Methods of the Theory of Functions of Many Complex Variables

VASILIY SERGEYEVICH VLADIMIROV

Methods of the Theory of Functions of Many Complex Variables

with a preface by
Academician N. N. Bogolyubov

translated by
SCRIPTA TECHNICA, INC.

edited by
LEON EHRENPREIS
Courant Institute of Mathematical Sciences
New York University

THE M.I.T. PRESS
Massachusetts Institute of Technology
Cambridge, Massachusetts, and London, England

Originally published under the title

METODY TEORII FUNKTSIY MNOGIKH
KOMPLEKSNYKH PEREMENNYKH
by
Nauka Press, Moscow, 1964

Library of Congress Catalog Card Number: 66-28672
Printed in the United States of America

Preface

The present monograph of V.S. Vladimirov is devoted to a systematic exposition of the fundamentals of the theory of single sheeted (schlicht) domains of holomorphy and its applications to quantum field theory, to the theory of functions, and to differential equations with constant coefficients.

In recent years, the theory of functions of several complex variables, which previously had not found much application in natural science, unexpectedly found numerous and fruitful applications in quantum field theory, especially in connection with questions regarding the basis of the so-called dispersion relations. The successes achieved in quantum field theory by this route have in turn had influence on the theory of functions of several complex variables. As it happened, a number of results and methods that were originally found for the solutions of particular problems in quantum field theory can be generalized to the theory of functions of several complex variables, enriching it with new and far-reaching theorems and methods. This applies, for example, to the theorem on the "edge of a wedge" to the theorem of a C-convex envelope, and to the Jost-Lehmann-Dyson integral representation.

The second portion of the monograph, containing results discovered partly by the author and partly by a number of other investigators in this field, is devoted to a study of this series of questions.

It should be noted that the theory of generalized functions is used extensively in the monograph, especially in the second portion. Generalized functions appear here as boundary values of holomorphic functions. This point of view enabled the author to single out the class of generalized functions that possess the distinctive property of quasianalyticity and to indicate applications of the theory of functions of this class to a number of problems in mathematics and quantum field theory.

For the convenience of the reader (especially, the theoretical physicist), the first portion of the monograph includes a detailed exposition of the theory of schlicht domains of holomorphy (pseudoconvex domains and integral representations. This is all the more necessary since the majority of the results and methods of the theory of functions of several complex variables that have found application in physics are either widely dispersed in a large number of periodical articles, not always easily accessible, or are insufficiently represented in monograph literature. Therefore, it is to be hoped that the present monograph will meet with the approval of theoretical physicists and mathematicians who are working with quantum field theory and who are interested in the applications of the methods of the theory of functions of several complex variables.

N. N. Bogolyubov

Introduction

In recent years, the theory of functions of several complex variables has found numerous applications to quantum field theory. This is due to the fact that the absence of any sort of satisfactory model in the contemporary theory of elementary functions that would enable us to explain and predict experimental results has lead to the development of an axiomatic (dispersion) approach to quantum field theory. The axiomatization of quantum field theory consists in a number of general principles, the most important of which are the following: Invariance under transformations of the inhomogeneous Lorentz groups, the existence of a complete system of (physical) states with positive energy, and microcausality. With this approach, physical quantities are treated as the boundary values of a function $f(z)$ of several complex variables that is holomorphic in some "primitive" (sometimes rather complicated) domain D defined by the axioms. However, in contrast with the situation with a single complex variable, in the space of several complex variables the following remarkable fact is true: not every domain is a domain of holomorphy. Therefore, in the absence of the specific form of the function f, the problem arises as to the construction of an envelope of holomorphy $H(D)$ of the domain D. Once the envelope $H(D)$ is constructed, we may hope in principle to write the integral representation (the dispersion relationship) expressing the value of an arbitrary function f that is holomorphic in D in terms of its values on a portion of the boundary $H(D)$. If these last values are connected with experimentally observed values, the integral representation obtained opens a path to the experimental checking of the theory. At the present time, the program envisioned is far from completed. In large measure, this is due to the fact that the methods of actually constructing envelopes of holomorphy for regions of complicated structure have been worked out only to a slight degree and are not standard methods.* In each individual case, this requires the application of some technique, often rather complicated, that is peculiar to the problem at hand. Thus, methods of constructing envelopes of holomorphy and methods of obtaining integral representations are of supreme interest in quantum field theory.

---

*The method of successive approximations (see section 21.6) might be considered an exception. However, actual application of this technique requires numerical methods.

In the present book, we shall give a systematic exposition of these methods and shall illustrate their usefulness with a number of examples and applications. Chapter 1 is introductory. In it, we define the basic concepts and the elementary propositions. As a rule these are trivial generalizations of the corresponding propositions from the theory of functions of a single complex variable to the case of two or more complex variables. The chapter also contains the more salient facts from the theory of functions of real variables (see section 2) and the theory of generalized functions (see section 3). The material of section 3 is used primarily in Chapter V. Chapter II is also of an auxiliary nature. It is devoted to the theory of plurisubharmonic functions and pseudoconvex domains. Chapter III treats various characteristics of domains of holomorphy: holomorphy: holomorphic convexity, the principles of continuity, local and global pseudoconvexity (Oka's theorem). The concept of an envelope of holomorphy is introduced and its properties are studied. The results of this chapter are of more than trivial value in a space of several complex variables. The case of a single variable is treated as a degenerate case since, for it, every domain is a domain of holomorphy. For greater clarity, the exposition is made for the simplest and, at the same time, most important case of single sheeted (schlicht) domains. Generalization to many-sheeted (plane covering) domains does not as a rule entail significant difficulties. (Appropriate remarks are made at certain points in the text.) The results of Chapters II and III are illustrated with examples of four types of domains: multiple-circular, tubular, semitubular, and Hartogs domains. Integral representations are expounded in Chapter IV. Three types of representations are examined: Martinelli-Bochner, Bergman-Weil, and Bochner representations: Chapter V is devoted to actual applications, especially in quantum field theory. However, the exposition is developed under more general hypotheses than are required for this theory. Therefore, the results given have an independent purely mathematical value. In this chapter, a study is made of the properties of the boundary values of functions that are holomorphic in tubular cones; Bogolyubov's theorem on the "edge of a wedge" and the theorem of a C-convex envelope are proven; a derivation is given of the Jost-Lehmann-Dyson integral representation; certain applications of these results to quantum field theory and the theory of functions are presented. It should be noted that the theory of generalized functions is used along with the theory of functions of several complex variables in this chapter.

The bibliography at the end of the book makes no claim at all to completeness. The references to sources mentioned in the text, especially in Chapters I-IV, are for the most part rather random and should by no means be regarded as material on the history of the theory of functions of several complex variables. In the preparation of this monograph, the following books proved of especial help: B. A. Fuks [1,2], S. Bochner and W. T. Martin [3], H. Behnke and P. Thullen [4], and also the surveys by H. Bremermann [13] and A. Wightman [61].

This book is an expansion of lectures given by me to my colleagues in the Laboratory of Theoretical Physics in Dubna in 1958-59 and 1961-62 and to students of the Department of Physics in Moscow State University during the academic year 1962-63.

I should like to take this opportunity to thank my colleagues and pupils in the division of quantum field theory at the Mathematical Institute of the Academy of Sciences, the Laboratory of Theoretical Physics, and the Department of Physics at Moscow State University for a number of comments that have contributed to the improvement of this book. In particular, I wish to thank N. N. Bogolyubov, Yu. N. Drozhzhinov, A. B. Zhizhchenko, B. V. Medvedev, M. K. Polivanov, B. M. Stepanov, I. T. Todorov, B. A. Fuks, B. V. Shabat, and M. Shirinbekov for valuable comments and attention to my work.

Contents

CHAPTER I

The Basic Properties of Holomorphic Functions

1. NOTATIONS AND DEFINITIONS

1. Notations

We denote by C^n and R^n the n-dimensional complex and real spaces respectively: $C^n = R^n + iR^n$. We denote points in the space R^n by $x = (x_1, x_2, \ldots, x_n)$, y, ξ, etc. Correspondingly, we denote points in C^n by $z = (z_1, z_2, \ldots, z_n) = x + iy$, $\zeta, \ldots$. We denote by $\bar{z}$ the complex conjugate of the point z; thus, $\bar{z} = x - iy$. The symbols $|z|$ and $|x|$ denote Euclidean lengths (norms):

$$|z| = \sqrt{z\bar{z}},$$

and the symbols $z\zeta$, $z\xi$, etc., denote scalar products:

$$z\zeta = z_1\zeta_1 + z_2\zeta_2 + \ldots + z_n\zeta_n.$$

If $\alpha = (\alpha_1, \alpha_2, \ldots, \alpha_n)$ is an n-dimensional vector with nonnegative integral components, we shall use the following abbreviations:

$$D = \left(\frac{\partial}{\partial z_1}, \frac{\partial}{\partial z_2}, \ldots, \frac{\partial}{\partial z_n}\right),$$

$$D^\alpha = \frac{\partial^{|\alpha|}}{\partial z_1^{\alpha_1} \partial z_2^{\alpha_2} \ldots \partial z_n^{\alpha_n}}, \quad |\alpha| = \alpha_1 + \alpha_2 + \ldots + \alpha_n,$$

$$\alpha! = \alpha_1!\alpha_2! \ldots \alpha_n!, \qquad z^\alpha = z_1^{\alpha_1} z_2^{\alpha_2} \ldots z_n^{\alpha_n}.$$

The symbol z^α will also be used for negative integral values of α_j. We shall denote the vector $(1, 1, \ldots, 1)$ by I. When there is no danger of confusion, we shall write z^I instead of $z_1 z_2 \ldots z_n$, dz instead of $dz_1 dz_2 \ldots dz_n$, and $\tilde{z}$ instead of $(z_2, \ldots, z_n)$.

2. Some concepts and generalizations from point-set theory

We shall denote by the symbol $[x : C]$ the set of all points x possessing property C. It is assumed that the reader is familiar with the basic concepts in the point-set theory of a finite-dimensional (Euclidean) space, such as the neighborhood of a point, an interior point, a cluster point, closure, open and closed sets, connectedness, etc.

Let A and B denote two arbitrary sets. The notation $x \in A$ (the notation $x \notin A$) indicates that x is (is not) an element of A. We denote by int A the interior (that is, the set of interior points) of A, by $\bar{A}$ the closure of A, by $U(A)$ a neighborhood of A (the union of neighborhoods of all points $x \in A$), by $e_A(x)$ the characteristic function of the set A (that is, $e_A(x) = 1$ for $x \in A$ and $e_A(x) = 0$ for $x \notin A$). We indicate by $A \subset B$ the fact that A is a subset of B. We denote by $A \cup B$ the union of A and B, by $A \cap B$ the intersection of A and B, by $\bigcup_\alpha A_\alpha$ and $\bigcap_\alpha A_\alpha$ the union and intersection respectively of a family of sets $\{A_\alpha\}$, by $A \setminus B$ the complement of B with respect to $A \cup B$, by $A \times B$ the Cartesian product of A and B (that is, the set of ordered pairs (a, b) as a and b range over the sets A and B respectively). The notation $A \Subset B$ indicates that A is a compact subset of B (that is, A is bounded and $\bar{A} \subset B$). We denote by ∂A the boundary of A (thus, $\partial A = \bar{A} \setminus \text{int } A$), by $\varnothing$ the empty set, and by $\{x^{(1)}, x^{(2)}, \ldots, x^{(k)}\}$ the set consisting of the elements $x^{(1)}, x^{(2)}, \ldots, x^{(k)}$.

A set is said to be bounded if all coordinates of its points are uniformly bounded. A domain is any open connected set. Every closed bounded set is compact.

We shall usually denote an open set by the symbol O. The set of cluster points of O that do not belong to O constitutes the boundary ∂O of the set O: $\partial O = \bar{O} \setminus O$. We note that int $\bar{O} \supset O$. Every open set can be uniquely decomposed into countably many disjoint domains, known as the components of the original open set.

We have the following

THEOREM (Heine-Borel). *If a compact set K is covered by a collection of neighborhoods, this collection contains a finite subcollection that also covers K.*

It follows from this theorem that if A is a bounded set, then $A \Subset U(A)$.

In what follows, we shall usually deal with neighborhoods of a point $z^0 \in C^n$ of a particular type, namely, polydiscs of radius $r = (r_1, r_2, \ldots, r_n)$ with center at that point:

$$S(z^0, r) = S(z_1^0, r_1) \times \ldots \times S(z_n^0, r_n),$$
$$S(z_j^0, r_j) = [z_j : |z_j - z_j^0| < r_j],$$

or balls

$$U(z^0, r) = [z : |z - z^0| < r].$$

Analogously, for a neighborhood of a point $x^0 \in R^n$, we shall usually take a poly-interval $S(x^0, r)$ or a hypercube $U(x^0, r)$.

We denote by A_ε and A^ε the particular ε-neighborhoods of the set A that are generated by the polydiscs or balls

$$A_\varepsilon = \bigcup_{z' \in A} S(z', \varepsilon I), \quad A^\varepsilon = \bigcup_{z' \in A} U(z', \varepsilon), \quad \varepsilon > 0.$$

Finally, we shall say that a sequence of sets $A_1, A_2, \ldots$ converges to a set A and we shall write $\lim_{k\to\infty} A_k = A$ if the set A consists of the limits of all possible convergent sequences of points $x^{(k)}$ (for $k=1, 2, \ldots$) in A_k.

A set X is called a topological space if a collection L of subsets of X exists that possesses the following two properties: (1) $\varnothing \in L$ and $X \in L$; (2) the union of an arbitrary number and the intersection of a finite number of members of L also belong to L. We then say that this collection L of subsets of X defines a topology on X and that these subsets are open in that topology. It is in terms of a topology that we usually define the concepts of convergence, continuity, connectedness, etc. (see, for example, Aleksandrov [14], Chapter 1).

A mapping of a topological space X onto a topological space Y is called a homeomorphism of X onto Y if it is one-to-one and bicontinuous.

3. *Distance functions and their properties*

Suppose that O is an open set in R^n. We denote by $\Delta_O(x)$ the function denoting the (Euclidean) distance from the point $x \in O$ to the boundary ∂O (see Fig. 1):

$$\Delta_O(x) = \sup r, \quad \text{if} \quad U(x, r) \subset O.$$

Sometimes, we shall use the distance generated by the norm $\|x\| = \max_{1 \leqslant j \leqslant n} |x_j|$. We denote this distance by $\delta_O(x)$:

$$\delta_O(x) = \sup r,$$

if

$$S(x, rI) \subset O.$$

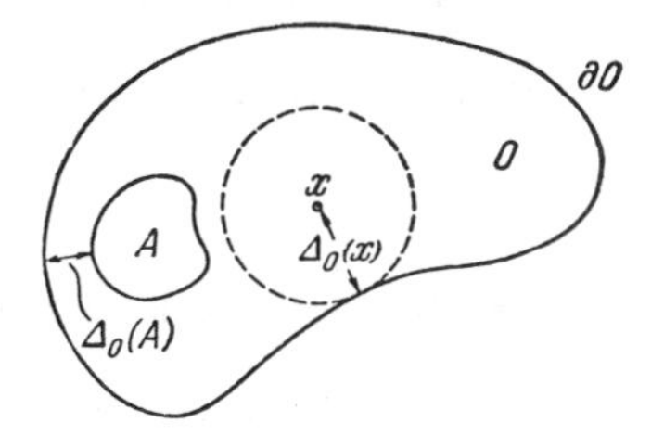

Fig. 1

If $O \neq R^n$, *mo* $\Delta_O(x)$ is a continuous function in O.

To see this, note that the function $\Delta_O(x)$ is bounded in O. Suppose that $|x'-x''| < \varepsilon$. Then, $|x - x'| < |x-x''| + \varepsilon$. If $|x-x''| < \Delta_O(x') - \varepsilon$, it follows that $|x - x'| < \Delta_O(x')$. From this it follows that $x \in O$ and, consequently, by definition, $\Delta_O(x'') \geqslant \Delta_O(x') - \varepsilon$. From symmetry considerations, we also have $\Delta_O(x') \geqslant \Delta_O(x'') - \varepsilon$. Therefore, $|\Delta_O(x') - \Delta_O(x'')| \leqslant \varepsilon$ if $|x'-x''| < \varepsilon$, which proves the assertion.

If $A \subset O$, we denote by $\Delta_O(A)$ the distance from A to O (see Fig. 1):

$$\Delta_O(A) = \inf_{x \in A} \Delta_O(x).$$

Clearly,

$$\Delta_O\left(\bigcup_\alpha A_\alpha\right) = \inf_\alpha \Delta_O(A_\alpha). \tag{1}$$

From this and from the Heine-Borel theorem, it follows that $A \Subset O$ implies $\Delta_O(A) > 0$. If A is bounded and $\Delta_O(A) > 0$, then $A \Subset O$.

Let O_α (for $\alpha = 1, 2, \ldots$) denote an increasing sequence of open sets and suppose that $\bigcup_\alpha O_\alpha = O$. Then, for an arbitary set $A \Subset O$, there exists a number $N = N(A)$ such that, for all $\alpha \geq N$,

$$\Delta_{O_\alpha}(A) \leqslant \Delta_{O_{\alpha+1}}(A) \leqslant \Delta_O(A) \text{ and } \lim_{\alpha \to \infty} \Delta_{O_\alpha}(A) = \Delta_O(A). \tag{2}$$

The proof follows immediately from the definitions.

Suppose that O is the interior of the intersection of open sets O_α. Then, for an arbitrary set $A \subset O$,

$$\Delta_O(A) = \inf_\alpha \Delta_{O_\alpha}(A). \tag{3}$$

To prove Eq. (3), we need only show that it holds for sets consisting of a single element $A = \{x\}$ and that

$$\Delta_B(x) = \inf_\alpha \Delta_{O_\alpha}(x)$$

for every component B of the open set $O = \operatorname{int} \bigcap_\alpha O_\alpha$.

Since $x \in B \subset O_\alpha$, we have $\Delta_B(x) \leqslant \Delta_{O_\alpha}(x)$ and therefore

$$\Delta_B(x) \leqslant \inf_\alpha \Delta_{O_\alpha}(x).$$

On the other hand, since $U(x, \Delta_{O_\alpha}(x)) \subset O_\alpha$ and $x \in B$, we have

$$U\left(x, \inf_\alpha \Delta_{O_\alpha}(x)\right) \subset B.$$

This implies the inequality

$$\inf_\alpha \Delta_{O_\alpha}(x) \leqslant \Delta_B(x),$$

which, in conjunction with the preceding one, completes the proof.

Since the norms $|x|$ and $\|x\|$ are equivalent, that is, since $\|x\| \leqslant |x| \leqslant \sqrt{n}\|x\|$, the function $\delta_O(x)$ possesses the properties enumerated.

Suppose that G is a domain in C^n, that a is a complex vector such that $|a|=1$, and that λ is a complex parameter. We denote by $\Delta_{a,G}(z)$ the distance from the point z to the boundary of the open (two-dimensional) set which is the intersection of the two-dimensional analytic plane $z'=z+\lambda a$ with the domain G. Thus,

$$\Delta_{a,G}(z)=\sup r, \quad \text{if} \quad [z':\ z'=z+\lambda a,\ |\lambda|<r]\subset G.$$

The function $\Delta_{a,G}(z)$ is lower-semicontinuous in G for all a (see section 2.1).

Introducing the open set $G_{z,a}=[\lambda: z+\lambda a\in G]$ in the plane of the complex variable λ, we may write

$$\Delta_{a,G}(z)=\Delta_{G_{z,a}}(0).$$

Obviously,

$$\Delta_G(z)=\inf_{|a|=1}\Delta_{a,G}(z). \tag{4}$$

4. *Surfaces*

We shall say that a function $f(x)$ belongs to the class $C^{(m)}=C^{(m)}(O)$ (for $m=0, 1, \ldots$) in an open set O if its first m derivatives are all continuous in O.

A set $S\subset R^n$ is called a k-dimensional surface (for $1\leqslant k\leqslant n-1$) of class $C^{(m)}$ (for $m=1, 2, \ldots$) if, in some neighborhood of every point $x^0\in S$, it is defined by the equations

$$x_j=x_j(x^0;\ t),\quad j=1, 2, \ldots, n,\quad t=(t_1, t_2, \ldots, t_k)$$

where the functions $x_j(x^0;\ t)\in C^{(m)}$ and the rank of the matrix

$$\frac{\partial x_j(x^0;\ t)}{\partial t_s}\qquad (j=1, 2, \ldots, n,\quad s=1, 2, \ldots, k)$$

is equal to k in the corresponding neighborhood of the point $t=0$, $x^0=x(x^0;\ 0)$.

It follows from this definition and from the implicit-function theorem that, in a neighborhood of a point $x^0\in S$, a k-dimensional surface S of class $C^{(m)}$ is defined by the equations

$$f_s(x^0;\ x)=0,\quad s=1, 2, \ldots, n-k,$$

where the functions $f_s(x^0;\ x)\in C^{(m)}$ and the rank of the matrix $\partial f_s/\partial x_j$ is $n-k$ in this neighborhood.

Thus, we can define a topology on a k-dimensional surface S of class $C^{(1)}$ in terms of the collection of those neighborhoods of the points in S that are homeomorphic to the k-dimensional ball. Suppose that a sequence $x^{(1)}, x^{(2)}, \ldots$ of points in S is a Cauchy sequence in S. Then, the limit $\lim_{k\to\infty} x^{(k)} = x^0 \in R^n$ exists. If $x^0 \notin S$, we shall call the point x^0 a boundary point of the surface S. The set of all boundary points of S constitutes the boundary ∂S of the surface S. If $\partial S = \emptyset$, we shall say that the surface S is *closed*.

Suppose that an open set S lies on a surface S' of class $C^{(m)}$. We assume that the topology in S is induced by the topology in S'. Then, obviously, S is also a surface of class $C^{(m)}$. We shall call the boundary ∂S of the surface S the boundary of the open set S. If the set $\bar{S} = S \cup \partial S$ also lies on the surface S', we shall say that $\bar{S}$ is a surface with a class $C^{(m)}$ boundary. In this case, we shall also say that the open set (or domain) S lies on the surface S' together with its boundary ∂S, $\bar{S} \subset S'$. (If, in addition, $\bar{S}$ is bounded, we shall say, in accordance with section 1.2, that S is compact in S' and we shall write $S \Subset S'$ (see Fig. 2).)

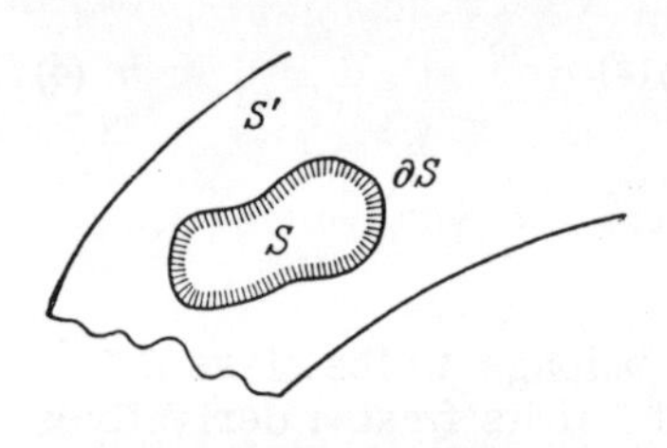

Fig. 2

We shall say that a surface is *piecewise smooth* if it consists of a finite number of connected surfaces with class $C^{(1)}$ boundary. One-dimensional surfaces are called *curves*; $(n-1)$-dimensional surfaces are called *hypersurfaces*. A continuous curve $x = x(t)$, where $0 \leqslant t \leqslant 1$, that has no multiple points is called a *Jordan curve*. Thus, on a Jordan curve, $x(t') \neq x(t'')$ for all $t' \neq t''$ except possibly for the case in which $t' = 0$ and $t'' = 1$. In the case in which $x(0) = x(1)$, the curve is said to be *closed*; otherwise, it is *open*.

5. *Simply connected regions*

Two Jordan curves that have common end points and that lie in some domain are said to be *homotopic* in that domain if, by means of a continuous deformation, it is possible to displace one of these curves into the other without leaving the domain anywhere in the process and such that the fixed end points are not moved during the displacement. A domain is said to be **simply connected** if any two such curves in it are homotopic.

We have the following theorems:

THEOREM (**the Jordan curve theorem**). *A closed Jordan curve L lying in the extended plane $\bar{C}^1$ separates this plane into two simply connected domains having L as their common boundary. One of these domains the interior of L, is bounded; the other, the exterior of L, contains the point ∞. The complement of an open Jordan curve L consists of a single simply connected domain containing the point ∞ and having L as its boundary.*

THEOREM (**Schönflies**). *All points of a closed Jordan curve in C^1 are accessible from both sides* (see Goluzin [5], 41).

A point $z^0 \in \partial G$ is said to be accessible from a domain G if it is possible to draw a Jordan curve one end point of which is z^0 and all other points of which lie in G.

It follows from the theorems of Jordan and Schönflies that all interior points of an open Jordan curve in C^1 are accessible.

6. *Other definitions*

We shall say that a series (sequence) converges uniformly in an open set O if this series (sequence) converges uniformly on every set that is compact in O.

If a function $f(x)$ is bounded (resp. bounded above, resp. bounded below) or is summable on every compact subset of the open set O, we shall say that this function is locally bounded (resp. bounded above, resp. bounded below) or locally summable in O.

We shall say that a function $f(x)$ defined on an open set $O \subset R^n$ is *unbounded* at a point $x^0 \in \partial O$ if there exists in O a sequence of points $\{x^{(k)}\}$ such that $x^{(k)} \to x^0$ and $|f(x^{(k)})| \to \infty$ as $k \to \infty$.

We shall say that a real function defined on an open set O approaches $+\infty$ everywhere on ∂O if, for arbitrary M, the set $[x : f(x) < M, x \in O]$ is compact in O.

Clearly, if $f(x)$ approaches $+\infty$ everywhere on ∂O, it is unbounded at all points of ∂O.

We shall say that a *maximum principle* applies to sets S and T with respect to a certain class of real functions if, for an arbitrary function $f(x)$ in that class,

$$\sup_{x \in T} f(x) = \sup_{x \in S \cup T} f(x).$$

2. SOME FACTS FROM THE THEORY OF FUNCTIONS OF n REAL VARIABLES

Most of the material of this section is contained in the book by Natanson [8].

1. *Definition of a semicontinuous function*

Suppose that a real function $f(x)$ is defined on a set $A \subset R^n$. If, for a point $x^0 \in A$,

$$f(x^0) = \overline{\lim_{x \to x^0}} f(x) = \lim_{\delta \to 0} \sup_{U(x^0, \delta) \cap A} f(x),$$

we say that $f(x)$ is *upper-semicontinuous* at x^0. Analogously, if

$$f(x^0) = \underline{\lim}_{x \to x^0} f(x),$$

we say that $f(x)$ is *lower-semicontinuous* at the point x^0. In particular, if $f(x^0) = +\infty$ (resp. $(f(x^0) = -\infty)$), then $f(x)$ is upper- (lower-) semicontinuous at the point x^0.

We note that if $f(x)$ is both upper- and lower-semicontinuous at the point x^0 and if $|f(x^0)| \neq \infty$, it is continuous at x^0.

We shall say that $f(x)$ is upper-semicontinuous on a set A if it is upper-semicontinuous at every point of A.

If a function $f(x)$ is upper-semicontinuous, the function $-f(x)$ is lower-semicontinuous. Therefore, we need study only, for example, upper-semicontinuous functions.

2. Tests for semicontinuity

Suppose that a function $f(x)$ is defined on a set A and that $x^0 \in A$.

For $f(x)$ to be upper-semicontinuous at a point x^0 such that $f(x^0) < \infty$, it is necessary and sufficient that for every $a > f(x^0)$ there exists a $\delta > 0$ such that, for all x in $U(x^0, \delta) \cap A$, $f(x) < a$.

The necessity follows from the fact that

$$\lim_{\delta \to 0} \sup_{U(x^0, \delta) \cap A} f(x) = f(x^0) < a.$$

Let us prove the sufficiency. If $f(x) < a$ in $U(x^0, \delta) \cap A$, then

$$\overline{\lim_{x \to x^0}} f(x) \leqslant a.$$

If we now let a approach $f(x^0)$, we obtain

$$\overline{\lim_{x \to x^0}} f(x) \leqslant f(x^0) \leqslant \overline{\lim_{x \to x^0}} f(x).$$

As a consequence of this test, we may make the following assertions:

If $-\infty < f(x^0) < \infty$, the function $f(x)$ is upper-semicontinuous at the point x^0 if and only if, for every $\varepsilon > 0$, there exists a $\delta > 0$ such that, for all $x \in U(x^0, \delta) \cap A$, the inequality $f(x) < f(x^0) + \varepsilon$ holds.

For a function $f(x)$ to be upper-semicontinuous in an open set O, it is necessary and sufficient that, for arbitrary $a < +\infty$, the set of points x in O for which $f(x) < a$ be open. In particular, semicontinuous functions are measurable.

For $f(x)$ to be upper-semicontinuous in an open set O, it is necessary and sufficient that the set

$$[(x, y): \ f(x) < y, \quad x \in O]$$

be open in R^{n+1}.

Suppose that a function $f(x)$ is upper-semicontinuous in an open set O. Then, the corresponding function $f(x)$ defined on any curve $x=x(t)$, where $0 \leqslant t \leqslant 1$, contained in O is upper-semicontinuous on that curve.

To see this, let $t_0 \in [0, 1]$ and $x^0 = x(t_0)$. Then,

$$f(x^0) = f[x(t_0)] \leqslant \overline{\lim_{t \to t_0}} f[x(t)] \leqslant \overline{\lim_{x \to x^0}} f(x) = f(x^0).$$

3. *Properties of semicontinuous functions*

A function $f(x)$ that is upper-semicontinuous on a compact set K attains its maximum value on K.

Proof: Define

$$M = \sup_{x \in K} f(x).$$

There exists a sequence $\{x_\alpha\} \subset K$ such that

$$M = \lim_{\alpha \to \infty} f(x_\alpha).$$

According to the Bolzano-Weierstrass theorem, there exists a subsequence $x_{\alpha_k} \to x_0 \in K$. Then,

$$f(x^0) = \overline{\lim_{x \to x^0}} f(x) \geqslant \lim_{k \to \infty} f\left(x_{\alpha_k}\right) = M \geqslant f(x^0).$$

If the value of a function that is upper-semicontinuous on K is never $+\infty$, the function is bounded from above on K. This follows from the preceding assertion.

The lower envelope

$$f(x) = \inf_\alpha f_\alpha(x)$$

of a family $\{f_\alpha(x)\}$ of functions that are upper-semicontinuous on a set A is upper-semicontinuous in A.

Proof: Suppose that $x^0 \in A$ and that $f(x^0) < a < +\infty$. Then, there exists an α_0 such that $f_{\alpha_0}(x^0) < a - \varepsilon$, where $\varepsilon = 1/2[a - f(x^0)]$. Since the function $f_{\alpha_0}(x)$ is upper-semicontinuous in A, there exists a $\delta > 0$ such that $f_{\alpha_0}(x) < a$ for all $x \in U(x^0, \delta) \cap A$ (see section 2.2). Then, $f(x) < a$ for all x in $U(x^0, \delta) \cap A$; that is, the function $f(x)$ is upper-semicontinuous at the point x^0. If $f(x^0) = +\infty$, $f(x)$ is also upper-semicontinuous at x^0, which completes the proof.

We note that the upper envelope of a finite number of upper-semicontinuous functions is also upper-semicontinuous. For an infinite number of functions, this may not be the case.

4. *Monotonic sequences of semicontinuous functions*

The limit of a decreasing sequence of upper-semicontinuous functions defined on a set A is upper-semicontinuous in A.

This follows from the preceding property since

$$f(x) = \lim_{\alpha \to \infty} f_\alpha(x) = \inf_\alpha f_\alpha(x),$$

if

$$f_1(x) \geqslant f_2(x) \geqslant \ldots, \qquad x \in A.$$

The converse is also valid:

If $f(x)$ is upper-semicontinuous on a compact set K and $f(x) < +\infty$ in K, there exists a decreasing sequence of continuous functions that converges to $f(x)$.

To prove this, we set

$$f_\alpha(x) = \max_{x' \in K} [f(x') - \alpha |x - x'|]. \tag{5}$$

The functions $f_\alpha(x)$ are continuous, nonincreasing, and uniformly founded from above (see section 2.3) and they are not less than $f(x)$:

$$f(x) \leqslant f_{\alpha+1}(x) \leqslant f_\alpha(x) \leqslant M, \qquad x \in K.$$

From this, it follows that

$$\lim_{\alpha \to \infty} f_\alpha(x) \geqslant f(x). \tag{6}$$

Let us assume first that $f(x) > -\infty$ on K. Let $z_\alpha(x)$ denote a point at which the maximum in Eq. (5) is attained. Then $z_\alpha(x) \to x$ as $\alpha \to \infty$. On the basis of the test given in section 2.2, for arbitrary $\varepsilon > 0$, there exists a $\delta > 0$ such that

$$f[z_\alpha(x)] < f(x) + \varepsilon \text{ and } |z_\alpha(x) - x| < \delta.$$

Then,

$$f_\alpha(x) = f[z_\alpha(x)] - \alpha |z_\alpha(x) - x| < f(x) + \varepsilon.$$

From this it follows that equality holds in conditional inequality (6).

If $f(x)$ assumes the value $-\infty$, we define the nonincreasing sequence

$$g_N(x) = \begin{cases} f(x), & \text{if } \quad f(x) \geqslant -N, \\ -N, & \text{if } \quad f(x) < -N. \end{cases}$$

These functions $g_N(x)$ are each bounded from below and upper-semicontinuous and the sequence converges to $f(x)$. If we apply the result obtained to each of the functions $g_N(x)$, we can see, by the familiar diagonal process that the assertion holds in this case also.

5. *Taking the limit under the Lebesgue integral sign*

The reader is assumed to be familiar with the basic concepts of Lebesgue integration: measurable sets, measurable functions, Lebesgue measure, the Lebesgue integral, "almost everywhere" (see [8]).

THEOREM (Lebesgue). *If a sequence of functions $\{f_\alpha(x)\}$ that are summable on a compact set K converges almost everywhere to $f(x)$ and if there exists a summable function $g(x)$ such that*

$$|f_\alpha(x)| \leqslant g(x), \qquad x \in K,$$

then $f(x)$ is summable and

$$\lim_{\alpha\to\infty} \int_K f_\alpha(x)\,dx = \int_K f(x)\,dx. \tag{7}$$

THEOREM (B. Levi). *If a sequence $\{f_\alpha(x)\}$ is such that*

$$f_{\alpha+1}(x) \leqslant f_\alpha(x) \text{ and } \left|\int_K f_\alpha(x)\,dx\right| < C,$$

the limit function $f(x)$ is summable and Eq. (7) holds.

6. *Extension of the concept of the Lebesgue integral*

For measurable functions that are bounded above (or below) on a compact set K, we shall extend the concept of the Lebesgue integral as follows: If the function is summable, we shall consider the extended Lebesgue integral as coinciding with the ordinary Lebesgue integral. If the function is not summable, we assign to the extended Lebesgue integral the value $-\infty$ (resp. $+\infty$). For example, if $f(x) < +\infty$ is upper-semicontinuous on K, it is bounded above on K (see section 2.3). Therefore, it is *integrable* on K (in the sense that the extended Lebesgue interval of $f(x)$ over K exists). Henceforth, all integrals of functions that are bounded above or below will be understood in the extended sense.

For the extended Lebesgue integral, Levi's theorem, and Fatou's lemma are valid in the following forms:

7. *Theorem (B. Levi)*

If a sequence of measurable functions $f_\alpha(x)$ *(for* $\alpha = 1, 2, \ldots$*) is such that*

$$-\infty \leqslant f_{\alpha+1}(x) \leqslant f_\alpha(x) \leqslant C < +\infty, \qquad \alpha = 1, 2, \ldots,$$

then,

$$\lim_{\alpha\to\infty} \int_K f_\alpha(x)\,dx = \int_K \lim_{\alpha\to\infty} f_\alpha(x)\,dx. \tag{8}$$

Proof: If

$$\lim_{\alpha\to\infty} \int_K f_\alpha(x)\,dx > -\infty,$$

Eq. (8) is valid by virtue of the theorem of Levi in section 2.5. On the other hand, if

$$\lim_{\alpha\to\infty} \int_K f_\alpha(x)\,dx = -\infty,$$

Eq. (8) follows from the inequality

$$\lim_{\alpha\to\infty} \int_K f_\alpha(x)\,dx \geqslant \int_K \lim_{\alpha\to\infty} f_\alpha(x)\,dx.$$

8. *Fatou's lemma*

If a sequence of measurable functions $\{f_\alpha(x)\}$ *is such that* $-\infty \leq f_\alpha(x) \leq C < +\infty$,

$$\overline{\lim_{\alpha\to\infty}} \int_K f_\alpha(x)\,dx \leqslant \int_K \overline{\lim_{\alpha\to\infty}} f_\alpha(x)\,dx. \tag{9}$$

Proof: We set

$$\varphi_\alpha(x) = \sup_{m\geqslant 0} f_{\alpha+m}(x).$$

Then,

$$\varphi_{\alpha+1}(x) \leqslant \varphi_\alpha(x) \leqslant C, \quad \lim_{\alpha\to\infty} \varphi_\alpha(x) = \overline{\lim_{\alpha\to\infty}} f_\sigma(x).$$

From this, we obtain by virtue of (8)

$$\lim_{\alpha\to\infty} \int_K \varphi_\alpha(x)\,dx = \int_K \overline{\lim_{\alpha\to\infty}} f_\alpha(x)\,dx. \tag{10}$$

But

$$f_\alpha(x) \leqslant \varphi_\alpha(x), \quad \int_K f_\alpha(x)\,dx \leqslant \int_K \varphi_\alpha(x)\,dx,$$

so that

$$\overline{\lim_{\alpha\to\infty}} \int_K f_\alpha(x)\,dx \leqslant \lim_{\alpha\to\infty} \int_K \varphi_\alpha(x)\,dx,$$

which, together with (10) yields inequality (9).

9. *Fubini's theorem*

If $f(x, y)$ is a measurable nonnegative function defined on a compact set $A \times B$, and if one of the iterated integrals

$$\int_A \left[\int_B f(x, y)\,dy\right] dx, \quad \int_B \left[\int_A f(x, y)\,dx\right] dy,$$

has a finite value, then $f(x, y)$ is summable and the other iterated integral also has a finite value. Furthermore,

$$\int_A \left[\int_B f(x, y)\,dy\right] dx = \int_B \left[\int_A f(x, y)\,dx\right] dy = \int_{A\times B} f(x, y)\,dx\,dy. \tag{11}$$

For the extended Lebesgue integral (see section 2.6), Fubini's theorem takes the following form: *If a measurable function $f(x,y) \leq C < +\infty$ is defined on a compact set $A \times B$, all three integrals exist and Eqs.* (11) *are valid.*

Proof: By applying Fubini's theorem to the nonnegative function $C - f(x, y)$, we see that all three integrals are finite or that all three are equal to $-\infty$. In both cases, Eqs. (11) hold.

3. FACTS TAKEN FROM THE THEORY OF GENERALIZED FUNCTIONS

Generalized functions found their first application in quantum mechanics when Dirac introduced the now-famous delta-function. In 1936, Sobolev introduced generalized functions in an explicit and nowadays commonly adopted form [77].*

---

*Editor's note: Actually the form in which generalized functions are used nowadays is due to L. Schwartz [11].

Most of the information from the theory of generalized functions that we shall present below is contained in the books by Schwartz [11], Gel'fand and Shilov [38], and Kantorovich and Akilov [95] and also in the lectures of Gårding and Lions [79]. References to results borrowed from other sources will be given from time to time in the exposition.

1. *The space of basic functions D (O)*

The closure of the set of points x for which a continuous function $\varphi(x)\neq 0$ is called the *carrier** of $\varphi(x)$ and is denoted by S_φ. If $S_\varphi \Subset O$, the function is said to be of compact carrier in O, where O is an is an open set.

We shall refer to the (linear) set of functions in the class $C^{(m)}(O)$ that are of compact carrier in O as the space $D^{(m)}(O)$ (for $m= 0, 1, 2, \ldots$). When we say that a sequence of functions in $D^{(m)}(O)$ converges, we mean that the sequence of functions and the sequences of the first m derivatives of these functions converge uniformly and that the carriers of the functions in the original sequence are contained in some compact subset of O. We shall refer to the space $D^{(\infty)}(O)=D(O)$ as the space of basic functions. We shall denote the space $D(R^n)$ simply by D.

An example of a basic function is

$$\omega(|x|)=\begin{cases} C_n \exp\left(-\frac{1}{1-|x|^2}\right), & \text{if} \quad |x|<1, \\ 0 \quad , & \text{if} \quad |x|\geqslant 1, \end{cases}$$

where the constant C_n is chosen so that

$$\int \omega(|x|)\,dx = C_n\sigma_n \int_0^1 e^{-\frac{1}{1-\rho^2}}\rho^{n-1}\,d\rho = 1.$$

Here, σ_n is the area of the surface of a unit sphere R^n.

The function $\omega(|x|)$ can be used to construct other basic functions by assigning to each function $f\in D^{(0)}(O)$ its *mean* function (regularization):

$$f_\varepsilon(x)=\frac{1}{\varepsilon^n}\int f(\xi)\,\omega\left(\left|\frac{x-\xi}{\varepsilon}\right|\right)d\xi, \qquad \varepsilon>0,$$

which, obviously, belongs to $D(O)$ if $\varepsilon<\Delta_O(S_f)$. Here, $f_\varepsilon(x)\rightarrow f(x)$ as $\varepsilon\rightarrow +0$ in the topology of $D^{(0)}(O)$.

Thus, $D(O)$ is dense in $D^{(0)}(O)$.

---

*This often called *support*.

Suppose that a set $A \Subset O$. Then, there exists a function $e \in D(O)$ such that $e(x) = 1$ for $x \in A$ and $0 \leq 1$ for $x \in O$.

Specifically, if $\Delta_O(A) < 3\varepsilon$, the mean function $(e_{A^\varepsilon})_\varepsilon(x)$, where $e_{A^\varepsilon}(x)$ is the characteristic function of the set $A^\varepsilon = \bigcup_{x \in A} U(x, \varepsilon)$, is the required function.

From this, we have the following assertion regarding the partition of unity. *Let O denote an open set of the form $O = \bigcup_\alpha O_\alpha$, where the open sets O_α are compact in O and each compact set $K \subset O$ has a nonempty intersection with only a finite number of sets O_α in the collection $\{O_\alpha\}$. Then, there exist functions $e_\alpha(x)$ such that $0 \leq e_\alpha \in D(O_\alpha)$ and u $\sum_\alpha e_\alpha(x) = 1$ for $x \in O$.*

We shall say that a set $M \subset D(O)$ is bounded in $D(O)$ if $S_\varphi \subset O' \Subset O$ and $|D^\alpha \varphi(x)| \leqslant K_\alpha$ for all $\varphi \in M$. It follows from this and from Arzelà's theorem that every closed bounded subset of $D(O)$ is compact in $D(O)$.

We shall say that an operator T that maps $D(O')$ into $D(O)$ is continuous if $T\varphi_\alpha \to 0$ as $\alpha \to \infty$ in $D(O)$ wherever $\varphi_\alpha \to 0$ as $\alpha \to \infty$ in $D(O')$.

An operator T is continuous if and only if it is bounded, that is, if it maps every bounded set in $D(O')$ into a set that is bounded in $D(O)$.

For example, the differential operators $D^\alpha \varphi$ and the operators of the form $\psi\varphi$ indicating multiplication by a function $\psi \in C^{(\infty)}(O)$ are continuous from $D(O')$ into $D(O)$.

2. *The space of generalized functions $D^*(O)$*

We shall refer to the set of all linear continuous functionals f defined on $D(O)$ as the space of generalized functions $D^*(O)$. The value of the functional $f \in D^*(O)$ when applied to a basic function $\varphi \in D(O)$ will be denoted by (f, φ), so that

$$(f, \alpha\varphi + \beta\psi) = \alpha(f, \varphi) + \beta(f, \psi)$$

where $f \in D^*(O)$ and $\varphi, \psi \in D(O)$; $(f, \varphi_k) \to 0$ if $\varphi_k \to 0$ as $k \to \infty$ in $D(O)$.

Sometimes, to show the argument of basic functions clearly, we shall use the notation $f(x)$ for a generalized function f and we shall write the quantity (f, φ) in the form of an "integral":

$$(f, \varphi) = \int f(x)\varphi(x)\,dx. \tag{12}$$

If a function $f(x)$ is locally summable in O, this formula* defines a generalized function $f \in D^*(O)$. We shall say that such

*In a number of works (see, for example, [38, 30, 40]), instead of [12], the definition

$$(f, \varphi) = \int \bar{f}(x)\varphi(x)\,dx,$$

is used. This means that the space $D^*(O)$ is antilinear.

generalized functions are *regular.* We shall say that all other generalized functions are *singular.* An example of a singular generalized function is Dirac's delta-function

$$(\delta, \varphi) = \int \delta(x) \varphi(x) \, dx = \varphi(0).$$

For a linear functional f over $D(O)$ to be continuous, that is, $f \in D^*(O)$, it is necessary and sufficient that it be bounded on every bounded set in $D(O)$.

Convergence of a sequence of functionals f_α, for $\alpha = 1, 2, \ldots$, in $D^*(O)$ to a functional f should be understood as weak convergence in the space $D^*(O)$; that is,

$$\lim_{\alpha \to \infty} (f_\alpha, \varphi) = (f, \varphi).$$

The space $D^*(O)$ *is complete with respect to (weak) convergence*; that is, if a sequence f_α (for $\alpha = 1, 2, \ldots$) of functions $D^*(O)$ is such that, for an arbitrary $\varphi \in D(O)$, the numerical sequence (f_α, φ) has a limit, there exists a function $f \in D^*(O)$ such that $f_\alpha \to f$ as $\alpha \to \infty$.

A generalized function $f \in D^*(O)$ *vanishes in an open set* $O' \subset O$ if $(f, \varphi) = 0$ for all $\varphi \in D(O')$. For $f \in D^*(O)$ to vanish in O', it is necessary and sufficient that it vanish in a neighborhood of every point of the set O'. This assertion follows immediately from the partition of unity (see section 3.1). It then follows that if two generalized functions coincide in a neighborhood of every point of the set O', they coincide everywhere in O'.

Suppose that O' is the largest open set in which a given generalized function $f \in D^*(O)$ vanishes. We shall call the complement of O' with respect to O the *carrier* of f and we shall write $S_f = O \setminus O'$. Obviously, S_f is closed in O. If $S_f \Subset O$, the generalized function f is said to be of compact carrier in O. If f is of compact carrier in R^n, we shall simply call it of compact carrier. For example, $\delta(x)$ is of compact carrier and its carrier is the point $x = 0$.

The set $M^* \subset D^*(O)$ is said to be bounded in $D^*(O)$ if, for an arbitrary $\varphi \in D(O)$, the set of numbers (f, φ) is uniformly bounded with respect to $f \in M^*$. It can be shown that *the set of numbers* (f, ϕ) *is uniformly bounded for all* $f \in M^*$ *and* $\phi \in M$, *where* M *is an arbitrary bounded set in* $D(O)$.

From this it follows that, if $f_\alpha \to f$ in $D^*(O)$ and $\varphi_\alpha \to \varphi$ in $D(O)$, then

$$(f_\alpha, \varphi_\alpha) \to (f, \varphi) \text{ as } \alpha \to \infty^*.$$

If $\varphi_\alpha \to 0$ in $D(O)$ as $\alpha \to \infty$, then $(f, \varphi_\alpha) \to 0$ uniformly with respect to f as $\alpha \to \infty$, where f belongs to an arbitrary bounded set in $D^*(O)$.

---

*Editor's Note: The author is somewhat sloppy here and does not distinguish between convergence and sequential convergence.

If $\varphi_\alpha \to 0$ in $D(O)$ as $\alpha \to \infty$, then $(f, \varphi_\alpha) \to 0$ uniformly with respect to f as $\alpha \to \infty$, where f belongs to an arbitrary bounded set in $D^*(O)$. If $f_\alpha \to 0$ in $D^*(O)$ as $\alpha \to \infty$, then $(f_\alpha, \varphi) \to 0$ uniformly with respect to φ as $\alpha \to \infty$, where φ belongs to an arbitrary bounded set of $D(O)$.

For the set M to be bounded in $D(O)$, it is necessary and sufficient that, for arbitrary $f \in D^*(O)$, the set of numbers (f, φ) be uniformly bounded with respect to $\varphi \in M$.

3. *Continuous linear operators in the space $D^*(O)$*

Linear operators in the space of generalized functions can be constructed from operators in the space of basic functions by taking adjoints.

Let T denote a continuous linear operator that maps the space $D(O')$ into $D(O)$. For the operator T, we define the adjoint operator T^*, which maps $D^*(O)$ into $D^*(O')$ according to the formula

$$(T^*f, \varphi) = (f, T\varphi), \quad f \in D^*(O), \quad \varphi \in D(O'). \tag{13}$$

The operator T^* is continuous and hence maps a bounded set in $D^*(O)$ into a bounded set in $D^*(O')$. Thus, the operator ψf in $D^*(O)$ which indicates multiplication by a function $\psi \in C^{(\infty)}(O)$ is defined, in accordance with (13), by the formula

$$(\psi f, \varphi) = (f, \psi\varphi), \quad f \in D^*(O), \quad \varphi \in D(O).$$

We define the derivative $D^\alpha f$ in the space $D^*(O)$ by the formula

$$(D^\alpha f, \varphi) = (-1)^{|\alpha|}(f, D^\alpha \varphi), \quad f \in D^*(O), \quad \varphi \in D(O).$$

From this it is clear that the differential operator $D^\alpha f$ in $D^*(O)$ is conjugate to the operator $(-1)^{|\alpha|}D^\alpha\varphi$ in $D(O)$.

Let $f_0 \in D^*(O)$ be a convolutor; that is, suppose that f_0 possesses the property that the convolution operator (see section 3.1)

$$f_0 * \varphi = (f_0(\xi), \varphi(x+\xi))$$

is continuous from $D(O')$ into $D(O)$. For example, if $S_{f_0} \subset U(0, r)$, the functional f_0 is a convolutor (from $D(O')$ into $D(O)$ if $\Delta_0(O') > 2r$). The convolution operator $f_0 * f$ in the space of generalized functions is defined as the adjoint operator to the convolution operator defined on the space of basic functions by formula (13):

$$(f_0 * f, \varphi) = (f, f_0 * \varphi), \quad f \in D^*(O), \quad \varphi \in D(O').$$

It follows from this definition that if $f_0(x)$ is a finite summable function and $f(x)$ is a locally summable function, then

$$(f_0 * f)(x) = \int f(\xi) f_0(x - \xi)\, d\xi.$$

If $\varphi \in D$, then

$$(\varphi * f)(x) = (f(\xi), \varphi(x-\xi)).$$

The operator $\varphi * f$ maps $D^*(O)$ into $C^{(\infty)}(O')$ if $\varphi \in D$, $S_\varphi \subset U(0, r)$, and $\Delta_O(O') > r$.

We have the following equations:

$$D^\alpha(f_0 * f) = D^\alpha f_0 * f = f_0 * D^\alpha f. \tag{14}$$

Suppose that a function $\omega(x)$ is such that (see section 3.1)

$$0 \leqslant \omega(x) \in D, \quad S_\omega \subset U(0, 1), \quad \int \omega(x)\, dx = 1.$$

Then,

$$\omega_\varepsilon(x) = \varepsilon^{-n} \omega\left(\frac{x}{\varepsilon}\right) \to \delta(x), \quad \varepsilon \to +0 \quad (\text{in } D^*).$$

Therefore, for an arbitrary generalized function $f \in D^*(O)$, we have the limit relation

$$\omega_\varepsilon * f \to f \text{ as } \varepsilon \to +0 \text{ (in an arbitrary } D^*(O'), \text{ where } O' \Subset O). \tag{15}$$

The function $(\omega_\varepsilon * f)(x)$ is called a *regularization* of the generalized function $f \in D^*(O)$. Since $S_{\omega_\varepsilon} \subset U(0, \varepsilon)$, we have $\omega_\varepsilon * f \in C^{(\infty)}(O')$ if $\Delta_O(O') > \varepsilon$.

Let us show that $D(O)$ is dense in $D^*(O)$.

Suppose that $f \in D^*(O)$. Consider a sequence of functions $\varphi_\varepsilon(x)$ in $D(O)$ such that $\varphi_\varepsilon(x) = 1$ for $\Delta_O(x) > \varepsilon$ and $|x| < 1/\varepsilon$. Let $\Delta_O(S_{\varphi_\varepsilon}) = 3\delta_\varepsilon$. Then, the sequence of functions $\omega_{\delta_\varepsilon} \times (f\varphi_\varepsilon)$ in $D(O)$ approaches f in $D^*(O)$ as $\varepsilon \to +0$, q.e.d.

4. *Measures*

A *measure* on a set A is a completely additive (complex) function that has a finite value on all Borel sets that are compact in A. A measure μ in an open set O defines a generalized function $\mu \in D^*(O)$ by the formula

$$(\mu, \varphi) = \int \varphi(x)\, d\mu(x).$$

Furthermore, this formula gives the general form of a continuous linear functional on the space $D^{(0)}(O)$. This assertion follows from the following theorems:

THEOREM (F. Riesz). *Every continuous linear function f on the space of continuous functions defined on a closed set K can be represented in the form*

$$(f, \varphi) = \int_K \varphi(x)\, d\mu(x), \quad \|f\| = \int_K |d\mu(x)|,$$

where μ is a measure on K.

THEOREM (**Hahn-Banach**). *Suppose that f is a continuous linear functional defined on a subspace m of a linear normed space M. Then, there exists a continuous linear functional F on M such that*

$$(F, \varphi) = (f, \varphi), \quad \varphi \in m; \quad \|F\|_M = \|f\|_m = \sup_{\|\varphi\| \leqslant 1, \ \varphi \in m} (f, \varphi).$$

Suppose that the measure $\mu(\varepsilon)$ is *absolutely continuous* in O, that is, that there exists a locally summable "density" $f(x)$ in O that, for an arbitrary Borel set $\varepsilon \Subset O$,

$$\mu(\varepsilon) = \int_\varepsilon f(x)\, dx.$$

In this case, let us agree to identify the generalized functions (μ, φ) and (f, φ) and to write $\mu = f$.

We shall say that a generalized function $f \in D^*(O)$ is positive in O and to write $f \geqslant 0$ if $(f, \varphi) \geqslant 0$ for all $\varphi \geqslant 0$ in $D(O)$. Thus, a positive measure in O defines a positive generalized function in O. The converse is also true: *Every positive generalized function in $D^*(O)$ defines a unique positive measure.*

The simplest example of a positive measure is the delta-function.

Suppose that a measure μ is defined on the entire space R^n. We shall say that this measure is of *power increase* if

$$\int \frac{|d\mu(x)|}{(1 + |x|)^\alpha} < \infty$$

for some $\alpha \geqslant 0$.

5. *The structure of generalized functions of finite order*

Suppose that $f \in D^*(O)$. The smallest integer m in $[0, \infty]$ such that $f \in D^{(m)*}(O)$ is called the *order* of the generalized function f in O.

For example, the order of a measure is equal to 0. If $f \in D^*(O)$ and $O' \Subset O$, then $f \in D^*(O')$ and the order of f in O' is always finite. In particular, if the carrier of f is compact in O, then f is of finite order in O.

If the order of $\in D^*(O)$ is equal to $m < \infty$ in O, then

$$f = \sum_{|\alpha| \leqslant m} D^\alpha \mu_\alpha$$

where the μ_α are measures in O. This result follows from the Riesz and Hahn-Banach theorems.

If the order of $f \in D^*(O)$ is equal to m in O, then $(f,\phi) = 0$ whenever $\phi \in D(O)$ and $D^\alpha \phi(x) = 0$, where $x \in S_f$ and $|\alpha| \leq m$.

However, if a sequence of functions φ_k in $D(O)$ is such that $D^\alpha\varphi_k(x) \to 0$ uniformly on S_f for all $|\alpha| \leqslant m$ as $k \to 0$, the quantity (f, φ_k) does not in general approach zero as $k \to \infty$ even if S_f is compact in O. On the other hand, if S_f is a regular set, such inconveniences cannot occur (for more details, see section 3.8).

A closed set F is said to be *regular* if there exist positive numbers d, ω, q (where $q \leqslant 1$) such that any two points x and x' in F at a distance $|x - x'| \leqslant d$ from each other can be connected in F by a rectifiable curve of length $l \leqslant \omega |x - x'|^q$.

For example, every convex closed set F is regular. In this case, we may take d as equal to the diameter F and $\omega = q = 1$.

Finally, we note that every generalized function f whose carrier is the point x^0 has a unique representation of the form

$$f(x) = \sum_{|\alpha| \leqslant m} c_\alpha D^\alpha \delta(x - x^0) \tag{16}$$

for a nonnegative number m and complex numbers c_α.

For a generalized function $f \in D^*(O)$ to be of compact carrier in O, it is necessary and sufficient that it be extended as a continuous linear functional on the space $C^{(\infty)}(O)$ with uniform convergence of each derivative on an arbitrary set that is compact in O.

6. *The space of basic functions S*

We shall refer to the (linear) set of functions belonging to the class $C^{(\infty)}(R^n)$ that, together with all their derivatives, decrease at infinity faster than an arbitrary power of $|x|^{-1}$ as the *set of basic functions S*. We introduce a topology in S by means of the countably many norms

$$\|\varphi\|_m = \sup_{x,\, |\alpha| \leqslant m} (1 + |x|)^m |D^\alpha \varphi(x)|, \quad \varphi \in S, \quad m = 0, 1, \ldots$$

Convergence in S is defined in terms of this topology: $\varphi_\alpha \to 0$ in S if $\|\varphi_\alpha\|_m \to 0$ for arbitrary m.

We denote by $S^{(m)}$ the completion of S with respect to the mth norm. The spaces $S^{(m)}$ are separable Banach spaces. D is dense in $S^{(m)}$.

For ϕ to belong to $S^{(m)}$, it is necessary and sufficient that ϕ belong to $C^{(m)}(R^n)$ and that $|x|^m D^\alpha \phi(x) \to 0$ as $|x| \to \infty$ for all $|\alpha| \leq m$.

Since the norms $\|\varphi\|_m$ increase with increasing m $\|\varphi\|_m \leqslant \|\varphi\|_{m+1}$ and are pairwise consistent (that is, every sequence that is a Cauchy sequence with respect to both of the norms and that converges to zero with respect to one of them must also converge to zero with respect to the other norm), it follows that the $S^{(m)}$ constitute a decreasing chain of spaces, each contained in the preceding one:

$$S^{(0)} \supset S^{(1)}, \ldots .$$

Here,

$$S = \bigcap_{m \geqslant 0} S^{(m)}.$$

There, S is a *complete space.*

We shall say that the set M is bounded in S if $\|\varphi\|_m \leqslant K_m$, where $\varphi \in M$ and $m = 0, 1, \ldots$. *Every closed set that is bounded in S is also compact in S.* (S is a Montel space.)

Let us denote by θ_M the class of multipliers in the space S, that is, the set of all functions in $C^{(\infty)}(R^n)$ multiplication by which is a continuous operator from S into S. It can be shown that *a function ψ in $C^{(\infty)}(R^n)$ belongs to θ_M if and only if all its derivatives are of power increase.*

Linear differential operators $D^\alpha \varphi(x)$ and displacement operators $\varphi(s+h)$ are continuous from S into S. The Fourier-transform operator

$$F[\varphi](x) = \int e^{ix\xi} \varphi(\xi) \, d\xi$$

performs a linear homemorphic mapping of S onto S. Here, the inverse operator is of the form

$$F^{-1}[\varphi](\xi) = \frac{1}{(2\pi)^n} \int e^{-ix\xi} \varphi(x) \, dx = \frac{1}{(2\pi)^n} F[\varphi(-x)](\xi).$$

7. *The space of generalized functions S**

We shall refer to the set of all continuous linear functionals on S as the *space of generalized functions* S^* (and we shall call them generalized functions of slow growth or *tempered distributions*).

Every measure of power increase (see section 3.4) and also every generalized function of compact carrier defines a functional in S^*. If $f \in S^*$, obviously $f \in D^*(O)$ for every S^*. Therefore, all the assertions made regarding generalized functions in the space $D^*(O)$ remain valid for generalized functions in the space S^*.

The spaces $S^{(m)*}$ dual to the $S^{(m)}$ constitute an increasing chain $S^{(0)*} \subset S^{(1)*} \subset \ldots$. The space S^* is the union of the spaces $S^{(m)*}$:

$$S^* = \bigcup_{m \geqslant 0} S^{(m)*}.$$

Therefore, every generalized function $f \in S^*$ belongs to $S^{(m)*}$ for some

$m \geqslant 0$. The smallest N such that $f \in S^{(N)*}$ is called the *order* of f. Thus, every $f \in S^*$ is of finite order $N = N(f)$. For each f, we may introduce the decreasing sequence of norms

$$\|f\|_{-m} = \sup_{\|\varphi\|_m = 1,\ \varphi \in S^{(m)}} (f, \varphi), \qquad m = N, N+1, \ldots .$$

From this it follows that *if $f(x)$ is a locally summable function and $f(x) \in S^*$, it does not increase faster than a polynomial* *.

We define convergence in S^* as weak convergence: $f_\alpha \to f$ in S^* if $(f_\alpha, \varphi) \to (f, \varphi)$ for all $\varphi \in S$. Since S is a perfect space, weak convergence in S^* implies strong convergence, that is, convergence with respect to the norm in some space $S^{(m)*}$: $\{f_\alpha\} \subset S^{(m)*}$ and $\|f_\alpha - f\|_{-m} \to 0$ as $\alpha \to \infty$.

The space S^* is complete with respect to weak (and, consequently, strong) *convergence.*

We shall say that a set $M^* \subset S^*$ is (weakly) bounded if $|(f, \varphi)| \leqslant c(\varphi)$ for all $f \in M^*$. In this case the set M^* is (strongly) bounded, that is, $M^* \subset S^{(m)*}$ and $\|f\|_{-m} \leqslant K$, where $f \in M^*$ for some m.

For a generalized function f in D^* to belong to S^*, it is necessary and sufficient that all its regularizations $\varphi * f$ for $\varphi \in D$ belong to θ_M.

Finally, we note that the space S is reflexive: $S = S^{**}$.

In the proof of these assertions, an important role is played by the

THEOREM (Banach-Steinhaus). *For a sequence $\{f_\alpha\}$ of continuous linear functionals on a Banach space B to converge weakly to a continuous linear functional on B, it is necessary and sufficient that*

(1) *the sequence of norms $\{\|f_\alpha\|\}$ be bounded and*

(2) *the limit $\lim_{\alpha \to \infty} (f_\alpha, \phi)$ exist for all ϕ in some set that is dense in B.*

8. *The structure of generalized functions in the space S**

If $f \in S^*$,

$$f = \sum_{|\alpha| \leqslant M} D^\alpha g_\alpha, \tag{17}$$

where the $g_\alpha(x)$ are continuous functions of power increase with carrier contained in a arbitrary neighborhood S_f^ϵ on the set S_f.

If the carrier S_f is a regular set, this theorem can be strengthened. The strengthening is based on the following theorem on the extension of differentiable functions.

---

*Editor's Note: This is not true; e.g., f could be the derivative of a function of polynomial growth.

Suppose that F *is a regular set (characterized by the numbers* $d > 0$, $\omega > 0$, *and* $0 < q \leq 1$ [see section 3.5]) *and that* m' *is a nonnegative integer. Then, a number* K *exists such that, for an arbitrary function* ϕ *in* $C^{(m)}(R^n)$, *where* m *is the smallest integer verifying the inequality* $mq \geq m'$, *there exists a function* ψ *in* $C^{(m)}(R^n)$ *with the properties*

$$D^\alpha\varphi(x) = D^\alpha\psi(x), \quad x \in F, \quad |\alpha| \leqslant m; \quad \|\psi\|_{m'} \leqslant K\|\varphi\|_{m,F}, \tag{18}$$

where

$$\|\varphi\|_{m,F} = \sup_{x \in F,\ |\alpha| \leqslant m} (1+|x|)^m |D^\alpha\varphi(x)|.$$

This theorem is proven in the author's article [42] by means of Whitney's construction [80] in the form given by Hörmander [81].

THEOREM. *Suppose that* $f \in S^*$ *and* $S_f \subset F$, where F is *a regular set. Then,*

$$f = \sum_{|\alpha| \leqslant M} D^\alpha \mu_\alpha, \tag{19}$$

where the $\mu_\alpha(x)$ *are measures of power increase with carriers in* F. (We note that the representation (19) is a generalization of formula (16).)

We indicate the proof of this theorem. Let N denote the order of f and let m denote the smallest integer that verifies the inequality $mq \geqslant N+1$ (where q is determined from F). According to the theorem on the extension of differentiable functions for arbitrary $\varphi \in S$, there exists a $\psi \in C^{(m)}(R^n)$ with properties (18):

$$D^\alpha\varphi(x) = D^\alpha\psi(x), \quad |\alpha| \leqslant m, \quad x \in F; \quad \|\psi\|_{N+1} \leqslant K\|\varphi\|_{m,F},$$

from which it follows that $\psi \in S^{(N)}$ (see section 3.6), that $(f, \varphi) = (f, \psi)$ (see section 3.5), and that

$$|(f, \varphi)| \leqslant \|f\|_{-N}\|\psi\|_N \leqslant \|f\|_{-N}\|\psi\|_{N+1} \leqslant$$
$$\leqslant K\|f\|_{-N}\|\varphi\|_{m,F}, \quad \varphi \in S.$$

From this inequality and the Riesz and Hahn-Banach theorems (see section 3.4), we obtain the required representation (19) for some $M \geqslant m$, q.e.d.

9. *Fourier transformations and convolutions*

As was pointed out in section 3.6, the Fourier transformation of basic functions in S is a continuous operator from S onto S with a continuous inverse operator. In accordance with section 3.3, we define the Fourier transformation $F[f]$ of a generalized function f in S^* as the operator adjoint to the operator F in the basic space S:

$$(F[f],\ \varphi)=(f,\ F[\varphi]),\quad f\in S^*,\quad \varphi\in S.$$

If $f\in L_2$, that is, if

$$\|f\|=\left[\int|f(x)|^2\,dx\right]^{\frac{1}{2}}<+\infty,$$

this operator $F[f]$ coincides with the ordinary Fourier transformation:*

$$F[f](x)=\int f(\xi)\,e^{i\xi x}\,d\xi=\operatorname{l.i.m}_{R\to\infty}\int_{|\xi|\leqslant R} f(\xi)\,e^{i\xi x}\,d\xi,$$

where l. i. m. should be understood in the L_2 sense. Then, Parseval's equality

$$\|F[f]\|^2=(2\pi)^n\,\|f\|^2$$

holds.

The carrier S_f is called the spectrum of the Fourier transformation $F[f]$.

The operator F performs a homeomorphic mapping of S^* onto S^* with the inverse operator $F^{-1}[f]=(2\pi)^{-n}F[f(-\xi)]$.

The following formulas are immediate consequences of the definition:

$$D^\alpha F[f]=F[(i\xi)^\alpha f];\quad F[D^\alpha f]=(-ix)^\alpha F[f]. \tag{20}$$

Furthermore, since $F[\delta]=1$ and $F[1]=(2\pi)^n\delta$, we obtain from formulas (20)

$$F\left[\xi^\alpha\right]=(2\pi)^n(-i)^{|\alpha|}D^\alpha\delta;\quad F\left[D^\alpha\delta\right]=(-ix)^\alpha. \tag{21}$$

Finally, we note the diagonalization formulas,

$$F[f(\xi-h)]=e^{ixh}F[f];\quad F[e^{i\xi h}f]=F[f](x+h), \tag{22}$$

where the symbol $f(\xi-h)$ denotes the displacement operator in the space S^* dual to the displacement operator $\varphi(\xi+h)$ in the basic space S.

If the functional f is of compact carrier, then

$$F[f](x)=(f(\xi),\ e^{ix\xi}). \tag{23}$$

---

*The theory of the Fourier transformation of functions in L_2 is expounded in the book by Bochner [154].

Let f_0 be a generalized function in S^* (e.g., a finite one) such that, for arbitrary $\varphi \in S$, the convolution

$$(f_0 * \varphi)(x) = (f_0(\xi), \ \varphi(\xi + x))$$

belongs to S and this operation is continuous from S into S, so that f_0 is the convolutor in S. We denote by θ_C^* the class of all convolutors in S (the generalized functions of rapid decrease).

THEOREM. *For* F_0 *to belong to* θ_C^*, *it is necessary and sufficient that* $F[f_0]$ *belong to* θ_M, *where* θ_M *is the class of multipliers in the space* S (*and, consequently, in the space* S^*).

Examples: $e^{i\xi^2} \in \theta_C^* \cap \theta_M$ since

$$F[e^{i\xi^2}] = \sqrt{\frac{\pi}{2}}(1+i)\,e^{-i\frac{x^2}{4}}.$$

If $f \in S^* \cap C^{(\infty)}$, then $f \in \theta_M$.

We define the convolution $f_0 * f$ in the space S^* (with convolutor $f_0 \in \theta_C^*$) as the operator adjoint to the convolution operator in the space S (see section 3.3):

$$(f_0 * f, \ \varphi) = (f, \ f_0 * \varphi), \quad f \in S^*, \ \varphi \in S.$$

For the convolution $f_0 * f$, the differentiation formulas (14) are valid.

We note the important formulas

$$F[f_0 * f] = F[f_0] \cdot F[f] \text{ if } f_0 \in \theta_C^* \text{ and } f \in S^*; \tag{24}$$

$$(2\pi)^n F[\psi f] = F[\psi] * F[f] \text{ if } \psi \in \theta_M \text{ and } f \in S^*. \tag{25}$$

It is easy to verify by use of formulas (24) and (25) that

$$\varphi * f \in \theta_M \quad \text{for all} \quad f \in S^* \text{ and } \varphi \in S; \tag{26}$$

$$\varphi f \in \theta_C^* \quad \text{for all} \quad f \in S^* \text{ and } \varphi \in S; \tag{27}$$

$$f_0 * g_0 \in \theta_C^* \quad \text{for all} \quad f_0 \in \theta_C^* \text{ and } g_0 \in \theta_C^*. \tag{28}$$

10. Differential equations with constant coefficients

Such equations are of the form

$$P(iD)\,u = f(x), \tag{29}$$

where $P(\xi)$ is a polynomial with constant coefficients (independent of x), f is a known and u an unknown generalized function in the space $D^*(O)$. If $f = 0$, Eq. (29) is said to be *homogeneous*.

Every solution $u \in D^*(O)$ of the homogenous equation (29) is the limit (in the sense of $D^*(O)$ of a sequence $u_\alpha(x)$, where $\alpha = 1, 2, \ldots$, of infinitely differentiable solution of the equation in the open sets

$$O_\alpha = \left[x : \Delta_O(x) > \frac{1}{\alpha},\ x \in O\right].$$

Specifically, by virtue of what was said in section 3.3, such a sequence is defined by

$$u_\alpha = \omega_{\frac{1}{\alpha}} * u, \quad \alpha = 1, 2, \ldots .$$

A *fundamental solution* $E(x)$ of Eq. (29) is a solution of this equation for $f = \delta(x)$:

$$P(iD)E(x) = \delta(x). \tag{30}$$

Every equation of the form (29) has a fundamental solution $E \in S^*$. Specifically, on the basis of (30), we have

$$P(\xi)F[E] = 1.$$

But the possibility of dividing by a polynomial in the space S^* has been established by Hörmander [81] and Loyasevich [82].

For example, Laplace's equation $\Delta u = 0$ has the fundamental solution

$$E(x) = \begin{cases} -\dfrac{1}{(n-2)\sigma_n} r^{2-n}, & n \geqslant 3, \\ \dfrac{\ln r}{2\pi}, & n = 2, \end{cases} \qquad \sigma_n = \frac{2\pi^{\frac{n}{2}}}{\Gamma\left(\frac{n}{2}\right)}.$$

Suppose that the right member f of Eq. (29) is a positive measure μ in O (see section 3.4). Then, we have the following

THEOREM. *Every solution $u \in D^*(O)$ of the equation $P(iD)\, u = \mu$, where the measure $\mu \geq 0$ in O, is represented in an arbitrary $O' \Subset O$ by the expansion formula of F. Riesz:*

$$u = E * \mu_{O'} + u_0, \tag{31}$$

where $\mu_{O'}(\epsilon) = \mu(\epsilon \cap O')$ is the restriction of the measure μ to O' and u_0 is a solution in $D^*(O')$ of the homogenous equation corresponding to Eq. (29).

This is true because, by virtue of formulas (14) and (30), we have in O'

$$P(iD)\, u = P(iD)\, E * \mu_{O'} + P(iD)\, u_0 = \delta * \mu_{O'} = \mu.$$

11. *Hypoelliptic differential operators*

A differential operator $P(iD)$ is said to be *hypoelliptic* if every solution u in $D^*(O)$ of the homogeneous equation $P(iD)u=0$ belongs to $C^{(\infty)}(O)$.

THEOREM (**Hörmander** [85]). *For an operator $P(iD)$ to be hypoelliptic, it is necessary and sufficient that*

$$\frac{D^\alpha P(\xi)}{P(\xi)} \to 0, \quad |\xi| \to \infty, \quad |\alpha| > 0. \tag{32}$$

It follows from this that, in particular, every solution of Laplace's equation $\Delta u = 0$ in $D^*(O)$ is an infinitely differentiable function (in fact, an analytic function).

Suppose that $x=(x_1, \ldots, x_m)$, $y=(y_1, \ldots, y_n)$, $O \subset R^m \times R^n$, and $f(x, y) \in D^*(O)$. We shall say that f is *infinitely differentiable* with respect to x if

$$(f(x, y), \varphi(y)) \in C^{(\infty)}(O_1)$$

for all $O_1 \subset R^m$ and $O_2 \subset R^n$ such that $O_1 \times O_2 \subset O$ and for all $\varphi \in D(O_2)$.

An operator $P(iD_x, iD_y)$ is said to be hypoelliptic with respect to x if every solution of the homogeneous equation

$$P(iD_x, iD_y)u=0$$

is infinitely differentiable with respect to x.

THEOREM (**Gårding and Malgrange** [83, 84]).* *For an operator $P(iD_x, iD_y)$ to be hypoelliptic with respect to x, it is necessary and sufficient that*

$$P(\xi, \eta) = P_0(\xi) + \sum_{j \geqslant 1} P_j(\xi) Q_j(\eta), \tag{33}$$

where $P_0(iD_x)$ is a hypoelliptic operator and

$$\frac{P_j(\xi)}{P_0(\xi)} \to 0, \quad |\xi| \to \infty, \quad j > 0.$$

It follows from this theorem, that, in particular, every solution in D^* of the equation

$$i^k \frac{\partial^k u}{\partial x_1^k} = P(iD)u, \tag{34}$$

that can be solved for the highest derivative with respect to x_1 is

---

*Editor's note: This is also proven by L. Ehrenpreis, "Solution of some problems of division, Part IV," American Jour. of Math, Vol. 82 (1962), pp. 522-588.

infinitely differentiable with respect to x_1. Therefore, for every fixed a, there exists a generalized function $u(a, \tilde{x}) \in D^*$, where $\tilde{x} = (x_2, x_3, \ldots, x_n)$. Here,

$$\lim_{a' \to a} u(a', \tilde{x}) = u(a, \tilde{x})$$

in the sense of convergence in D^*.

On the other hand, if a solution u of Eq. (34) belongs to S^*, then $u(a, \tilde{x}) \in S^*$.

To see this, note that for every $\varphi(\tilde{x}) \in D$, the convolution $\varphi * u = (u(x_1, \tilde{x}), \varphi(\tilde{x} - \tilde{x}'))$ is infinitely differentiable with respect to all arguments and belongs to S^*. Consequently, $\varphi * u \in \theta_M$ (see section 3.6). Then, $(u(a, \tilde{x}'), \varphi(\tilde{x} - \tilde{x}')) \in \theta_M$ for every a. Therefore, by virtue of the criterion in section 3.7, it follows that $u(a, \tilde{x}) \in S^*$, as asserted.

12. *Schwartz' theorem of the kernel*

For a linear operator T to be continuous from $D(O_x)$ into $D^*(O'_y)$, it is necessary and sufficient that there exist a generalized function $f_T \in D^*(O_x \times O'_y)$, such that

$$(T\varphi, \psi) = (f_T, \varphi\psi), \qquad \varphi \in D(O_x), \quad \psi \in D(O'_y).$$

An analogous theorem holds for the space S^*.

4. DEFINITIONS AND SIMPLEST PROPERTIES OF HOLOMORPHIC FUNCTIONS

1. *Definitions of holomorphic functions*

A function $f(z)$ is said to be holomorphic at a point $z^0 \in C^n$ if, in some neighborhood of z^0, it is the sum of an absolutely convergent power series

$$f(z) = \sum_{|\alpha| \geqslant 0} a_\alpha (z - z^0)^\alpha. \tag{35}$$

(This is Weierstrass' definition.)

A function $f(z)$ is said to be holomorphic at a point $z^0 \in C^n$ if, in some neighborhood of z^0, all the first partial derivatives $\partial f / \partial \bar{z}_j = 0$, for $j = 1, 2, \ldots, n$, that is, if the Cauchy-Riemann conditions

$$\frac{\partial u}{\partial x_j} = \frac{\partial v}{\partial y_j}, \qquad \frac{\partial u}{\partial y_j} = -\frac{\partial v}{\partial x_j}, \qquad j = 1, 2, \ldots, n, \tag{36}$$

where $f = u + iv$ and $z_j = x_j + iy_j$ are satisfied. (This is Riemann's definition.)

Thus, a function is holomorphic at a point z^0 according to Riemann's definition if it is holomorphic with respect to each variable z_j separately (with the other variables fixed) in some neighborhood of that point.

The first definition is also suitable for defining holomorphic functions of real variables whereas the second definition is not since there exist infinitely differentiable nonholomorphic functions of a real variable, for example,

$$f(x)=e^{-\frac{1}{x^2}} \quad \text{for} \quad x \neq 0, \; f(0)=0).$$

Suppose that a function $f(z)$ is holomorphic at a point z^0 in the sense of Weierstrass. Then, the series (35) converges absolutely in the closed polycircle $\bar{S}(z^0, r)$ and, hence,

$$|a_\alpha| \leqslant \frac{M}{r^\alpha}, \qquad |\alpha|=0, 1, \ldots \tag{37}$$

for some $M>0$. (Here, $r=(r_1, r_2, \ldots, r_n)$ [cf. the notations in section 1.2].) Consequently, in $S(z^0, r)$, the series (35) is dominated by the series

$$M \sum_{|\alpha| \geqslant 0} \left(\frac{z-z^0}{r}\right)^\alpha = M\left(1-\frac{z-z^0}{r}\right)^{-I};$$

it can be termwise differentiated infinitely many times and all its derivatives are continuous functions. In particular,

$$a_\alpha = \frac{1}{\alpha!} D^\alpha f(z^0). \tag{38}$$

Thus, *every function that is holomorphic in the sense of Weierstrass is continuous and infinitely differentiable, and all its derivatives are also holomorphic functions.* In particular, every function that is holomorphic in the sense of Weierstrass is holomorphic in the sense of Riemann also.

From this and from the Cauchy-Riemann conditions (36), it follows that the real and imaginary parts u and v of a holomorphic function (these are referred to as *conjugate functions*) are infinitely differentiable and satisfy the system of differential equations

$$\frac{\partial^2 u}{\partial x_j \partial x_k} = -\frac{\partial^2 u}{\partial y_j \partial y_k}, \quad \frac{\partial^2 u}{\partial x_j \partial y_k} = \frac{\partial^2 u}{\partial y_j \partial x_k}. \tag{39}$$
$$j, k = 1, 2, \ldots, n.$$

Every generalized function $u(x, y) \in D^*(G)$ that satisfies the system of equations (39) is said to be pluriharmonic (harmonic in the case in which $n=1$) in the domain G.

Every pluriharmonic function is infinitely differentiable.

This is true because Eqs. (29) imply that the pluriharmonic function $u \in D^*(G)$ satisfies Laplace's equation

$$\Delta_x u + \Delta_y u = 0,$$

in the domain G and hence is infinitely differentiable in G (see section 3.11).

2. *The equivalence of the definitions of holomorphic functions*

In the preceding section, it was shown that a function that is holomorphic in the sense of Weierstrass is also holomorphic in the sense of Riemann. The converse assertion, constituting the fundamental theorem of Hartogs [6], is also valid:

If a function is holomorphic with respect to each variable individually, it is also holomorphic with respect to the entire set of variables (in the sense of Weierstrass).

The proof of this theorem is not trivial, and we shall postpone it until section 15.8. We note that this theorem would not be valid for functions of real variables. For example, the function

$$f(x, y) = \frac{xy}{x^2 + y^2} \quad \text{as} \quad (x, y) \neq (0, 0), \; f(0, 0) = 0$$

is holomorphic for each variable individually in a neighborhood of the point (0, 0), but it is not even continuous at that point.

Here, we shall prove this theorem under the additional hypothesis that the function in question is bounded.

Suppose that a function $f(z)$ is holomorphic with respect to each variable z_j individually and that it is bounded in the closed* polycircle $\bar{S}(z^0, r)$. Then, if we apply Cauchy's formula (for a single variable) n times, we obtain for all $z \in S(z^0, r)$

$$a_\alpha = \frac{1}{(2\pi)^n} \int_0^{2\pi} \cdots \int_0^{2\pi} \frac{f(z^0 + re^{i\theta})}{r^\alpha} e^{-i\theta\alpha} \, d\theta = \frac{1}{(2\pi i)^n} \int\limits_{\partial S(z_1^0, r_1) \times \cdots \times \partial S(z_n^0, r_n)} \frac{f(\zeta) \, d\zeta}{(\zeta - z^0)^{\alpha + I}} \tag{40}$$

where we take the positive direction on $\partial S(z_j^0, r_j)$ counterclockwise. By hypothesis, the integrand

*By virtue of the Heine-Borel theorem, this means that $f(z)$ is holomorphic with respect to each variable individually in the somewhat larger polycircle $S(z^0, r + \varepsilon I)$ for some $\varepsilon > 0$.

$$f(\zeta)=f(z_1^0+r_1e^{i\theta_1},\ z_2^0+r_2e^{i\theta_2},\ \ldots,\ z_n^0+r_ne^{i\theta_n})\equiv$$
$$\equiv f(z^0+re^{i\theta})$$

in Eq. (40) is uniformly bounded in absolute value for $0\leqslant\theta_j\leqslant 2\pi$, where $j=1,\ 2,\ \ldots,\ n$. It is easy to show that this function is measurable. Therefore, it is summable and, from Fubini's theorem (see section 2.9), the iterated integral (40) may be replaced with the multiple integral

$$\begin{aligned} f(z)&=\frac{r^I}{(2\pi)^n}\int_0^{2\pi}\cdots\int_0^{2\pi}\frac{f(z^0+re^{i\theta})\,e^{i\theta I}\,d\theta}{(z^0+re^{i\theta}-z)^I}\equiv\\ &\equiv\frac{1}{(2\pi i)^n}\int\limits_{\partial S(z_1^0,\ r_1)\times\ldots\times\partial S(z_n^0,\ r_n)}\frac{f(\zeta)\,d\zeta}{(\zeta-z)^I},\quad z\in S(z^0,\ r).\end{aligned}\tag{41}$$

Suppose that $z\in S(z^0,\ r')$, where $r'_j<r_j$. Then, the series

$$\frac{1}{(z^0+re^{i\theta}-z)^I}=\sum_{|\alpha|\geqslant 0}\frac{(z-z^0)^\alpha}{r^{\alpha+I}}\,e^{-i\theta\,(\alpha+I)}\tag{42}$$

converges absolutely and uniformly with respect to z and θ. Since $f(\zeta)$ is bounded, the series (42) can be substituted into formula (41), and the order of integration and summation can be reversed. This leads us to Taylor's series (35), which converges absolutely in the polycircle $S(z^0,\ r)$. The coefficients in this series are

$$\begin{aligned} a_\alpha&=\frac{1}{(2\pi)^n}\int_0^{2\pi}\cdots\int_0^{2\pi}\frac{f(z^0+re^{i\theta})}{r^\alpha}\,e^{-i\theta\alpha}\,d\theta\equiv\\ &\equiv\frac{1}{(2\pi i)^n}\int\limits_{\partial S(z_1^0,\ r_1)\times\ldots\times\partial S(z_n^0,\ r_n)}\frac{f(\zeta)\,d\zeta}{(\zeta-z^0)^{\alpha+I}}.\end{aligned}\tag{43}$$

This proves the assertion.

Thus, on the basis of this theorem, we no longer need to specify whether a function is holomorphic in the sense of Riemann or Weierstrass and we may treat these two concepts of holomorphy as equivalent.

We shall say that a function $f(z)$, defined in a domain $G\subset C^n$ is holomorphic in G if it is holomorphic at every point in G.

It follows from this definition that holomorphy of a function in a domain G implies its single-valuedness in G.

We shall denote by H_G the set of functions that are holomorphic in a region G.

We are now in a position to prove the following assertion: *If $f\in D^*(G)$ satisfies the Cauchy-Riemann conditions, $f(z)$ is holomorphic in G.*

Proof: In this case, the real and imaginary parts belong to $D^*(G)$ and satisfy the system of equations (39). Therefore, $f(z)\in C^{(\infty)}(G)$

(cf. section 4.1). But then the function $f(z)$ satisfies the Cauchy-Riemann conditions (36) in the usual sense and, hence, is holomorphic with respect to each individual variable in the domain G. This means that $f(z)$ is holomorphic in G, q.e.d.

3. *Cauchy's formula*

Suppose that $f(z)$ is holomorphic in a domain $G = G_1 \times G_2 \times \ldots \times G_n$ where each G_j is a domain in the z_j-plane with a piecewise-smooth boundary ∂G_j, and is continuous in $\bar{G}$. Then, by following the reasoning of the preceding section, we can verify that Cauchy's formula is valid for the function $f(z)$:

$$\frac{1}{(2\pi i)^n} \int\limits_{\partial G_1 \times \ldots \times \partial G_n} \frac{f(\zeta)\, d\zeta}{(\zeta - z)^I} = \begin{cases} f(z), & \text{if} \quad z \in G, \\ 0, & \text{if} \quad z \bar{\in} \bar{G}, \end{cases} \tag{44}$$

where the positive direction on ∂G_j is taken in such a way that the domain G_j is on one's left. We note that formula (41) is a particular case of formula (44) with $G = S(z^0, r)$.

Cauchy's formula (44) expresses the values of $f(z)$ in a $2n$-dimensional domain $G = G_1 \times G_2 \times \ldots \times G_n$ in terms of its values on the n-dimensional oriented manifold $\partial G_1 \times \partial G_2 \times \ldots \times \partial G_n$—that is, the hull of the domain G. The hull constitutes a portion of the boundary ∂G of the domain G. We note that ∂G consists of points z of the form

$$z_j \in \partial G_j, \ (z_1, \ldots, z_{j-1}, z_{j+1}, \ldots, z_n) \in \bar{G}_1 \times \ldots$$
$$\ldots \times \bar{G}_{j-1} \times \bar{G}_{j+1} \times \ldots \times \bar{G}_n, \quad j = 1, 2, \ldots, n.$$

For generalizations of Cauchy's formula for different regions G, see Chapter IV.

Formula (44) can be differentiated an arbitrary number of times:

$$\frac{\alpha!}{(2\pi i)^n} \int\limits_{\partial G_1 \times \ldots \times \partial G_n} \frac{f(\zeta)\, d\zeta}{(\zeta - z)^{\alpha + I}} = \begin{cases} D^\alpha f(z), & z \in G, \\ 0, & z \bar{\in} \bar{G}. \end{cases} \tag{45}$$

4. *The simplest properties of holomorphic functions*

Suppose that a function $f(z)$ is holomorphic in a polycircle $S(z^0, r)$ and continuous in $\bar{S}(z^0, r)$. Then, in $S(z^0, r)$, this function can be represented in the form of Cauchy's integral (41) and expounded in an absolutely convergent power series with coefficients given by (43). From (43), we have Cauchy's inequality

$$|a_\alpha| \leqslant \frac{1}{r^\alpha} \max_{z \in \bar{S}(z^0, r)} |f(z)|. \tag{46}$$

From this, in turn, we have the following assertion:

If a function $f(z)$ is holomorphic in a domain G, then each of its representations in the form of a power series (35) is unique and is valid in the largest polycircle $S(z^0, r)$ contained in G. Here, the coefficients of the series (35) are calculated from the equivalent formulas (38) and (43).

We note that the second portion of this assertion is not valid for holomorphic functions of real variables.

Suppose that a sequence of functions $f_k(z)$, for $k = 1, 2, \ldots$, that are holomorphic in a domain G is such that, for arbitrary $\phi \in D(G)$, the sequence of numbers $\int f_k(z)\phi(x,y)\,dx\,dy$ has a limit as $k \to \infty$. Then, there exists a function $f(x)$ that is holomorphic in G such that $D^\alpha f_k(z) \to D^\alpha f(z)$ uniformly in G as $k \to \infty$, where $|\alpha| = 0, 1, \ldots$.

Proof: Because of the (weak) completeness of the space $D^*(G)$ (cf. section 3.2), there exists a generalized function $f \in D^*(G)$ such that

$$\lim_{k\to\infty} \int f_k(z)\varphi(x, y)\,dx\,dy = (f, \varphi),$$
$$\varphi \in D(G).$$

Since the functions $f_k(z)$ satisfy the Cauchy-Riemann conditions and the differential operators are continuous from $D^*(G)$ into $D^*(G)$ (cf. section 3.3), it follows that the generalized function f also satisfies these conditions and, consequently, f is a holomorphic function in the domain G (see section 4.2).

Let us show that $D^\alpha f_k(z) \to D^\alpha f(z)$ uniformly in G. Suppose that $S(z^0, rI) \Subset G$. Then, if we apply formula (41) to the functions $f_k(z)$ and remember (see Fig. 3) that

$$S\left(z^0, \frac{r}{4} I\right) \cup S\left(z', \frac{r}{2} I\right) \subset S(z^0, rI),$$

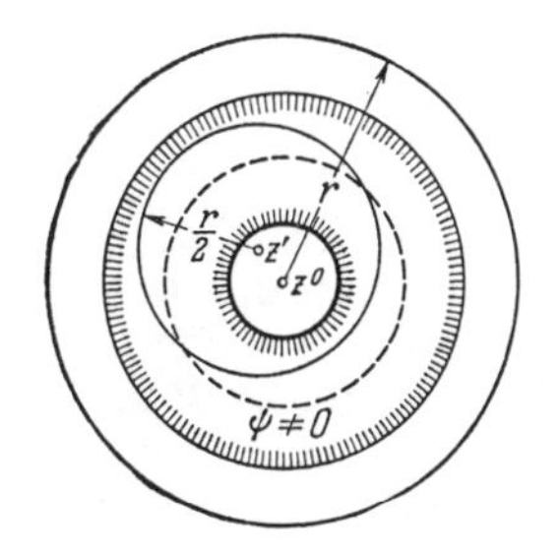

Fig. 3

we obtain for all $(z, z') \in S\left(z^0, \frac{r}{4} I\right) \times S\left(z^0, \frac{r}{4} I\right)$

$$D^\alpha f_k(z) = \frac{\alpha!}{(2\pi i)^n} \int_{\partial S\left(z^0, \frac{r}{2} I\right)} \frac{f_k(\zeta - z^0 + z')\,d\zeta}{(\zeta - z^0 + z' - z)^{\alpha+I}}.$$

Suppose that the function $\varphi(z') \equiv \varphi(x', y')$ belongs to $D[S(0, r/4I)]$ and $\int \varphi(z')\,dx'\,dy' = 1$. If we multiply the preceding equation by $\varphi(z' - z^0)$ and integrate with respect to z', we obtain the following representation for all z in $S(z^0, r/4\, I)$:

$$D^\alpha f_k(z) = \\ = \frac{\alpha!}{(2\pi i)^n} \int \frac{f_k(z')}{(z' - z)^{\alpha+I}} \int_{\partial S\left(z^0, \frac{r}{2} I\right)} \varphi(z' - \zeta)\,d\zeta\,dx'\,dy'. \qquad (47)$$

The function

$$\psi(z') = \int_{\partial S\left(z^0, \frac{r}{2} I\right)} \varphi(z' - \zeta)\, d\zeta$$

is infinitely differentiable and finite. Its carrier is contained in the domain $[z\colon 1/4\ r < |z_j - z_j^0| < 3/4\ r,\ j = 1, 2, \ldots, n]$ (see Fig. 3). Therefore, the set of functions

$$\left\{ \frac{\psi(z')}{(z' - z)^{\alpha + I}},\quad z \in S\left(z^0, \frac{r}{8} I\right) \right\}$$

is bounded in $D(G)$ (see section 3.1). Since $f_k(z) \to f(z)$ in $D^*(G)$, we conclude on the basis of (47) (see section 3.2) that

$$D^\alpha f_k(z) \to$$
$$\to \frac{\alpha!}{(2\pi i)^n} \int \frac{f(z')}{(z' - z)^{\alpha + I}} \int_{\partial S\left(z^0, \frac{r}{2} I\right)} \varphi(z' - \zeta)\, d\zeta\, dx'\, dy' =$$
$$= \frac{\alpha!}{(2\pi i)^n} \int_{\partial S(z^0, rI)} \frac{f(\zeta)}{(\zeta - z)^{\alpha + I}}\, d\zeta = D^\alpha f(z)$$

uniformly with respect to $z \in S(z^0, r/8\, I)$. Since an arbitrary subdomain $G' \Subset G$ can be covered by a finite number of the polycircles $S(z^0, r/8\, I)$ such that $S(z^0, rI) \Subset G$, it follows that $D^\alpha f_k(z) \to D^\alpha f(z)$ uniformly in G' as $k \to \infty$, q.e.d.

We note that the proposition just proven is not valid for holomorphic functions of real variables. For example, the function $\sqrt{x^2 + \varepsilon} \to |x|$ uniformly as $\varepsilon \to +0$ for $|x| \leqslant 1$, although the function $|x|$ is not holomorphic at 0.

5. *Formal derivatives*

Suppose that a function $f(x, y) = u(x, y) + iv(x, y)$ possesses all first-order derivatives. Then,

$$df = \sum_{j=1}^{n} \left(\frac{\partial f}{\partial x_j} dx_j + \frac{\partial f}{\partial y_j} dy_j \right). \tag{48}$$

We define the complex variables z_j and $\bar{z}_j$ in accordance with the formulas

$$z_j = x_j + iy_j,\quad \bar{z}_j = x_j - iy_j;$$
$$x_j = \frac{z_j + \bar{z}_j}{2},\quad y = \frac{z_j - \bar{z}_j}{2i},\quad j = 1, 2, \ldots, n,$$

so that

$$dz_j = dx_j + i\,dy_j, \quad d\bar{z}_j = dx_j - i\,dy_j;$$
$$dx_j = \frac{1}{2}(dz_j + d\bar{z}_j), \quad dy_j = \frac{1}{2i}(dz_j - d\bar{z}_j).$$

If we differentiate the equation

$$f(x, y) = f\left(\frac{z+\bar{z}}{2}, \frac{z-\bar{z}}{2i}\right)$$

formally with respect to z_j and $\bar{z}_j$, we obtain

$$\frac{\partial f}{\partial z_j} = \frac{1}{2}\left(\frac{\partial f}{\partial x_j} - i\frac{\partial f}{\partial y_j}\right), \quad \frac{\partial f}{\partial \bar{z}_j} = \frac{1}{2}\left(\frac{\partial f}{\partial x_j} + i\frac{\partial f}{\partial y_j}\right) \tag{49}$$
$$j = 1, 2, \ldots, n.$$

The quantities $\partial f/\partial z_j$ and $\partial f/\partial \bar{z}_j$ defined by formulas (49) are known as *formal derivatives*. These should not be confused with partial derivatives with respect to z_j and $\bar{z}_j$ since the complex variables z_j and $\bar{z}_j$ are not independent

We note that the Cauchy-Riemann conditions (36) can be written in the form

$$\frac{\partial f}{\partial \bar{z}_j} = 0, \qquad j = 1, 2, \ldots, n. \tag{50}$$

Equations (39) in formal derivatives become

$$\frac{\partial^2 u}{\partial z_j\,\partial \bar{z}_k} = 0, \qquad 1 \leqslant j, \quad k \leqslant n. \tag{51}$$

Formula (48) in formal derivatives becomes

$$df = \sum_{j=1}^{n}\left(\frac{\partial f}{\partial z_j}\,dz_j + \frac{\partial f}{\partial \bar{z}_j}\,d\bar{z}_j\right). \tag{52}$$

For a holomorphic function f, this equation is simplified to

$$df = \sum_{j=1}^{n}\frac{\partial f}{\partial z_j}\,dz_j. \tag{53}$$

These formulas enable us to assert that all the rules of differential calculus remain in force for formal derivatives.

6. *Holomorphy of a composite function*

Suppose that a function $f(z)$ is holomorphic in a domain G and that functions $z_j(\lambda)$, for $j = 1, 2, \ldots, n$, are holomorphic in a domain $B \subset C^m$. Suppose that a *vector-valued function $z(\lambda)$ maps the domain B into the domain G. Then, the function $f[z(\lambda)]$ is holomorphic in B.*

Proof: Suppose that $\lambda^0 \in B$. Then, $z(\lambda^0) \in G$ and the power series

$$f(z) = \sum_{|\alpha| \geqslant 0} a_\alpha [z - z(\lambda^0)]^\alpha \tag{54}$$

converges absolutely in some polycircle $S[z(\lambda^0), r]$. Furthermore, there exists a polycircle $S(\lambda^0, \delta)$ such that $z(\lambda) \in S[z(\lambda^0), r]$ for all $\lambda \in S(\lambda^0, \delta)$, and the Taylor series

$$z_j(\lambda) = z_j(\lambda^0) + \sum_{|\beta| \geqslant 1} b_\beta^{(j)} (\lambda - \lambda^0)^\beta, \quad j = 1, 2, \ldots, n \tag{55}$$

converge absolutely in this polycircle. If we substitute these absolutely convergent series into the absolutely convergent series (54), we obtain a new power series that will converge absolutely in $S(\lambda^0, \delta)$ and will coincide with the function $f[z(\lambda)]$ in that polycircle.

The holomorphy of the function $f[z(\lambda)]$ can be proven even more simply if we use the technique of formal derivatives (see section 4.5). Specifically, by using the Cauchy-Riemann conditions (50), we have

$$\frac{\partial f}{\partial \bar{\lambda}_k} = \sum_{j=1}^{n} \left(\frac{\partial f}{\partial z_j} \frac{\partial z_j}{\partial \bar{\lambda}_k} + \frac{\partial f}{\partial \bar{z}_j} \frac{\partial \bar{z}_j}{\partial \bar{\lambda}_k} \right) = 0, \qquad k = 1, 2, \ldots, m.$$

5. HOLOMORPHIC FUNCTIONS AT INFINITELY DISTANT POINTS

1. *Extension of the space C^1*

Consider the group of fractional linear transformations

$$\zeta = \frac{az + b}{cz + d}, \quad ad - cb \neq 0. \tag{56}$$

The transformation (56) maps an arbitrary point of the plane C^1, deleted by the point $-d/c$ if $c \neq 0$, into a finite point.

We shall say that the point $-d/c$ is mapped by the transformation (56) into the infinitely distant point, which we denote by ∞. Here, the inverse transformation

$$z = \frac{-\zeta d + b}{\zeta c - a}$$

maps the point ∞ into the point $-d/c$.

By adding the infinitely distant point ∞ to the plane C^1, we obtain the *extended plane* $\bar{C}^1$. We shall use the term *neighborhood of the point* ∞ to refer to the exterior of every circle with center at 0. Thus, the transformation $\zeta = 1/z$ defines a one-to-one mapping of a neighborhood $|z| < r$ of the point 0 onto the neighborhood $|\zeta| > 1/r$ of the point ∞. From this it follows that the fractional linear transformation (56) defines a homeomorphic (that is, one-to-one and bicontinuous) mapping of the extended plane $\bar{C}^1$ onto itself.

Geometrically, the extended plane $\bar{C}^1$ can be thought of as a sphere (the so-called Riemann sphere). Specifically, a stereographic projection of the unit sphere onto the plane C^1 (see Fig. 4) establishes a homeomorphism between points M of this sphere and points z of the extended plane $\bar{C}^1$. Here, the pole P of the projection corresponds to the infinitely distant point of the plane. In this projection, to each rotation of the sphere there correspond two fractional linear transformations of the plane $\bar{C}^1$ with unit matrices whose determinants are each 1 and which differ in sign. Conversely, to each fractional linear transformation of the plane $\bar{C}^1$ there corresponds a single rotation of the sphere.

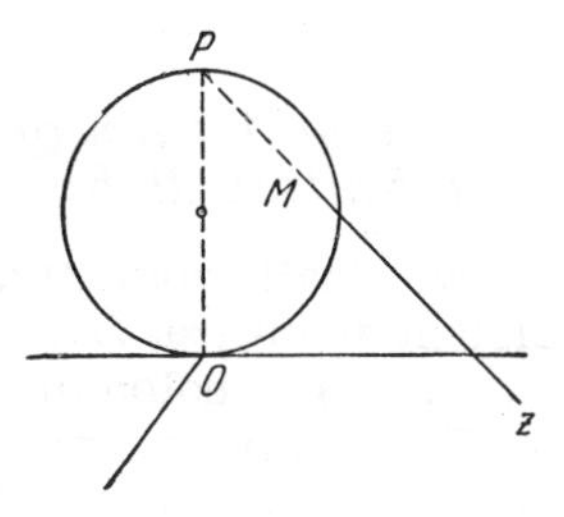

Fig. 4

2. *Extension of the plane* C^n

We define the extended $\bar{C}^n$ as the Cartesian product of the spaces $\bar{C}^1$:

$$\bar{C}^n = \underbrace{\bar{C}^1 \times \bar{C}^1 \times \ldots \times \bar{C}^1}_{n \text{ times}}.$$

Here, the topology of the space $\bar{C}^n$ is induced by the topology of the space $\bar{C}^1$. $\bar{C}^n$ is called the *space of the theory of functions*.*

It follows from this definition that every two-dimensional analytic plane $z_j = z_j^0$, for $j \neq k$, in $\bar{C}^n$ has one infinitely distant point, which we denote by $(z_1^0, \ldots, z_{k-1}^0, \infty, z_{k+1}^0, \ldots, z_n^0)$. The transformation

$$\zeta_j = z_j, \quad j \neq k, \quad \zeta_k = \frac{1}{z_k}$$

maps this point into the finite point $(z_1^0, \ldots, z_{k-1}^0, 0, z_{k+1}^0, \ldots, z_k^0)$. Every four-dimensional analytic plane, for example, $z_j = z_j^0$, for $j = 3, \ldots, n$, has two infinitely distant (two-dimensional) planes $(\infty, a, z_3^0, \ldots, z_n^0)$ and $(a, \infty, z_3^0, \ldots, z_n^0)$ and one infinitely distant point $(\infty, \infty, z_3^0, \ldots, z_n^0)$, etc. Therefore, the set of infinitely distant points $\bar{C}^n$ constitutes a $(2n-2)$-dimensional manifold.

*Other ways of extending the space C^n are expounded in the book by Fuks [1], section 5.

In $\overline{C}^n$, every infinite sequence of points has a convergent subsequence.

A set $A \subset C^n$ is said to be *bounded* if its closure does not contain an infinitely distant point. If it does, it is said to be *unbounded*. Henceforth, unless the contrary is explicitly stated, all the sets that we shall consider are assumed to be bounded.

3. *The definition of a function that is holomorphic at infinitely distant points*

We shall say that a function $f(z)$ is holomorphic at an infinitely distant point $(\infty, \ldots, \infty z_k^0, \ldots, z_n^0)$ if the function $f(1/z_1 \ldots, 1/z_{k-1}, z_k, \ldots, z_n)$ is holomorphic at the point $(0, \ldots, 0, z_k^0, \ldots, z_n^0)$.

To illustrate, consider the case in which $n=2$. The manifold of infinitely distant points in $\overline{C}^2$ is two-dimensional and consists of points of the forms (a, ∞), (∞, a) and (∞, ∞). Corresponding to these are the transformations

$$\zeta_1 = z_1, \zeta_2 = \frac{1}{z_2}; \; \zeta_1 = \frac{1}{z_1}, \; \zeta_2 = z_2; \; \zeta_1 = \frac{1}{z_1} \;\; \zeta_2 = \frac{1}{z_2}$$

which map these infinitely distant points into the points $(a, 0)$, $(0, a)$ and $(0, 0)$, respectively. The function $f(\zeta)$ is holomorphic at these infinitely distant points if the functions

$$f\left(z_1, \frac{1}{z_2}\right), \; f\left(\frac{1}{z_1}, z_2\right), \; f\left(\frac{1}{z_1}, \frac{1}{z_2}\right)$$

are holomorphic at the points $(a, 0)$, $(0, a)$ and $(0, 0)$, respectively. Therefore, the expression "$f(z)$ is holomorphic at the point (a, ∞)," for example, means that there exists a bicircle $|z_1 - a| < r_1$, $|z_2| < r_2$ at which the function $f(z_1 \, 1/z_2)$ can be expanded in absolutely convergent Taylor series and, hence, that the function $f(z)$ can be expanded in a Laurent series

$$f(z) = \sum_{|\alpha| \geqslant 0} a_\alpha (z_1 - a)^{\alpha_1} z_2^{-\alpha_2}$$

In the neighborhood $|z_1 - a| < r_1$, $|z_2| > 1/r_2$ of the point (a, ∞).

4. *The Laurent expansion*

Suppose that a function $f(z)$ is holomorphic in the Cartesian product of the closed circular rings $r_j^- \leqslant |z_j| \leqslant r_j^+$ for $j = 1, 2, \ldots, n$. By applying Cauchy's formula (44), we obtain

$$f(z) = \frac{1}{(2\pi i)^n} \sum \int_{|\zeta_1| = r_1^{\epsilon_1}} \ldots \int_{|\zeta_n| = r_n^{\epsilon_n}} \frac{f(\zeta) \, d\zeta}{(\zeta - z)^I}, \tag{57}$$

where the summation is carried out over all $\varepsilon=(\varepsilon_1, \ldots, \varepsilon_n)$, where ε_j is $+$ or $-$ for $j = 1, 2, \ldots, n$. By expanding $(\zeta - z)^{-I}$ in the corresponding series, substituting into (57), and replacing the integrals over the contours $|z_j| = r_j^-$ with integrals over the contours $|z_j| = r_j^+$ (with the corresponding change in sign), we obtain the expansion of $f(z)$ in a Laurent series

$$f(z) = \sum_{\alpha} a_\alpha z^\alpha \tag{58}$$

with coefficients

$$a_\alpha = \frac{1}{(2\pi i)^n} \int\limits_{|\zeta_1| = r_1^+} \cdots \int\limits_{|\zeta_n| = r_n^+} \frac{f(\zeta)\, d\zeta}{\zeta^{\alpha+I}}. \tag{59}$$

Special cases of formulas (58) and (59) are the Taylor expansion (35) in a neighborhood of $z^0 = 0$ and the Laurent expansions in neighborhoods of the infinitely distant points of the form $(\ldots, 0, \ldots, \infty, \ldots)$. In particular, if $r_j^- = 0$, it follows from (59) that $a_\alpha = 0$ for $\alpha_j + 1 \leqslant 0$.

6. HOLOMORPHIC CONTINUATION

1. A theorem on holomorphic continuation

Suppose that two functions $f_1(z)$ and $f_2(z)$ are holomorphic in domains G_1 and G_2 respectively and that $G_1 \cap G_2$ is a domain. Suppose that the functions $f_1(z)$ and $f_2(z)$ concide in a real neighborhood $S(x^0, r)$, $y = y^0$ of the point $z^0 = x^0 + iy^0$ in $G_1 \cap G_2$. Then, these functions are holomorphic continuation each of the other; that is, there exists a unique function $f(z)$ that is holomorphic in $G_1 \cup G_2$ and that coincides with $f_1(z)$ in G_1 and with $f_2(z)$ in G_2.

Proof: It will be sufficient to show that $f_1(z) \equiv f_2(z)$ for $z \in G_1 \cap G_2$. In the largest polycircle $S(z^0, r_0 I)$ contained in $G_1 \cap G_2$, we have the expansions (see section 4.4)

$$f_k(z) = \sum_{|\alpha| \geqslant 0} a_\alpha^{(k)} (z - z^0)^\alpha, \quad a_\alpha^{(k)} = \frac{1}{\alpha!} D^\alpha f_k(z^0), \quad k = 1, 2. \tag{60}$$

If we take the derivatives in the real neighborhood $S(x^0, r)$, $y = y^0$, where, by hypothesis, $f_1(x + iy^0) = f_2(x + iy^0)$, we see that $a_\alpha^{(1)} = a_\alpha^{(2)}$. From this and from (60), it follows that $f_1(z) = f_2(z)$ in $S(z^0, r_0 I)$.

Suppose now that z^* is an arbitrary point in $G_1 \cap G_2$. Since, by hypothesis $G_1 \cap G_2$ is a domain, the point z^* can be connected with z^0 by a piecewise-smooth curve L lying entirely in $G_1 \cap G_2$ (see Fig. 5). Let us take a point z' in the polycircle $S(z^0, r_0 I)$ that lies on the curve L, and let us expand the functions $f_k(z)$ in a power series in the largest polycircle $S(z', r_1 l)$ contained in $G_1 \cap G_2$. From what we proved above, $f_1(z) \equiv f_2(z)$ in $S(z^0, r_1 l)$. By continuing this process, after a finite number of steps we arrive at the point z^*, at which $f_1(z^*) = f_2(z^*)$. The finiteness of the number of steps follows from the Heine-Borel theorem. To see this, note that since L is compact, the distance from L to $\partial(G_1 \cap G_2)$ is positive. Therefore, the radii $r_0, r_1, \ldots$ of the polycircles referred to above are bounded below by some positive number. But the length of the curve L is finite. From this it follows that the number of steps is finite, which completes the proof.

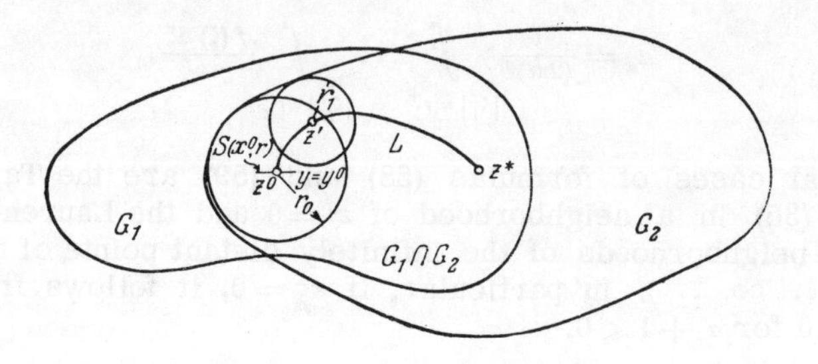

Fig. 5

Remark: If $G_1 \cap G_2$ is not a domain, $f(z)$ may not be single-valued in $G_1 \cup G_2$. In this case also, we shall say that f_1 (resp. f_2) is holomorphically continued into G_2 (resp. G_1).

2. *Consequences of the holomorphic continuation theorem*

If a function $f(x)$ of a real n-tuple is holomorphic in a domain $B \subset R^n$, there exists a unique function $F(z)$ that is holomorphic in some domain $G \subset C^n$ and that coincides with $f(x)$ for $x \in B$.

Proof: If $x^0 \in B$, we have

$$f(x) = \sum_{|\alpha| \geqslant 0} a_\alpha (x - x^0)^\alpha$$

in some poly-interval $S(x^0, r)$. Therefore, the function

$$F(z) = \sum_{|\alpha| \geqslant 0} a_\alpha (z - x^0)^\alpha$$

is holomorphic in the polycircle $S(x^0, r)$ and possesses the required properties.

This property enables us to obtain a principle of holomorphic continuation for holomorphic functions of real as well as complex variables.

If $f(z)$ is holomorphic in a region G and if $D^\alpha f(z^0)=0$ at a point $z^0 \in G$ for all $|\alpha|>m$, then $f(z)$ is a polynomial of degree not exceeding m. This assertion is obvious.

Suppose that $f(z,w)$ is a function of two complex n-tuples z and w. Suppose that f is holomorphic in a neighborhood of the point $z=0$, $w=0$ and that f vanishes on the manifold $w=\bar{z}$ (inside that neighborhood.) Then, $f(z,w)$ vanishes identically in that neighborhood.

Proof: We introduce new variables Z and W defined by

$$Z=\frac{z+w}{2}, \quad W=\frac{z-w}{2i}; \quad z=Z+iW, \quad w=Z-iW.$$

The function $f(Z+iW, Z-iW)$ is holomorphic in the corresponding neighborhood of the point 0 (see section 4.6) and it vanishes for real Z and W. According to the holomorphic continuation theorem, $f(Z+iW, Z-iW)=f(z,w)=0$.

3. *Liouville's theorem*

If $f(z)$ is holomorphic in C^n (i.e., is an entire function) and if $|f(z)|\le c(1+|z|)^m$, then $f(z)$ is a polynomial of degree not exceeding m.

Proof: If we apply formula (45) to the polycircle $S(0, RI)$, we obtain

$$D^\alpha f(0)=\frac{\alpha!}{(2\pi i)^n}\int\limits_{|z_1|=R}\cdots\int\limits_{|z_n|=R}\frac{f(\zeta)\,d\zeta}{\zeta^{\alpha+I}}. \tag{61}$$

Therefore,

$$|D^\alpha f(0)|\leqslant \alpha!\frac{c(1+\sqrt{n}R)^m}{R^{|\alpha|}}\to 0 \text{ as } R\to\infty \quad \text{if} \quad |\alpha|>m.$$

Therefore, $|D^\alpha f(0)|=0$ for $|\alpha|>m$. Application of the assertion of the preceding section completes the proof of the theorem.

4. *The maximum modulus theorem*

If $f(z)$ is holomorphic in a domain G and if $f(z)\not\equiv$ const, then $|f(z)|$ cannot have a maximum value in G.

Proof: Suppose that $|f(z)|$ does have a maximum value M at a point $z^0\in G$. Let us take an arbitrary polycircle $S(z^0, r)\Subset G$ and apply Cauchy's formula (44) to $f(z)$:

$$f(z^0)=\frac{1}{(2\pi i)^n}\int\limits_{\partial S(z_1^0, r_1)}\cdots\int\limits_{\partial S(z_n^0, r_n)}\frac{f(z)\,dz}{(z-z^0)^I}. \tag{62}$$

Let us show that $|f(z)| = M$ on the hull $\partial S(z_1^0, r_1) \times \ldots \partial S(z_n^0, r_n)$. If $|f(z')|$ were less than M at some point z' of this set, there would, because of the continuity, be a neighborhood of this point throughout which $|f(z)|$ would be less than M. But this contradicts Eq. (62) because we would then have

$$|f(z^0)| = M \leqslant \frac{1}{(2\pi)^n} \int_0^{2\pi} \ldots \int_0^{2\pi} |f(z^0 + re^{i\theta})| \, d\theta_1 \ldots d\theta_n < M.$$

Since the numbers r_j can be chosen arbitrarily (though sufficiently small), it follows that $|f(z)| = M$ in the polycircle $S(z^0, r)$.

Let us show that $f(z) \equiv$ const in $S(z^0, r)$. This is obvious for $M = 0$ since in that case, $f(z) \equiv 0$. Suppose $M > 0$. Then, if we differentiate the equation $f(z)\overline{f}(z) = M^2$ and apply the Cauchy-Riemann conditions $\partial f / \partial \bar{z}_j$, for $j = 1, 2, \ldots, n$, we obtain

$$\overline{f}\frac{\partial f}{\partial z_j} + f\frac{\partial \overline{f}}{\partial z_j} = \overline{f}\frac{\partial f}{\partial z_j} + f\frac{\overline{\partial f}}{\partial \bar{z}_j} = \overline{f}\frac{\partial f}{\partial z_j} = 0, \quad j = 1, 2, \ldots, n.$$

Since $|\overline{f}| = M > 0$, this last relationship implies the equations $\partial f / \partial z_j = 0, \quad j = 1, 2, \ldots, n$. These equations, together with the Cauchy-Riemann conditions yield $f(z) \equiv$ const in $S(z^0, r)$.

By virtue of the holomorphic continuation theorem $f(z) \equiv$ const in G. But this is ruled out by the hypotheses of the theorem. Consequently, $|f(z)|$ cannot assume a maximum value within G, which completes the proof.

Thus, by virtue of a well-known theorem of Weierstrass, we may say that the maximum theorem holds for bounded domains G and their boundaries ∂G as applied to the absolute values of functions that are holomorphic in G and continuous in $\overline{G}$ (see section 1.6).

5. *Construction of a holomorphic continuation*

The theorem on holomorphic continuation enables us to construct the holomorphic continuation of a holomorphic function as follows: Suppose that a function $f(z)$ is holomorphic in a polycircle $S(z^0, r^0)$. We choose a point z' in this polycircle and expand $f(z)$ in a power series about that point. This series will converge absolutely in some polycircle $S(z', r')$. This means that $f(z)$ can be continued holomorphically (see section 6.1) to points of $S(z', r') \setminus S(z^0, r^0)$. Thus, we define a new function [which we still denote by $f(z)$] that is holomorphic in $S(z^0, r^0) \cup S(z', r')$. We repeat this process for other points in the region $S(z^0, r^0) \cup S(z', r')$. As a result of this continuation, we obtain a function $f(z)$ (not, in general, single-valued) that is holomorphic at every point of the set of polycircles that we have obtained. The extended function $f(z)$ may be nonsingle-valued for the simple reason that we may reach a point z^0 from the given point z^* by two or more different paths and in the process obtain different power series representing the function $f(z)$ in a neighborhood of the point z^*.

6. *The monodromy theorem*

What was said in the preceding section naturally leads to the problem of finding sufficient conditions to ensure that the result of a holomorphic continuation is independent of the steps by which this continuation is achieved. One such condition is given by the following

THEOREM (on monodromy). *Suppose that a function $f(z)$ that is holomorphic in some neighborhood of a point z^0 is holomorphically continued outside this neighborhood along every path lying entirely in some domain G. Then, the result of continuing $f(z)$ to an arbitrary point $z^* \in G$ along all homotopic paths in G connecting the points z^0 and z^* will be the same. In particular, if the domain G is simply connected, $f(z)$ will be single-valued in G.*

Proof: By use of power-series expansions, we continue the function $f(z)$ from the point z^0 to the point z^* along any piecewise-smooth path L. Since L is compact, there will, by virtue of the Heine-Borel theorem, be a finite number of polycircles covering L in which the corresponding power series for $f(z)$ converge. Therefore, there will exist neighborhood $U(L)$ of the curve L in which the function $f(z)$ is holomorphic and single-valued.

Now suppose that L^* is an arbitrary path in G homotopic to L (see Fig. 6). Since the surface F along which a continuous deformation of the curve L takes the curve L into the curve L^* is compact, it follows, on the basis of the preceding reasoning and the Heine-Borel theorem that this surface is covered by a finite number of neighborhoods $U(L), U(L_1)$, ..., $U(L^*)$ such that (1) the function $f(z)$ is single-valued in each of these neighborhoods and (2) the values of the function $f(z)$ coincide in the common portion of two adjacent neighborhoods. This means that the function $f(z)$ is single-valued in some neighborhood of the surface F and, thus, the continuations of $f(z)$ into the point z^* along the two paths L and L^* coincide, which completes the proof.

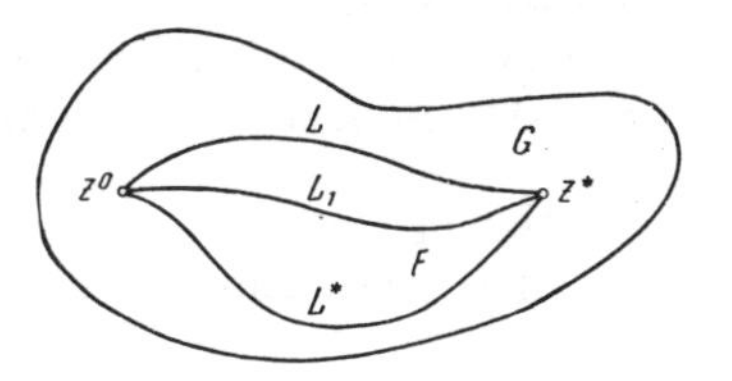

Fig. 6

7. *Representation of a holomorphic function by means of a line integral*

Suppose that $f(z)$ is holomorphic at some point z^0 and is holomorphically continued from the point z^0 along every path lying entirely in the domain G. Then, *the following formula holds for the continuation of* $f(z)$ *along any piecewise-smooth curve L lying in G that connects the points* z^0 *and z:*

$$f(z)=f(z^0)+\int_{z^0}^{z}\sum_{j=1}^{n}\frac{\partial f}{\partial z'_j}\,dz'_j, \tag{63}$$

where the integration is carried out over the curve L. Here, the result does not change if we replace the curve L in formula (63) *with an arbitrary piecewise-smooth curve in G that is homotopic to it. In particular, if the domain G is simply connected, the integral in* (63) *is independent of the path of integration.*

Proof: If we integrate Eq. (53) over the path L, we obtain formula (63).

Necessary and sufficient conditions for the value of the line integral in (63) to be independent of the path of integration are conditions (39). By virtue of the monodromy theorem, these conditions are satisfied in a neighborhood of that surface along which a continuous deformation of the path L transforms that path into another path in G that is homotopic to L. This completes the proof.

It follows from this theorem that conjugate functions (see section 4.1) can be expressed each in terms of the other. For example,

$$v(x,y)=v(x^0,y^0)+\int_{z^0}^{z}\sum_{j=1}^{n}\left(-\frac{\partial u}{\partial y'_j}\,dx'_j+\frac{\partial u}{\partial x'_j}\,dy'_j\right). \tag{64}$$

Conversely, *if* $u(x,y)$ *is a pluriharmonic function in a simply connected domain G, the function* $f(z)=u+iv$, *where the function* $v(x,y)$ *is defined by formula* (64), *is holomorphic in G.*

Proof: The function $u(x,y)$ is infinitely differentiable (see section 4.1). Equations (39) ensure that the integral (64) is independent of the path of integration in the simply connected domain G. By differentiating the integral (64), we see that the Cauchy-Riemann conditions (36) are satisfied in the domain G, which completes the proof.

7. HOLOMORPHIC MAPPINGS

1. *Definition of a holomorphic mapping*

A mapping $\zeta=\zeta(z)$ of a domain G onto a domain G_1 is said to be holomorphic* if the vector-valued function $\zeta(z)=[\zeta_1(z),\ldots,\zeta_n(z)]$ is holomorphic in G and

$$\frac{\partial(\zeta_1,\ldots,\zeta_n)}{\partial(z_1,\ldots,z_n)}\neq 0,\qquad z\in G. \tag{65}$$

*A holomorphic mapping is sometimes called "pseudoconformal."

For $n=1$, such a mapping is conformal. In contrast with the case in which $n=1$, holomorphic mappings for $n \geqslant 2$ do not in general preserve angles between directions.

A holomorphic mapping is locally homeomorphic; that is, a sufficiently small neighborhood of any point is mapped in a one-to-one manner and bicontinuously into the corresponding neighborhood of the image of that point.

Proof: If we set $\zeta_j = u_j + iv_j$ and $z_j = x_j + iy_j$ and use the Cauchy-Riemann conditions, we obtain from (69)

$$\frac{\partial(u_1, v_1, \ldots, u_n, v_n)}{\partial(x_1, y_1, \ldots, x_n, y_n)} = \frac{\partial(\zeta_1, \bar{\zeta}_1, \ldots, \zeta_n, \bar{\zeta}_n)}{\partial(z_1, \bar{z}_1, \ldots, z_n, \bar{z}_n)} =$$
$$= \frac{\partial(\zeta_1, \ldots, \zeta_n, \bar{\zeta}_1, \ldots, \bar{\zeta}_n)}{\partial(z_1, \ldots, z_n, \bar{z}_1, \ldots, \bar{z}_n)} = \left| \frac{\partial(\zeta_1, \ldots, \zeta_n)}{\partial(z_1, \ldots, z_n)} \right|^2 \neq 0, \qquad z \in G.$$

By virtue of a classical theorem of analysis, the assertion made follows from this.

2. *Biholomorphic mappings*

A holomorphic mapping of a domain G onto a domain G_1 is not in general one-to-one. For example, the conformal mapping $\zeta = (z-1)^n$ of the circle $|z| < 1$ onto the corresponding domain is not one-to-one for $n \geqslant 3$.

If a holomorphic mapping of a domain G onto a domain G_1 is one-to-one, such a mapping is said to be *biholomorphic* (*single-sheeted*). (schlicht). In this case, we say that the domains G and G_1 are equivalent.

Let us show that the *mapping $z = z(\xi)$, that is, the inverse of a biholomorphic mapping $\zeta = \zeta(z)$, is also a biholomorphic mapping.*

Proof: It follows from what was said in the preceding section that the mapping $z = z(\zeta)$ is infinitely differentiable and that the corresponding Jacobian is nonzero in the domain G_1 (on the basis of classical theorems). It remains to show that the vector-valued function $z(\zeta) = [z_1(\zeta), \ldots, z_n(\zeta)]$ is holomorphic in G_1. On the basis of section 4.2, to do this we need only show that

$$\frac{\partial z_j(\zeta)}{\partial \bar{\zeta}_k} = 0, \qquad \zeta \in G_1; \qquad 1 \leqslant j, \qquad k \leqslant n. \tag{66}$$

If we differentiate the equations

$$\zeta_j = \zeta_j[z(\zeta)]$$

with respect to $\bar{\zeta}_k$, we obtain

$$\sum_{l=1}^{n} \frac{\partial \zeta_j}{\partial z_l} \frac{\partial z_l}{\partial \bar{\zeta}_k} = 0, \qquad 1 \leqslant j, \qquad k \leqslant n,$$

from which, on the basis of condition (65), relations (66) follow.

Thus, *a biholomorphic mapping is a homeomorphic mapping not just locally but globally.*

Remark: For an arbitrary holomorphic mapping $\zeta = \zeta(z)$ of a domain G onto a domain G_1, it is possible to define a covering domain G^* with projection G_1 (see section 8.6) such that the mapping will be a homeomorphism of G onto G^*, that is, such that the vector-valued function $z = z(\zeta)$ of the inverse mapping will be holomorphic (and single-valued) in G^* (see Fuks [1], p. 171).

3. *Boundary points*

Let us make more precise the concept of a boundary point of a finite domain G. Suppose that $z^0 \in \partial G$. This means that there exists a sequence of points $z^{(k)}$, $k = 1, 2, \ldots$ such that $z^{(k)} \to z^0$ for $z^{(k)} \in G$. It may happen that the points of this sequence lie in different components of the open set $G \cap U(z^0, r)$. Therefore, it is sometimes convenient (for example, in the theory of holomorphic mappings) to assign to a single geometric point $z^0 \in \partial G$ several (possible infinitely many) boundary points $R_1(z^0)$, $R_2(z^0)$, ... lying over z^0 and corresponding to different sequences of points of the domain G that converge to z^0.

Suppose that the point $z^0 \in \partial G$ is accessible from the domain G (see section 1.5).

We shall say that this sequence $z^{(k)} \in G$, for $k = 1, 2, \ldots$, converges to the boundary point $R(z^0)$ (cf. Section 8.7) lying over the point $z^0 \in \partial G$ if (1) $z^{(k)} \to z^0$ as $k \to \infty$ and (2) for every r, there exists an N such that all points $z^{(k)}$, for $k \geqslant N$, are contained in the same component of the open set $G \cap U(z^0, r)$. We shall identify two boundary points $R_1(z^0)$ and $R_2(z^0)$ if the corresponding sequences $\{z^{(k)}\}$ and $\{\zeta^{(k)}\}$ are such that the sequence $z^{(1)}, \zeta^{(1)}, z^{(2)}, \zeta^{(2)}, \ldots$ converges to some boundary point lying over z^0.

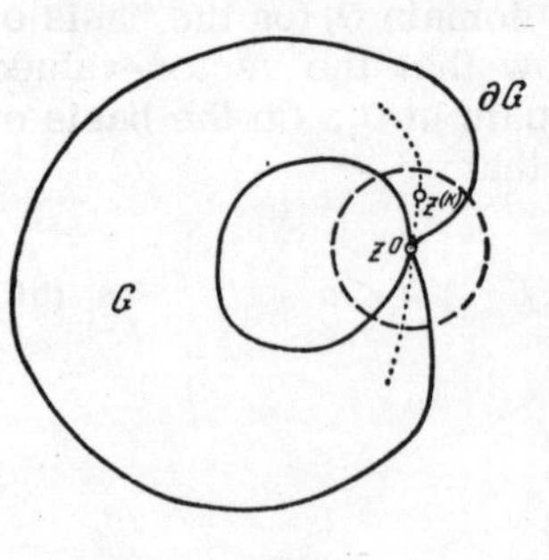

Fig. 7

If exactly p (where $1 \leqslant p \leqslant \infty$) distinct boundary points lie over a geometric point $z^0 \in \partial G$, we shall say that the point z^0 is *p-multiple*. For example, in Fig. 7, the point z^0 is 2-multiple. Every point on the boundary of a plane domain bounded by a closed Jordan curve is 1-multiple. Every interior point of a plane nonclosed Jordan curve is 2-multiple. These assertions follow from Jordan's and Schönflies' theorems (see section 1.5).

For an unbounded infinite domain G, we assign only finite points to its boundary ∂G. However, under biholomorphic mappings of a domain G, finite points of ∂G may be mapped into infinitely distant ones. Therefore, it is natural to consider an unbounded domain G in some sort of extension of the space C^n. Here, it is necessary to assign to its boundary those infinitely distant points that are the limits of unbounded sequences of points of this region (where convergence of these sequences must be understood in the extended space [see section 5.2]).

The set of boundary points of the domain G in this sense (that is, with consideration of their multiplicities and with the addition of suitable infinitely distant points) will be called the boundary $\partial' G$ (as distinct from the set-theoretic boundary $\partial G = \overline{G} \setminus G$). In what follows, in absence of explicit mention to the contrary, by the boundary of a domain G, we shall mean ∂G.

4. *Biholomorphic in C^1*

The following theorems are valid in the plane C^1 (see Goluzin [5], Chapter II; Lavrent'yev and Shabat [94], Chapter II).

THEOREM (Riemann). *Let G denote an arbitrary simply connected domain boundary consisting of more than one point. Then, there exists a unique biholomorphic mapping $\zeta = \zeta(z)$ of this domain onto the unit circle $|\zeta| < 1$ such that a given point $z^0 \in G$ and a direction assigned at that point are mapped respectively into the point $\zeta = 0$ and the direction of the positive real axis.*

THEOREM (Carathéodory). *If all the boundary points of a bounded simply connected domain G are accessible, then, under a biholomorphic mapping $\xi = \xi(z)$ of this domain onto the open unit circle $|\zeta| < 1$, the correspondence of the points $\partial' G$ and the points of the circle $|\zeta| = 1$ will be one-to-one. Here, the inverse function $z = z(\zeta)$ is continuous for $|\zeta| \leq 1$ if we understand for its value on the circle $|\zeta| = 1$ the limit values from the open circle $|\zeta| < 1$. In particular, if the domain G is bounded by a closed Jordan curve, the function $\zeta = \zeta(z)$ maps $\overline{G}$ onto $|\zeta| \leq 1$ in a one-to-one bicontinuous manner.*

THEOREM (Koebe). *If a function $z = z(\zeta)$, where $z(0) = 0$, maps the circle $|\zeta| < 1$ biholomorphically onto the domain G and does not assume a value $c > 0$ in that circle, then $|z'(0)| \leq 4c$.*

5. *The group of automorphisms*

A biholomorphic mapping of a domain onto itself is called an *automorphism* of that region. The set of all automorphisms of a domain obviously constitutes a group, which we shall call the group of automorphisms of the given domain.

Let us give some examples of domains and their automorphisms.

Multiple-circular domains (Reinhardt domains). A domain G possessing the property that $z^0 \in G$ implies $z \in G$ for all points $z = (z^0 - a)e^{i\theta} + a$, where $\theta = (\theta_1, \theta_2, \ldots, \theta_n)$ is an arbitrary real vector, is called a multiple-circular domain (or a Reinhardt domain). The point a is called the *center* of such a domain. Multiple-circular domains possess automorphisms of the form

$$\zeta_j = (z_j - a_j)\, e^{i\theta_j} + a_j, \qquad j = 1, 2, \ldots, n.$$

If the relation $z^0 \in G$, where G is a multiple-circular domain, implies that all the points

$$[z:\ z_j = \lambda_j (z_j^0 - a_j) + a_j, \qquad |\lambda_j| \leqslant 1, \quad j = 1, 2, \ldots, n],$$

also belong to G, such a domain is called a complete multiple-circular domain.

Hartogs domains (polycircular domains). If $z^0 \in G$ implies that all points of the form

$$[(z_1^0 - a_1)\, e^{i\theta_1} + a_1,\ \tilde{z}^0], \qquad \tilde{z}^0 = (z_2^0, \ldots, z_n^0),$$

where θ_1 is an arbitrary real number, also belong to G, such a domain is called a *Hartogs domain*. The plane $z_1 = a_1$ is called the *plane of symmetry* of that domain. Hartogs domains possess automorphisms of the form

$$\zeta_1 = (z_1 - a_1)\, e^{i\theta_1} + a_1, \qquad \tilde{\zeta} = \tilde{z}.$$

If the relation $z \in G$, where z^0 is a Hartogs domain, implies that all points of the form

$$[z:\ z_1 = \lambda_1 (z_1^0 - a_1) + a_1, \qquad \tilde{z} = \tilde{z}^0, \qquad |\lambda_1| \leqslant 1],$$

also belong to G, this domain G is called a *complete* Hartogs domain.

Tubular domains are domains of the form

$$T_B = B + iR^n = [z = x + iy : x \in B,\ |y| < \infty]$$

or

$$T^B = R^n + iB,$$

where B is a domain in R^n known as the *base* of the domain T^B or T_B. The mapping $\zeta = iz$ maps the domain T_B onto the domain T^B. The transformations $\zeta = z + a$, where a is an arbitrary real vector, are automorphisms of the region T^B.

Semitubular domains are domains of the form

$$G = [z = (x_1 + iy_1, \tilde{z}) : (x_1, \tilde{z}) \in D, \quad y_1 - \text{arbitrary}]$$

where the base D is a domain in $R^1 \times C^{n-1}$. The transformations $\zeta_1 = z_1 + ia_1$ and $\tilde{\zeta} = \tilde{z}$, where a_1 is an arbitrary real number, are automorphisms of the domain G.

The problem arises as to how to construct the group of automorphisms of a given domain. As an example, let us construct the group of automorphisms of the tubular domain T^{Γ^+}, where $\Gamma^+ = [y : y_1 > \sqrt{y_2^2 + \ldots + y_n^2}]$ is a future light cone. (This example is of particular interest in quantum field theory (see Bros [63].) We can verify directly that the following three mappings belong to the group of automorphisms of the domain T^{Γ^+}:

(1) real translations $T_a : \zeta = z + a$ (n parameters);

(2) similarity transformations $\zeta = \lambda z$, where $\lambda > 0$ (one parameter);

(3) transformations defined by the elements L of the orthochronous Lorentz group $\zeta = Lz$, where $L = \|g_{kj}\|$ for $k, j = 1, 2, \ldots, n$ $[n^2 - n)/2$ parameters];

(4) the "inversion" transformation J:

$$\zeta = \frac{-z}{z_1^2 - z_2^2 - \ldots - z_n^2}.$$

When we examine the transformation $\lambda T_a LJT_b$, we see that the set of automorphisms thus constructed depends on no fewer than

$$1 + n + \frac{1}{2}(n^2 - n) + n = \frac{1}{2}(n + 1)(n + 2)$$

parameters. On the other hand, Cartan [123] has shown that the group of automorphisms of the domain $T^{\Gamma^+} - 1/2\,(n+1)(n+2)$ is a parametric group. Therefore, this domain has no other automorphisms and the automorphisms listed (1)-(4) generate its group of automorphisms.

If T maps a domain G biholomorphically into a domain G_1 and if A is the group of automorphisms of the domain G, the group of automorphisms of the domain G_1 is

$$\{TaT^{-1},\ a \in A\}.$$

This is true because $TaT^{-1}G_1 = TaG = TG = G_1$.

It follows from this that, in particular, multiple-circular domains $S(0, 1)$ and $U(0, 1)$ in the space C^n (for $n \geqslant 2$) are not equivalent since they have different groups of automorphisms. Thus, in the space C^n, for $n \geqslant 2$, there exist simply connected domains that cannot be mapped biholomorphically onto each other. On the other hand, in the space C^1, a theorem of Riemann (see section 7.4) tells us that all simply connected domains are equivalent. Herein lies a significant difference between the spaces C^1 and C^n for $n \geqslant 2$.

8. DOMAINS OF HOLOMORPHY

1. *Definition of a domain of holomorphy*

A domain* G is said to be a *domain of holomorphy* of a function $f(z)$ if $f(z)$ is holomorphic in G but is not holomorphic in a larger domain (more precisely, if f cannot be continued holomorphically outside G). The domain of holomorphy of a function is also called the *domain of existence* of that function.

For example, the domain of holomorphy of the function

$$f_0(z) = \sum_{\alpha=0}^{\infty} z^{\alpha!} \tag{67}$$

is the unit circle $S(0, 1)$.

To see this, note that $f_0(z)$ is unbounded (see section 1.6) at the points $z = e^{i\varphi} \in \partial S(0, 1)$ for rational $\varphi = p/q$ (where p and q are integers) since

$$|f_0(re^{i\varphi})| \geqslant \left| \sum_{\alpha=q+1}^{\infty} r^{\alpha!} e^{i\frac{p}{q}\alpha!} \right| - \left| \sum_{\alpha=0}^{q} r^{\alpha!} e^{i\frac{p}{q}\alpha!} \right| =$$
$$= \sum_{\alpha=q+1}^{\infty} r^{\alpha!} - \left| \sum_{\alpha=0}^{q} r^{\alpha!} e^{i\frac{p}{q}\alpha!} \right| \to \infty \quad \text{for } r \to 1-0$$

but the set of rational values for φ is everywhere dense in the interval $[0, 2\pi]$ and, therefore, $f_0(z)$ has singularities at all points of $\partial S(0, 1)$.

If every component of an open set $O \subset C^n$ is a domain of holomorphy, we shall say that O is an *open set of holomorphy.*

2. *Domains of holomorphy in* C^1

The question arises as to whether every domain is a domain of holomorphy of some function. The answer to this question is quite different for $n = 1$ and $n \geqslant 2$.

In the space C^1, *every domain is a domain of holomorphy.*

At the moment, we shall prove this assertion with the hypothesis that the domain G in question is bounded by a closed Jordan curve. It will be proven in its general form in sections 19.2 and 16.8 (see also Stoilow [117], Vol. I, Chapter IX).

Indeed, there exists a biholomorphic mapping $\zeta = \omega(z)$ of the domain G onto unit circle $|\zeta| < 1$ that maps $\overline{G}$ onto $|\zeta| \leqslant 1$ in a one-to-one and bicontinuous manner (see section 7.4). Therefore, the

*We recall that we are considering only bounded domains.

function $f_0[\omega(z)]$, where the function f_0 is defined by (67), is holomorphic in G (see section 4.6) and is not bounded at every point of ∂G. Consequently, G is a domain of holomorphy.

3. *Domains of holomorphy in C^n for $n \geqslant 2$*

In contrast with the space C^1, *not every domain in the space C^n for $n \geq 2$ is a domain of holomorphy.*

This is one of the more surprising features that sharply distinguish the theory of functions of several complex variables $(n \geqslant 2)$ from the theory of functions of a single complex variable.

To prove this assertion, we shall exhibit a single counterexample in C^2 which is a slight modification of the example presented by Bochner and Martin ([3], p. 91).

Let us show that every function $f(z)$ that is holomorphic in the semihollow sphere

$$G: \quad \left[z=(z_1, z_2): \frac{1}{2} < |z| < 1\right],$$

must necessarily be holomorphic in a larger domain, namely, the open sphere $|z| < 1$, (more precisely, that it admits a holomorphic continuation into the sphere $|z| < 1$).

This assertion means that G *cannot be the domain of holomorphy of any function.*

To prove it, consider the function

$$\varphi(z)=\frac{1}{2\pi i} \int\limits_{|\zeta|=\frac{3}{4}} \frac{f(\zeta, z_2)}{\zeta - z_1}\, d\zeta. \tag{68}$$

This function is holomorphic with respect to z_1 and z_2 individually in the bicircle $|z_1| < 3/4$, $|z_2| < \sqrt{7}/4$ since $[(\zeta, z_2): |\zeta| = \frac{3}{4}, |z_2| < \sqrt{7}/4] \subset G$ (see Fig. 8). According to the fundamental theorem of Hartogs (see section 4.2), $\varphi(z)$ is holomorphic in that bicircle. Furthermore, the function $\varphi(z)$ coincides with $f(z)$ in the domain $|z_1| \leqslant 3/4$, $1/2 < |z_2| < \sqrt{7}/4$ since this domain is contained in G, so that Cauchy's formula

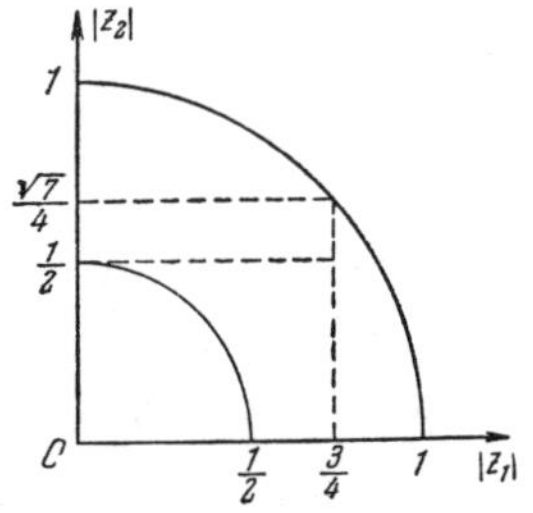

$$f(z)=\frac{1}{2\pi i} \int\limits_{|\zeta|=\frac{3}{4}} \frac{f(\zeta, z_2)}{\zeta - z_1}\, d\zeta,$$

Fig. 8

is applicable. On the basis of (68), it follows from this that $f(z) = \varphi(z)$ in the domain mentioned. Since the set

$$\left[z: \ |z_1| < \frac{3}{4}, \quad |z_2| < \frac{\sqrt{7}}{4}\right] \cap G$$

is a domain, according to the holomorphic-continuation theorem (see section 6.1), the function $f(z)$ is holomorphic in the sphere

$$[|z|<1]=\left[z: |z_1|<\frac{3}{4}, \quad |z_2|<\frac{\sqrt{7}}{4}\right]\cup G,$$

as asserted.

4. *Holomorphic extension of domains*

The preceding example leads us to introduce the following definition. Consider a domain $G\subset C^n$, where $n\geqslant 2$. If there exists a domain $G^*\supset G$ such that $H_G=H_{G^*}$ (see section 4.2), we shall call G^* a *holomorphic extension* of the domain G.

Thus, the sphere $|z|<1$ is a holomorphic extension of the domain $1/2<|z|<1$. Furthermore, it is clear that if G is a domain of holomorphy, then $G^*=G$.

In what follows, we shall repeatedly encounter examples of domains possessing nontrivial holomorphic extensions.

5. *Construction of domains of holomorphy*

A domain of holomorphy of a function $f(z)$ that is holomorphic in some given neighborhood $S(z^0, r)$ can be constructed by means of its holomorphic continuation (see section 6.5). Let us expand $f(z)$ in a power series in the polycircle $S(z^0, r)$. It may turn out that this series converges absolutely in a larger polycircle. Let $S(z^0, r^0)$ be the largest of these polycircles. Let us take a countable everywhere-dense set of points $\{z^{(k)}\}$ of this polycircle and expand the function $f(z)$ in a power series in the largest neighborhood $S(z^{(k)}, r^{(k)})$ of each point $z^{(k)}$. It may happen that some of the polycircles $S(z^{(k)}, r^{(k)})$ will extend beyond the polycircle $S(z^0, r^0)$. From the holomorphic-continuation theorem (see section 6.1), this means that the function $f(z)$ is holomorphic (more precisely, it can be holomorphically continued) beyond the boundary of the polycircle $S(z^0, r^0)$. Now, in each of the polycircles $S(z^{(k)}, r^{(k)})$, let us choose a countable everywhere-dense set of points and let us expand $f(z)$ in a power series in the largest neighborhood of each such point. We continue this process (see Fig. 9).

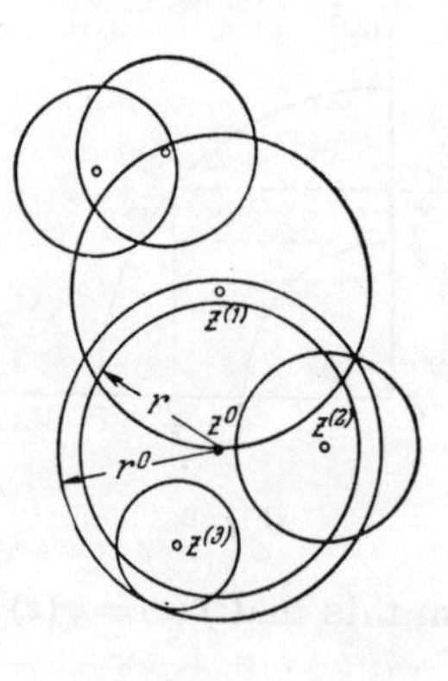

Fig. 9

Let us suppose that as the result of a countable number of such steps, we have obtained a *single-valued* function $f(z)$ in the domain G which is the union of the (countably many) polycircles constructed at each step in the holomorphic continuation. By construction, this function is holomorphic in G and nonholomorphic at all finite points lying outside G. This means that the domain G that we have constructed is the domain of holomorphy of the function $f(z)$.

Another, more complicated case results when the process of holomorphic continuation described above yields a *nonsingle-valued* function $f(z)$. To make the function $f(z)$ single-valued, it will be necessary to broaden the concept of a domain. This is done by introducing so-called covering domains (nonsingle-sheeted domains) on C^n (see Fuks [1], Chapter II, Behnke and Thullen [4]).

6. *Construction of nonsingle-sheeted domains of holomorphy*

In the holomorphic continuation of the function $f(z)$ in the preceding section, we obtained a countable number of pairs (S_k, f_k), for $k = 1, 2, \ldots$, the first member of which is a maximum polycircle S_k and the second member is the corresponding power series f_k in which the function $f(z)$ is expanded in that polycircle. By combining all the pairs (S_k, f_k), we obtain a point set G over C^n. Here, we identify two points of the set G if they (1) have the same coordinates and (2) belong to the same pair. Thus, every (*analytic*) point $P(z) \in G$ represents the coordinates of the (*geometric*) point z and the corresponding power series of the function $f(z)$.

It follows from the construction of the system of pairs $\{(S_k, f_k)\}$ that the set G is connected in the sense that, for arbitrary points P and Q in G, there exists a finite sequence of pairs $(S_{\alpha_1}, f_{\alpha_1}) \left(S_{\alpha_N}, f_{\alpha_N}\right)$ such that $P \in (S_{\alpha_1}, f_{\alpha_1})$, $Q \in \left(S_{\alpha_N}, f_{\alpha_N}\right)$, and any two adjacent pairs in this sequence have at least one common point. This means that G is a (covering) domain.

Obviously, no more than a countable number of distinct analytic points $P(z)$ in the covering domain G lying over the geometric point z can exist. We denote this number by $p(z)$, where $1 \leqslant p(z) \leqslant \infty$. Furthermore, the set of projections of all analytic points of the covering domain G constitutes a domain in C^n, which we shall call the *projection* of G. The number $p = \sup\limits_{z \in \operatorname{pr} G} p(z)$ characterizes the degree of branching of the domain G. Here, we say that G is a p-sheeted (covering) domain. If G is a single-sheeted (schlicht) domain ($p = 1$), it is identified with its projection and, in this case, we may take $P(z) \equiv z$.

We shall say that a sequence of points $P_k(z^{(k)})$ for $k = 1, 2, \ldots$, in the covering domain G converges to a point $P_0(z^0) \in G$ if (1) $z^{(k)} \to z^0$ and (2) for an arbitrary pair (S_α, f_α) containing the point $P_0(z^0)$, there exists a number N such that $P_k(z^{(k)}) \in (S_\alpha, f_\alpha)$ for all $k \geqslant N$. The concepts of continuity, curve, etc., are defined in G in the usual manner in terms of this convergence.

The function $f(z)$ is single-valued and holomorphic in the covering domain G that we have just constructed but is not holomorphic in any larger domain. Therefore, G is the domain of holomorphy (existence) of the function $f(z)$. In view of the uniqueness of a holomorphic continuation (see section 6.1), the domain G is independent of the choice of the original neighborhood $S(z^0, r)$.

Thus, *to every holomorphic function there corresponds a unique domain of holomorphy, namely, the domain of existence of the function.*

7. Boundary points of a domain of holomorphy

A sequence of points $P_k(z^{(k)})$, $k = 1, 2, \ldots$, in a domain G is called a fundamental sequence if (1) $z^{(k)} \to z^0$ and (2) for every polycircle $S(z^0, r)$, there exists an N such that the inequalities $l \geqslant N$ and $k \geqslant N$ imply that the points $P_l(z^{(l)})$ and $P_k(z^{(k)})$ can be connected by a Jordan curve in G the projection of which is contained in that polycircle (see Fig. 7). Two fundamental sequences $\{P_k\}$ and $\{Q_k\}$ are said to be equivalent if the sequence $P_1, Q_1, P_2, Q_2, \ldots$ is a fundamental sequence. If a fundamental sequence $\{P_k(z^{(k)})\}$ does not have a limit in G, we shall call this sequence a *boundary point* of the domain G and shall denote it by $R(z^0)$ (cf. section 7.3). We shall identify boundary points corresponding to equivalent fundamental sequences. For example, the domain of holomorphy of the function $\ln z$ possesses a single boundary point $R(z^0)$, at which $z^0 = 0$.

The set of all boundary points of a domain of holomorphy G constitutes its *boundary* $\partial' G$. The boundary points of a domain of holomorphy of a function are called the *singular points* of that function.

8. Remark

We defined covering domains in terms of holomorphic continuation. However, the class of covering domains thus introduced turns out to be insufficient, for example, in the case in which we need to consider the holomorphic continuation of several functions simultaneously. This leads us to the definition of intersection of domains of holomorphy. In this case, the concept of a covering domain needs to be extended by using the abstract definition of such a domain. (For details, see Fuks [1] and Behnke and Thullen [4].) In the last decade, the concept of a covering domain has been considerably generalized, which has lead to the rapidly developing theory of complex spaces by use of present-day topological and algebraic methods (see Fuks [1], Chapter III; Grauert and Remmert [60]). The basic concepts of this theory that deal with domains of holomorphy are also expounded in Wightman [61].

For our purposes, such general concepts are unnecessary since we shall deal primarily with single-sheeted domains and hence with single-valued functions. Henceforth, unless the contrary is explicitly stated, we shall assume that all domains are *bounded* and *single-sheeted*.

CHAPTER II

Plurisubharmonic Functions and Pseudoconvex Domains

This chapter is devoted to plurisubharmonic (in particular, convex) functions and the pseudoconvex (in particular, convex) domains corresponding to them. We shall see that plurisubharmonic functions constitute a basic analytic instrument in the study of domains of holomorphy.

9. SUBHARMONIC FUNCTIONS.

Plurisubharmonic functions of (complex) variables are defined in terms of plurisubharmonic functions of a single variable. These last coincide with subharmonic functions of two real variables. Therefore, it will be natural for us to study first subharmonic functions of two variables. The corresponding results (except for the material of section 9.16) remain valid without significant changes for subharmonic functions of an arbitrary number of variables. A detailed exposition of subharmonic functions is to be found in the book by Privalov [9].

1. Definition of a subharmonic function

A function $u(z)=u(z, \bar{z})\equiv u(x, y)$ is said to be *subharmonic* in a domain $G\subset C^1$ if (1) $-\infty\leqslant u(z)<+\infty$ in G, (2) $u(z)$ is upper-semicontinuous in G, and (3) for an arbitrary subdomain $G'\Subset G$ and an arbitrary function $U(z)$ that is harmonic in G' and continuous in $\bar{G}'$, the inequality $u(z)\leqslant U(z)$ on $\partial G'$ implies the inequality $u(z)\leqslant U(z)$ in G'. A function u such that $-u$ is a subharmonic function is called a *superharmonic function*. A harmonic function is both a subharmonic and a superharmonic function.

A subharmonic function in a domain G is locally bounded above in G (see section 2.3 and 1.6) and hence is locally integrable in G (in the sense of section 2.6). A subharmonic function considered on an arbitrary continuous curve contained in G is upper-semi-continuous on that curve (see section 2.2). Therefore, if this curve is piecewise smooth and bounded, the subharmonic function is integrable on it.

2. *Poisson's formula*

Suppose that a function $U(z)$ is harmonic in $S(z^0, r)$ and continuous in $\overline{S}(z^0, r)$. Then, for $U(z)$, Poisson's formula is valid:

$$U(z) = \int_0^{2\pi} P(z - z^0, re^{i\theta}) U(z^0 + re^{i\theta})\, d\theta, \qquad z \in S(z^0, r), \tag{1}$$

where $P(z, \zeta)$ is Poisson's kernel:

$$P(z, \zeta) = \frac{1}{2\pi} \frac{r^2 - \rho^2}{r^2 - 2r\rho\cos(\varphi - \theta) + \rho^2}, \qquad \zeta = re^{i\theta}, \quad z = \rho e^{i\varphi}. \tag{2}$$

The kernel $P(z, \zeta)$ possesses the properties

$$P(0, \zeta) = \frac{1}{2\pi}, \qquad \int_0^{2\pi} P(z, re^{i\theta})\, d\theta = 1, \tag{3}$$

$$\frac{1}{2\pi} \frac{r - \rho}{r + \rho} \leqslant P(z, \zeta) \leqslant \frac{1}{2\pi} \frac{r + \rho}{r - \rho}. \tag{4}$$

In particular, from (1) and (3), we have

$$U(z) = \frac{1}{2\pi} \int_0^{2\pi} U(z + re^{i\theta})\, d\theta. \tag{5}$$

3. *Harnack's Theorem*

THEOREM (Harnack's) *A decreasing sequence $\{U_\alpha\}$ of functions that are harmonic in a domain G converges uniformly in G either to a harmonic function or to $-\infty$.* Suppose that $\overline{S}(z^0, r) \Subset G$. If we apply Poisson's formula (1) to the nonnegative harmonic functions $U_\alpha - U_{\alpha+\beta}$ and use properties (4) and (5), we obtain Harnack's inequality

$$\begin{aligned} \frac{r-\rho}{r+\rho} [U_\alpha(z^0) - U_{\alpha+\beta}(z^0)] \leqslant U_\alpha(z) - U_{\alpha+\beta}(z) \leqslant \\ \leqslant \frac{r+\rho}{r-\rho} [U_\alpha(z^0) - U_{\alpha+\beta}(z^0)], \end{aligned} \tag{6}$$

which is valid for all $z \in S(z^0, r)$.

Suppose that $U_\alpha(z^0) \to -\infty$. Then it follows from inequality (6) that $U_\alpha(z) \to U(z)$ uniformly in $S\left(z^0, r/2\right)$. On the basis of the Heine-Borel theorem, we conclude from this that $U_\alpha(z) \to -\infty$ uniformly in every subdomain $G' \Subset G$, that is, uniformly in G.

Suppose now that $U_\alpha(z^0) \to a > -\infty$. Then, from what we have just proven, $U(z) = \lim U_\alpha(z) > -\infty$ in G. It follows from inequality

(6) $U_\alpha(z) \to U(z)$ uniformly in $S(z^0\ r/2)$. On the basis of the Heine-Borel theorem, $U_\alpha(z)$ converges uniformly to $U(z)$ in every subdomain $G' \Subset G$. Therefore, $U(z)$ is a harmonic function in G, which completes the proof.

4. *A test for subharmonicity*

Suppose that $u(z) < +\infty$ is a subharmonic function in $S(z^0, r)$ and that $u(z)$ is upper-semicontinuous in $\overline{S}(z^0, r)$. Then, for all $z \in S(z^0, r)$,

$$u(z) \leqslant \int_0^{2\pi} P(z - z^0,\ re^{i\theta})\, u(z^0 + re^{i\theta})\, d\theta. \tag{7}$$

Conversely, if $u(z) < +\infty$ is upper-semicontinuous in the domain G and satisfies the inequality

$$u(z) \leqslant \frac{1}{2\pi} \int_0^{2\pi} u(z + re^{i\theta})\, d\theta \tag{8}$$

for all $z \in G$ and for all sufficiently small $r \leq r_0(z)$, then $u(z)$ is a subharmonic function in G.

Proof: Suppose that $u(z) < +\infty$ is subharmonic in $S(z^0, r)$ and upper-semicontinuous in $\overline{S}(z^0, r)$. On the basis of section 2.4, there exists a decreasing sequence of continuous functions $U_\alpha(z)$ for $\alpha = 1, 2, \ldots$, that converges to $u(z)$ in $S(z^0, r)$. Let $u_\alpha(z)$ be a harmonic function $S(z^0, r)$ that assumes the value $u_\alpha(z)$ on $\partial S(z^0, r)$. Since

$$U_{\alpha+1}(z) = u_{\alpha+1}(z) \leqslant u_\alpha(z) = U_\alpha(z), \qquad z \in \partial S(z^0,\ r),$$

we have, by virtue of the maximum principle,

$$U_{\alpha+1}(z) \leqslant U_\alpha(z), \qquad z \in \overline{S}(z^0,\ r). \tag{9}$$

In accordance with Harnack's theorem (see section 9.3), we conclude from this that the function

$$U^*(z) = \lim_{\alpha\to\infty} U_\alpha(z) \tag{10}$$

is either identically equal to $-\infty$ or is a harmonic function in $S(z^0, r)$. But the function $u(z)$ is subharmonic in G and

$$u(z) \leqslant u_\alpha(z) = U_\alpha(z), \qquad z \in \partial S(z^0,\ r). \tag{11}$$

Therefore,

$$u(z) \leqslant U_\alpha(z), \qquad z \in \bar{S}(z^0, r)$$

and, consequently, on the basis of (10),

$$u(z) \leqslant U^*(z), \qquad z \in \bar{S}(z^0, r). \tag{12}$$

If we now use (1), (9)-(11), and the theorem of Levi (see section 2.7), we obtain the chain of equations

$$\begin{aligned} U^*(z) = \lim_{\alpha\to\infty} U_\alpha(z) = \lim_{\alpha\to\infty} \int_0^{2\pi} P(z-z^0, re^{i\theta}) U_\alpha(z^0+re^{i\theta})\, d\theta = \\ = \int_0^{2\pi} P(z-z^0, re^{i\theta}) \lim_{\alpha\to\infty} U_\alpha(z^0+re^{i\theta})\, d\theta = \\ = \int_0^{2\pi} P(z-z^0, re^{i\theta}) \lim_{\alpha\to\infty} u_\alpha(z^0+re^{i\theta})\, d\theta = \\ = \int_0^{2\pi} P(z-z^0, re^{i\theta}) u(z^0+re^{i\theta})\, d\theta, \end{aligned} \tag{13}$$

which, together with (12) yields the desired inequality (7).

Conversely, suppose that the function $u(z) < +\infty$ is upper-semicontinuous in G and satisfies inequality (8). Suppose, furthermore, that $U(z)$ is a harmonic function in $G' \Subset G$ which is continuous in $\bar{G}'$ and that $U(z) \geqslant u(z)$ on $\partial G'$. We assume that there exists a point $z' \in G'$ such that $U(z') < u(z')$. Then, the function $f(z) = u(z) - U(z)$ is upper-semicontinuous in $\bar{G}'$, nonpositive on $\partial G'$, and positive at the point $z' \in G'$. Therefore, it attains its maximum $M > 0$ (see section 2.3) at the point $z^0 \in G'$. But, by virtue of Eq. (5), the function $f(z)$ satisfies inequality (8) for all sufficiently small $r \leqslant r_0(z^0)$. This means that $f(z) \equiv M$ in $S(z^0, r)$ because, if $f(z)$ were less than M at some point $z' \in S(z^0, r)$, this inequality would, in view of the upper-semicontinuity of the function $f(z)$, be retained in some neighborhood of the point z' (see section 2.2), which would contradict inequality (8).

By virtue of the Heine-Borel theorem, we can exhibit a number $r_0 = r_0(G') > 0$ such that inequality (8) will be satisfied for all $z \in \bar{G}'$ and $r \leqslant r_0$. Then, if we apply this inequality to all points of the domain G' at which $f(z) = M$, we see that $f(z) \equiv M > 0$ in $\bar{G}'$, which contradicts the fact that $f(z) \leqslant 0$ on $\partial G'$. This contradiction proves the inequality $U(z) \geqslant u(z)$ in G', so that $u(z)$ is a subharmonic function in G. This completes the proof of the theorem.

5. *The smallest harmonic majorant*

If a function $u(z)$, such that $-\infty \not\equiv u(z) < +\infty$, is subharmonic in $S(z^0, r)$ and upper-semicontinuous in $\bar{S}(z^0, r)$, the function $U^*(z)$

constructed in the proof of the direct part of the theorem of section 9.4 is harmonic $S(z^0, r)$ and upper-semicontinuous in $S(z^0, r)$. According to Eqs. (13), this function is independent of the choice of decreasing sequence $u_\alpha \to u$. The function $U^*(z)$ is called the *least harmonic majorant* of the function $u(z)$ in $S(z^0, r)$. It possesses the property that the inequality $u(z) \leqslant U(z)$ on $\partial S(z^0, r)$, where $U(z)$ is harmonic in $S(z^0, r)$ and continuous in $\bar{S}(z^0, r)$, implies the inequality $U^*(z) \leqslant U(z)$ in $S(z^0, r)$. Specifically, from (1) and (13), we have

$$U^*(z) = \int_0^{2\pi} P(z - z^0, re^{i\theta})\, u(z^0 + re^{i\theta})\, d\theta \leqslant$$

$$\leqslant \int_0^{2\pi} P(z - z^0, re^{i\theta})\, U(z^0 + re^{i\theta})\, d\theta = U(z).$$

If the boundary ∂G is such that the Dirichlet problem is solvable for an arbitrary continuous function defined on ∂G, we define in an analogous manner the least harmonic majorant of a function $u(z)$ such that $-\infty \not\equiv u(z) < +\infty$, that is subharmonic in G and upper-semicontinuous in G.

6. *The simplest properties of subharmonic functions*

The following properties of subharmonic functions follow from the test of section 9.4:

A function defined in a domain is subharmonic in that domain if and only if it is subharmonic in a neighborhood of every point of that domain.

A linear combination, with positive coefficients, of subharmonic functions is a subharmonic function.

The limit of a uniformly convergent sequence of subharmonic functions is a subharmonic function.

A monotonically decreasing sequence $\{u_\alpha\}$ of subharmonic functions converges to a subharmonic function.

Proof: On the basis of section 2.3, for an arbitrary subregion $G' \Subset G$, there exists a number $C < +\infty$ such that $u_{\alpha+1}(z) \leqslant u_\alpha(z) \leqslant C$. Therefore (see section 2.4), the function $\lim\limits_{\alpha\to\infty} u_\alpha(z) < +\infty$ is upper-semicontinuous. Furthermore, according to section 2.7, it is possible to take the limit under the integral sign in inequality (8):

$$\lim_{\alpha\to\infty} u_\alpha(z) \leqslant \frac{1}{2\pi} \int_0^{2\pi} \lim_{\alpha\to\infty} u_\alpha(z + re^{i\theta})\, d\theta.$$

From the theorem of section 9.4, the function $\lim\limits_{\alpha\to\infty} u_\alpha(z)$ is subharmonic.

A nonconstant subharmonic function $u(z)$ in a domain G cannot have a maximum value within G.

To see this, if the function $u(z)$ did have a maximum value M at a point $z^0 \in G$, we should have, just as in the proof of the converse portion of the theorem of section 9.4, $u(z) \equiv M$ in G, which is impossible.

The upper envelope

$$u(z) = \overline{\lim_{z' \to z}} \sup_{\alpha} u_{\alpha}(z')$$

and the limit superior

$$v(z) = \overline{\lim_{z' \to z}} \ \overline{\lim_{\alpha \to \infty}} u_{\alpha}(z')$$

of the family $\{u_{\alpha}\}$ of functions that are subharmonic and locally uniformly bounded above in the domain G are subharmonic functions in G.

Proof: The function $u(z) < +\infty$ and is upper-semicontinuous in G. If we apply inequality (8) (for $r < \Delta_G(z)$) to each function $u_{\alpha}(z)$, we obtain by use of Levi's theorem (see section 2.7)

$$u(z) = \overline{\lim_{z' \to z}} \sup_{\alpha} u_{\alpha}(z') \leqslant \overline{\lim_{z' \to z}} \frac{1}{2\pi} \int_0^{2\pi} \sup_{\alpha} u_{\alpha}(z' + re^{i\theta})\, d\theta \leqslant$$

$$\leqslant \frac{1}{2\pi} \int_0^{2\pi} \overline{\lim_{z' \to z}} \sup_{\alpha} u_{\alpha}(z' + re^{i\theta})\, d\theta = \frac{1}{2\pi} \int_0^{2\pi} u(z + re^{i\theta})\, d\theta.$$

On the basis of the test of section 9.4, this proves that the function $v(z)$ is subharmonic. We can treat the function $u(z)$ in an analogous fashion.

7. *The greatest subharmonic minorant*

Suppose that a function $r(z)$ is upper-semicontinuous in G. A subharmonic function $v(z)$ is called the *greatest subharmonic minorant* of the function $r(z)$ in the domain G if (1) $v(z) \leqslant r(z)$ and (2) $v(z) \geqslant u(z)$ for an arbitrary subharmonic function $u(z)$ that is subharmonic

From this definition, it is clear that the function $v(z)$ is unique. Let us show that if $r(z) < +\infty$, then $v(z)$ always exists and is given by the formula

$$v(z) = \overline{\lim_{z' \to z}} \sup_{[u(z') \leqslant r(z')]} u(z'),$$

where the supremum is over all functions $u(z)$ that are subharmonic in G and that do not exceed $r(z)$.

Since the function $r(z)$ is upper-semicontinuous in G, the set of subharmonic functions $u(z)$ that do not exceed $r(z)$ are uniformly locally bounded above in G (see section 2.3). Therefore, the function $v(z)$ is subharmonic in G and

$$v(z) \leqslant \overline{\lim_{z' \to z}} \, r(z') = r(z).$$

Finally, if $u(z)$ is a subharmonic function and $u(z) \leqslant r(z)$, then $u(z) \leqslant v(z)$. This means that the function $v(z)$ that we have constructed is the greatest subharmonic minorant of the function $r(z)$ in the domain G.

The least superharmonic majorant $V(z)$ of a lower-semicontinuous function $R(z) > -\infty$ in a domain G is defined analogously:

$$V(z) = \underline{\lim_{z' \to z}} \inf_{[u(z') \geqslant R(z')]} u(z'),$$

where the infimum is over all functions $u(z)$ that are superharmonic in G and that are not less than $R(z)$.

8. *The mean value of a subharmonic function*

Suppose that $u(z)$ is a subharmonic function in a domain G. The function

$$J(r, z^0; u) = \frac{1}{2\pi} \int_0^{2\pi} u(z^0 + re^{i\theta}) \, d\theta$$

is called the *mean value* of $u(z)$. Obviously, it is defined for all r such that $[z : |z - z^0| = r] \subset G$.

If $u(z)$ is a subharmonic function in G, its mean value increases with respect to r in $[0, R)$ if $[|z - z^0| < R] \subset G$, and is convex with respect to $\ln r$ in (r, R) if $[\rho < |z - z^0| < R] \subset G$.

To see this, let r_1 and r_2, where $r_1 < r_2$, be arbitrary numbers in $[0, R)$ and let $U^*(z)$ be the smallest harmonic majorant of the function $u(z)$ in the circle $|z - z^0| < r_2$ (see section 9.5). Then,

$$J(r_1, z^0; u) \leqslant J(r_1, z^0; U^*) = J(r_2, z^0; U^*) = J(r_2, z^0; u),$$

which shows that the function $J(r, z^0; u)$ is an increasing function of r in $[0, R)$.

Let r_1 and r_2, where $r_1 < r_2$, be arbitrary numbers in (ρ, R) and let $V(z)$ be the least harmonic majorant of the function $u(z)$ in the annulus $r_1 < |z - z^0| < r_2$ (see Fig. 10). By using Green's formula, we obtain, for all $r \in (r_1, r_2)$ and all $\varepsilon \in (0, r - r_1)$,

$$-\ln(r_1+\varepsilon)\int\limits_{|z-z^0|=r_1+\varepsilon}\frac{\partial V}{\partial\rho}\bigg|_{\rho=r_1+\varepsilon}ds+\ln r\int\limits_{|z-z^0|=r}\frac{\partial V}{\partial\rho}\bigg|_{\rho=r}ds=$$
$$=-\frac{1}{r_1+\varepsilon}\int\limits_{|z-z^0|=r_1+\varepsilon}V\,ds+\frac{1}{r}\int\limits_{|z-z^0|=r}V\,ds. \tag{14}$$

Since the function V is harmonic, we have

$$\int\limits_{|z-z^0|=r_1+\varepsilon}\frac{\partial V}{\partial\rho}\bigg|_{\rho=r_1+\varepsilon}ds=\int\limits_{|z-z^0|=r}\frac{\partial V}{\partial\rho}\bigg|_{\rho=r}ds=a,$$

where the number a is independent of r and ε. Therefore, if we take the limit in (14) as $\varepsilon\to 0$, we see that the function

$$J(r,\ z^0;\ V)=\frac{1}{2\pi r}\int\limits_{|z-z^0|=r}V\,ds$$

is linear with respect to $\ln r$ in $(r_1,\ r_2)$. Furthermore, the function $J(r,\ z^0;\ u)$ does not exceed $J(r,\ z^0;\ V)$ in $(r_1,\ r_2)$ and it coincides with

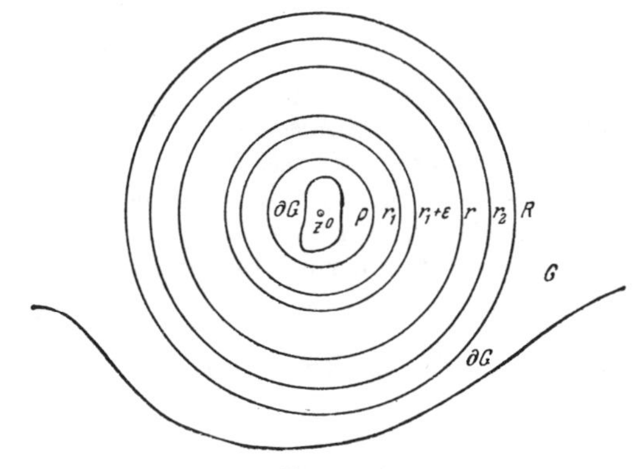

Fig. 10

$J(r,\ z^0;\ V)$ at the points r_1 and r_2. These statements prove that $J(r,\ z^0;\ u)$ is convex with respect to $\ln r$ in $(r_1,\ r_2)$. Because r_1 and r_2 were arbitrarily chosen (except that $\rho<r_1<r_2<R$), we conclude from this that this function is convex with respect to $\ln r$ for all $r\in(\rho,\ R)$, which completes the proof.

Remark: This theorem holds also in R^n for arbitrary $n\geqslant 3$. Here, the function $\ln r$ must be replaced by r^{-n+2}.

9. *The summability of a subharmonic function*

If $u(z)\not\equiv-\infty$ is a subharmonic function in a domain G, it is locally summable in G.

Proof: The function $u(z)$ is integrable over an arbitrary subdomain $G' \Subset G$ and its integral $< +\infty$. Let us show that this integral cannot be equal to $-\infty$. Suppose that it is equal to $-\infty$. Then, there exists a point $z^0 \in G'$ and a positive number r_0 such that

$$\int\limits_{S(z_0,\ r_0)} u(z)\,dx\,dy = -\infty, \quad S(z_0,\ 3r_0) \Subset G. \tag{15}$$

If we integrate inequality (8)

$$u(z^0) \leqslant \frac{1}{2\pi}\int\limits_0^{2\pi} u(z^0 + re^{i\theta})\,d\theta,$$

with respect to $r\,dr$ from 0 to r_0 and apply Fubini's theorem (see section 2.9), we obtain by virtue of (15),

$$\frac{1}{2}u(z^0)\,r_0^2 \leqslant$$

$$\leqslant \frac{1}{2\pi}\int\limits_0^{r_0} r\,dr \int\limits_0^{2\pi} u(z^0 + re^{i\theta})\,d\theta = \frac{1}{2\pi}\int\limits_{S(z_0,\ r_0)} u(z)\,dx\,dy = -\infty.$$

From this it follows that $u(z^0) = -\infty$.

Suppose that $z' \in S(z^0,\ r_0)$. Since $S(z',\ 2r_0) \subset S(z^0,\ 3r_0)$, it follows from (15) that the integral of the function $u(z)$ with respect to $S(z',\ 2r_0)$ is equal to $-\infty$. But then, according to what we have just proven, $u(z') = -\infty$. Thus, $u(z) \equiv -\infty$ in $S(z^0,\ r_0)$. This implies that $u(z) \equiv -\infty$ in G, which contradicts our assumption.

10. *The uniqueness of a subharmonic function*

LEMMA. *Suppose that $u_1(z)$ and $u_2(z)$ are subharmonic functions in a domain G and that they coincide almost everywhere in G. Then, $u_1(z) = u_2(z)$ everywhere in G.*

This lemma follows from the following more general proposition.

Suppose that a subharmonic function $u_1(z)$ coincides almost everywhere in G with an upper-semicontinuous function $u_2(z) < +\infty$. Then, $u_1(z) \leq u_2(z)$ everywhere in G.

Proof: Suppose, on the contrary, that $u_2(z^0) < u_1(z^0)$, where $z^0 \in G$. Since the function $u_2(z) < +\infty$ is upper-semicontinuous at the point z^0, there exists a positive number δ such that, for all $|z - z^0| < \delta$, the inequality $u_2(z) < u_1(z^0)$ is satisfied (see section 2.2). If we apply inequality (8) to the function $u_1(z)$, we obtain

$$u_2(z) < \frac{1}{2\pi}\int\limits_0^{2\pi} u_1(z^0 + re^{i\theta})\,d\theta, \quad z \in S(z^0,\ \delta), \quad r \leqslant \delta.$$

If we set $z = re^{i\varphi} + z^0$ and integrate the above inequality with respect to $r\,dr$ from 0 to δ and with respect to $d\varphi$ from 0 to 2π, we obtain, by use of Fubini's theorem (see section 2.9)

$$\int_{S(z^0,\,\delta)} u_2(z)\,dx\,dy < \int_{S(z^0,\,\delta)} u_1(z)\,dx\,dy,$$

which is impossible.

11. Examples of subharmonic functions

If $u(z) \geqslant 0$, is a subharmonic function, the function $u^p(z)$ (for $p \geqslant 1$) is also subharmonic.

This assertion follows from Holder's inequality

$$u^p(z) \leqslant \left[\frac{1}{2\pi}\int_0^{2\pi} u(z + re^{i\theta})d\theta\right]^p \leqslant \frac{1}{2\pi}\int_0^{2\pi} u^p(z + re^{i\theta})\,d\theta$$

and the test of section 9.4.

If $u(z)$ is a subharmonic function, the function $e^{u(z)}$ is also subharmonic.

This assertion follows from the inequality between the geometric and arithmetic means:

$$e^{\int p \ln f\,dx} \leqslant \int pf\,dx, \quad \text{if } \int p\,dx = 1, \quad p \geqslant 0, \quad f \geqslant 0$$

(see [10], p. 167) and section 9.4. Specifically,

$$e^{u(z)} \leqslant \exp\left\{\frac{1}{2\pi}\int_0^{2\pi} u(z + re^{i\theta})\,d\theta\right\} \leqslant$$
$$\leqslant \frac{1}{2\pi}\int_0^{2\pi} \exp\{u(z + re^{i\theta})\}\,d\theta.$$

The converse is not true (see section 9.15).

12. Jensen's inequality

Suppose that $f(z)$ is holomorphic in G. Then, the functions $\ln|f(z)|$, $\ln^+|f(z)|$ and $|f(z)|^p$ $(p \geq 0)$ are subharmonic in G.

Proof: Since

$$|f(z)|^p = e^{p \ln|f(z)|}, \quad \ln^+|f(z)| = \max(0, \ln|f(z)|),$$

it will, by virtue of section 9.11, be sufficient to prove the above assertion for the function $\ln|f(z)|$. But this function is subharmonic according to the test of section 9.4 since it does not take the value $+\infty$; it is upper-semicontinuous in G; and, it satisfies inequality (8) for all sufficiently small $r \leqslant r_0(z)$ for $z \in G$. The last assertion follows from the following considerations: Either $\ln|f(z)| = -\infty$ and inequality (8) is trivally satisfied or $\ln|f(z)| = \operatorname{Re} \ln f(z) > -\infty$ is a harmonic function in some neighborhood of the point z and inequality (8) is satisfied by virtue of Eq. (5).

Thus, if a function $f(z)$ is holomorphic in $\bar{S}(0, r)$, Jensen's inequality

$$\ln|f(z)| \leqslant \int_0^{2\pi} P(z, re^{i\theta}) \ln|f(re^{i\theta})| \, d\theta$$

follows from inequality (7).

13. The approximation of subharmonic functions

For $u(z)$ to be a subharmonic function in a domain G, it is necessary and sufficient that $u(z)$ be the limit of a decreasing sequence of subharmonic functions $u_\alpha(z)$, for $\alpha = 1, 2, \ldots$, in the class $C^{(\infty)}(G_\alpha)$, where the domains G_α are such that $G_\alpha \subset G_{\alpha+1}$ and $\bigcup_\alpha G_\alpha = G$.

Proof: Sufficiency of the above condition follows from section 9.6.

Let us prove the necessity. If $u(z) \equiv -\infty$, the required sequence is $u_\alpha(z) = -\alpha$. If $u(z) \not\equiv -\infty$, $u(z)$ is locally summable in G (see section 9.9). Suppose that a function $\omega(|z|)$ defined in D possesses the properties: $0 \leqslant \omega(|z|)$ $\omega(|z|) = 0$ for $|z| \geqslant 1$;

$$2\pi \int_0^1 \omega(\rho) \rho \, d\rho = 1 \tag{16}$$

(see section 3.1). We construct a sequence of mean functions:

$$\begin{aligned} u_\alpha(z) &= \int u\left(z + \frac{z'}{\alpha}\right) \omega(|z'|) \, dx' \, dy' = \\ &= \int u(z') \, \omega(\alpha|z - z'|) \, \alpha^2 \, dx' \, dy' = \\ &= \int_0^1 r\omega(r) \left[\int_0^{2\pi} u\left(z + \frac{r}{\alpha} e^{i\theta}\right) d\theta \right] dr = \\ &= 2\pi \int_0^1 J\left(\frac{r}{\alpha}, \ z; \ u\right) r\omega(r) \, dr, \qquad \alpha = 1, 2, \ldots . \end{aligned} \tag{17}$$

Since the function $u(z)$ is subharmonic in G, inequalities (17) and the properties of the function $\omega(|z|)$ imply, by virtue of the test of section 9.4, that the functions $u_\alpha(z)$ are subharmonic and infinitely differentiable in the open sets

$$O_\alpha = \left[z:\ \Delta_G(z) > \frac{1}{\alpha},\quad z \in G\right],$$

where

$$O_\alpha \subset O_{\alpha+1} \text{ and } \bigcup_\alpha O_\alpha = G.$$

As our set G_α, we take that component O_α to which the fixed point of the domain G belongs.

Let us show that the monotonically decreasing sequence of functions $\{u_\alpha\}$ converges to $u(z)$ at every point $z \in G$. If $z \in G$, then $S(z, \rho) \Subset G$ for some $\rho > 0$. But then, there exists a number $N \geqslant 1$ such that $S(z, \rho) \Subset G_\alpha$ for all $\alpha \geqslant N$. Since the function $J(r/\alpha, z; u)$ decreases monotonically with increasing α (see section 9.8), it follows from the last term of Eq. (17) that the sequence $u_\alpha(z)$, for $\alpha \geqslant N$, decreases monotonically. On the other hand, the test of section 2.2 tells us that, for arbitrary positive ε, there exists a positive number δ (where $(\delta \leqslant \rho)$ such that $u(z') < u(z) + \varepsilon$ for all $|z' - z| < \delta$. From this and from (16 and 17), it follows that $u_\alpha(z) < u(z) + \varepsilon$ for all sufficiently large α. This, together with the inequality $u(z) \leqslant u_\alpha(z)$, proves that $\lim_{\alpha\to\infty} u_\alpha(z) = u(z)$ for all $z \in G$, which completes the proof.

14. *A property of positivity of subharmonic functions*

If $u(z) \not\equiv -\infty$ is a subharmonic function in a domain G, it satisfies the inequality.

$$4\frac{\partial^2 u(z)}{\partial z\, \partial \bar{z}} = \Delta u(x, y) \geqslant 0, \qquad z \in G. \tag{18}$$

Conversely, if $u \in D^*(G)$ and satisfies inequality (18) in G, then $u(z)$ is a measurable function and there exists a unique subharmonic function in G that coincides with $u(z)$ almost everywhere in G.

Remark: Differentiability and positivity in inequality (18) must be understood in the sense of generalized functions: $\Delta u \geqslant 0$ means that $\int u \Delta \varphi \, dx\, dy \geqslant 0$ for arbitrary $\varphi(z) \geqslant 0$ in $D(G)$ (see section 3.4).

Proof: Suppose that $u(z)$ is a subharmonic function in G. Since $u(z) \not\equiv -\infty$, $u(z)$ must, because of section 9.9, be locally summable in G. Hence, $u(z)$ defines a continuous linear functional u over $D(G)$ by the formula $(u, \varphi) \doteq \int u\varphi \, dx\, dy$ (see section 3.2); that is, $u \in D^*(G)$. Inequality (8) for the function $u(z)$ can be rewritten in terms of generalized functions in the form of the convolution (see section 3.3)

$$(\mu_r - \delta) * u \geqslant 0, \tag{19}$$

where the generalized function μ_r is defined by the equation

$$(\mu_r, \varphi) = \frac{1}{2\pi} \int_0^{2\pi} \varphi(re^{i\theta})\, d\theta, \qquad \varphi \in D(G).$$

But since

$$\left(\frac{\mu_r - \delta}{r^2}, \ \varphi\right) = \frac{1}{r^2}\left[\frac{1}{2\pi}\int_0^{2\pi} \varphi(re^{i\theta})\, d\theta - \varphi(0)\right] \to \frac{1}{4}\Delta\varphi(0)$$

for arbitrary $\varphi \in D(G)$, it follows that

$$\frac{\mu_r - \delta}{r^2} \to \frac{1}{4}\Delta\delta, \quad \text{as} \quad r \to +0. \tag{20}$$

Using the continuity of the convolution (see section 3.3), we derive inequality (18) from (19) and (20):

$$0 \leqslant \frac{\mu_r - \delta}{r^2} * u \to \frac{1}{4}\Delta\delta * u = \frac{1}{4}\Delta u.$$

Conversely, suppose that $u(z) \in D^*(G)$ and satisfies inequality (18) in the domain G. Then, Δu defines a positive measure in G (see section 3.4). We denote this measure by μ. Thus, u satisfies Poisson's equation $\Delta u = \mu$. We denote by $E(z) = (1/2\pi)\ln|z|$ the fundamental solution of this equation, $\Delta E = \delta$, and we denote by $\mu_{G'}$ the restriction of the measure μ to an arbitrary subdomain $G' \Subset G$. Then, $u(z)$ can be represented by Riesz' formula in G' (see section 3.10):

$$u = E * \mu_{G'} + U_{G'}, \tag{21}$$

where $U_{G'}(z)$ is a harmonic function in the region G'.

Furthermore, since the function

$$\int_{G'} |\ln|z - z'||\, dx\, dy$$

is continuous with respect to z' in $\overline{G}'$ and since the measure $\mu_{G'}$ defines a continuous linear functional over the space of functions that are continuous in $\overline{G}'$, the iterated integral

$$\int_{G'}\left[\int_{G'} |\ln|z - z'||\, dx\, dy\right] d\mu(z')$$

exists and is finite. Then, we conclude from Fubini's theorem (see section 2.9) that, for almost all $z \in G'$, the integral

$$\frac{1}{2\pi}\int_{G'} \ln|z-z'|\, d\mu(z') = E * \mu_{G'},$$

exists and is finite. This integral is a summable function on G'. From this and from (21), it follows that the generalized function u is locally summable in G' and can, for almost all $z \in G'$, be represented in the form

$$u(z) = \frac{1}{2\pi}\int_{G'} \ln|z-z'|\, d\mu(z') + U_{G'}(z). \tag{22}$$

Let us show now that the right member of Eq. (22) is a subharmonic function in the domain G'. We denote it by $v(z)$. Since, for every $z \in G'$, the sequence of continuous functions $\ln(|z-z'|+\varepsilon)$ approaches $\ln|z-z'|$ monotonically from above as $\varepsilon \to +0$, it follows on the basis of Levi's theorem (see section 2.7) that

$$v(z) = \lim_{\varepsilon \to +0} u_\varepsilon(z) + U_{G'}(z), \tag{23}$$

where

$$u_\varepsilon(z) = \frac{1}{2\pi}\int_{G'} \ln(|z-z'|+\varepsilon)\, d\mu(z').$$

The functions $u_\varepsilon(z)$ are continuous in G'. Furthermore, since the measure $\mu \geqslant 0$ and the function

$$\ln(|z|+\varepsilon) = \sup_{|\zeta| \leqslant \varepsilon} \ln|z+\zeta|$$

is subharmonic (see sections 9.6 and 9.12), we conclude, on the basis of the test of section 9.4, that the function $u_\varepsilon(z)$ is subharmonic in G' and that it decreases monotonically as $\varepsilon \to +0$. Remembering that $U_{G'}(z)$ is a harmonic function in G', we conclude from equation (23) (see section 9.6) that the function $v(z)$ is subharmonic in G'.

Thus, the function $u(z)$ coincides (almost everywhere) in an arbitrary subdomain $G' \Subset G$ with a subharmonic function. This means that $u(z)$ coincides almost everywhere in G with a subharmonic function. From section 9.10, this subharmonic function is unique, which completes the proof.

15. *Logarithmically subharmonic functions*

A nonnegative function $u(z)$ is said to be logarithmically subharmonic in a domain G if the function $\ln u(z)$ is subharmonic in G.

On the basis of section 9.11, the function $u(z)$ is also subharmonic. Examples of logarithmically subharmonic functions are the absolute values of holomorphic functions (see section 9.12).

THEOREM. *For a nonnegative function $u(z)$ to be logarithmically subharmonic in a domain G, it is necessary and sufficient that the function $u(z)\ |e^{az}|$ be subharmonic in G for all complex a.*

Proof: The necessity of the condition is obvious since the function $\ln u(z) + \mathrm{Re}\,(az)$ is subharmonic for all a and, by virtue of section 9.11, the function $u(z)|e^{az}|$ is also subharmonic.

To prove the sufficiency, note that the function $\ln u(z) < +\infty$ and is upper-semicontinuous in G since

$$\overline{\lim_{z\to z^0}}\, u(z) = u(z^0) = 0 = \underline{\lim_{z\to z^0}}\, u(z)$$

and hence the nonnegative function $u(z)$ is continuous at points z^0 at which $u(z^0) = 0$.

We introduce the decreasing sequence

$$u_\alpha(z) = \int u\left(z + \frac{z'}{\alpha}\right)\omega(|z'|)dx'\,dy' + \frac{1}{\alpha}, \qquad \alpha = 1, 2, \ldots$$

of positive infinitely differentiable subharmonic functions in the corresponding domains G_α, where $G_\alpha \subset G_{\alpha+1}$ and $\bigcup_\alpha G_\alpha = G$ (see section 9.13). It follows from the equation

$$u_\alpha(z)\ |e^{az}| =$$
$$= \int u\left(z + \frac{z'}{\alpha}\right)\left|e^{a\left(z+\frac{z'}{\alpha}\right)}\right|\omega(|z'|)\,e^{-\mathrm{Re}\left(\frac{az'}{\alpha}\right)}dx'\,dy' + \frac{1}{\alpha}|e^{az}|$$

and the hypothesis of the theorem that the function $u_\alpha(z)|e^{az}|$ is subharmonic in G_α for all α. Then, by virtue of section 9.14, we have

$$\Delta(u_\alpha|e^{az}|) = |e^{az}|\left[\Delta u_\alpha + 2b\frac{\partial u_\alpha}{\partial x} - 2c\frac{\partial u_\alpha}{\partial y} + (b^2+c^2)\,u_\alpha\right] \geqslant 0$$

for all b and c, where $a = b + ic$. From this, taking the minimum over b and c and using the fact that $u_\alpha \geqslant 1/\alpha$ is positive, we obtain the inequality

$$u_\alpha \Delta u_\alpha - \left(\frac{\partial u_\alpha}{\partial x}\right)^2 - \left(\frac{\partial u_\alpha}{\partial y}\right)^2 \geqslant 0.$$

On the other hand, we have

$$\Delta \ln u_\alpha = \frac{1}{u_\alpha^2}\left[u_\alpha \Delta u_\alpha - \left(\frac{\partial u_\alpha}{\partial x}\right)^2 - \left(\frac{\partial u_\alpha}{\partial y}\right)^2\right].$$

Therefore, $\Delta \ln u_\alpha \geqslant 0$ in G_α. From this, we conclude on the basis of the test in section 9.14 that the function $\ln u_\alpha(z)$ is subharmonic in G_α. Since $u_\alpha \to u$, monotonically decreasing (see section 9.13), as $\alpha \to \infty$, it follows that $\ln u_\alpha \to \ln u$, also decreasing monotonically. Therefore, the function $\ln u$ is subharmonic in G (see section 9.6), which completes the proof.

16. The trace of a subharmonic function on a Jordan curve

Suppose that a function $u(z)$ is subharmonic in a domain G and that $L = [z: z = z(t), 0 \leq t \leq 1]$ is a Jordan curve lying entirely in G. Then,

$$\overline{\lim_{t\to 0,\ t\neq 0}}\ u[z(t)] = u[z(0)]. \tag{24}$$

(See Oka [75] and Rothstein [66].)

Proof: Without loss of generality, we may assume that $z(0) = 0$. Let us denote the left member of Eq. (24) by a. Since the function $u[z(t)]$ is upper-semicontinuous for $t \in [0, 1]$ (see section 9.1), we have

$$a \leqslant \overline{\lim_{t\to 0}}\ u[z(t)] = u(0) < +\infty.$$

Let us suppose that $a < u(0)$. Define $\varepsilon = 1/2[u(0) - a]$. Then, since $0 \in G$ and the function $u(z)$ is upper-semicontinuous in G, there exist positive numbers $r \leqslant 1$ and $t_0 \leqslant 1$ such that $S(0, r) \Subset G$ and

$$u[z(t)] \leqslant u(0) - \varepsilon, \quad t \in (0, t_0]; \tag{25}$$

Here, we may assume that the point $z(t_0)$ lies on the circle $|z| \leqslant r$ and that the curve $z = z(t)$, for $0 \leqslant t \leqslant t_0$, lies entirely inside the closed circle $|z| = r$. Suppose that $\rho < r$. We denote by L_ρ that portion of the curve L that lies between the points $z(t_0)$ and $z(t_\rho)$, where $z(t_\rho)$ is the last point of intersection of L with the circle $|z| = \rho$ (as we move along L to the point 0 [see Fig. 11]).

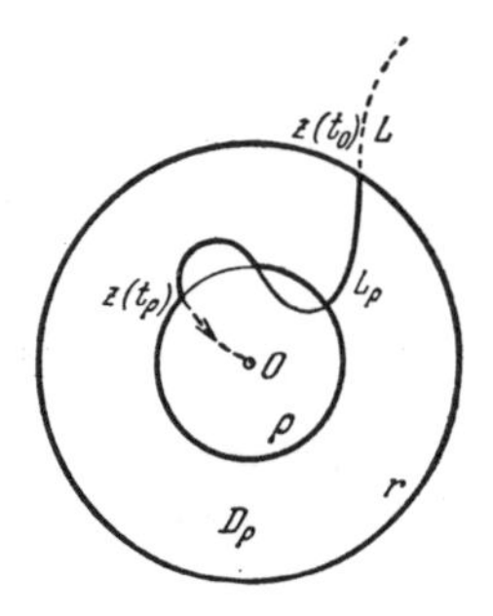

Fig. 11

We denote by D_ρ the circle $|\zeta| < r$ deleted by the curve L_ρ. Since L_ρ is a Jordan curve, D_ρ is a simply connected domain and all points of its boundary ∂D_ρ are accessible from D_ρ (see section 1.5). Here, points of curve L_ρ are double points. We denote by $\partial' D_\rho$ the boundary of D_ρ with the multiplicity of its points taken into consideration [see section 7.3].

Let us consider the function $z=\omega_\rho(\zeta)$ that maps the open circle $|\zeta|<1$ single-sheetedly and conformally onto D_ρ in such a way that $\omega_\rho(0)=0$ and $\omega_\rho'(0)>0$. According to Carathéodory's theorem (see section 7.4), ω_e maps the circle $|\zeta|\leqslant 1$ in a one-to-one manner onto $\partial' G_\rho$. Here, the function $\frac{\omega_\rho(\zeta)}{\zeta}$ is continuous in $|\zeta|<1$.

Thus, the function $\omega_\rho(\zeta)/\zeta$ is holomorphic in $|\zeta|<1$ and is continuous and nonzero in $|\zeta|=1$. Therefore, the function $\ln\left|\frac{\omega_\rho(\zeta)}{\zeta}\right|$ is harmonic in $|\zeta|<1$ and continuous in $|\zeta|\leqslant 1$. If we apply formula (5) to it, we obtain

$$\ln|\omega_\rho'(0)|=$$

$$=\frac{1}{2\pi}\int_0^{2\pi}\ln\left|\frac{\omega_\rho(e^{i\theta})}{e^{i\theta}}\right|d\theta=\frac{1}{2\pi}\int_0^{2\pi}\ln|\omega_\rho(e^{i\theta})|\,d\theta. \tag{26}$$

We denote by l_ρ the preimage of the curve $L_\rho\subset\partial' D_\rho$ under the mapping $z=\omega_\rho(\zeta)$ and we denote by $2\pi a_\rho$ the measure of the set l_ρ (on the circle $|\zeta|=1$). Obviously,

$$|\omega_\rho(\zeta)|\geqslant\rho,\quad \zeta\in l_\rho;\quad |\omega_\rho(\zeta)|\leqslant r\leqslant 1,\quad |\zeta|=1.$$

From Eq. (26) and the above inequalities, we obtain the inequality

$$\ln|\omega_\rho'(0)|\geqslant a_\rho\ln\rho. \tag{27}$$

On the other hand, since the function $\omega_\rho(\zeta)$ does not assume the value ρ in the circle $|\zeta|<1$, it follows, on the basis of Koebe's theorem (see section 7.4) that $|\omega_\rho'(0)|\leqslant 4\rho$. From this and from inequality (27), it follows that

$$\rho^{1-a_\rho}\geqslant\frac{1}{4}.$$

Since $a_\rho\leqslant 1$, this inequality means that $a_\rho\to 1$ as $\rho\to+0$.

The function $u[\omega_\rho(\zeta)]$ is subharmonic in the open circle $|\zeta|<1$ (see section 10.11) and upper-semicontinuous in the closed circle $|\zeta|\leqslant 1$. Therefore, it satisfies inequality (8):

$$u(0)=u[\omega_\rho(0)]\leqslant\frac{1}{2\pi}\int_0^{2\pi}u[\omega_\rho(e^{i\theta})]\,d\theta. \tag{28}$$

Furthermore, since $u(z)\leqslant M<+\infty$ for $|z|\leqslant r$, we obtain on the basis of inequalities (25) and (28)

$$u(0)\leqslant[u(0)-\varepsilon]\,a_\rho+M(1-a_\rho),$$

which is impossible as $r\to+0$ (since $a_\rho\to 1$). This contradiction proves Eq. (24).

10. PLURISUBHARMONIC FUNCTIONS

1. *The definition of a plurisubharmonic function*

A function $u(z) \equiv u(z, \bar{z}) \equiv u(x, y)$ is said to be plurisubharmonic (pseudoconvex) in a domain $G \subset C^n$ if (1) $u(z)$ is upper-semicontinuous in G and (2) for arbitrary $z^0 \in G$ and $a z^0 \in G$, the function $u(z^0 + \lambda a)$ is subharmonic with respect to λ in every component of the open set

$$G_{z^0, a} = [\lambda : z^0 + \lambda a \in G].$$

A function u is said to be plurisuperharmonic if $-u$ is a plurisubharmonic function. We note that a plurisubharmonic function does not assume the value $+\infty$ and therefore is bounded from above in G (see sections 2.3 and 1.6) and is locally integrable in G (see section 2.6).

For the behavior of plurisubharmonic functions, see the works by Lelong [12, 49], Bremermann [13], and Oka [19, 20].

2. *A test for plurisubharmonicity*

Suppose that $u(z)$ is a plurisubharmonic function in a domain G and that $[z' : z' = z + \lambda a, |\lambda| < r] \Subset G$. Then,

$$u(z) \leqslant \frac{1}{2\pi} \int_0^{2\pi} u(z + rae^{i\theta})\, d\theta. \tag{29}$$

Conversely, if $u(z) < +\infty$ is upper-semicontinuous in G and satisfies inequality (29) for all $z \in G$, $|a| = 1$ and $r \leq r_0(z, a)$, then $u(z)$ is a plurisubharmonic function in G.

This theorem follows from the definition of section 10.1 and the test in section 9.4.

3. *The simplest properties of plurisubharmonic functions*

The following properties of plurisubharmonic functions follow from the preceding test and from section 9.6.

A function $u(z)$ is plurisubharmonic in a domain G if and only if it is plurisubharmonic in a neighborhood of each point of G.

A linear combination, with positive coefficients, of plurisubharmonic functions is a plurisubharmonic function.

The limit of a uniformly convergent sequence of plurisubharmonic functions is a plurisubharmonic function.

A monotonically decreasing sequence of plurisubharmonic functions is a plurisubharmonic function.

A nonconstant plurisubharmonic function in a domain G cannot attain a maximum inside G.

Thus, by use of the theorem in section 2.3, we may assert that the maximum principle (see section 1.6) holds for bounded regions G and their boundaries ∂G with respect to functions that are plurisubharmonic in G and upper-semicontinuous in $\bar{G}$.

The upper envelope

$$u(z) = \overline{\lim_{z' \to z}} \sup_{\alpha} u_\alpha(z')$$

and the limit superior

$$v(z) = \overline{\lim_{z' \to z}} \overline{\lim_{\alpha \to \infty}} u_\alpha(z')$$

of the family $\{u_\alpha\}$ of functions that are plurisubharmonic in G and locally uniformly bounded above in G are plurisubharmonic functions in G.

4. *The greatest plurisubharmonic minorant*

The greatest (resp. least) plurisubharmonic (resp. plurisuperharmonic) minorant $v(z)$ (resp. majorant $V(z)$) of an upper- (resp. lower-) semicontinuous functions $r(z) < \infty$ (resp. $R(z) > -\infty$) is defined exactly as in section 9.7. We have the corresponding formulas

$$v(z) = \overline{\lim_{z' \to z}} \sup_{[u(z') \leqslant r(z')]} u(z'),$$
$$V(z) = \underline{\lim_{z' \to z}} \inf_{[u(z') \geqslant R(z')]} u(z'),$$

where the supremum (resp. infimum) is over all plurisubharmonic (resp. plurisuperharmonic) functions in G such that $u(z) \leqslant r(z)$ (resp. $(u(z) \geqslant R(z))$.

5. *Examples of plurisubharmonic functions*

If $u(z)$ is a plurisubharmonic function, the function $e^{u(z)}$ is also a plurisubharmonic function. If, in addition, $u(z) \geqslant 0$, then $u^p(z)$ is a plurisubharmonic function for all $p \geqslant 1$ (see section 9.11).

If $f(z)$ is holomorphic in G, the functions $\ln|f(z)|$, $\ln^+|f(z)|$ and $|f(z)|^p$ (where $(p \geqslant 0)$ are plurisubharmonic in G (see section 9.12).

A function $u(z) \geqslant 0$ is said to be logarithmically plurisubharmonic in a domain G if the function $\ln u(z)$ is plurisubharmonic in G. It follows from this that the function $u(z)$ must itself be plurisubharmonic. Examples of logarithmically plurisubharmonic functions are the absolute values of holomorphic functions.

THEOREM. *For a function $u(z) \geq 0$ to be logarithmically plurisubharmonic, it is necessary and sufficient that the function $u(z)\,|e^{az}|$ be plurisubharmonic for all complex vectors a.*

This follows from the theorem of section 9.15.

We conclude from this theorem that the property of logarithmic plurisubharmonicity is conserved under addition and multiplication by a positive number. Therefore, the function

$$V(z)=\int u^p(z;\ \xi)\,d\mu(\xi), \qquad 0<p<\infty$$

is logarithmically plurisubharmonic if the function $u(z;\ \xi)$ is logarithmically plurisubharmonic with respect to z for every ξ and u^p is summable with respect to the measure $\mu \geqslant 0$ (under the assumption that $V(z)$ is upper-semicontinuous).

6. *Jensen's inequality*

If a function $u(z)$ is plurisubharmonic in G and $S(z^0,\ r) \Subset G$, then, for all $z \in S(z^0,\ r)$,

$$u(z) \leqslant \int_0^{2\pi} \cdots \int_0^{2\pi} P_n(z-z^0,\ re^{i\theta})\,u(z^0+re^{i\theta})\,d\theta_1 \ldots d\theta_n, \tag{30}$$

where

$$re^{i\theta}=(r_1e^{i\theta_1},\ \ldots,\ r_ne^{i\theta_n}), \quad P_n(z,\ \zeta)=\prod_{1\leqslant j\leqslant n} P(z_j,\ \zeta_j).$$

Inequality (30) is easily proven by induction on n since a plurisubharmonic function is subharmonic with respect to each variable individually and this inequality was proven in section 9.4 for $n=1$.

Inequality (30) implies Jensen's inequality

$$\ln|f(z)| \leqslant \int_0^{2\pi} \cdots \int_0^{2\pi} P_n(z,\ re^{i\theta})\ln|f(re^{i\theta})|\,d\theta,$$

which is valid for all functions $f(z)$ that are holomorphic in $\bar{S}(0,\ r)$. The following properties of the kernel $P_n(z,\ \zeta)$ follow from (3) and (4):

$$P_n(0,\ \zeta)=\frac{1}{(2\pi)^n}, \quad \int_0^{2\pi} \cdots \int_0^{2\pi} P_n(z,\ re^{i\theta})\,d\theta=1, \tag{31}$$

$$m_n(r, \rho) = \frac{1}{(2\pi)^n} \prod_{1 \leqslant j \leqslant n} \frac{r_j - \rho_j}{r_j + \rho_j} \leqslant P_n(z, \zeta) \leqslant \leqslant \frac{1}{(2\pi)^n} \prod_{1 \leqslant j \leqslant n} \frac{r_j + \rho_j}{r_j - \rho_j} = M_n(r, \rho). \tag{32}$$

7. *A generalization of Harnack's theorem*

From inequality (30), we get the following generalization of Harnack's theorem (see, for example, Bochner and Martin [3], p. 189):

Suppose that a sequence of functions $u_\alpha(z)$, for $\alpha = 1, 2, \ldots$, of functions that are plurisubharmonic in a domain G and locally uniformly bounded above in G is such that

$$\overline{\lim_{\alpha \to +\infty}}\, u_\alpha(z) \leqslant A < +\infty, \qquad z \in G. \tag{33}$$

Then, for arbitrary $\epsilon \subset 0$ and arbitrary subdomain $G' \Subset G$, there exists a number such that, for all $\alpha \geq N$ and all $z \in G'$, the inequality

$$u_\alpha(z) \leqslant A + \varepsilon \tag{34}$$

is valid.

Proof: On the basis of the Heine-Borel theorem, the compact set $\bar{G}'$ can be covered by a finite number of polycircles that are compact in G. Therefore, it will be sufficient to prove the conclusion of the theorem for just one of these polycircles $S(z^0, r^0)$. For simplicity in writing, we shall assume that $z^0 = 0$. Since $S(0, r^0) \Subset G$, there exists a greatest polycircle $S(0, r)$, where $r_j > r_j^0$ for $j = 1, 2, \ldots, n$, which is also compact in G. From the hypothesis of the theorem,

$$u_\alpha(z) \leqslant M, \quad z \in \bar{S}(0, r). \tag{35}$$

If we apply inequality (30) to the functions $u_\alpha(z)$ and to the polycircle $S(0, r)$:

$$u_\alpha(z) \leqslant \int_0^{2\pi} \ldots \int_0^{2\pi} P_n(z, re^{i\theta})\, u_\alpha(re^{i\theta})\, d\theta, \tag{36}$$

we obtain

$$u_\alpha^*(\theta) = \sup_{m \geqslant 0} u_{\alpha+m}(re^{i\theta}),$$

$$u_0(\theta) = \overline{\lim_{\alpha \to +\infty}}\, u_\alpha(re^{i\theta}).$$

Then, on the basis of (33) and (35), we have, for $z \in \bar{S}(0, r)$,

$$\left.\begin{aligned} &u_\alpha(re^{i\theta}) \leqslant u_\alpha^*(\theta) \leqslant M, \\ &u_{\alpha+1}^*(\theta) \leqslant u_\alpha^*(\theta) \to u_0(\theta) \leqslant A. \end{aligned}\right\} \tag{37}$$

From this and from inequality (36), we conclude that, for $z \in \bar{S}(0, r^0)$,

$$u_\alpha(z) \leqslant \int_0^{2\pi} \ldots \int_0^{2\pi} P_n(z, re^{i\theta}) u_\alpha^*(\theta)\, d\theta \leqslant$$
$$\leqslant \int_0^{2\pi} \ldots \int_0^{2\pi} P_n(z, re^{i\theta}) u_0(\theta)\, d\theta + \tag{38}$$
$$+ M_n(r, r^0) \int_0^{2\pi} \ldots \int_0^{2\pi} [u_\alpha^*(\theta) - u_0(\theta)]\, d\theta \leqslant$$
$$\leqslant A + M_n(r, r^0)\, \varepsilon_\alpha,$$

where the numbers $M_n(r, r^0)$ are defined in (32) and

$$\varepsilon_\alpha = \int_0^{2\pi} \ldots \int_0^{2\pi} [u_\alpha^*(\theta) - u_0(\theta)]\, d\theta.$$

Suppose that $\int_0^{2\pi} \ldots \int_0^{2\pi} u_0(\theta)\, d\theta > -\infty$. It then follows from (37) and from section 2.8 that $\varepsilon_\alpha \to 0$, monotonically decreasing, as $\alpha \to \infty$. We choose for $N = N(\varepsilon, r^0)$ the smallest integer such that $M_n(r, r^0)\, \varepsilon_N \leqslant \varepsilon$. From this and from (38), we have the desired inequality (34) in the polycircle $\bar{S}(0, r^0)$ for $\alpha \geqslant N$.

On the other hand, if

$$\int_0^{2\pi} \ldots \int_0^{2\pi} u_0(\theta)\, d\theta = -\infty,$$

then

$$\eta_\alpha = \int_0^{2\pi} \ldots \int_0^{2\pi} u_\alpha^*(\theta)\, d\theta \to -\infty, \quad \alpha \to \infty,$$

monotonically decreasing, as $a \to \infty$. By using inequalities (36), (37), and (32) and Eq. (31), we obtain, for all $z \in S(0, r^0)$,

$$u_\alpha(z) \leqslant M \int_0^{2\pi} \ldots \int_0^{2\pi} P_n(z, re^{i\theta})\, d\theta +$$
$$+ \int_0^{2\pi} \ldots \int_0^{2\pi} P_n(r, e^{i\theta}) [u_\alpha^*(\theta) - M]\, d\theta \leqslant$$
$$\leqslant M - m_n(r, r^0) \int_0^{2\pi} \ldots \int_0^{2\pi} [M - u_\alpha^*(\theta)]\, d\theta =$$
$$= M[1 - (2\pi)^n m_n(r, r^0)] + m_n(r, r^0)\, \eta_\alpha.$$

From this it follows that $u_\alpha(z) \to -\infty$ uniformly for all $z \in \bar{S}(0, r^0)$. Consequently, inequality (34) remains valid in this case also in the polycircle $\alpha \to \infty$ for all sufficiently large α. This completes the proof of the theorem.

8. *The mean value of a plurisubharmonic function*

Suppose that $u(z)$ is a plurisubharmonic function in a domain G and that $z^0 \in G$. The function

$$J(r, z^0; u) = \frac{1}{\sigma_{2n}} \int\limits_{|a|=1} u(z^0 + ra)\, da$$

is called the *mean value* of $u(z)$, where σ_{2n} is the area of the surface of the unit sphere in R^{2n}:

$$\sigma_{2n} = \frac{2\pi^n}{(n-1)!}.$$

If $u(z)$ is a plurisubharmonic function in G, then

$$u(z^0) \leqslant J(r, z^0; u),$$

also the function $J(r, z^0; u)$ is an increasing function with respect to $r \in [0, R)$ if $U(z^0, R) \subset G$ and is convex with respect to $\ln r$ in (ρ, R) if $[z : \rho < |z - z^0| < R] < G$ (see section 9.8).

Proof: If we integrate inequality (29) over all a such that $|a| = 1$ and use the properties of integrals over a unit sphere, we obtain

$$\begin{aligned} u(z^0)\,\sigma_{2n} &\leqslant \frac{1}{2\pi} \int\limits_{|a|=1} da \int\limits_0^{2\pi} u(z^0 + rae^{i\theta})\, d\theta = \\ &= \frac{1}{2\pi} \int\limits_0^{2\pi} d\theta \int\limits_{|a|=1} u(z^0 + rae^{i\theta})\, da = \\ &= \int\limits_{|a|=1} u(z^0 + ra)\, da = \sigma_{2n} J(r, z^0; u). \end{aligned} \tag{39}$$

Furthermore, since the function $u(z^0 + \lambda a)$ is subharmonic with respect to λ, it follows on the basis of the theorem of section 9.8 that the function

$$\frac{1}{2\pi} \int\limits_0^{2\pi} u(z^0 + rae^{i\theta})\, d\theta$$

is an increasing function and is convex with respect to $\ln r$ for all a. But then it follows from (39) that the function $J(r, z^0, u)$ is an increasing function and is convex with respect to $\ln r$ under analogous conditions, which completes the proof.

It follows from this theorem and from the test in section 9.4 (for $2n$ variables) that plurisubharmonic functions are subharmonic functions. In particular, if $u(z) \not\equiv -\infty$ is a plurisubharmonic function in G, it is locally summable in G (see section 9.9). If two plurisubharmonic functions in G coincide almost everywhere in G, they are identical in G (see section 9.10).

9. *Approximation of plurisubharmonic functions*

For a function $u(z)$ to be plurisubharmonic in a domain G, it is necessary and sufficient that it be the limit of a decreasing sequence of plurisubharmonic functions $u_\alpha(z)$, for $\alpha = 1, 2, \ldots$, of the class $C^{(\infty)}(G_\alpha)$, where the domains G_α are such that $G_\alpha \subset G_{\alpha+1}$ and $U_\alpha G_\alpha = G$.

The proof is analogous to the proof of the theorem in section 9.13. Suppose that a function $\omega(|z|) \in D$ possesses the properties

$$0 \leqslant \omega(|z|); \quad \omega(|z|) = 0, \quad |z| \geqslant 1;$$
$$\sigma_{2n} \int_0^1 \omega(\rho)\, \rho^{2n-1}\, d\rho = 1$$

(see section 3.1). We define the required sequence of functions $u_\alpha(z)$ for $\alpha = 1, 2, \ldots$, as in (17):

$$u_\alpha(z) = \int u\left(z + \frac{z'}{\alpha}\right) \omega(|z'|)\, dx'\, dy' =$$
$$= \int u(z')\, \omega(\alpha\, |z - z'|)\, \alpha^{2n}\, dx'\, dy' =$$
$$= \int_0^1 r^{2n-1} \omega(r) \left[\int_{|a|=1} u\left(z + \frac{r}{\alpha} a\right) da \right] dr =$$
$$= \sigma_{2n} \int_0^1 J\left(\frac{r}{\alpha},\, z;\, u\right) r^{2n-1} \omega(r)\, dr.$$

The monotonic decrease of the functions $u_\alpha(z)$ follows from the theorem in section 10.8.

Remark: If a function $u(z)$ is logarithmically plurisubharmonic, the sequence $u(z)$ constructed in the theorem consists of logarithmically plurisubharmonic functions. This follows from the theorem in section 10.5 (cf. also the proof of the theorem in section 9.15).

10. The property of positivity for plurisubharmonic functions

If $u(z) \not\equiv -\infty$ is a plurisubharmonic function in a domain G, the Hermitian form

$$\sum_{j,k} \frac{\partial^2 u}{\partial z_j \partial \bar{z}_k} a_j \bar{a}_k = (H(z;\ u)\, a,\ \bar{a})$$

is positive in Gl that is, for an arbitrary nonnegative function $\phi \in D(G)$ and, arbitrary vectors a,

$$\sum_{j,k} \int u(z) \frac{\partial^2 \varphi(z)}{\partial z_j \partial \bar{z}_k}\, dx\, dy\, a_j \bar{a}_k \geqslant 0.$$

Conversely, if $u \in D^*(G)$ is such that the Hermitian form $(H(z; u)\, a, \bar{a})$ is positive in G, there exists a unique function that is plurisubharmonic in G and that coincides with $u(z)$ almost everywhere (cf. section 9.14).

Proof: If $u(z) \not\equiv -\infty$ is a plurisubharmonic function in G, there exists, by virtue of the theorem in section 10.9, a sequence of plurisubharmonic functions $u_\alpha(z)$, for $\alpha = 1, 2, \ldots,$ in the class $C^{(\infty)}(G_\alpha)$, where $G_\alpha \subset G_{\alpha+1}$ and $\bigcup_\alpha G_\alpha = G$ such that $u_\alpha(z) \to u(z)$, decreasing monotonically. Therefore, if we use the definition of plurisubharmonic functions and the theorem in section 9.14, we conclude that

$$(H(z;\ u_\alpha)\, a,\ \bar{a}) = \frac{\partial^2 u_\alpha(z + \lambda a)}{\partial \lambda\, \partial \bar{\lambda}} \bigg|_{\lambda=0} \geqslant 0, \qquad z \in G_\alpha.$$

Since $u_\alpha(z) \to u(z)$ in the sense of $D^*(G)$, if we take the limit as $\alpha \to \infty$ in the inequality that we have just obtained and make use of the continuity of the differentiation operator in $D^*(G)$ (see section 3.3), we see that the form $(H(z;\ u)\, a,\ \bar{a})$ is positive in G.

Conversely, suppose that $u \in D^*(G)$ and that the form $(H(z;\ u)\, a,\ \bar{a})$ is positive in G. Then, $\Delta u \geqslant 0$ in the sense of $D^*(G)$. In accordance with the theorem in section 9.14 (more precisely, in accordance with the analogous theorem for the $2n$-dimensional case), the function $u(z)$ coincides almost everywhere in G with the unique subharmonic function $v(z)$. It then becomes clear that $(H(z;\ v)\, a,\ \bar{a}) \geqslant 0$.

Suppose that the function $\omega(|z|)$ satisfies the conditions of section 10.9. We construct a sequence of mean functions

$$v_\alpha(z) = \int v\left(z + \frac{z'}{\alpha}\right) \omega(|z'|)\, dx'\, dy' = \\ = \int v(z')\, \omega(\alpha\, |z - z'|)\, \alpha^{2n}\, dx'\, dy'. \tag{40}$$

The functions $v_\alpha(z) \in C^{(\infty)}(G_\alpha)$, where $G_\alpha \subset G_{\alpha+1}$ and $\bigcup_\alpha G_\alpha = G$, converge to $v(z)$, decreasing monotonically, (see section 9.13). Furthermore, it follows from (40) and from the properties of the function

$\omega(|z|)$ that, for all a,

$$0 \leqslant (H(z;\ v_\alpha)\,a,\ \bar{a}) = \frac{\partial^2 v_\alpha(z^0+\lambda a)}{\partial\lambda\,\partial\bar{\lambda}}, \quad z^0+\lambda a = z \in G_\alpha.$$

From this we conclude on the basis of the theorem in section 9.14 that the functions $v_\alpha(z)$ are plurisubharmonic in G_α. But then, the function $v(z)$ is plurisubharmonic in G (see section 10.3), which completes the proof.

A consequence of this theorem is the following: For a function to be pluriharmonic (see section 4.1), it is necessary and sufficient that it be both plurisubharmonic and plurisuperharmonic.

11. *Invariance under biholomorphic mappings*

Plurisubharmonic functions are mapped under biholomorphic transformation into plurisubharmonic functions.

Proof: Consider first a plurisubharmonic function $u(z) \in C^{(2)}$ in a domain G. If $\zeta = \zeta(z)$ is a biholomorphic mapping of G onto G_1, the function $u_1(\zeta) = u[z(\zeta)]$ where $z = z(\zeta)$ is the inverse transformation (see section 7.2), satisfies the inequality $(H(\zeta;\ u_1)\,a,\ \bar{a}) \geqslant 0$ since

$$(H(\zeta;\ u_1)\,a,\ \bar{a}) = (H(z;\ u)\,Aa,\ \overline{Aa}) \geqslant 0,$$

where

$$A = \left\| \frac{\partial z_j}{\partial \zeta_k} \right\|, \qquad j,\ k = 1,\ 2,\ \dots,\ n.$$

On the basis of the theorem in section 10.10, the function $u_1(\zeta)$ is plurisubharmonic in G_1.

Suppose now that $u(z) \bar{\in} C^{(2)}$. If we apply the result that we have just obtained to the decreasing sequence of plurisubharmonic functions of the class $C^{(\infty)}$ that converges to $u(z)$ (see section 10.9), we easily establish the assertion made in the general case.

Remark: Since a holomorphic mapping is locally biholomorphic (see section 7.2) and since plurisubharmonic functions are defined locally (see section 10.3), the assertion made above remains in force for holomorphic mappings.

Finally, we note that if $u \in C^{(2)}$, the inequality $(H(z;\ u)\,a,\ \bar{a}) \geqslant 0$ when $\sum_j \frac{\partial u}{\partial z_j} a_j = 0$ remains valid for a biholomorphic transformation.

To see this, note that, for $u_1(\zeta) = u[z(\zeta)]$,

$$\sum_j \frac{\partial u_1(\zeta)}{\partial \zeta_j} a_j = \sum_j \frac{\partial u}{\partial z_j} (Aa)_j = 0$$

and hence

$$(H(\zeta;\ u_1)\,a,\ \bar{a}) = (H(z;\ u)\,Aa,\ \overline{Aa}) \geqslant 0, \quad \text{q. e. d.}$$

12. *Analytic surfaces*

A set $F \subset C^n$ is called a $2k$-dimensional analytic surface if it is defined in some neighborhood of every point $z^0 \in F$ by equations of the form

$$z_j = z_j(z^0;\ \lambda), \qquad j = 1,\ 2,\ \dots,\ n, \quad \lambda = (\lambda_1, \dots, \lambda_k),$$

where the functions $z_j(z^0;\ \lambda)$ are holomorphic and the rank of the matrix

$$\frac{\partial z_j(z^0;\ \lambda)}{\partial \lambda_s} \qquad (j = 1,\ 2,\ \dots,\ n, \quad s = 1,\ 2,\ \dots,\ k)$$

is equal to k in the corresponding neighborhood of the point $\lambda = 0$, $z^0 = z(z^0;\ 0)$.

It follows from this definition that, in a neighborhood of a point $z^0 \in F$, a $2k$-dimensional analytic surface F is defined by the equations

$$f_s(z^0;\ z) = 0, \qquad s = 1,\ 2,\ \dots,\ n - k,$$

where the functions $f_s(z^0;\ z)$ are holomorphic and the rank of the matrix of the $\partial f_s / \partial z_j$ is equal to $n-k$ in that neighborhood (cf. 7.2).

In accordance with the definitions of section 1.4, an analytic surface is a surface of class $C^{(\infty)}$. Therefore, the concepts expounded in section 1.4 are applicable to analytic surfaces.

If a manifold F of dimension $2k$ is given by the equations $f_s(z) = 0$ for $s = 1,\ 2,\ \dots,\ n - k$, where the functions $f_s(z)$ are holomorphic in a neighborhood of F, then F is called a $2k$-*dimensional analytic set*.

13. *Plurisubharmonic functions on analytic surfaces*

Suppose that a function $u(z)$ is defined on an analytic surface F. We shall say that $u(z)$ is plurisubharmonic on F if, for all $z^0 \in F$, the function $u[z(z^0;\ \lambda)]$ is plurisubharmonic with respect to λ in some neighborhood of the point $\lambda = 0$. We note that this definition is of a local nature (cf. section 10.3).

A holomorphic function on an analytic surface is defined analogously.

If a function $u(z)$ is plurisubharmonic in a neighborhood $U(F)$ of a $2k$-dimensional analytic surface F, it is also a plurisubharmonic function of F.

Proof: If we set $\varphi(\lambda) = u[z(z^0;\lambda)]$, we obtain (cf. section 10.11)

$$(H(\lambda;\varphi)a, \bar{a}) = (H(z;u)Aa, \overline{Aa}) \geqslant 0,$$

where

$$A = \left\| \frac{\partial z_j}{\partial \lambda_s} \right\|, \qquad a = (a_1, a_2, \ldots, a_k).$$

It remains only to use the theorem of section 10.10.

An analogous assertion holds for functions that are holomorphic in $U(F)$.

14. *The maximum principle*

The maximum principle holds for bounded domains S lying on an analytic surface F and for their boundaries ∂S *such that* $\bar{S} = S \cup \partial S \subset D$, *and with respect to functions that are plurisubharmonic in S and upper-semicontinuous on* $\bar{S}$ (cf. section 10.3).

Proof: The assertion clearly holds for $u \equiv$ const. Let us now suppose that $u(z) \not\equiv$ const is a plurisubharmonic function on S that is upper-semicontinuous on $\bar{S}$. Let us suppose that the assertion formulated above does not hold for this function. We define

$$A = [z : u(z) = M, \quad z \in \bar{S}],$$

where

$$M = \sup_{z \in \bar{S}} u(z).$$

Since $u(z)$ is upper-semicontinuous on $\bar{S}$, the set A is nonempty (see section 2.3). By hypothesis, $A \subset S$. But A is a closed set on z^0 (see section 9.4) and since it is bounded, we have $A \Subset S$. Let z^0 denote a point in ∂A. By definition (see section 10.12), in a neighborhood of the point z^0, an analytic surface F is defined in the form $z_j = z_j(\lambda)$, for $j = 1, 2, \ldots, n$ where the $z_j(\lambda)$ are holomorphic in some neighborhood U of the point $\lambda = 0$, $z^0 = z(0)$. Let us assume that the neighborhood U is sufficiently small that $[z : z = z(\lambda), \lambda \in U] \subset S$ (such a neighborhood always exists since $A \Subset S$). But then, the function $u[z(\lambda)]$ is plurisubharmonic in U (see section 10.13) and it attains its maximum value M at an interior point $\lambda = 0$ of the domain U. Consequently, $u(z) \equiv M$ in $[z : z = z(\lambda), \lambda \in U]$ (see section 10.3), which contradicts the fact that z^0 is a boundary point of the set A. This completes the proof.

15. *Hartogs functions*

The class of Hartogs functions F_G (see [16, 3]) is defined as the smallest class of real-valued functions defined in a domain G

that contains all functions $\ln|f(z)|$, where $f(z)$ is a holomorphic function in G, and that is closed under the following operations:

(1) if $u_1, u_2 \in F_G$ and $\lambda_1 \geqslant 0$, $\lambda_2 \geqslant 0$, then $\lambda_1 u_1 + \lambda_2 u_2 \in F_G$;

(2) if $\{u_\alpha\} \subset F_G$ where $u_\alpha \leqslant C_{G'} < +\infty$ for every subdomain $G' \Subset G$, then $\sup u_\alpha \in F_G$;

(3) if $\{u_\alpha\} \subset F_G$ and $u_\alpha \geqslant u_{\alpha+1}$ then $\lim\limits_{\alpha \to \infty} u_\alpha \in F_G$;

(4) if $u \in F_G$, then $\overline{\lim\limits_{z' \to z}}\, u(z') \in F_G$;

(5) if $u \in F_{G'}$ for all subdomains $G' \Subset G$, then $u \in F_G$.

From the properties of plurisubharmonic functions, it follows that Hartogs' functions that are upper-semicontinuous are plurisubharmonic functions. However, not every plurisubharmonic function is a Hartogs function. A counterexample has been constructed by Bremermann [16, 17]. On the other hand, for the case in which G is a domain of holomorphy, the classes of upper-semicontinuous Hartogs' functions and plurisubharmonic functions coincide (see [16]).

11. CONVEX FUNCTIONS

Convex functions constitute an important special case of plurisubharmonic functions. Therefore, a number of propositions in the theory of convex functions follow from the corresponding assertions regarding plurisubharmonic functions. For more detailed information on convex functions, see, for example, Bonnesen and Fenchel [121], Hardy, Littlewood, and Pólya [10], and Bremermann [13].

1. *Definition of a convex function*

A real-valued $u(x) < +\infty$ of a real variable x is said to be convex in an interval (a, b) if, for all x and x' in (a, b) and for all $\lambda \in [0, 1]$, it satisfies the inequality (see Fig. 12)

$$u[\lambda x + (1-\lambda)x'] \leqslant \lambda u(x) + (1-\lambda)u(x'). \quad (41)$$

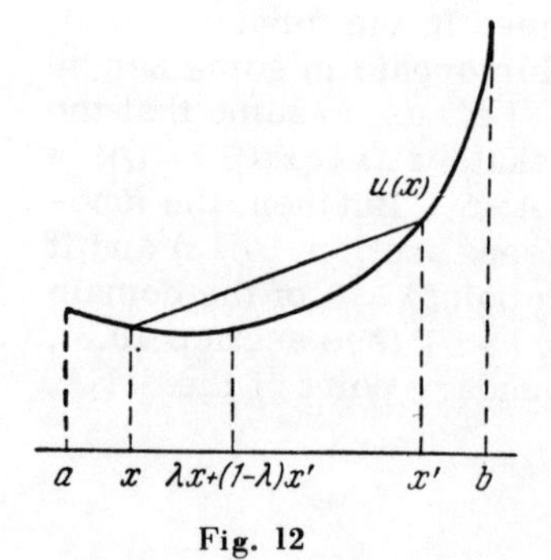

Fig. 12

A function $u(x)$, where $x = (x_1, \ldots, x_n)$ is said to be convex in a domain $B \subset R^n$ if, for all $x^0 \in B$ and all b such that $|b| = 1$, the function $u(x^0 + tb)$ is convex with respect to t in every interval contained in the open set

$$B_{x^0, b} = [t : x^0 + bt \in B].$$

It follows from these definitions that if a function $u(x)$ is convex in a domain B, either $u(x) \equiv -\infty$ or $u(x)$ is finite at all points in B.

2. *Continuity of convex functions*

If a function $u(x) \not\equiv -\infty$ is convex in a domain B, it is continuous in B.

Proof: Let us show first that $u(x)$ is bounded above on every compact set $K \subset B$. On the basis of the Heine-Borel theorem, it will be sufficient to prove this assertion for closed bounded convex polyhedra contained in B. We denote by $x^{(k)}$, for $k = 1, 2, \ldots, N$, the vertices of the polyhedron π (see Fig. 13). An arbitrary point $x \in \pi$ can be represented in the form

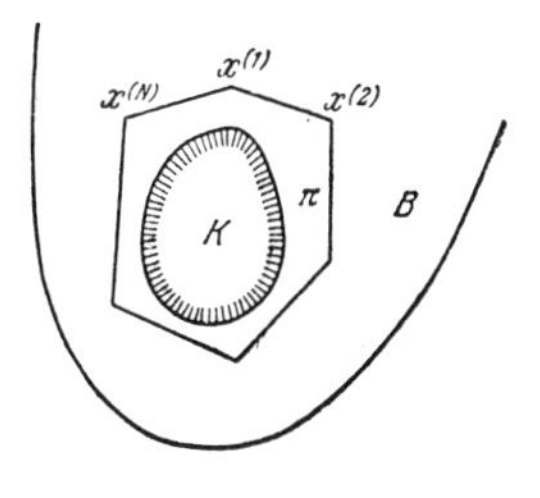

Fig. 13

$$x = \sum_{1 \leqslant k \leqslant N} t_k x^{(k)}, \quad t_k \geqslant 0, \quad \sum_{1 \leqslant k \leqslant N} t_k = 1.$$

Then, the inequality

$$u(x) \leqslant \sum_{1 \leqslant k \leqslant N} t_k u(x^{(k)}), \quad x \in \pi \tag{42}$$

is valid.

To see this, note that inequality (42) is valid for $N = 1$. Suppose that it is valid for $N - 1$. Then, by using inequality (41), we obtain

$$\begin{aligned} u(x) &= u\left(\sum_{1 \leqslant k \leqslant N-1} t_k x^{(k)} + t_N x^{(N)}\right) \leqslant \\ &\leqslant (1 - t_N)\, u\left(\sum_{1 \leqslant k \leqslant N-1} \frac{t_k}{1 - t_N} x^{(k)}\right) + \\ &\qquad + t_N u(x^{(N)}) \leqslant \sum_{1 \leqslant k \leqslant N} t_k u(x^{(k)}). \end{aligned}$$

Inequality (42) implies that the function $u(x)$ is bounded above on π since

$$u(x) \leqslant \max_{1 \leqslant k \leqslant N} \{u(x^{(k)})\}, \qquad x \in \pi.$$

Consider any point $x^0 \in B$. We shall show that $u(x)$ is continuous at x^0. From what we have already proven, $u(x) \leqslant a$ in a sufficiently small sphere $|x - x^0| \leqslant r$. Let $x_k \to x^0$. Then, $1 \geqslant \varepsilon_k = |x_k - x^0| / r \to 0$. If we apply inequality (41) taking $\lambda = \varepsilon_k$, $x = x_k - x^0 / \varepsilon_k + x^0$, and $x' = x^0$, we obtain

$$u(x_k) \leqslant \varepsilon_k u\left(\frac{x_k - x^0}{\varepsilon_k} + x^0\right) + (1 - \varepsilon_k)\, u(x^0).$$

From this it follows that

$$u(x_k) - u(x^0) \leqslant \varepsilon_k [a - u(x^0)]. \tag{43}$$

If we apply inequality (41), this time taking $\lambda = 1/1 + \epsilon_k$ $x = x_k$ and $x' = x^0 - x_k/\epsilon_k + x^0$, we obtain

$$u(x^0) \leqslant \frac{1}{1+\varepsilon_k} u(x_k) + \frac{\varepsilon_k}{1+\varepsilon_k} u\left(\frac{x^0 - x_k}{\varepsilon_k} + x^0\right),$$

so that

$$u(x^0) - u(x_k) \leqslant \varepsilon_k [a - u(x^0)].$$

This, together with (43) proves that $u(x)$ is continuous at the point x^0.

3. *Properties of convex functions*

The properties of convex functions are analogous to the properties of continuous plurisubharmonic functions and they follow from the properties of convex functions of a single variable just as the properties of plurisubharmonic functions follow from the properties of subharmonic functions of a single (complex) variable (see section 10).

Let us note some of these properties.

A function $u(x)$ is convex in a region B if and only if the quadratic form

$$\sum_{j,k} \frac{\partial^2 u}{\partial x_j \partial x_k} b_j b_k = (H(x;\ u)b,\ b)$$

is positive semidefinite in B (in the sense of $D^*(B)$) (cf. section 10.10).

The upper envelope

$$\sup_{\alpha} u_\alpha(x)$$

and the limit superior

$$\overline{\lim_{\alpha \to \infty}}\ u_\alpha(x)$$

of the family $\{u_\alpha(x)\}$ of functions that are convex in B and that are locally uniformly bounded above in B are convex functions in B (cf. section 10.4).

If a function $u(z) = u(x, y)$ is convex in a domain $G \subset R^{2n}$, it is plurisubharmonic in that domain.

Proof: If we take $a_j = b_j + ic_j$, we have

$$\sum_{j,k} \frac{\partial^2 u}{\partial z_j \partial \bar{z}_k} a_j \bar{a}_k = \frac{1}{4} \sum_j \left(\frac{\partial^2 u}{\partial x_j^2} + \frac{\partial^2 u}{\partial y_j^2} \right) (b_j^2 + c_j^2) + \\ + \frac{1}{4} \sum_{j \neq k} \left[\left(\frac{\partial^2 u}{\partial x_j \partial x_k} + \frac{\partial^2 u}{\partial y_j \partial y_k} \right) (b_j b_k + c_j c_k) + \right. \\ \left. + \left(\frac{\partial^2 u}{\partial x_j \partial y_k} - \frac{\partial^2 u}{\partial x_k \partial y_j} \right) (c_j b_k - c_k b_j) \right], \tag{44}$$

that is, if we define $B = (b, c) = (b_1, \ldots, b_n, c_1, \ldots, c_n)$ and $C = (c, -b)$, we have

$$(H(z; u)a, \bar{a}) = \frac{1}{4}(H(x, y; u)B, B) + \frac{1}{4}(H(x, y; u)C, C),$$

from which the assertion follows.

For a function $u(x)$ to be convex in a domain $B \subset R^n$, it is necessary and sufficient that it be plurisubharmonic in the tubular domain $B + iR^n \subset C^n$.

This is true because $\partial u / \partial y_j = 0$ for $j = 1, 2, \ldots, n$ and, hence, it follows from (44) that

$$(H(z; u)a, \bar{a}) = \frac{1}{4}(H(x; u)b, b) + \frac{1}{4}(H(x; u)c, c), \text{ q.e.d.}$$

4. *Convex functions with respect to* $(\ln r_1, \ldots, \ln r_n)$

For a function $R(r)$, where $r = (r_1, r_2, \ldots, r_n)$, to be convex with respect to the variables $(\ln r_1, r_2, \ldots, \ln r_n)$ in a domain B contained in the octant $r_j > 0$, for $j > 1, 2, \ldots, n$, it is necessary and sufficient that the function $u(z) = R(|z_1|, \ldots, |z_n|)$ be plurisubharmonic in the multiple-circular domain

$$G = [z : (|z_1|, \ldots, |z_n|) \in B].$$

Proof: Either the function $R(r)$ is continuous in B or $R(r) \equiv -\infty$ (see section 11.1). Furthermore, taking

$$z_j = r_j e^{i\varphi_j}, \quad a_j = |a_j| e^{i\theta_j},$$

we obtain

$$\frac{\partial^2 u}{\partial z_j \partial \bar{z}_j} = \frac{1}{4} \frac{\partial^2 R}{\partial r_j^2} + \frac{1}{4r_j} \frac{\partial R}{\partial r_j}, \quad \frac{\partial^2 u}{\partial z_j \partial \bar{z}_k} = \frac{1}{4} \frac{\partial^2 R}{\partial r_j \partial r_k} e^{i(\varphi_k - \varphi_j)}$$
$$(j \neq k).$$

It follows from this that

$$\sum_{j,k} \frac{\partial^2 u}{\partial z_j \partial \bar{z}_k} a_j \bar{a}_k = \frac{1}{4} \sum_{j,k} \frac{\partial^2 R}{\partial r_j \partial r_k} e^{i(\varphi_k - \varphi_j)} a_j \bar{a}_k +$$
$$+ \frac{1}{4} \sum_j \frac{\partial R}{\partial r_j} \frac{|a_j|^2}{r_j} = \frac{1}{4} \sum_{j,k} \frac{\partial^2 R}{\partial r_j \partial r_k} |a_j||a_k| \times$$
$$\times [\cos(\varphi_j - \theta_j)\cos(\varphi_k - \theta_k) + \sin(\varphi_j - \theta_j)\sin(\varphi_k - \theta_k)] +$$
$$+ \frac{1}{4} \sum_j \frac{\partial R}{\partial r_j} \frac{|a_j|^2}{r_j} [\cos^2(\varphi_j - \theta_j) + \sin^2(\varphi_j - \theta_j)] =$$
$$= \frac{1}{4} \sum_{j,k} \frac{\partial^2 R}{\partial \ln r_j \partial \ln r_k} r_j |a_j| r_k |a_k| [\cos(\varphi_j - \theta_j)\cos(\varphi_k - \theta_k) +$$
$$+ \sin(\varphi_j - \theta_j)\sin(\varphi_k - \theta_k)].$$

If we define

$$\alpha = (r_1|a_1|\cos(\varphi_1 - \theta_1), \ \ldots, \ r_n|a_n|\cos(\varphi_n - \theta_n)),$$
$$\beta = (r_1|a_1|\sin(\varphi_1 - \theta_1), \ \ldots, \ r_n|a_n|\sin(\varphi_n - \theta_n)),$$

we obtain

$$(H(z;\, u)\,a,\ \bar{a}) = \frac{1}{4}(H(\ln r;\ R)\,\alpha,\ \alpha) + \frac{1}{4}(H(\ln r;\ R)\,\beta,\ \beta),$$

from which the assertion made follows.

5. *Functions that are plurisubharmonic in multiple-circular domains*

For a function $u(z) = R(|z_1|, \ldots, |z_n|)$ to be plurisubharmonic in a multiple-circular domain G, it is necessary and sufficient that the function $R(r)$ possess the following properties:

(1) *$R(r) < +\infty$ and is upper-semicontinuous on the set*

$$B = [r : r_j = |z_j|, \quad j = 1, 2, \ldots, n;\ z \in G];$$

2) *$R(r_1^0, \ldots, r_j, \ldots, r_n^0)$ increases with respect to each variable r_j (with the others fixed) in the interval $[0, R_j^0)$ whenever the interval $(r_1^0, \ldots, r_j, \ldots, r_n^0)$, where $0 \le r_j < R_j^0$, is contained in B;*

(3) *$R(r)$ is convex with respect to $(\ln r_1, \ln r_2, \ldots, \ln r_n)$ at all interior points of the set B;*

(4) *$R(r)$ possesses property (3) at all interior points of every (nonempty) face*

$$r_{j_1} = r_{j_2} = \ldots = r_{j_k} = 0, \quad 1 \leqslant j_1 < j_2 < \ldots < j_k \leqslant n,$$
$$k = 1, 2, \ldots, n-1$$

of the set B (with respect to the other variables).

Proof: The necessity of conditions (1) is obvious. Let us prove condition (2). The function $R(r_1^0, \ldots, r_j, \ldots, r_n^0)$ is subharmonic in the circle $|z_j| < R_j^0$ (see Fig. 14). Then, the function

$$R(r_1^0, \ldots, r_j, \ldots, r_n^0) = \frac{1}{2\pi}\int_0^{2\pi} R(r_1^0, \ldots, |r_j e^{i\theta}|, \ldots, r_n^0)\, d\theta$$

increases with respect to r_j in the interval $[0, R_j^0)$ (see section 9.8). The necessity of conditions (3) and (4) follows from the theorem in section 11.4 and from the fact that the function $u(z)$ on an arbitrary analytic plane $z_{j_1} = z_{j_2} = \ldots = z_{j_k} = 0$ is plurisubharmonic with respect to the remaining variables in the intersection of the domain G by that plane (see section 10.13).

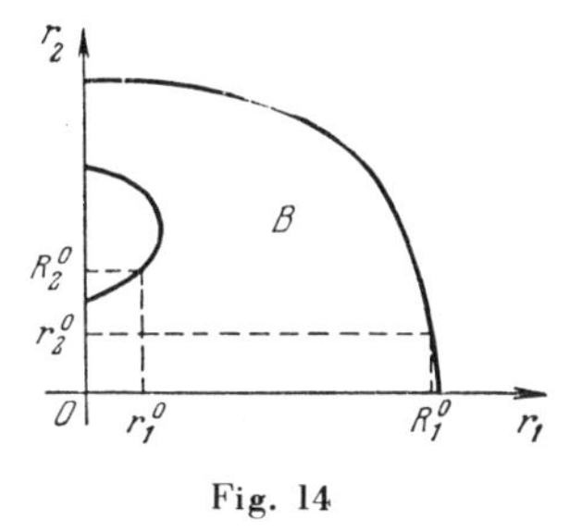

Fig. 14

Proof of the sufficiency: It follows easily from condition (1) that the function $u(z)$ $u(z) = R(|z_1|, \ldots, |z_n|) < +\infty$ and is upper-semicontinuous in G. Let us show that the function $u(z)$ is plurisubharmonic in G. From condition (3) and the lemma of section 11.4, we conclude that $u(z)$ is plurisubharmonic in G if $z_1 \ldots z_n \neq 0$. It remains for us to consider the points of the manifold $z_1 \ldots z_n = 0$ belonging to the region G. Suppose, for example, that

$$z^0 = (z_1^0, \ldots, z_k^0, 0, \ldots, 0) \in G,$$
$$z_1^0 \ldots z_k^0 \neq 0 \qquad (0 \leqslant k \leqslant n-1).$$

Then, for some $r^0 > 0$, we have $S(z^0, r^0 I) \Subset G$ and $z_1 \ldots z_k \neq 0$ for $z \in \bar{S}(z^0, r^0 I)$. According to what we have proven, it follows from condition (4) that the function $R(|z_1|, \ldots, |z_k|, 0, \ldots, 0)$ is plurisubharmonic at those points of the intersection of the domain G by the analytic plane $z_{k+1} = \ldots = z_n = 0$ at which $z_1 \ldots z_k \neq 0$. In particular, this function is plurisubharmonic in the polydisk $|z_j - z_j^0| \leqslant r^0$, for $j = 1, 2, \ldots, k$. Therefore, for all $a = (a_1, a_2, \ldots, a_n)$, we have (see section 10.2)

$$u(z^0) = R(|z_1^0|, \ldots, |z_k^0|, 0, \ldots, 0) \leqslant$$
$$\leqslant \frac{1}{2\pi}\int_0^{2\pi} R(|z_1^0 + r^0 a_1 e^{i\theta}|, \ldots, |z_k^0 + r^0 a_k e^{i\theta}|, 0, \ldots, 0)\, d\theta.$$

Furthermore, it follows from condition (2) that

$$R(|z_1^0 + r^0 a_1 e^{i\theta}|, \ldots, |z_k^0 + r^0 a_k e^{i\theta}|, 0, \ldots, 0) \leqslant$$
$$\leqslant R(|z_1^0 + r^0 a_1 e^{i\theta}|, \ldots, |z_k^0 + r^0 a_k e^{i\theta}|, |r^0 a_{k+1} e^{i\theta}|, \ldots, |r^0 a_n e^{i\theta}|).$$

From this and from the preceding inequality, we obtain

$$u(z^0) \leqslant \frac{1}{2\pi} \int_0^{2\pi} R(|z_1^0 + r^0 a_1 e^{i\theta}|, \ldots, |z_k^0 + r^0 a_k e^{i\theta}|, |r^0 a_{k+1} e^{i\theta}|, \ldots, |r^0 a_n e^{i\theta}|)\, d\theta = \frac{1}{2\pi} \int_0^{2\pi} u(z^0 + r^0 e^{i\theta})\, d\theta.$$

Thus, inequality (29) (see section 10.2) is satisfied for all points of the manifold $z_1 \ldots z_n = 0$. Therefore, the function $u(z)$ is plurisubharmonic in G, which completes the proof.

6. *Logarithmically convex functions*

A function $u(x) \geqslant 0$ is said to be logarithmically convex in a domain B if the function $\ln u(z)$ is convex in B. Here, the function $u(x) = e^{\ln u(x)}$ itself is also convex in B. This follows from the fact that $u(x)$ is a logarithmically plurisubharmonic function in the tubular domain $B + iR^n$ (see sections 10.5 and 11.3).

THEOREM. *For a nonnegative function $u(x)$ to be logarithmically convex in a domain B, it is necessary and sufficient that the function $u(x)e^{bx}$ be convex in B for all b.*

Proof: The necessity of the condition is obvious (cf. section 9.15). Let us prove the sufficiency. Since the function $u(x)e^{bx}$ is convex in B for all b, it follows that $u(x)|e^{az}|$, where $a = b + ic$, is convex and hence plurisubharmonic in $B + iR^n$. Then, on the basis of the theorem of section 10.5, $\ln u(x)$ is plurisubharmonic in $B + iR^n$ and hence is convex in B (see section 11.3), which completes the proof.

The sufficiency of the condition of this theorem is also easily established directly (without use of the corresponding theorem for plurisubharmonic functions) by use of the following lemma, which is of interest in its own right.

LEMMA. *For a function $-\infty \leq u(x) < +\infty$ to be convex in a domain B, it is necessary and sufficient that, for all b and for every rectangular interval lying in B, the function $u(x) + bx$ attain its maximum value on the boundary of that interval.*

The necessity of the condition is obvious. Let us show that it is sufficient. If $u(x_0) = -\infty$ for $x_0 \in B$, then, obviously, $u(x) \equiv -\infty$ for $x \in B$. Therefore, let us assume $u(x) > -\infty$ for $x \in B$. Suppose that an interval $x = \lambda x' + (1-\lambda)x''$, for $0 \leqslant \lambda \leqslant 1$, is contained in B. We choose the vector b so that

$$u(x') + bx' = u(x'') + bx''.$$

Therefore, by using the maximum theorem, we obtain for all $\lambda \in [0, 1]$,

$$u[\lambda x' + (1-\lambda)x''] + \lambda bx' + (1-\lambda)bx'' \leqslant u(x') + bx',$$

that is,

$$u[\lambda x' + (1-\lambda)x''] \leqslant \lambda u(x') + (1-\lambda)u(x''),$$

which means that the function $u(x)$ is convex in B, which completes the proof.

7. *Examples of logarithmically convex functions*

(1) *Suppose that a function $u(z)$ is logarithmically plurisubharmonic in a complete circular domain G. Then, its mean value with respect to the hull of the polycircle $S(0,r) \Subset G$ of order p*

$$I_p(r;\, u) = \begin{cases} \left[\dfrac{1}{(2\pi)^n}\displaystyle\int_0^{2\pi}\cdots\int_0^{2\pi} u^p(re^{i\theta})\,d\theta\right]^{\frac{1}{p}}, & 0 < p < \infty; \\ \max\limits_{|z_j|=r_j,\ j=1,2,\ldots,n} u(z), & p = \infty, \end{cases}$$

where $re^{i\theta} = (r_1e^{i\theta_1}, \ldots, r_ne^{i\theta_n})$, is an increasing function with respect to each variable r_j and is logarithmically convex with respect to $(\ln r_1, \ldots, \ln r_n)$ in the domain

$$\operatorname{int} B = [r : r_j = |z_j|,\ j = 1, 2, \ldots, n;\ z_1 \ldots z_n \neq 0,\ z \in G].$$

Proof: For every $\theta = (\theta_1, \ldots, \theta_n)$, the function $u(ze^{i\theta})$ is logarithmically plurisubharmonic in the domain G. Therefore, the function

$$I_p(|z_1|, \ldots, |z_n|;\, u) = \left[\frac{1}{(2\pi)^n}\int_0^{2\pi}\cdots\int_0^{2\pi} u^p(ze^{i\theta})\,d\theta\right]^{\frac{1}{p}}$$

is logarithmically plurisubharmonic in the domain G and depends on z through $(|z_1|, \ldots, |z_n|)$ (see section 10.5). Then, it satisfies conditions (1)-(4) of the theorem in section 11.5. In particular, it increases with respect to each r_j and is logarithmically convex with respect to $(\ln r_1, \ldots, \ln r_n)$ in the domain int B.

Suppose now that $p = \infty$. If $u(z)$ is a continuous function, then

$$I_p(r;\, u) \to I_\infty(r;\, u), \text{ as } p \to \infty,$$

and, consequently, $I_\infty(r;\, u)$ is an increasing function with respect to each r_j and is logarithmically convex with respect to $(\ln r_1, \ldots, \ln r_n)$. On the other hand, if $u(z)$ is not a continuous function, it must, according to the remark following the theorem in section 10.9, be the limit of a decreasing sequence of continuous logarithmically plurisubharmonic functions u_α for $\alpha = 1, 2, \ldots$. Then, the function $I_\infty(r; u)$ is also the limit of a decreasing sequence of functions $I_\infty(r; u_\alpha)$ for $\alpha = 1, 2, \ldots$. From what we have proven, the functions $I_\infty(r; u_\alpha)$ are increasing functions with respect to the r_j and are logarithmically convex with respect to $(\ln r_1, \ldots, \ln r_n)$. Our assertion follows from this for $p = \infty$.

For $n = 1$ and $u = |f|$, where f is a holomorphic function in the circle $|z| < R$, this assertion includes Hardy's theorem (for $(0 < p < \infty)$ and Hadamard's three-circle theorem (for $(p = \infty)$.

(2) *Suppose that the function $u(z)$ is logarithmically plurisubharmonic in the hypersphere $|z| < R$. Then, its mean value on a sphere of radius $r < R$ of order p*

$$J_p(r; u) = \begin{cases} \left[\dfrac{1}{2\pi} \displaystyle\int_{|a|=1} u^p(ra)\, da \right]^{\frac{1}{p}}, & 0 < p < \infty; \\ \max\limits_{|z|=r} u(z), & p = \infty \end{cases}$$

is an increasing logarithmically convex function of $\ln r$ *in the interval* $(0, R)$.

Proof: We note that for every a such that $|a| = 1$, the function $u(za) = u(za_1, \ldots, za_n)$ is logarithmically subharmonic in the open circle $|z| < R$. Therefore, the function

$$J_p(z; u) = \left[\frac{1}{2\pi} \int_{|a|=1} u^p(za)\, da \right]^{\frac{1}{p}}$$

is logarithmically subharmonic (see section 10.5). Because of the group properties of integrals over a unit sphere (see section 10.8), the function $J_p(z; u)$ does in fact depend on $|z|$. According to the theorem of section 11.5, the function $J_p(r; u)$ is an increasing logarithmically convex function of $\ln r$ in $(0, R)$. The case of $p = \infty$ is treated in a manner analogous to that of example (1).

(3) *Suppose that the function $u(z)$ is logarithmically plurisubharmonic in the tubular domain $B + iR^n$ and that, for each $x \in B$, its pth power is summable with respect to y (for $p = \infty$, $u(x + iy)$ is bounded above with respect to y). Then, the function*

$$\lambda_p(x; u) = \begin{cases} \left[\displaystyle\int u^p(x + iy)\, dy \right]^{\frac{1}{p}}, & 0 < p < \infty; \\ \sup\limits_{y} u(x + iy), & p = \infty \end{cases}$$

is logarithmically convex in the domain B if it is locally bounded from above in B.

Proof: For every ξ, the function $u(x + iy + i\xi)$ is logarithmically plurisubharmonic with respect to $z = x + iy$, and its pth power is summable with respect to ξ. Therefore (see section 10.5), the function

$$\lambda_{p,R}(x+iy;\ u)=\left[\int\limits_{|\xi|<R} u^p(x+iy+i\xi)\,d\xi\right]^{\frac{1}{p}}$$

is logarithmically plurisubharmonic in $B+iR^n$ and approaches 0 as $|y|\to\infty$. Suppose that x_1 and x_2 are arbitrary points in the domain B such that the straight-line segment connecting them is also contained in B. Then, from the maximum theorem for subharmonic functions, we have

$$\begin{aligned}&\xi bx_1+(1-\xi)\,bx_2+\ln\lambda_{p,R}(\zeta x_1+(1-\zeta)\,x_2;\ u)\leqslant\\&\leqslant\sup_{\xi=0,\,1;\,|\eta|<\infty}[\xi bx_1+(1-\xi)\,bx_2+\ln\lambda_{p,R}(\zeta x_1+(1-\zeta)\,x_2;\ u)]\end{aligned}$$

for all b and for all $\zeta=\xi+i\eta$ in the strip $0\leqslant\xi\leqslant 1$. If we now let R approach $+\infty$, we deduce, since $\lambda_{p,R}(z;\ u)\to\lambda_p(x;\ u)$, monotonically increasing, that, for all b, the function $\ln\lambda_p(x;\ u)+bx$ defined on every interval lying entirely in B attains its maximum value on the boundary of that interval. According to the lemma in section 11.6, the function $\ln\lambda_p(x;\ u)$ is convex in B.

Now consider the case in which $p=\infty$. Following Phragmén and Lindelöf, we introduce the logarithmically plurisubharmonic function in the domain $B+iR^n$

$$u_\varepsilon(z)=\left|e^{\varepsilon\left(z_1^2+\dots+z_n^2\right)}\right|u(z),\qquad \varepsilon>0.$$

For every $x\in B$ and for all real b, we have

$$\lim_{|y|\to\infty}|e^{bz}|\,u_\varepsilon(z)=\lim_{|y|\to\infty}e^{\varepsilon(|x|^2-|y|^2)+bx}\,u(x+iy)=0.$$

Therefore, we conclude on the basis of the maximum theorem for plurisubharmonic functions (see section 10.3), that the function $\lambda_\infty(x;\ u_\varepsilon)\,e^{bx}$ defined on any interval lying entirely in the domain B attains its maximum value on the boundary of that interval. According to the lemma of section 11.6, the function $\ln\lambda_\infty(x;\ u_\varepsilon)$ is convex in B. But $\lambda_\infty(x;\ u_\varepsilon)\to\lambda_\infty(x;\ u)$ as $\varepsilon\to+0$. Therefore, the function $\lambda_\infty(x;\ u)$ is logarithmically convex, which completes the proof.

8. *Remarks*

We point out some relationships between subharmonic, plurisubharmonic, and convex functions in (real) spaces of various dimensions n: for $n=1$, subharmonic and convex functions coincide; for $n=2$, subharmonic and plurisubharmonic functions, but convex functions constitute only a portion of subharmonic functions; for $n \geqslant 3$ and odd, convex functions constitute a portion of all subharmonic functions; for $n \geqslant 4$ and even, convex functions constitute a portion of plurisubharmonic functions, which in turn constitute a portion of all subharmonic functions.

12. PSEUDOCONVEX DOMAINS

For pseudoconvex domains, see, for example, Oka [19, 20], Lelong [159], and also the survey by Bremermann [13].

1. *The definition of a pseudoconvex domain*

Pseudoconvex domains are defined in terms of plurisubharmonic functions in a manner similar to the way in which convex domains are defined in terms of convex functions. Pseudoconvexity is a generalization of the concept of convexity defined in the real space R^n to the case of a complex space C^n.

Consider a point $Z \in C^n$ and a domain $G \subset C^n$. Let $\Delta_G(z)$ denote the distance from the point z to the boundary of the domain G (see definition in section 1.3). The domain G is said to be pseudoconvex if the function $\ln \Delta_G(z)$ is plurisubharmonic in G.

The function $\ln \Delta_G(z)$, which is continuous in the domain G, assumes the value $+\infty$ at all finite points of the boundary ∂G. Therefore, if G is a pseudoconvex domain, the continuous function

$$\max\left[-\ln \Delta_G(z),\ |z|^2\right]$$

is plurisubharmonic in G and approaches $+\infty$ everywhere on ∂G (in the sense of the definition in section 1.6).

2. *The simplest properties of pseudoconvex domains*

An arbitrary component of the interior of the intersection of pseudoconvex domains is a pseudoconvex domain.

Proof: Let G_α denote pseudoconvex domains. Then, the functions $\ln \Delta_{G_\alpha}(z)$ are plurisubharmonic in G_α. Let G denote any component of $\bigcap_\alpha G_\alpha$. The functions $\ln \Delta_{G_\alpha}(z)$ are locally uniformly bounded from above in G since $\Delta_{G_\alpha}(z) \geqslant \Delta_G(G') > 0$ for all z in $G' \Subset G$. From this, we conclude, on the basis of formula (3) of section 1.3, that

$$-\ln \Delta_G(z) = \sup_\alpha\left[-\ln \Delta_{G_\alpha}(z)\right]$$

is a plurisubharmonic function in G (see section 10.3). Thus, G is a pseudoconvex domain, which completes the proof.

If the domains $G \subset C^n$ and $D \subset C^m$ are pseudoconvex, the domain $G \times D \subset C^{n+m}$ is also pseudoconvex.

Proof: Since $G \times D = (G \times C^m) \cap (C^n \times D)$, it will, by virtue of the preceding results, be sufficient to show that the domains $G \times C^m$ and $C^n \times D$ are pseudoconvex in C^{n+m}. But this assertion follows from the relation

$$-\ln \Delta_G(z) = -\ln \Delta_{G \times C^m}(z, w),$$

according to which the function $\ln \Delta_{G \times C^m}(z, w)$ is plurisubharmonic in $G \times C^m$; that is, the domain $G \times C^m$ is pseudoconvex, which completes the proof.

The union of an increasing sequence of pseudoconvex domains is a pseudoconvex domain.

Proof: Suppose that $G_\alpha \subset G_{\alpha+1}$ and $G = \bigcup_\alpha G_\alpha$. Then, on the basis of (2) (see section 1.3),

$$-\ln \Delta_{G_\alpha}(z) \geqslant -\ln \Delta_{G_{\alpha+1}}(z) \to -\ln \Delta_G(z) \quad \text{as} \quad \alpha \to \infty$$

is an arbitrary subdomain $G' \Subset G$. The functions $\ln \Delta_{G_\alpha}(z)$ are plurisubharmonic in G'. Therefore, on the basis of section 10.3, the function $\ln \Delta_G(z)$ is plurisubharmonic in G'. Since G' is an arbitrary compact subdomain of G, this function is necessarily plurisubharmonic in G, which completes the proof.

3. *The weak continuity principle*

We shall say that the weak continuity principle applies to a domain G if the following assertion holds: Let $\{S_\alpha\}$ denote a sequence of domains that lie, together with their boundaries ∂S_α, on the two-dimensional analytic surfaces F_α and suppose that

$$S_\alpha \cup \partial S_\alpha \Subset G, \qquad \lim_{\alpha\to\infty} S_\alpha = S_0, \qquad \lim_{\alpha\to\infty} \partial S_\alpha = T_0 \Subset G.$$

Then, if S_0 is bounded, $S_0 \Subset G$ (see Fig. 15).

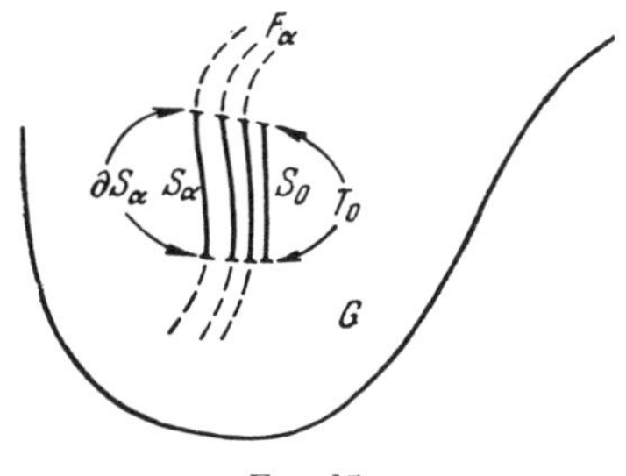

Fig. 15

Clearly, the weak continuity principle holds for every domain in C^1.

LEMMA (**Bremermann** [13]). *If the weak continuity principle applies to a domain G, the following equations hold:*

$$\inf_{z \in S \cup \partial S} |\chi(z)| \Delta_{a, G}(z) = \inf_{z \in \partial S} |\chi(z)| \Delta_{a, G}(z), \quad |a| = 1; \tag{45}$$

$$\inf_{z \in S \cup \partial S} |\chi(z)| \Delta_G(z) = \inf_{z \in \partial S} |\chi(z)| \Delta_G(z), \tag{46}$$

where S is an arbitrary domain lying on an arbitrary two-dimensional analytic surface $z = z^0 + \lambda b$ such that $S \cup \partial S \Subset G$ and $\chi(z)$ is an arbitrary function that is holomorphic and nonvanishing in the domain G.

Proof: Since $\partial S \Subset G$ and $\chi(z) \neq 0$ in G, the number

$$m = \inf_{z \in \partial S} \Delta_{a, G}(z) |\chi(z)| > 0$$

(see section 1.3) for every vector a such that $|a| = 1$. From this and the fact that the set ∂S is closed, it follows that the sets

$$T_\alpha = [z : z = z' + a\alpha\, \chi(z')^{-1}, \quad z' \in \partial S]$$

are such that $T_\alpha \Subset G$ for arbitrary $\alpha < m$.

Suppose that the vectors a and b are linearly independent. Then, for $\alpha \in [0, m)$, the two-dimensional set

$$F_\alpha = [z : z = z^0 + \lambda b + a\alpha\, \chi(z^0 + \lambda b)^{-1}, \quad \lambda \in G_{z^0, b}],$$

where (see section 1.3)

$$G_{z^0, b} = [\lambda : z^0 + \lambda b \in G],$$

is an analytic surface because (see section 10.12)

$$\frac{\partial z}{\partial \lambda} = b + a\alpha \sum_{1 \leqslant k \leqslant n} \frac{\partial}{\partial z_k} \chi^{-1}(z^0 + \lambda b)\, b_k \neq 0.$$

Furthermore, for each α in the interval indicated, the set

$$S_\alpha = [z : z = z' + a\alpha\, \chi(z')^{-1}, \quad z' \in S]$$

is a domain that lies together with its boundary $\partial S_\alpha = T_\alpha$ on F_α. However, by hypothesis, $S(0) = S \Subset G$. Therefore, there exists a number α_0 in $(0, m)$ such that $S_\alpha \Subset G$ for all $\alpha \in [0, \alpha_0]$. Thus, since $\partial S_\alpha = T_\alpha \Subset G$ for arbitrary $\alpha < m$, we conclude from the assumption that the weak continuity principle holds that $S_\alpha \Subset G$ for arbitrary $\alpha < m$. This means that

$$\Delta_{a,G}(z) \geqslant \alpha |\chi(z)|^{-1}, \qquad z \in S.$$

If we take the limit as $\alpha \to m - 0$ in this inequality, we obtain

$$\Delta_{a,G}(z) \geqslant m |\chi(z)|^{-1}, \qquad z \in S,$$

from which Eq. (45) follows.

In particular, we have shown that Eq. (45) holds for an arbitrary domain in C^1. In this case, it reduces to Eq. (46).

Let us assume now that the vectors a and b are linearly dependent. Without loss of generality, we may assume that $a = b$. In this case, if we use the formula

$$\Delta_{a,G}(z^0 + a\lambda) = \Delta_{G_{z^0,a}}(\lambda)$$

(see section 1.3), Eq. (45) reduces to Eq. (46) for $n = 1$ (in the λ-plane) with G replaced by $G_{z^0,a}$, $\chi(z)$ by $\chi(z^0 + \lambda a)$, and S and ∂S by the hulls of these sets in the λ-plane. But, for $n = 1$, Eq. (46) is already proven. Therefore, Eq. (45) is established for all a such that $|a| = 1$.

By use of formula (4) (see section 1.3), we can obtain Eq. (46) from Eq. (45). Specifically,

$$\inf_{|a|=1} \inf_{z \in \partial S} |\chi(z)| \Delta_{a,G}(z) = \inf_{|a|=1} \inf_{z \in S \cup \partial S} |\chi(z)| \Delta_{a,G}(z) =$$
$$= \inf_{z \in \partial S} \inf_{|a|=1} |\chi(z)| \Delta_{a,G}(z) = \inf_{z \in S \cup \partial S} \inf_{|a|=1} |\chi(z)| \Delta_{a,G}(z) =$$
$$= \inf_{z \in \partial S} |\chi(z)| \Delta_G(z) = \inf_{z \in S \cup \partial S} |\chi(z)| \Delta_G(z), \text{ q. e. d.}$$

4. *A condition for pseudoconvexity, I*

For a domain G to be pseudoconvex, it is necessary and sufficient that a weak continuity principle apply to it.

Proof of the sufficiency: Since a weak continuity principle applies to the region G, Eqs. (45) are valid for it. In the proof of the necessity of the second condition for pseudoconvexity (see section 12.5), it will be established that Eqs. (45) imply that the function $\ln \Delta_{a,G}(z)$ is plurisubharmonic in G for all a such that $|a| = 1$. The function $\ln \Delta_{a,G}(z)$ is uniformly bounded above for all a such that $|a| = 1$ and all $z \in G' \Subset G$. We then conclude on the basis of formula (4) in section 1.3, that the function

$$-\ln \Delta_G(z) = \sup_{|a|=1} [-\ln \Delta_{a,G}(z)]$$

is plurisubharmonic in G (see section 10.3). This means that the domain G is pseudoconvex.

Proof of the necessity: Suppose that G is a pseudoconvex domain. Consider a sequence of domains S_α, for $\alpha = 1, 2, \ldots$, that lie together with their boundaries ∂S_α on the two-dimensional analytic surfaces F_α. Suppose that $S_\alpha \cup \partial S_\alpha \Subset G$, that $\lim S_\alpha = S_0$ is bounded, and that $\lim \partial S_\alpha = T_0 \Subset G$. Since the function $\ln \Delta_G(z)$ is plurisubharmonic, we have, on the basis of the maximum theorem (see section 10.14)

$$\sup_{z \in \partial S_\alpha} [-\ln \Delta_G(z)] = \sup_{z \in S_\alpha \cup \partial S_\alpha} [-\ln \Delta_G(z)].$$

If we take the limit as $\alpha \to \infty$ in this equation and use the continuity of the function $\ln \Delta_G(z)$ (see section 1.3) and the boundedness of the sets S_0 and T_0, we obtain the equation

$$\sup_{z \in T_0} [-\ln \Delta_G(z)] = \sup_{z \in S_0 \cup T_0} [-\ln \Delta_G(z)].$$

It follows from this equation that $\Delta_G(S_0) \geqslant \Delta_G(T_0)$. But $T_0 \Subset G$ and hence $\Delta_G(S_0) \geqslant \Delta_G(T_0) > 0$. In view of the boundedness of the set S_0, this inequality implies that $S_0 \Subset G$, which completes the proof.

COROLLARY. *Every domain in C^1 is pseudoconvex.*

We note in passing that every domain in R^1 is convex.

5. *A condition for pseudoconvexity, II*

For a domain G to be pseudoconvex, it is necessary and sufficient that the function $-\ln \Delta_{a,G}(z)$ be plurisubharmonic in G for all a such that $|a| = 1$.

The sufficiency of the condition was proven when the preceding condition for pseudoconvexity was proven (see section 12.4).

To prove the necessity, note that since the domain G is pseudoconvex, it follows from the necessity of the preceding condition for pseudoconvexity that a weak continuity principle applies for G and hence Eqs. (45) are valid. Let us show that in this case, the function $\ln \Delta_{a,G}(z)$ is plurisubharmonic in G for arbitrary a such that $|a| = 1$.

Suppose that this is not the case. Then, there exist a vector a such that $|a| = 1$ and an analytic plane $z = z^0 + \lambda b$, where $|b| = 1$, such that the function $V(\lambda) = -\ln \Delta_{a,G}(z^0 + \lambda b)$ is not subharmonic in $G_{z^0, b}$ (see section 10.1). Therefore, there exists a function $h(\lambda)$ that is harmonic in the open circle $[\lambda : |\lambda| < r] \Subset G_{z^0, b}$ and continuous in the closed circle $|\lambda| \leqslant r$ such that

$$V(\lambda) \leqslant h(\lambda) \quad \text{for} \quad |\lambda| = r, \tag{47}$$

and there exists a point λ^0 at which

$$V(\lambda^0) > h(\lambda^0), \qquad |\lambda^0| < r.$$

From this and from (47), it follows that

$$V(\lambda) < h(\lambda) + \eta \quad \text{for} \quad |\lambda| = r; \qquad V(\lambda^0) > h(\lambda^0) + \eta, \tag{48}$$

where $\eta = 1/2\,[V(\lambda^0) - h(\lambda^0)] > 0$. But the function $V(\lambda) < +\infty$ and is upper-semicontinuous in the circle $|\lambda| \leqslant r$. Therefore, it follows from (48) and from the Heine-Borel theorem that there exists a number r_1 (see section 2.2) such that $|\lambda^0| < r_1 < r$ and

$$V(\lambda) < h(\lambda) + \eta \text{ for } |\lambda| = r_1;\ V(\lambda^0) > h(\lambda^0) + \eta. \tag{49}$$

Since the function $h(\lambda)$ is harmonic in the circle $|\lambda| < r > r_1$, there exists a sequence of harmonic polynomials that converge uniformly to the function $h(\lambda)$ in the closed circle $|\lambda| \leqslant r_1$. (This assertion follows from Poisson's formula (1); see Stoilow [117], Vol. II, p. 23 [p. 36 of the Russian translation].) Therefore, on the basis of (49), we can exhibit a harmonic polynomial $p(\lambda)$ that satisfies the inequalities

$$p(\lambda) \geqslant V(\lambda) \quad \text{for} \quad |\lambda| = r_1; \qquad p(\lambda^0) < V(\lambda^0). \tag{50}$$

Consider the harmonic polynomial $p(\lambda)$ that is the harmonic conjugate of the harmonic polynomial $q(\lambda)$ (see section 6.7). Let us form the polynomial $g(\lambda) = p(\lambda) + iq(\lambda)$ in the complex variable λ. We continue the polynomial $g(\lambda)$ over the entire space C^n by setting, for example, $g(z) = g(\lambda)$ at the points $z = z^0 + \lambda b + \rho_2 b^{(2)} + \ldots + \rho_n b^{(n)}$, where $|\rho_2| < \infty, \ldots, |\rho_n| < \infty$ and the vectors $(b, b^{(2)}, \ldots, b^{(n)})$ constitute an orthogonal basis on C^n.

We define

$$S = [z : z = z^0 + \lambda b,\ |\lambda| < r_1],$$
$$T = [z : z = z^0 + \lambda b,\ |\lambda| = r_1].$$

Then, on the basis of Eq. (45), we have

$$\Delta_{a, G}(z)\,|e^{g(z)}| \geqslant \inf_{z \in T} \Delta_{a, G}(z)\,|e^{g(z)}|, \qquad z \in S. \tag{51}$$

But, from the first of inequalities (50), we have, for $z \in T$,

$$V(z) \leqslant \operatorname{Re} g(z) = p(z),$$

that is,

$$\Delta_{a, G}(z)\,|e^{g(z)}| \geqslant 1, \qquad z \in T.$$

From this and from (51), it follows that

$$\Delta_{a, G}(z)\,|e^{g(z)}| \geqslant 1, \qquad z \in S. \tag{52}$$

On the other hand, from the second of inequalities (50), we have, at the point $z' = z^0 + \lambda^0 b \in S$,

$$p(z') = \operatorname{Re} g(z') < V(z'),$$

that is,

$$\Delta_{a,G}(z')|e^{g(z')}| < 1,$$

which contradicts inequality (52). This contradiction proves that the function $-\ln \Delta_{a,G}(z)$ is plurisubharmonic in G.

6. *A condition for pseudoconvexity, III*

For a domain G to be pseudoconvex, it is necessary and sufficient that the function $-\ln \Delta_G(z)$ be plurisubharmonic in the boundary strip $G \cap U(\partial G)$ of the domain G, where $U(\partial G)$ is a neighborhood of ∂G.

The necessity of the condition is obvious. To prove the sufficiency, we first suppose that G is a bounded domain. Then,

$$\Delta_G[G \setminus U(\partial G)] = \rho > 0 \tag{53}$$

(see section 1.3). From the first condition for pseudoconvexity, it will be sufficient to show that the weak continuity principle applies to the domain G.

Suppose that a sequence of surfaces S_α for $\alpha = 1, 2, \ldots$, lying on two-dimensional analytic surfaces F_α is such that $S_\alpha \cup \partial S_\alpha \Subset G$, $\lim S_\alpha = S_0$ is bounded, and $\lim \partial S_\alpha = T_0 \Subset G$. We introduce the sequence of open sets $S'_\alpha = S_\alpha \cap U(\partial G)$ lying on F_α. Obviously,

$$\left.\begin{array}{c} S'_\alpha \cup \partial S'_\alpha \Subset G, \quad \partial S'_\alpha \subset \bar{S}_\alpha \setminus S'_\alpha \subset \partial S_\alpha \cup [G \setminus U(\partial G)], \\ S_\alpha \subset S'_\alpha \cup [G \setminus U(\partial G)]. \end{array}\right\} \tag{54}$$

From the maximum theorem for the sets S'_α and $\partial S'_\alpha$ and the function $-\ln \Delta_G(z)$ (see section 10.14), we have

$$\sup_{z \in S'_\alpha} [-\ln \Delta_G(z)] \leqslant \sup_{z \in \partial S_\alpha} [-\ln \Delta_G(z)].$$

From this it follows that $\Delta_G(S'_\alpha) \geqslant \Delta_G(\partial S'_\alpha)$. From this inequality and from (53) and (54), we obtain the chain of inequalities

$$\begin{aligned} \Delta_G(S_\alpha) &\geqslant \min\{\Delta_G(S'_\alpha), \Delta_G[G \setminus U(\partial G)]\} \geqslant \\ &\geqslant \min[\Delta_G(\partial S'_\alpha), \rho] \geqslant \min\{\min[\Delta_G(\partial S_\alpha), \Delta_G[G \setminus U(\partial G)]], \rho\} = \\ &= \min[\Delta_G(\partial S_\alpha), \rho]. \end{aligned} \tag{55}$$

If we take the limit as $\alpha \to \infty$ in (55) and use the continuity of the function $\Delta_G(z)$ and the boundedness of the sets S_0 and T_0, we derive the inequality $\Delta_G(S_0) \geqslant \min[\Delta_G(T_0), \rho]$. It follows from this inequality and from the inclusion $T_0 \Subset G$ that $\Delta_G(S_0) > 0$. But S_0 is a set. Therefore, $S_0 \Subset G$. Thus, a weak continuity principle applies to the domain G and, hence, G is a pseudoconvex domain.

Suppose now that G is an unbounded domain. We introduce the open bounded sets $G_R = G \cap U(0, R)$ (see Fig. 16). On the basis of formula (3) of section 1.3, we have

$$-\ln \Delta_{G_R}(z) = \max[-\ln \Delta_G(z), \ -\ln \Delta_{U(0, R)}(z)].$$

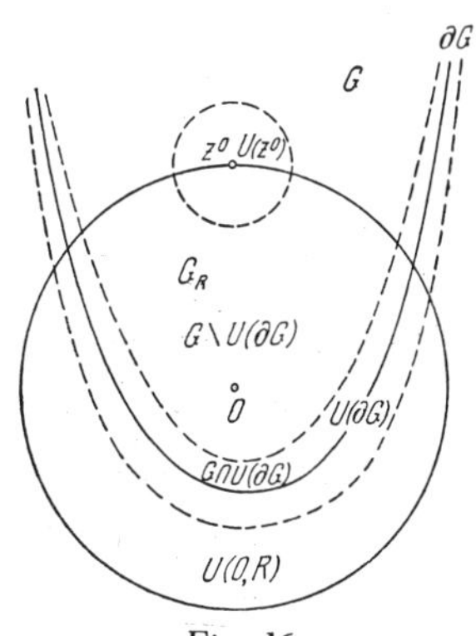

Fig. 16

But the function

$$-\ln \Delta_{U(0, R)}(z) = -\ln(R - |z|)$$

is plurisubharmonic in $U(0, R)$. Therefore, the function $-\ln \Delta_{G_R}(z)$ is plurisubharmonic in $G \cap U(\partial G) \cap U(0, R)$. Let z^0 be a point such that $z^0 \in \partial U(0, R)$ and $z^0 \bar{\in} G \cap U(\partial G)$. Then $|z^0 - z| \geqslant m > 0$ for all $z \in \partial G$. Consequently, there exists a neighborhood $U(z^0)$ such that

$$-\ln \Delta_{G_R}(z) = -\ln(R - |z|),$$
$$z \in U(z^0) \cap U(0, R).$$

Therefore, the function $-\ln \Delta_{G_R}(z)$ is plurisubharmonic in $U(z^0) \cap U(0, R)$. Thus, the function $-\ln \Delta_{G_R}(z)$ is plurisubharmonic in the boundary strip of the open bounded set G_R. From what has been proven, each component of this set is a pseudoconvex domain. But G_R increases monotonically and $\bigcup_{R>0} G_R = G$. Therefore, the domain G is a pseudoconvex (see section 12.2), which completes the proof.

7. *A condition for pseudoconvexity, IV*

Each component of the open set

$$G = [z: \ V(z) < 0, \quad z \in U(\bar{G})],$$

where $V(z)$ is a plurisubharmonic function in a neighborhood $U(\bar{G})$ of the set G, is a pseudoconvex domain.

Proof: Since the function $V(z)$ is upper-semicontinuous in $U(\bar{G})$, the set G is open (see section 2.2). Let us suppose first that G is a bounded set. Then, $G \Subset U(\bar{G})$ (see section 1.2). Let us suppose further that the function $V(z)$ is continuous in $U(\bar{G})$. Reasoning as in the proof of the necessity of the first condition for psuedoconvexity (see section 12.4) (with $-\ln \Delta_G(z)$ replaced by $V(z)$), we can show that the weak continuity principle applies to the open set G. Consequently, each component of the set G is a pseudoconvex domain.

Let us now drop the assumption that the function $V(z)$ is continuous. According to the theorem in section 10.9, there exists a decreasing sequence $V_\alpha(z)$, for $\alpha = 1, 2, \ldots$, of continuous plurisubharmonic functions in the open set G', where $G \Subset G' \Subset U(\bar{G})$, that converges to the function $V(z)$. Therefore, the sequence of open sets

$$G_\alpha = [z: \ V_\alpha(z) < 0, \quad z \in G'], \quad \alpha = 1, 2, \ldots,$$

is an increasing sequence and $\bigcup_\alpha G_\alpha = G$ (see Fig. 17). From what we have proven, each component of the set G_α is a pseudoconvex domain. From the theorem in section 12.2, each component of the set G is a pseudoconvex domain.

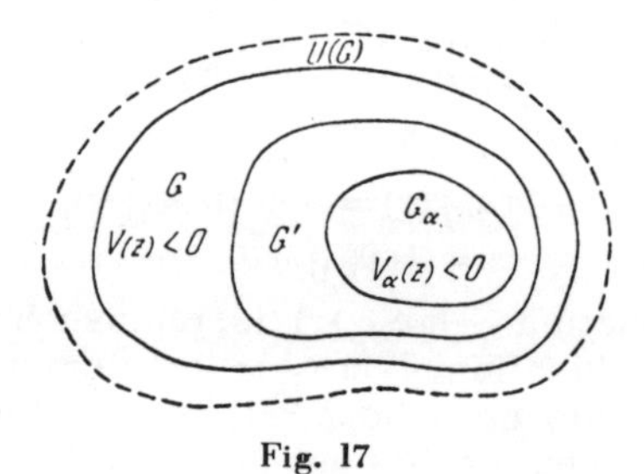

Fig. 17

Suppose now that the set G is unbounded. Let us consider the sequence of bounded open sets $G_R = G \cap U(0, R)$, where $R = 1, 2, \ldots$. Clearly,

$$G_R = [z: \ V_R(z) < 0, \quad z \in U(\bar{G})], \quad G_R \Subset U(\bar{G}),$$

where the function

$$V_R(z) = \max[V(z), \ |z|^2 - R^2]$$

is plurisubharmonic in $U(\bar{G})$ (see section 10.3). From what we have proven, each component of the set G_R is a pseudoconvex domain. But $G_R \subset G_{R+1}$ and $\bigcup_{R>0} G_R = G$. From the theorem in section 12.2, each component of the set G is a pseudoconvex domain, which completes the proof.

8. *A condition for pseudoconvexity,* V

For a domain G to be pseudoconvex, it is necessary and sufficient that there exists a function $V(z)$ that is plurisubharmonic in G and that approaches $+\infty$ everywhere on ∂G.

The necessity of this condition was established in section 12.1. Let us prove its sufficiency. Since the function $V(z)$ approaches $+\infty$ everywhere on ∂G, the sequence of open sets $G_\alpha = [z\colon V(z) - \alpha < 0,\ z \in G]$ for $\alpha = 1, 2, \ldots$ possesses the properties that $G_\alpha \subset G_{\alpha+1} \Subset G$ and $\bigcup_\alpha G_\alpha = G$ (see section 1.6). Since $G_\alpha \Subset G$ and the function $V(z) - \alpha$ is plurisubharmonic in G, it follows from the fourth condition for pseudoconvexity (see section 12.7) that each component of the open set G_α is a pseudoconvex domain. Then, according to the theorem in section 12.2, the domain G is pseudoconvex, which completes the proof.

9. *A condition for pseudoconvexity,* V_1

Suppose that G is a pseudoconvex domain and that a function $V(z)$ is plurisubharmonic in G. Then, every component of the open set $G' = [z\colon V(z) < 0,\ z \in G]$ is pseudoconvex domain.

Proof: According to the fifth condition for pseudoconvexity (see section 12.8), there exists a function $V^*(z)$ that is plurisubharmonic in G such that $[z\colon V^*(z) < \alpha,\ z \in G] \Subset G$ for all $\alpha = 1, 2, \ldots$. The functions $V_\alpha(z) = \max[V^*(z) - \alpha,\ V(z)]$ are plurisubharmonic in G and the sequence of open sets $G_\alpha = [z\colon V_\alpha(z) < 0$, for $\alpha = 1, 2, \ldots$, possesses the three properties: $G_\alpha \Subset G$, $G_\alpha \subset G_{\alpha+1}$, and $\bigcup_\alpha G_\alpha = G'$. From the fourth condition for pseudoconvexity (see section 12.7), each component of the open set G_α is a pseudoconvex domain and, consequently, according to the theorem in section 12.2, every component of the set G' is also a pseudoconvex domain, as asserted.

10. *Strictly pseudoconvex domains*

We shall say that a domain G is strictly pseudoconvex if

$$G = [z\colon\ V(z) < 0, \quad z \in U(\bar{G})],$$

where the function $V(z) \in C^{(2)}$ and satisfies the inequality

$$(H(z;\, V)a,\ \bar{a}) \geqslant \sigma |a|^2, \quad \sigma > 0,$$

in a neighborhood $U(\bar{G})$ of the domain G.

We note that, on the basis of section 10.10, the function $V(z)$ is plurisubharmonic in $U(\bar{G})$. Therefore, on the basis of the fourth condition for pseudoconvexity (see section 12.7), the domain G is pseudoconvex.

For a domain G to be pseudoconvex, it is necessary and sufficient that it be the union of an increasing sequence of strictly pseudoconvex domains

$$G_\alpha = [z: \quad V_\alpha(z) < 0, \quad z \in U(\bar{G}_\alpha)], \quad \alpha = 1, 2, \ldots$$

such that $G_\alpha \Subset G_{\alpha+1} \Subset G$, *where* $V_\alpha(z) \in C^{(\infty)}$ *in* $U(\bar{G})_\alpha$.

Proof: The sufficiency follows from section 12.2. Let us prove the necessity. Since G is a pseudoconvex domain, there exists a function $V(z)$ that is plurisubharmonic in G such that $[z: V(z) < \alpha] \Subset G$ for all $\alpha = 1, 2, \ldots$ (see section 12.1). Furthermore, there exists a decreasing sequence $\overset{*}{V}_k(z)$, for $k = 1, 2, \ldots$, of plurisubharmonic functions of class $C^{(\infty)}$ in open sets B_k, that converges to the function $V(z)$. The sets B_k can be chosen in such a way that $B_k \Subset B_{k+1} \Subset G$ and $\bigcup_k B_k = G$ (see section 10.9). From what has been said, it follows that, for every $\alpha \geqslant 1$, there exists a number $k(\alpha) \geqslant 1$ such that

$$[z: \quad V(z) < \alpha] \Subset B_{k(\alpha)}.$$

We may assume that the function $k(\alpha)$ is an increasing function. Thus, the sequence of open sets

$$G'_\alpha = \left[z: \quad \overset{*}{V}_{k(\alpha)}(z) + \frac{1}{\alpha}|z|^2 - \alpha < 0, \quad z \in D_\alpha\right],$$

where $D_\alpha = B_{k(\alpha)}$ possesses the properties

$$G'_\alpha \Subset D_\alpha \Subset G, \qquad G'_\alpha \Subset G'_{\alpha+1}, \qquad \bigcup_\alpha G'_\alpha = G.$$

The functions

$$V_\alpha(z) = \overset{*}{V}_{k(\alpha)}(z) + \frac{1}{\alpha}|z|^2 - \alpha$$

belong to the class $C^{(\infty)}$ and satisfy the inequality

$$(H(z; V_\alpha)a, \bar{a}) \geqslant \frac{1}{\alpha}|a|^2$$

in D_α. Therefore, every component of the open set G'_α is a strictly pseudoconvex domain. Thus, for the sequence of strictly pseudoconvex domains G_α referred to in the statement of the theorem, we may take the sequence of components of the open sets G'_α that contain a fixed point $z^0 \in G$. This completes the proof.

11. *Invariance under biholomorphic mappings*

We shall subsequently need the following

LEMMA. *Suppose that a biholomorphic mapping $\xi = \xi$ maps a domain G onto a domain G_1 and that a function $u(z)$ converges to $+\infty$ everywhere on ∂G. Then, the function $u[z(\xi)]$, where $z = z(\xi)$, representing the inverse transformation, approaches $+\infty$ everywhere on ∂G_1.*

Proof: Since

$$[z: \quad u(z) < M, \quad z \in G] \Subset G,$$

for arbitrary real M (see section 1.6) it follows from the fact that a biholomorphic mapping homeomorphic that

$$[\zeta: \quad u[z(\zeta)] < M, \quad \zeta \in G_1] \Subset G_1,$$

as asserted.

A biholomorphic mapping $\zeta = \zeta(z)$ maps a pseudoconvex domain G onto a pseudoconvex domain G_1.

Proof: Since the domain G is pseudoconvex, there exists a function $V(z)$ that is plurisubharmonic in G and that approaches $+\infty$ everywhere on ∂G (see section 12.1). But then, the function $V[z(\zeta)]$, where $z = z(\zeta)$, representing the inverse transformation, is plurisubharmonic in G_1 (see section 10.11) and, from the preceding lemma, it approaches $+\infty$ everywhere on ∂G_1. Therefore, the domain G_1 is pseudoconvex (see section 12.8).

12. *Intersections with pseudoconvex domains*

Every component contained in the intersection g of a pseudoconvex domain G with a $2k$-dimensional analytic plane F is a pseudoconvex domain in C^k.

On the basis of the preceding theorem, we may, by making a suitable (linear) biholomorphic mapping, assume that the plane F is defined by the equations $z_{n-k} = \ldots = z_n = 0$. For the pseudoconvex domain G, there exists a function $V(z)$ that is plurisubharmonic in G and that approaches $+\infty$ everywhere on ∂G. But then, the function $V(z_1, \ldots, z_k, 0, \ldots, 0)$ is plurisubharmonic in g and approaches $+\infty$ everywhere on ∂g (in the sense of section 1.6). Therefore, every component g is a pseudoconvex domain (see section 12.8), which completes the proof.

We shall now give some examples of pseudoconvex domains.

13. *Pseudoconvex Hartogs domains*

We may write a complete Hartogs domain G with plane of symmetry $z_1=0$ (see section 7.5) in the form

$$G=[z:\quad |z_1|<R(\tilde{z}),\quad \tilde{z}\in B],$$

where B is a domain in the space C^{n-1} and the function $R(\tilde{z})$ is upper-semicontinuous in B (see Fig. 18).

A complete Hartogs domain G is pseudoconvex if and only if B is a pseudoconvex domain and the function $-\ln R(\tilde{z})$ is plurisubharmonic in B.

Proof: If G is a pseudoconvex domain, then the intersection B of G and the analytic plane $z_1=0$ is also a pseudoconvex domain (see section 12.12). Define $a=(1, 0, \dots, 0)$. Then, $-\ln\Delta_{a,G}(z)$ is a plurisubharmonic function in G (see section 12.5). But

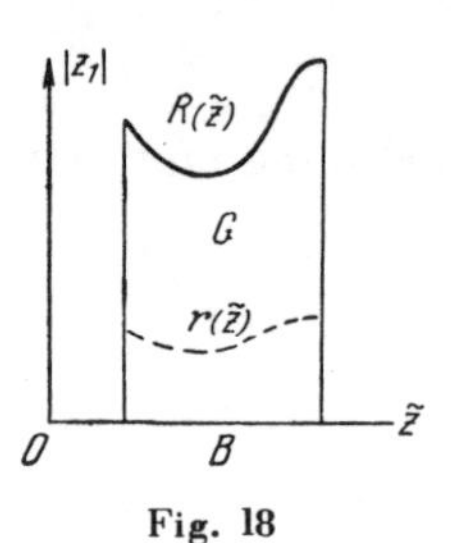

Fig. 18

$$-\ln\Delta_{a,G}(0, \tilde{z})=-\ln R(\tilde{z}).$$

Therefore, the function $-\ln R(\tilde{z})$ is plurisubharmonic in B.

Conversely, suppose that B is a pseudoconvex domain and that the function $-\ln R(\tilde{z})$ is plurisubharmonic in B. Then, the domain $C^1\times B$ is a pseudoconvex (see section 12.2) and the function $\ln|z_1|-\ln R(\tilde{z})$ is plurisubharmonic in $C^1\times B$. But

$$G=[z:\quad \ln|z_1|-\ln R(\tilde{z})<0,\quad z\in C^1\times B].$$

According to the sixth condition for pseudoconvexity (see section 12.9), the domain G is pseudoconvex.

Let us now consider a Hartogs domain that contains no points of its plane of symmetry. The simplest domain of this type is a domain of the form

$$G=[z:\quad r(\tilde{z})<|z_1|<R(\tilde{z}),\quad \tilde{z}\in B],$$

where B is a domain in C^{n-1} and the functions $R(\tilde{z})$ and $r(\tilde{z})>0$ are lower- and upper-semicontinuous in B, respectively (see section 2.2 and Fig. 18).

If B is a pseudoconvex domain and the functions $r(\tilde{z})$ and $-\ln R(\tilde{z})$ are plurisubharmonic in B, the domain G is pseudoconvex.

Proof: If we delete the point 0 from the plane C^1, we obtain the domain $C_0^1\subset C^1$. Therefore, the domain $C_0^1\times B$ is pseudoconvex in C^n (see section 12.2). The function

$$u(z)=\max\left[\ln|z_1|-\ln R(\tilde{z}),\ \ln\left|\frac{1}{z_1}\right|+\ln r(\tilde{z})\right]$$

is plurisubharmonic in $C_0^1 \times B$. But

$$G=\left[z:\quad u(z)<0,\quad z\in C_0^1\times B\right].$$

According to the sixth condition for pseudoconvexity (see section 12.9), the domain G is pseudoconvex, which completes the proof.

14. *Pseudoconvex semitubular domains*

Consider a semitubular domain (see section 7.5) of the form

$$G=[z=(x_1+iy_1,\tilde{z}):\ v(\tilde{z})<x_1<V(\tilde{z}),\ \tilde{z}\in B,\ |y_1|<\infty],$$

where B is a domain in the space C^{n-1} (see Fig. 19). The functions $v(\tilde{z})$ and $V(\tilde{z})$ must be upper- and lower-semicontinuous in B, respectively (see section (2.2).

A semitubular domain G is pseudoconvex if B is a pseudoconvex domain and the functions $v(\tilde{z})$ and $-V(\tilde{z})$ are plurisubharmonic in B.

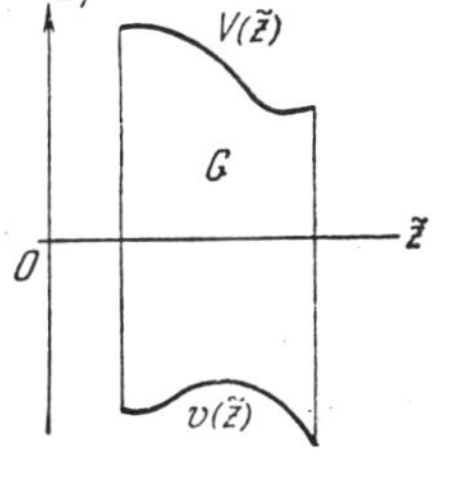

Fig. 19

Proof: Then the function

$$V^*(z)=\max[x_1-V(\tilde{z}),\ v(\tilde{z})-x_1]$$

is plurisubharmonic in the pseudoconvex domain $C^1\times B$ (see section 12.2). But, from the construction

$$G=[z:\quad V^*(z)<0,\quad z\in C^1\times B].$$

Consequently, G is a pseudoconvex domain (see section 12.9), which completes the proof.

An example of a semitubular pseudoconvex domain (see Bremermann [18]). Suppose that $n=2$, $R>2r$, and

$$G=\left[z:|x_1|<\sqrt{r^2-(|z_2|-R)^2},\quad R-r<|z_2|<R+r\right].$$

The domain G is a cylinder in C^2, the generator of which is directed along the y_1-axis and whose base is the torus $(|z_2|-R)^2+x_1^2<r^2$ in R^3 (see Fig. 20). Let us show that the function $V(z_2)=-\sqrt{r^2-(|z_2|-R)^2}$ is subharmonic in the annulus $R-r<|z_2|<R+r$. (This annulus is a pseudoconvex domain [see section 12.4].) We have

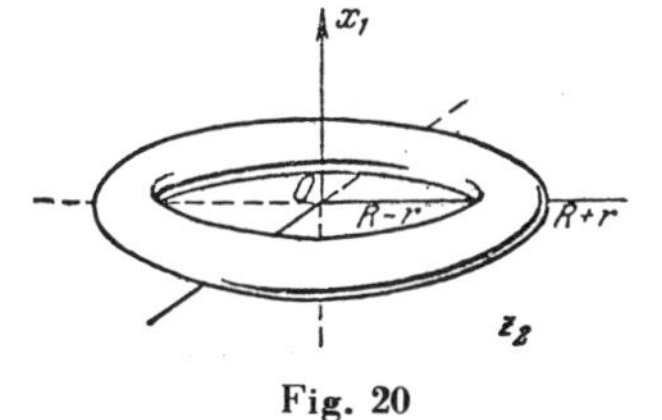

Fig. 20

$$\frac{\partial^2 V(z_2)}{\partial^2 \ln|z_2|}=-$$

$$\frac{(R-2|z_2|)|z_2|}{V(z_2)}+\frac{(R-|z_2|)^2|z_2|^2}{V^3(z_2)}>0,$$

since, by assumption $R > 2r$. On the basis of the theorem of section 11.4, the function $V(z_2)$ is subharmonic and hence the domain G is pseudoconvex.

15. *Analytic polyhedra*

The open set $\mathfrak{P} \subset C^n$ is called an analytic polyhedron if there exists N holomorphic functions $\chi_\alpha(z)$, for $\alpha = 1, 2, \ldots, N$, defined in $U(\overline{\mathfrak{P}})$ such that (see Fig. 21)

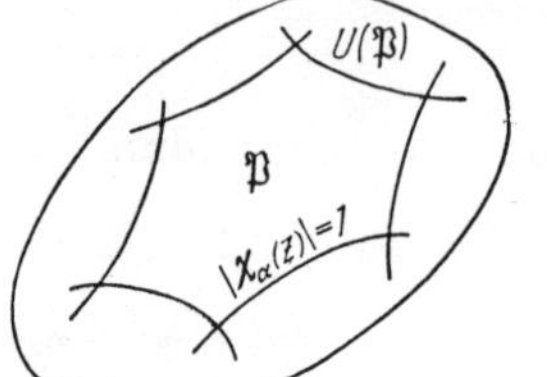

Fig. 21

$$\mathfrak{P} = [z: \ |\chi_\alpha(z)| < 1, \ \alpha = 1, 2, \ldots, N; \ z \in U(\overline{\mathfrak{P}})].$$

Every component of an analytic polyhedron is a pseudoconvex domain.

Proof: On the basis of section 10.3, the function

$$V(z) = \max_{1 \leqslant \alpha \leqslant N} \{\ln |\chi_\alpha(z)|\}$$

is plurisubharmonic in $U(\overline{\mathfrak{P}})$. But

$$\mathfrak{P} = [z: \ V(z) < 0, \ z \in U(\overline{\mathfrak{P}})].$$

Now, we need only use the fourth condition for pseudoconvexity, which completes the proof.

A connected bounded analytic polyhedron is called a Weil domain if $N \geqslant n$ and the intersection of arbitrary k hypersurfaces $|\chi_{\alpha_i}(z)| = 1$, where $1 \leqslant k \leqslant n$ and $i = 1, 2, \ldots, k$, is of dimension not exceeding $2n - k$.

13. CONVEX DOMAINS

As was mentioned in section 12.1, there is an analogy between convex and pseudoconvex domains similar to that between convex and plurisubharmonic functions. To pursue this analogy further and at the same time get a more thorough understanding of pseudoconvex domains, we shall now enumerate the basic properties of convex domains that are analogous to the properties of pseudoconvex domains. In this, we shall follow the work of Bremermann [13]. The theory of convex compact sets is expounded in detail in the monograph of Bonnesen and Fenchel [12].

1. *Definition of a convex domain*

A domain $B \subset R^n$ is said to be convex if, for arbitrary points $x' \in B$ and $x'' \in B$,

$$[x : x = tx' + (1 - t)x'', \quad 0 \leqslant t \leqslant 1] \subset B.$$

It is well known that a domain is convex if and only if, for every boundary point, there exists a supporting plane.

From this it follows that, for a domain B to be convex, it is necessary and sufficient that it be the interior of the intersection of the semispaces $a(x-x^0)>0$:

$$B = \operatorname{int} \bigcap_{x^0 \in \partial B} [x : a(x-x^0) > 0].$$

Here, $a(x-x^0)=0$ is the supporting plane to the domain B at the point $x^0 \in \partial B$ (see Fig. 22).

2. *A condition for convexity, I*

For a domain B to be convex, it is necessary and sufficient that the following weak continuity principle apply to it: If S_α, for $\alpha = 1, 2, \ldots$, is a sequence of intervals such that $S_\alpha \cup \partial S_\alpha \Subset B$, $\lim S_\alpha = S_0$ is bounded and $\lim \partial S_\alpha = T_0 \Subset B$, then $S_0 \Subset B$ (see sections 12.3 and 12.4).

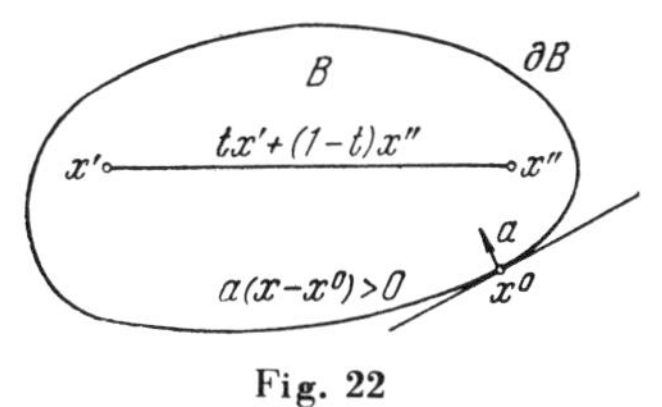

Fig. 22

Proof of the necessity: Suppose that B is a convex domain. Let us show that

$$\Delta_B(x) \geqslant \Delta_B(\partial S_\alpha), \qquad x \in S_\alpha. \tag{56}$$

If this were not the case, there would be a point $x^0 \in S_\alpha$ at which

$$\Delta_B(x^0) < \Delta_B(\partial S_\alpha). \tag{57}$$

Let us denote by x' some point on ∂B lying at a distance $\Delta_B(x^0)$ from x^0 (see Fig. 23). Let us draw a straight line L through x' parallel to the segment S_α. Let $\partial S_\alpha = \{a_\alpha, b_\alpha\}$. From the construction, there exist points a'_α and b'_α on the straight line L such that $|a_\alpha - a'_\alpha| = \Delta_B(x^0)$ and $|b_\alpha - b'_\alpha| = \Delta_B(x^0)$. It follows from this and inequality (57) that $a'_\alpha \in B$ and $b'_\alpha \in B$. Since the point x^0 lies between a_α and b_α, the point x' must lie between a'_α and b'_α. But this is impossible since the domain is convex. This contradiction proves inequality (57). Taking the limit as $\alpha \to \infty$ and making use of the continuity of the function $\Delta_B(x)$ and the boundedness of the sets S_0 and T_0, we obtain $\Delta_B(S_0) \geqslant \Delta_B(T_0)$. Since $T_0 \Subset G$, it follows from this that $\Delta_B(S_0) > 0$, which, together with the boundedness of S_0, implies that $S_0 \Subset B$. This means that the weak continuity principle is valid for the domain B.

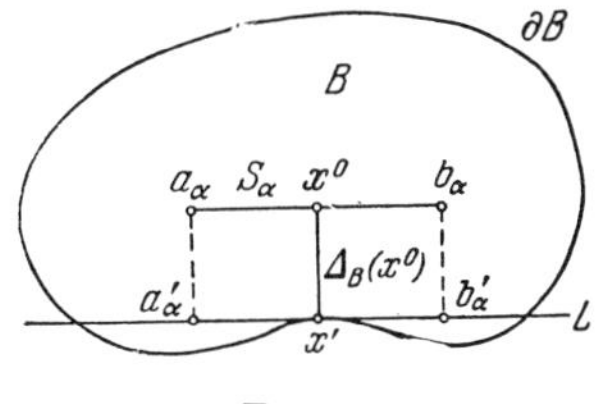

Fig. 23

Proof of the sufficiency: Suppose that $x' \in B$ and $x'' \in B$. Let us connect these points by a piecewise-smooth curve $[x : x = x(t),\ 0 \leqslant t \leqslant 1]$, where $x' = x(0)$ and $x'' = x(1)$, which lies entirely in the domain B. Since x' is an interior point of B, it follows that, for sufficiently small t, all straight line segments connecting the points $x' = x(0)$ and $x(t)$ lie entirely in B (see Fig. 24). Then, from the weak continuity principle, these segments are also contained in the domain B for all $t \in [0, 1]$ and, in particular, for $t = 1$. Consequently, B is a convex domain.

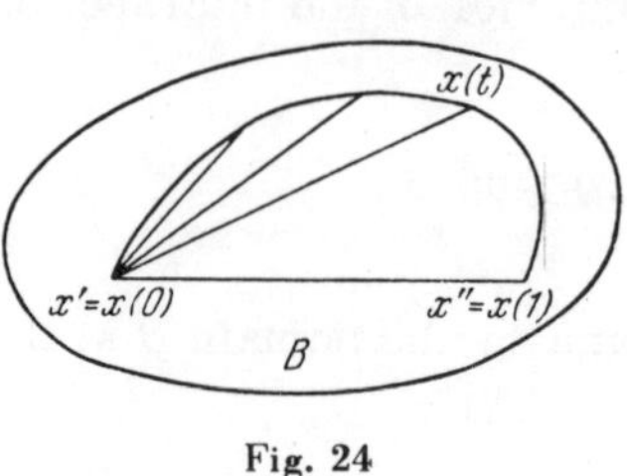

Fig. 24

3. *A condition for convexity, II*

For a domain B to be convex, it is necessary and sufficient that the function $-\ln \Delta_B(x)$ *be convex in* B (see section 12.1).

Proof of the necessity: Suppose that B is a convex domain. Through every point $x^0 \in \partial B$, let us pass the supporting plane $a(x - x^0) = 0$, where $|a| = 1$ (see section 13.1). As we know, the quantity $a(x - x^0)$ is the distance from the point $x \in B$ to this plane. Therefore [see section 1.3) formula (3)],

$$\Delta_B(x) = \inf_{x^0 \in \partial B} \{a(x - x^0)\}$$

and, consequently,

$$-\ln \Delta_B(x) = \sup_{x^0 \in \partial B} \{-\ln a(x - x^0)\}. \tag{58}$$

But the function $-\ln a(x - x^0)$ is convex in B since (see section 11.3)

$$(H(x;\ -\ln a(x - x^0))\,b,\ b) = \\ = \sum_{j,k} \frac{a_j a_k b_j b_k}{[a(x - x^0)]^2} = \frac{(ab)^2}{[a(x - x^0)]^2} \geqslant 0.$$

Furthermore, it is clear that the family of functions $\{-\ln a(x - x^0),\ x^0 \in \partial B\}$ is uniformly bounded above on every subdomain $B' \Subset B$. Therefore, on the basis of section 11.3, Eq. (58) implies that the function $-\ln \Delta_B(x)$ is convex in B.

Proof of the sufficiency: Suppose that the function $-\ln \Delta_B(x)$ is convex in B. Let us show that the weak continuity principle applies to the domain B. Let S_α, where $\alpha=1, 2, \ldots$, be a sequence of intervals such that $\bigcup \partial S_\alpha \Subset B$, $\lim S_\alpha = S_0$ is bounded, and $\lim \partial S_\alpha = T_0 \Subset B$. For the convex domain $-\ln \Delta_B(x)$, the maximum theorem holds:

$$-\ln \Delta_B(x) \leqslant \sup_{x\in\partial S_\alpha} [-\ln \Delta_B(x)], \qquad x \in S_\alpha,$$

which is equivalent to inequality (56). Just as in the proof of the necessity of the first condition for convexity (see section 13.2), it follows from this that the weak continuity principle applies to the domain B. From the first condition for convexity, the domain B is convex. This completes the proof.

COROLLARY. *Every convex domain $G \subset C^n$ is pseudoconvex.*

Proof: Since the function $-\ln \Delta_G(x, y)$ is convex in G, it is also plurisubharmonic in G (see section 11.3).

4. *Properties of convex domains*

The interior of the intersection of convex domains is a convex domain (see section 12.2).

The union of an increasing sequence of convex domains is a convex domain (see section 12.2).

For a domain B to be convex, it is necessary and sufficient that there exist a function that is convex in B and that approaches $+\infty$ everywhere on ∂B (see section 12.8).

If $V(x)$ is a convex function in $U(\overline{B})$ the domain

$$B = [x: \ V(x) < 0, \ x \in U(\overline{B})]$$

is convex (see section 12.7).

If B is a convex domain and the function $V(x)$ is convex in B, the domain $B' = [x: \ V(x) < 0, \ x \in B]$ is convex (see section 12.9).

For a domain B to be convex, it is necessary and sufficient that it be the union of an increasing sequence of strictly convex domains of the form

$$B_\alpha = [x; \ V_\alpha(x) < 0, \ x \in U(\overline{B}_\alpha)], \qquad \alpha = 1, 2, \ldots,$$

where $B_\alpha \Subset B_{\alpha+1} \Subset B$ in $V_\alpha(x) \in C^{(\infty)}$, such that $U(\overline{B}_\alpha)$ (see section 12.10). (The definition of a strictly convex domain is analogous to the definition of a strictly pseudoconvex domain [see section 12.10].)

These assertions are proved in the same way as the corresponding assertions were proven for pseudoconvex domains (with the corresponding simplifications).

Let us now consider some examples illustrating the general theorem on convex domains.

5. *Pseudoconvex tubular domains*

For a tubular domain T_B (see section 7.5) *to be pseudoconvex, it is necessary and sufficient that its base B be a convex domain.*

This assertion follows from the formula

$$-\ln \Delta_{T_B}(z) = -\ln \Delta_B(x)$$

and the theorem of section 11.3.

6. *Logarithmically convex domains*

We shall say that a multiple-circular domain G with center $a=0$ (see section 7.5) is logarithmically convex if the image Q of this domain in the space of logarithms of the absolute values is a convex domain:

$$Q = [\xi: \ \xi_j = \ln|z_j|, \ j = 1, 2, \ldots, n; \ \ z \in G, \ z_1 z_2 \ldots z_n \neq 0].$$

Suppose that G is a multiple-circular domain. We introduce the sequence

$$A_\alpha(G) = \sup_{z \in G} |z^\alpha|, \ \alpha = (\alpha_1, \ldots, \alpha_n), \tag{59}$$

where the a_j are nonzero integers.

If a multiple-circular domain G can be represented in the form

$$G = \operatorname{int} \bigcap_\alpha [z: \ \ |z^\alpha| < B_\alpha],$$

it is a logarithmically convex domain and $A_\alpha(G) \leqslant B_\alpha$.

This is true because

$$Q = \operatorname{int} \bigcap_\alpha [\xi: \ \ \alpha\xi < \ln B_\alpha]$$

is a convex domain (see section 13.1) and $\ln A_\alpha(G) = \sup_{\xi \in Q} \alpha\xi \leqslant \ln B_\alpha$.

If G is a bounded complete logarithmically convex domain, it can be represented in the form

$$G = \operatorname{int} \bigcap_{|\alpha| \geqslant 1} [z: |z^\alpha| < A_\alpha(G)]. \tag{60}$$

Proof: The image Q of G in the space of the logarithms of the absolute values is a convex domain possessing the property that, together with each point ξ^0, it contains the entire octant $\xi_j \leqslant \xi_j^0$, for $j = 1, 2, \ldots, n$ (see Fig. 25). Therefore, Q is the interior of the intersection of the semispaces (see section 13.1) such that all the direction cosines of their boundary are nonnegative. But the set of vectors with nonnegative rational coordinates is dense in the set of all vectors with nonnegative coordinates. Since G is a bounded domain, we have

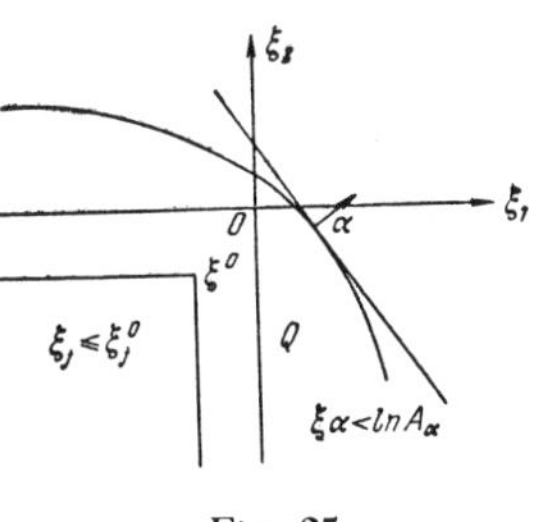

Fig. 25

$$Q = \operatorname{int} \bigcap_{|\alpha| \geqslant 1} [\xi : \xi\alpha < \ln A_\alpha(G)], \tag{61}$$

where α_j are nonnegative integers and the planes $\xi\alpha = \ln A_\alpha(G)$ are the supporting planes for the domain Q, so that

$$\ln A_\alpha(G) = \sup_{\xi \in Q} \xi\alpha; \qquad A_\alpha(G) = \sup_{z \in G} |z^\alpha|.$$

It follows from Eq. (61) that

$$[z : z \in G,\ z_1 \ldots z_n \neq 0] = \\ = \operatorname{int} \bigcap_{|\alpha| \geqslant 1} [z : |z^\alpha| < A_\alpha(G),\ z_1 \ldots z_n \neq 0].$$

But the domain G is complete. Therefore, the representation (60) follows from the last equation, as asserted.

If a multiple-circular domain G is pseudoconvex, it is logarithmically convex.

Proof: Every domain $D_j = [z : z_j \neq 0]$, for $j = 1, 2, \ldots, n$, is pseudoconvex since $-\ln \Delta_{D_j}(z) = -\ln |z_j|$ is a plurisubharmonic function in D_j. Therefore, the domain

$$G' = [z:\ z \in G,\ z_1 \ldots z_n \neq 0] = G \bigcap \Big\{ \bigcap_{1 \leqslant j \leqslant n} D_j \Big\},$$

being the intersection of pseudoconvex domains, is pseudoconvex (see section 12.2). Hence, the function $-\ln \Delta_{G'}(z)$ is plurisubharmonic in G'. But, G' is obviously a multiple-circular domain with center at zero. Therefore, the function $-\ln \Delta_{G'}(z)$ depends only on $|z_j|$. But then, the function

$$V(|z_1|, \ldots, |z_n|) = \max[-\ln \Delta_{G'}(z),\ |z|^2]$$

is plurisubharmonic in G' and converges to $+\infty$ everywhere on $\partial G'$. From this it follows that the function $V(e^{\xi_1}, \ldots, e^{\xi_n})$ is convex in Q (see section 11.4). Furthermore, since the mapping $r = e^\xi$ is a homeomorphism of $(-\infty, \infty)$ onto $(0, \infty)$, the function $V(e^{\xi})$ converges to $+\infty$ everywhere on ∂Q (cf. lemma in section 12.11). Therefore, the domain Q is convex (see section 13.4).

We shall prove the converse assertion for the two limiting cases of complete domains and domains not containing points of the manifold $z_1 \ldots z_n = 0$.

If G is a complete logarithmically convex domain or if it does not contain points of the manifold $z_1 \ldots z_n = 0$, it is a pseudoconvex domain.

Proof: Suppose that G is a complete logarithmically convex domain. Then, $G = \bigcup_{R>0} G_R$, where the $G_R = G \cap S(0, RI)$ are bounded complete logarithmically convex domains. On the basis of section 12.2, we need only prove that each of the domains G_R is pseudoconvex. From what we have already proven, G_R can be represented by formula (60):

$$G_R = \operatorname{int} \bigcap_{|\alpha| \geqslant 1} [z : |z^{\alpha}| < A_{\alpha}(G_R)].$$

Let us show that the domain $[z : |z^{\alpha}| < A_{\alpha}]$ is pseudoconvex if $|\alpha| \geqslant 1$. Note that

$$[z : |z^{\alpha}| < A_{\alpha}] = [z : \alpha_1 \ln |z_1| + \ldots + \alpha_n \ln |z_n| - \ln A_{\alpha} < 0].$$

But, for $\alpha_j \geqslant 0$, the function $\alpha_1 \ln |z_1| + \ldots + \alpha_n \ln |z_n| - \ln A_{\alpha}$ is plurisubharmonic in C^n (see section 10.5). Therefore, (see section 12.9), the domain $[z : |z^{\alpha}| < A_{\alpha}]$ is pseudoconvex. But then the domain G_R, being the interior of the intersection of pseudoconvex domains, is pseudoconvex (see section 12.2).

Suppose now that G is a logarithmically convex domain containing no points of the manifold $z_1 \ldots z_n = 0$. Since the corresponding Q is convex, there exists a function $V(\xi)$ that is convex in Q and that approaches $+\infty$ everywhere on ∂Q (see section 13.4). Therefore, the function $V(\ln |z_1|, \ldots, \ln |z_n|)$ is plurisubharmonic in G (see section 11.4). Furthermore, since the mapping $\xi = \ln r$ is a homeomorphism of $(0, \infty)$ onto $(-\infty, \infty)$, the function $V(\ln |z|)$ approaches $+\infty$ everywhere on ∂G (cf. lemma in section 12.11). Therefore, (see section 12.8), the domain G is pseudoconvex, which completes the proof.

In conclusion, let us take a look at complete logarithmically convex domains in the space C^2. Such a domain can be represented in one of the following two forms:

$$[z : |z_1| < R_1(|z_2|), \ |z_2| < R'],$$
$$[z : |z_2| < R_2(|z_1|), \ |z_1| < R],$$

where the function R_2 is the inverse of the function R_1 (see Fig. 26).

The functions $R_1(r)$ and $R_2(r)$ are lower-semicontinuous and decreasing functions in $[0, R')$ and $[0, R)$, respectively.

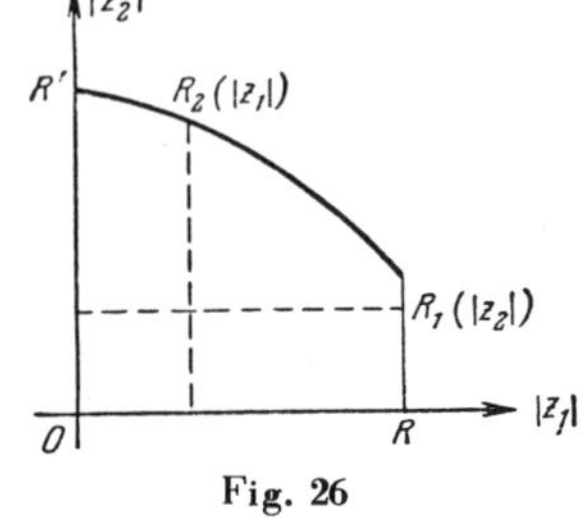

Fig. 26

For a complete multiple-circular domain G to be logarithmically convex, it is necessary and sufficient that the function $R_1(r)$ *possess the following properties:* (1) *It is lower-semicontinuous and decreasing in* $[0, R')$; (2) *the function* $-R_1(r)$ *is convex with respect to* $\ln r$ *in* $(0, R')$.

It follows from these properties that $R_1(r)$ is a continuous function in $(0, R')$ and that $R = R_1(0) = \lim\limits_{r \to +0} R_1(r)$.

The function $R_2(r)$ possesses analogous properties.

CHAPTER III

Domains and Envelopes of Holomorphy

In this chapter, we shall expound the general theory of (single-sheeted) domains of holomorphy and envelopes of holomorphy. We shall take up the following four types of properties of domains of holomorphy: (1) holomorphic convexity, (2) principles of continuity, (3) local pseudoconvexity, and (4) global pseudoconvexity. Although all these properties are closely interrelated, they characterize domains of holomorphy in quite different terms. In particular, according to the fundamental theorem of Oka, the class of (single-sheeted) domains of holomorphy coincides with the class of pseudoconvex domains. Therefore, study of domains of holomorphy amounts to a study of geometric objects, namely, the pseudoconvex domains examined in the preceding chapter. Before embarking on an exposition of the general theory of domains of holomorphy, let us examine in greater detail functions that are holomorphic in multiple-circular domains and Hartogs domains.

14. MULTIPLE-CIRCULAR DOMAINS AND POWER SERIES

1. *Holomorphic extension of multiple-circular domains*

Suppose that a multiple-circular domain G contains its center $a=0$ (see section 7.5). Let $f(z)$ be a function that is holomorphic in G. Let us take an arbitrary multiple-circular subdomain $G' \Subset G$ that contains the center 0. We define

$$R=\sup_{z\in G'}|z|, \quad \Delta_G(G')=\eta>0, \quad r=1+\frac{\eta}{2R}.$$

Let us construct the function (see Cartan [118])

$$\varphi(z)=\frac{1}{(2\pi i)^n}\int\limits_{|\lambda_1|=r}\cdots\int\limits_{|\lambda_n|=r}\frac{f(\lambda_1 z_1,\ldots,\lambda_n z_n)}{(\lambda-1)^I}\,d\lambda. \tag{1}$$

When $z \in G'$, the points $(\lambda_1 z_1, \ldots, \lambda_n z_n)$ are strictly contained in G for $|\lambda_j| = r$, where $j = 1, 2, \ldots, n$, since

$$|(\lambda_1 z_1, \ldots, \lambda_n z_n) - (e^{i \arg \lambda_1} z_1, \ldots, e^{i \arg \lambda_n} z_n)| \leqslant$$
$$\leqslant |z|(r-1) \leqslant \frac{\eta}{2}.$$

Therefore, the function $\varphi(z)$ is holomorphic in G' and $|f(\lambda_1 z_1, \ldots, \lambda_n z_n)| \leqslant M$ for all $z \in G'$, where $|\lambda_j| = r$, for $j = 1, 2, \ldots, n$.

Furthermore, since $0 \in G'$, the function $f(z)$ can be expanded in a Taylor series

$$f(z) = \sum_{|\alpha| \geqslant 0} a_\alpha z^\alpha, \qquad a_\alpha = \frac{1}{\alpha!} D^\alpha f(0) \tag{2}$$

in some polydisk $S(0, \rho)$. If we substitute this series into formula (1), we see that

$$\varphi(z) = f(z), \qquad z \in S\left(0, \frac{\rho}{r}\right).$$

On the basis of the holomorphic continuation theorem (see section 6.1), we conclude that $\varphi(z) \equiv f(z)$ in the domain G', that is, that

$$f(z) = \frac{1}{(2\pi i)^n} \int_{|\lambda_1| = r} \cdots \int_{|\lambda_n| = r} \frac{f(\lambda_1 z_1, \ldots, \lambda_n z_n)}{(\lambda - 1)^I} d\lambda.$$

From this, we get the expansion

$$f(z) = \sum_{\alpha = 0}^{\infty} \varphi_\alpha(z), \qquad z \in G', \tag{3}$$

where

$$\varphi_\alpha(z) = \frac{1}{(2\pi i)^n} \int_{|\lambda_1| = r} \cdots \int_{|\lambda_n| = r} \frac{f(\lambda_1 z_1, \ldots, \lambda_n z_n)}{\lambda^{\alpha + I}} d\lambda = z^\alpha f_\alpha(z),$$
$$f_\alpha(z) = \frac{1}{(2\pi i)^n} \int_{r|z_1| = |z_1'|} \cdots \int_{r|z_n| = |z_n'|} \frac{f(z')}{z'^{\,\alpha + I}} dz'. \tag{4}$$

The functions $\varphi_\alpha(z)$ are holomorphic in G' and the series (3) converges absolutely and uniformly in G'.

Let us show that $f_\alpha(z) = a_\alpha$ in G'. It follows from (4) and Cauchy's theorem that these equations are valid in $S\left(0, \frac{\rho}{r}\right)$. But then, the equations $\varphi_\alpha(z) = a_\alpha z^\alpha$, which are valid in $S\left(0, \frac{\rho}{r}\right)$, remain valid in G'. This proves our assertion.

Thus, the series in (3) is independent of r (and hence of G') and coincides with the power series (2).

Remembering that G' is an arbitrary compact subdomain of G, we can, by use of Abel's theorem, obtain the following result (see Hartogs [6], Cartan [118]):

If a function $f(z)$ is holomorphic in a multiple-circular domain G that contains its center $a = 0$, it is holomorphic (and, consequently, single-valued) in the smallest complete multiple-circular domain $\pi(G)$ containing the given domain G,

$$\pi(G) = [z\colon z_j = \lambda_j z'_j, \quad z' \in G, \quad |\lambda_j| \leqslant 1, \quad j = 1, 2, \ldots, n]$$

and it can be expanded in the absolutely convergent power series (2) in $\pi(G)$. Thus, $\pi(G)$ is a holomorphic extension of the domain G.

2. *Domains of absolute convergence of power series*

We shall say that a multiple-circular domain G is a domain of absolute convergence of a power series if there exists a power series that converges absolutely in the domain G but does not converge (absolutely) in any larger domain.

Thus, the preceding theorem reduces domains of absolute convergence of power series to complete multiple-circular domains. We shall see below that not every complete multiple-circular domain is a domain of absolute convergence of any power series. The question arises as to how domains of absolute convergence of power series may be characterized. The answer to this question for bounded domains was already known to Hartogs [6]. Here, following the work of Ayzenberg and Mityagin [62], we shall give a simple solution of this problem.

We introduce the sequence of numbers (see Eq. (59) in section 13.6)

$$A_\alpha(G) = \sup_{z \in G} |z^\alpha|, \quad |\alpha| \geqslant 0.$$

LEMMA 1. *If the power series*

$$\sum_{|\alpha| \geqslant 0} a_\alpha z^\alpha \tag{5}$$

converges absolutely in a closed bounded complete multiple-circular domain $\overline{G}$, then

$$|a_\alpha| \leqslant \frac{M}{A_\alpha(G)}, \quad \text{where} \quad M = \max_{z \in \overline{G}} \left| \sum_{|\alpha| \geqslant 0} a_\alpha z^\alpha \right|.$$

Proof: For an arbitrary polydisk $S(0, r) \subset G$, where $r_j = |b_j|$ for $b \in G$, the following inequality of Cauchy (see section 4.4) is valid:

$$|a_\alpha| \leqslant |b^\alpha|^{-1} \max_{z \in \bar{S}(0,r)} \left| \sum_{|\alpha| \geqslant 0} a_\alpha z^\alpha \right| \leqslant \frac{M}{|b^\alpha|}.$$

Since this inequality holds for all $b \in G$, we have

$$|a_\alpha| \leqslant \inf_{b \in G} \frac{M}{|b^\alpha|} = \frac{M}{A_\alpha(G)}, \quad \text{q. e. d.}$$

LEMMA 2. *If the series* (5) *converges absolutely in a complete multiple-circular domain* G, *the series*

$$\sum_{|\alpha| \geqslant 0} |a_\alpha| \, |A_\alpha(G')| \tag{6}$$

converges for an arbitrary subdomain $G' \Subset G$

Proof: Since $G' \Subset G$, there exists a bounded complete multiple-circular domain G_0 such that $G' \subset G_0 \Subset G$. Furthermore, since $G_0 \Subset G$, we have $rG_0 \Subset G$ for $r > 1$ sufficiently close to 1. By using Lemma 1, we then obtain

$$|a_\alpha| \leqslant \frac{M}{A_\alpha(rG_0)}, \quad \text{where } M = \max_{z \in r\bar{G}_0} \left| \sum_{|\alpha| \geqslant 0} a_\alpha z^\alpha \right|.$$

From this inequality and the equation

$$A_\alpha(rG_0) = r^{|\alpha|} A_\alpha(G_0) \tag{7}$$

we conclude that the series (6) is convergent:

$$\sum_{|\alpha| \geqslant 0} |a_\alpha| \, |A_\alpha(G')| \leqslant \sum_{|\alpha| \geqslant 0} |a_\alpha| \, |A_\alpha(G_0)| \leqslant M \sum_{|\alpha| \geqslant 0} r^{-|\alpha|} < +\infty.$$

THEOREM. *For the series* (5) *to converge absolutely in a bounded complete multiple-circular domain* G, *it is necessary and sufficient that the series*

$$\sum_{|\alpha| \geqslant 0} a_\alpha A_\alpha(G) \, z^\alpha \tag{8}$$

converge absolutely in the unit polydisk $S(0, I)$.

This sufficiency is obvious. Let us prove the necessity. On the basis of Lemma 2, absolute convergence of the series (5) in G implies convergence of the series

$$\sum_{|\alpha| \geqslant 0} |a_\alpha| \, A_\alpha(rG)$$

for arbitrary $0 < r < 1$; that is, on the basis of (7), it implies convergence of the series

$$\sum_{|\alpha|\geqslant 0} |a_\alpha| A_\alpha(G) r^\alpha.$$

But this means that the series (8) converges absolutely in $S(0, I)$.

3. *Logarithmically convex envelopes*

The smallest logarithmically convex domain G^* containing a given multiple-circular domain G is called the logarithmically convex envelope of G.

Let Q denote the image of the domain G in the space of logarithms of the absolute values:

$$Q = [\xi: \xi_j = \ln |z_j|, \ j = 1, 2, \ldots, n; \ z \in G, \ z_1 z_2 \ldots z_n \neq 0].$$

It follows from the above definition that the image of the domain G^* in this space is $O(Q)$, which is the convex envelope of Q:

$$O(Q) = [\xi: \xi_j = \ln |z_j|, \quad j = 1, 2, \ldots, n; \\ z \in G^*, \quad z_1 z_2 \ldots z_n \neq 0].$$

It follows from this and the results of section 13.6 that the logarithmically convex envelope G of a complete multiple-circular domain G is of the form

$$G^* = \bigcup_{R>0} G_R^*,$$

$$G_R^* = \text{int} \bigcap_{|\alpha|\geqslant 1} [z : |z^\alpha| < A_\alpha(G_R)], \quad G_R = G \bigcap S(0, RI).$$

In particular, if G is a bounded domain, then

$$G^* = \text{int} \bigcap_{|\alpha|\geqslant 1} [z : |z^\alpha| < A_\alpha(G)].$$

For unbounded domains, this formula is not in general valid. For example, suppose that

$$G = [z : |z_1|^p |z_2|^q < 1] \subset C^2, \ \frac{p}{q} \quad \text{irrational.}$$

Then, $G = G^*$, whereas

$$\bigcap_{|\alpha|\geqslant 1} [z : |z^\alpha| < A_\alpha(G) = +\infty] = C^2 \neq G^*.$$

This fact was discovered by Engibaryan.

4. *Holomorphic extension of complete multiple-circular domains*

Every function that is holomorphic in a complete multiple-circular domain G is holomorphic (and consequently single-valued) in its logarithmically convex envelope G^* and can be expanded in a power series that converges absolutely in G^*.

Thus, G^* is the holomorphic extension of the domain G. We note that G^* is a pseudoconvex domain (see section 13.6).

This assertion follows immediately from the following theorem of Hartogs [6].

For a complete multiple-circular domain to be the domain of absolute convergence of a power series, it is necessary and sufficient that it be logarithmically convex.

Proof: Suppose that a complete multiple-circular domain G is a domain of absolute convergence of the power series (5). Let us set

$$G_R = G \bigcap S(0,\ RI), \qquad R > 0.$$

Then, the bounded domain G_R is the domain of absolute convergence of the series

$$\sum_{|\alpha| \geqslant 0} \left[a_\alpha + \left(\frac{1}{R}\right)^{|\alpha|} e^{i \arg a_\alpha} \right] z^\alpha. \tag{9}$$

Let us show that G_R is a logarithmically convex domain: $G_R = \overset{*}{G}_R$. If we apply the theorems in section 14.2 and then use the equations $A_\alpha(G_R) = A_\alpha(\overset{*}{G}_R)$, we see that the series (9) converges absolutely in $\overset{*}{G}_R$. But then, $G_R = \overset{*}{G}_R$. Thus, the domain G is the union of an increasing sequence of logarithmically convex domains G_R. Consequently, it is itself a logarithmically convex domain (see section 13.6).

For the time being, we shall prove the converse only for bounded domains. The proof for unbounded domains will be given in section 19.3.

Suppose that G is a complete bounded logarithmically convex domain: $G = G^*$. According to the theorem of section 14.2, the series

$$\sum_{|\alpha| \geqslant 0} \frac{1}{A_\alpha(G)} z^\alpha \tag{10}$$

converges absolutely in G. Now if this series converges absolutely in a larger bounded complete multiple-circular domain G', the series

$$\sum_{|\alpha| \geqslant 0} \frac{A_\alpha(G')}{A_\alpha(G)} z^\alpha \tag{11}$$

must, on the basis of the theorem in section 14.2, converge

absolutely in $S(0, I)$. Let us suppose that there exists an α_0 such that $A_{\alpha_0}(G') > A_{\alpha_0}(G)$. It follows from the absolute convergence of the series (11) in $S(0, I)$ and from the relationship $A_{\alpha m}(G) = (A_\alpha(G))^m$ that the series

$$\sum_{m=0}^{\infty} \left(\frac{A_{\alpha_0}(G')}{A_{\alpha_0}(G)} \right)^m z^{\alpha_0 m},$$

also converges absolutely in $S(0, I)$, which is impossible. Therefore,

$$A_\alpha(G') \leqslant A_\alpha(G) \text{ for all } \alpha.$$

But the inclusion $G' \supset G$ implies the reverse inequalities. Therefore, $A_\alpha(G') = A_\alpha(G)$. It follows from this that $G' = G^* = G$, which is impossible. Consequently, G is a domain of absolute convergence of the series (10), which completes the proof.

15. HARTOGS' DOMAINS AND SERIES

In sections 6, 8 and 14, (multiple) power-series expansions of functions constituted our basic tool for holomorphic expansions. In this section, we shall introduce another method of holomorphic continuation, namely, the method of expanding functions in power series in a single variable (Hartogs' series). In contrast with the preceding method, this method can be applied only in the case of several complex variables ($n \geqslant 2$). The results that we shall present in connection with Hartogs' series are of independent interest but they also have a number of applications in the theory of several complex variables, for example, in the proof of the fundamental theorem of Hartogs (see section 4.2) and in the study of semitubular domains.

1. Expansion in Hartogs series

Suppose that a Hartogs domain G (see section 7.5) contains points of its plane of symmetry $z_1 = a_1$. For simplicity, we shall assume that $a_1 = 0$.

Suppose that a function $f(z)$ is holomorphic in the domain G. Let G' be an arbitrary domain that is compact in G and that contains points of the form $(0, \tilde{z})$. Without loss of generality, we may assume that G' is a Hartogs domain. We define

$$R = \sup_{z \in G'} |z_1|, \qquad \Delta_G(G') = \eta, \qquad r = 1 + \frac{\eta}{2R}. \tag{12}$$

Let us construct the function (see Cartan [118])

$$\varphi(z)=\frac{1}{2\pi i}\int_{|\lambda_1|=r}\frac{f(\lambda_1 z_1, \tilde{z})}{\lambda_1-1}\,d\lambda_1. \tag{13}$$

When $z\in G'$, the points $|\lambda_1|=r$ are strictly contained in G for all $(\lambda_1 z_1, \tilde{z})$ since, on the basis of (12),

$$|(\lambda_1 z_1, \tilde{z})-(e^{i\arg\lambda_1}z_1, \tilde{z})|\leqslant|\lambda z_1-e^{i\arg\lambda_1}z_1|\leqslant$$
$$\leqslant|z_1|(|\lambda_1|-1)\leqslant\frac{\eta}{2}.$$

Therefore, the function $\varphi(z)$ is holomorphic in G' (see section 4.6) and $|f(\lambda_1 z_1, \tilde{z})|\leqslant M$ for $|\lambda_1|=r$ and $z\in G'$.

Furthermore, since there exists a point $(0, \tilde{z}^0)$ belonging to G', the function $f(z)$ can be expanded in an absolutely convergent power series

$$f(z)=\sum_{|\alpha|\geqslant 0}a_\alpha z_1^{\alpha_1}(\tilde{z}-\tilde{z}^0)^{\tilde{\alpha}}, \qquad a_\alpha=\frac{1}{\alpha!}D^\alpha f(0, \tilde{z}^0)$$

in some polydisk $S((0, \tilde{z}^0), \rho)$. If we substitute this series into Eq. (13), we see that

$$\varphi(z)=\frac{1}{2\pi i}\int_{|\lambda_1|=r}\frac{1}{\lambda_1-1}\sum_{|\alpha|\geqslant 0}a_\alpha\lambda_1^{\alpha_1}z_1^{\alpha_1}(\tilde{z}-\tilde{z}^0)^{\tilde{\alpha}}\,d\lambda_1=$$
$$=\sum_{|\alpha|\geqslant 0}a_\alpha z_1^{\alpha_1}(\tilde{z}-\tilde{z}^0)^{\tilde{\alpha}}=f(z)$$

for

$$z\in S\left((0, \tilde{z}^0), \frac{\rho}{r}\right).$$

Therefore, on the basis of the holomorphic continuation theorem (see section 6.1), we conclude that $\varphi(z)\equiv f(z)$ in the domain G'; that is,

$$f(z)=\frac{1}{2\pi i}\int_{|\lambda_1|=r}\frac{f(\lambda_1 z_1, \tilde{z})}{\lambda_1-1}\,d\lambda_1, \qquad z\in G'.$$

From this we get the expansion

$$f(z)=\sum_{\alpha=0}^{\infty}\varphi_\alpha(z), \qquad z\in G', \tag{14}$$

where

$$\varphi_\alpha(z) = \frac{1}{2\pi i} \int\limits_{|\lambda_1| = r} f(\lambda_1 z_1, \tilde{z}) \frac{d\lambda_1}{\lambda_1^{\alpha+1}} = z_1^\alpha f_\alpha(z), \tag{15}$$

$$f_\alpha(z) = \frac{1}{2\pi i} \int\limits_{r|z_1| = |z_1'|} \frac{f(z_1', \tilde{z})}{z_1'^{\alpha+1}} dz_1', \quad \alpha = 0, 1, \ldots. \tag{16}$$

The series (14) converges absolutely and uniformly in G'.

According to Cauchy's theorem, the integral (16) defining the function f_α is independent of r. Therefore, we may take the limit in formula (16) as $r \to 1+0$. We then obtain

$$f_\alpha(z) = \frac{1}{2\pi i} \int\limits_{|z_1| = |z_1'|} \frac{f(z_1', \tilde{z})}{z_1'^{\alpha+1}} dz_1', \quad \alpha = 0, 1, \ldots. \tag{17}$$

The representation (17) is valid for $z \in G'$. But the right member in (17) is independent of $G' \Subset G$. Therefore, the functions $f_\alpha(z)$ are defined throughout the entire domain G. Furthermore, on the basis of Cauchy's theorem, these functions are independent of z_1 when z_1 varies in some annulus $r(\tilde{z}) < |z_1| < R(\tilde{z})$ (or in the circle $|z_1| < R(\tilde{z})$) lying entirely in the domain G (see Fig. 27).

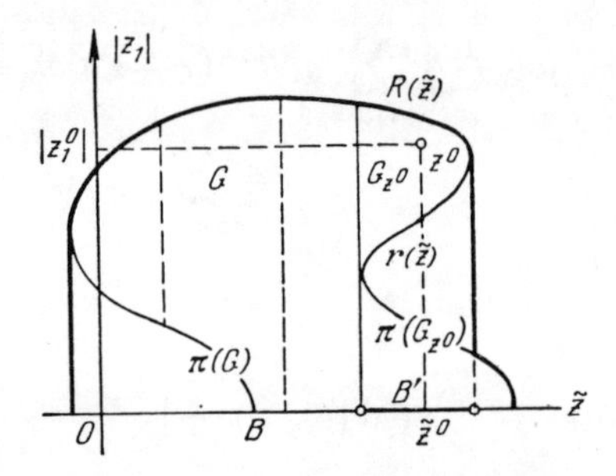

Fig. 27

In particular, if $(0, \tilde{z}) \in G$, by taking the limit in (17) as $z_1 \to 0$, we obtain

$$f_\alpha(z) = \frac{1}{\alpha!} D^\alpha f(0, \tilde{z}), \quad \text{if} \quad [z\colon |z_1| < R(\tilde{z})] \subset G. \tag{18}$$

Let us show that the functions $f_\alpha(z)$ are holomorphic in G. It follows from (15) that the functions $\varphi_\alpha(z)$ are holomorphic in G' and that $f_\alpha(z) = \varphi_\alpha(z) z_1^{-\alpha}$. Therefore, the functions $f_\alpha(z)$ are holomorphic in G' if $z_1 \neq 0$. On the other hand, if the point $(0, \tilde{z}) \in G'$, the functions $f_\alpha(z)$ will, by virtue of what we have proven, be independent of z_1 in some neighborhood of that point. Thus, the functions $f_\alpha(z)$ are also holomorphic at the points $(0, \tilde{z})$ in the domain G'. Since G' is an arbitrary subdomain that is compact in G, this means that the $f_\alpha(z)$ are holomorphic in G.

Summing up what we have said, we have the following result (see Hartogs [6], Cartan [118]).

Every function $f(z)$ that is holomorphic in a Hartogs domain G that contains points of its plane of symmetry $z_1 = 0$, can be expanded in an absolutely and uniformly convergent Hartogs series

$$f(z) = \sum_{\alpha=0}^{\infty} z_1^\alpha f_\alpha(z), \tag{19}$$

in that domain. Here, the functions $f_\alpha(z)$ are holomorphic in G and are independent of z_1 in every component (annulus or circle) of the open set

$$[z_1: (z_1, \tilde{z}) \in G].$$

2. *Holomorphic extension of Hartogs domains*

We denote by B the projection of the domain G onto the plane of symmetry $z_1 = 0$. Obviously, B is a domain in C^{n-1}. Let $z^0 = (z_1^0, \tilde{z}^0) \in G$. Then, $\tilde{z}^0 \in B$. Let G_{z^0} be any subdomain of the domain G that contains the point z^0 and that can be represented in one of the following forms (see Fig. 27):

$$G_{z^0} = [z: \ r_{z^0}(\tilde{z}) < |z_1| < R_{z^0}(\tilde{z}), \ \ \tilde{z} \in B'],$$
$$G_{z^0} = [z: \ |z_1| < R_{z^0}(\tilde{z}), \ \ \tilde{z} \in B'],$$

where B' is a subdomain of the domain B that depends on z^0. On the basis of section 2.2, the functions $r_{z^0}(\tilde{z})$ and $R_{z^0}(\tilde{z})$ must be upper- and lower-semicontinuous, respectively, in B'.

On the basis of the results of the preceding section, in the expansion (19), the functions $f_\alpha(z)$ in the domain G_{z^0} are independent of z_1. Therefore, the functions $\tilde{f}_\alpha(\tilde{z}) \equiv f_\alpha(z)$, being holomorphic in G, are also holomorphic in B'. Thus, the expansion (19) in the domain G_{z^0} can be written as follows:

$$f(z) = \sum_{\alpha=0}^{\infty} z_1^\alpha \tilde{f}_\alpha(\tilde{z}). \tag{20}$$

According to Abel's theorem, the series (20) converges absolutely and uniformly in the smallest complete Hartogs domain (see Fig. 27)

$$\pi(G_{z^0}) = [z: \ z_1 = \lambda z_1', \ \ \tilde{z} = \tilde{z}', \ \ z' \in G_{z^0}, \ \ |\lambda| \leqslant 1] =$$
$$= [z: \ |z_1| < R^*_{z^0}(\tilde{z}), \ \ \tilde{z} \in B'],$$

containing the domain G_{z^0} and thus defines a holomorphic function in $\pi(G_{z^0})$.

Thus, every function $f(z)$ that is holomorphic in a Hartogs domain G containing points of its plane of symmetry $z_1 = 0$ is holomorphic in every complete Hartogs domain $\pi(G_{z^0})$ (for $(z^0 \in G)$) and is represented in it by the series (20).

Therefore, by means of the series (20), the function $f(z)$ can be holomorphically continued to every point of the smallest complete Hartogs domain $\pi(G)$ containing the given domain G (see Fig. 27):

$$\pi(G) = \bigcup_{z^0 \in G} \pi(G_{z^0}) =$$
$$= [z: z_1 = \lambda z_1', \quad \tilde{z} = \tilde{z}', \quad z' \in G, \quad |\lambda| \leqslant 1] =$$
$$= [z: |z_1| < R(\tilde{z}), \ \tilde{z} \in B].$$

Clearly, the function $f(z)$ will not in general be single-valued in $\pi(G)$. However, by using the concepts of section 8, we may construct a covering domain on B in which the functions $f_\alpha(\tilde{z})$ will be holomorphic (and single-valued). Here, the series (20) defines a function $f(z)$ that is holomorphic in the corresponding covering domain (over $\pi(G)$). The exposition makes it possible to construct examples of single-sheeted Hartogs domains with nonsingle-sheeted envelopes of holomorphy (see section 20.2).

3. *Hartogs' theorem*

Here, we shall confine ourselves to a consideration of the simplest case, that in which the function $f(z)$ is single-valued in $\pi(G)$, that is, to the case of complete Hartogs domains.

Suppose that the function $f(z)$ is holomorphic in a complete Hartogs domain

$$G = [z: \ |z_1| < R(\tilde{z}), \ \tilde{z} \in B],$$

where the domain B is the projection G onto the plane of symmetry $z_1 = 0$ and the function $R(\tilde{z})$ is lower-semicontinuous in B (see section 2.2). (B coincides with the intersection of the domain G and the analytic plane $z_1 = 0$.) On the basis of the results of sections 15.1 and 15.2, the function $f(z)$ can be represented in G by a Hartogs series

$$f(z) = \sum_{\alpha=0}^{\infty} z_1^\alpha f_\alpha(\tilde{z}), \quad f_\alpha(\tilde{z}) = \frac{1}{\alpha!} D^\alpha f(0, \tilde{z}), \tag{21}$$

which converges absolutely and uniformly in the domain G. The functions $f_\alpha(\tilde{z})$ are holomorphic in B.

For each $\tilde{z} \in B$, we denote the radius of convergence of the series (21) by $R_f^0(\tilde{z})$. Since the series (21) converges for $|z_1| < R(\tilde{z})$, it follows that $R_f^0(\tilde{z}) \geqslant R(\tilde{z})$. Furthermore, on the basis of the Cauchy-Hadamard theorem, we have

$$\frac{1}{R_f^0(\tilde{z})} = \overline{\lim_{\alpha \to +\infty}} \sqrt[\alpha]{|f_\alpha(\tilde{z})|}. \tag{22}$$

The function $R_f^0(\tilde{z})$ does not need to be lower-semicontinuous. If it is not, we set

$$R_f(\tilde{z}) = \underline{\lim_{\tilde{z}' \to \tilde{z}}} R_f^0(\tilde{z}'). \tag{23}$$

Obviously, $R_f(\tilde{z}) \leqslant R_f^0(\tilde{z})$. Furthermore, since the function $R(\tilde{z})$ is assumed to be lower-semicontinuous, it follows from the inequality $R_f^0 \geqslant R$ and Eq. (23) that $R(\tilde{z}) \leqslant R_f(\tilde{z})$ (see Fig. 28).

The function $R_f(\tilde{z})$ is lower-semicontinuous in the domain B. Therefore, the set

$$G_f = [z: \ |z_1| < R_f(\tilde{z}), \ \tilde{z} \in B]$$

is a domain (see section 2.2). Clearly, $G_f \supset G$.

It turns out that the series (21) converges absolutely and uniformly in the domain G_f. We shall prove this statement in the following section. As a preliminary, we shall prove the following, less general, theorem, belonging to Hartogs (6).

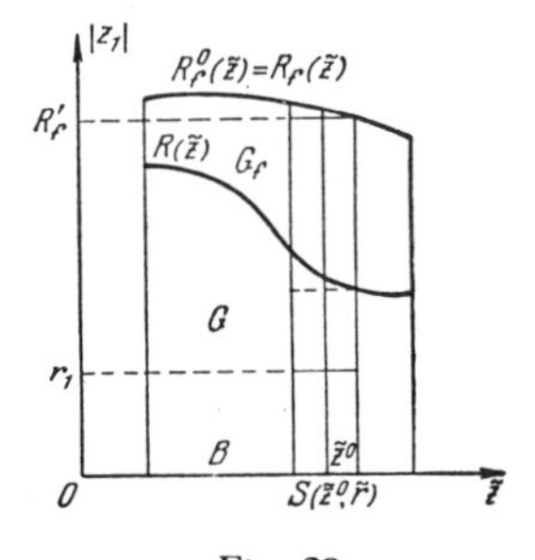

Fig. 28

If the Hartogs series (20) converges absolutely and uniformly in a closed polycircle $\overline{S}(0, r_1) \times \overline{S}(\tilde{z}^0, \tilde{r})$ and converges absolutely in a circle $|z_1| > R_1 < r_1$ for arbitrary fixed $\tilde{z}$ in $\overline{S}(\tilde{z}^0, \tilde{r})$, this series converges absolutely and uniformly in the polycircle $S(0, R_1) \times S(\tilde{z}^0, \tilde{r})$.

Proof: It follows from the conditions of the theorem and from Eq. (22) that

$$\left.\begin{array}{c} |f_\alpha(\tilde{z})| r_1^\alpha \leqslant M, \\ \overline{\lim\limits_{\alpha \to +\infty}} \frac{1}{\alpha} \ln |f_\alpha(\tilde{z})| \leqslant -\ln R_1, \quad \tilde{z} \in \overline{S}(\tilde{z}^0, \tilde{r}) \end{array}\right\} \tag{24}$$

for $\alpha = 1, 2, \ldots$. Suppose that $0 < R_1^0 < R < R_1$, and $0 < r_j^0 < r_j$ for $j = 2, \ldots, n$. Then, on the basis of (24), the functions $\varphi_\alpha(\tilde{z}) = f_\alpha(\tilde{z}) R^\alpha M^{-1}$ which are holomorphic in $S(\tilde{z}^0, \tilde{r})$, satisfy the conditions

$$\frac{1}{\alpha} \ln |\varphi_\alpha(\tilde{z})| \leqslant \ln \frac{R}{r_1}, \quad \overline{\lim_{\alpha \to +\infty}} \frac{1}{\alpha} \ln |\varphi_\alpha(\tilde{z})| \leqslant \ln \frac{R}{R_1}. \tag{25}$$

The functions $u_\alpha(\tilde{z}) = \frac{1}{\alpha} \ln |\varphi_\alpha(\tilde{z})|$ are plurisubharmonic in $S(\tilde{z}^0, \tilde{r})$ (see section 10.5) and, on the basis of (25), satisfy the conditions of the generalized Harnack's theorem (see section 10.7) with G replaced by $S(\tilde{z}^0, \tilde{r})$ and A replaced by $\ln R - \ln R_1$. If we set $\varepsilon = \ln R_1 - \ln R > 0$, we conclude from this theorem that there exists a number $N = N(\varepsilon, \tilde{r}^0)$ such that, for all $\alpha \geqslant N$ and $\tilde{z}$ in $\overline{S}(\tilde{z}^0, \tilde{r}^0)$,

$$u_\alpha(\tilde{z}) = \frac{1}{\alpha} \ln |\varphi_\alpha(\tilde{z})| \leqslant 0,$$

that is,

$$|f_\alpha(\tilde{z})| R^\alpha \leqslant M, \quad \alpha \geqslant N, \quad \tilde{z} \in \overline{S}(\tilde{z}^0, \tilde{r}^0).$$

This inequality implies the absolute and uniform convergence of the series (21) in the polydisk $\overline{S}(0, R_1^0) \times \overline{S}(\tilde{z}^0, \tilde{r}^0)$. Since the numbers R_1^0 and r_j^0 $(j = 2, \ldots, n)$ can be chosen arbitrarily close to R_1 and r_j $(j = 2, \ldots, n)$, respectively, it follows that the series (21) converges absolutely and uniformly in the polydisk $S(0, R_1) \times S(\tilde{z}^0, \tilde{r})$, which completes the proof.

4. *Domains of convergence of Hartogs' series*

The following generalization follows from Hartogs' theorem (see section 15.3): *If the Hartogs series* (21) *converges absolutely and uniformly in a complete Hartogs domain*

$$G = [z: \ |z_1| < R(\tilde{z}), \ \tilde{z} \in B],$$

it also converges absolutely and uniformly in the larger complete Hartogs domain

$$G_f = [z: \ |z_1| < R_f(\tilde{z}), \ \tilde{z} \in B]$$

where the function

$$-\ln R_f(\tilde{z}) = \overline{\lim_{\tilde{z}' \to \tilde{z}}} \ \overline{\lim_{\alpha \to +\infty}} \ \frac{1}{\alpha} \ln |f_\alpha(\tilde{z}')|$$

is plurisubharmonic in B and $R_f \geq R$.

Proof: Suppose that B' is an arbitrary relatively compact subdomain of B. On the basis of the Heine-Borel theorem, we can cover the compact set $\overline{B}'$ with a finite number of polydisks that are compact in B. Let $S(\tilde{z}^0, \tilde{r})$ be one of these. Since the function $R_f(\tilde{z})$ is lower-semicontinuous in B, it assumes its smallest value $\overline{S}(\tilde{z}^0, \tilde{r})$ on $R_f' > 0$ (see section 2.3). Therefore, on the basis of (22),

$$\overline{\lim_{\alpha \to +\infty}} \sqrt[\alpha]{|f_\alpha(\tilde{z})|} = \frac{1}{R_f^0(\tilde{z})} \leqslant \frac{1}{R_f(\tilde{z})} \leqslant \frac{1}{R_f'}, \quad \tilde{z} \in \overline{S}(\tilde{z}^0, \tilde{r}).$$

It follows from this that the series (21) converges absolutely in the circle $|z_1| < R'_f$ for every $\tilde{z} \in \bar{S}(\tilde{z}^0, \tilde{r})$ (see Fig. 28). Furthermore, this series converges absolutely and uniformly in the polydisk $\bar{S}(0, r_1) \times \bar{S}(\tilde{z}^0, \tilde{r})$, where $r_1 = 1/2 \min R(\tilde{z})$ for $\tilde{z} \in \bar{S}(\tilde{z}^0, \tilde{r})$. (The function $R(\tilde{z})$ is lower-semicontinuous and positive in B and hence $r_1 > 0$ [see section 2.3].) According to Hartogs' theorem (see section 15.3), the series (21) converges absolutely and uniformly in the polydisk $S(0, R'_f) \times S(\tilde{z}^0, \tilde{r})$. Since an arbitrary subdomain G_f that is compact in G' can be covered by a finite number of such polydisks, the series (21) converges absolutely and uniformly in G'. This means that the series (21) converges absolutely and uniformly in G_f (see section 1.6).

It follows from the absolute and uniform convergence of the series (21) in the domain G that, for an arbitrary subdomain $B' \Subset B$, there exist numbers M and r_1 such that

$$|f_\alpha(\tilde{z})| r_1^\alpha \leqslant M, \quad \tilde{z} \in B', \ \alpha = 1, 2, \ldots .$$

From this we conclude that the set $\left\{\frac{1}{\alpha} \ln |f_\alpha(\tilde{z})|\right\}$ of functions that are plurisubharmonic in B is uniformly bounded from above on B'. But then, the function $-\ln R_f(\tilde{z})$ is plurisubharmonic in B (see section 10.3), which completes the proof.

Proof: The function $-\ln R_f(\tilde{z})$ belongs to the class F_B of Hartogs' functions (see section 10.15).

COROLLARY. *Every function that is holomorphic in a complete Hartogs domain G is holomorphic (and single-valued) in the domain G_f.*

5. *Logarithmically plurisubharmonic envelopes*

Consider the complete Hartogs domain

$$G = [z: |z_1| < R(\tilde{z}), \ \tilde{z} \in B] .$$

We shall refer to a complete Hartogs domain of the form

$$G^* = [z: |z_1| < e^{V(\tilde{z})}, \ \tilde{z} \in B],$$

where $V(\tilde{z})$ is the least plurisuperharmonic majorant of the function $\ln R(\tilde{z})$ in the domain B, as the logarithmically plurisuperharmonic envelope of the complete Hartogs domain G. Since the functions $R(\tilde{z})$ is lower-semicontinuous (see section 2.2) and positive in B, the function $\ln R(\tilde{z}) < \infty$ and is lower-semicontinuous in B. Therefore, this function has a least plurisuperharmonic majorant $V(\tilde{z})$ (see section 10.4). Therefore, the domain G^* is pseudoconvex (see section 12.13). Clearly, $G^* \supset G$.

The logarithmically plurisuperharmonic envelope of the complete Hartogs domain is a holomorphic extension of that domain.

Proof: Let $f \in H_G$. Then, the function $f(z)$ is holomorphic in the larger domain (see section 15.4)

$$G_f = [z: |z_1| < e^{V_f(\tilde{z})}, \tilde{z} \in B],$$

where

$$V_f(\tilde{z}) = -\overline{\lim_{\tilde{z}' \to \tilde{z}}} \ \overline{\lim_{\alpha \to +\infty}} \frac{1}{\alpha} \ln |f_\alpha(\tilde{z}')|$$

is a plurisuperharmonic function in the domain B that satisfies the inequality $V_f(\tilde{z}) \geqslant \ln R(\tilde{z})$.

From this and from the definition of least plurisuperharmonic majorant (see sections 10.4 and 9.7), we get the inequality

$$V_f(\tilde{z}) \geqslant V(\tilde{z}), \ \tilde{z} \in B. \tag{26}$$

Inequality (26) means that $G^* \subset G_f$. Thus, every function $f(z)$ that is holomorphic in G is holomorphic in G^*, which completes the proof.

6. *Hartogs-Laurent series*

Let us now consider a Hartogs domain that does not contain a single point of its plane of symmetry (see section 12.13)

$$G = [z: r(\tilde{z}) < |z_1| < R(\tilde{z}), \ \tilde{z} \in B]$$

(see Fig. 18).

Every function $f(z)$ that is holomorphic in a Hartogs domain G is holomorphic in the domain

$$G_f = [z: e^{v_f(\tilde{z})} < |z_1| < e^{V_f(\tilde{z})}, \ \tilde{z} \in B]$$

and, throughout this domain, can be expanded in an absolutely and uniformly convergent Hartogs-Laurent series

$$f(z) = \sum_{\alpha=-\infty}^{\infty} z_1^\alpha f_\alpha(\tilde{z}). \tag{27}$$

Here, the functions $f_\alpha(\tilde{z})$ are holomorphic in B and

$$f_\alpha(\tilde{z}) = \frac{1}{2\pi i} \int_l \frac{f(z_1', \tilde{z})}{z_1'^{\alpha+1}} dz_1', \quad \alpha = 0, \pm 1, \ldots, \tag{28}$$

where the integration is carried out over an arbitrary closed piecewise-smooth contour lying entirely in the annulus

$$e^{v_f(\tilde{z})} < |z_1| < e^{V_f(\tilde{z})}$$

and containing the point $z_1 = 0$. *Finally,*

$$V_f(z) = -\overline{\lim_{\tilde{z}' \to \tilde{z}}}\ \overline{\lim_{\alpha \to +\infty}} \frac{1}{\alpha} \ln |f_\alpha(\tilde{z}')|, \quad V_f(\tilde{z}) \geqslant \ln R(\tilde{z}); \tag{29}$$

$$v_f(\tilde{z}) = \overline{\lim_{\tilde{z}' \to \tilde{z}}}\ \overline{\lim_{\alpha \to -\infty}} \frac{1}{-\alpha} \ln |f_\alpha(\tilde{z}')|, \quad v_f(\tilde{z}) \leqslant \ln r(\tilde{z}). \tag{30}$$

Proof: For every $\tilde{z} \in B$, let us expand the function $f(z)$ in a series of the form (27). Here, the functions $f_\alpha(\tilde{z})$ are given by formula (28) with the contour l lying in the annulus $r(\tilde{z}) < |z_1| < R(\tilde{z})$. We have

$$f(z) = f_+(z) + f_-(z),$$

where

$$f_+(z) = \sum_{\alpha=0}^{\infty} z_1^\alpha f_\alpha(\tilde{z}), \quad f_-(z) = \sum_{\alpha=1}^{\infty} z_1^{-\alpha} f_{-\alpha}(\tilde{z}).$$

It follows from this and from the representation (27) that the function $f_+(z)$ is holomorphic in the complete Hartogs domain

$$G^+ = [z:\ |z_1| < R(\tilde{z}),\quad \tilde{z} \in B].$$

On the basis of section 15.4, this function is holomorphic in the larger Hartogs domain

$$G_f^+ = \left[z:\ |z_1| < e^{V_f(\tilde{z})},\quad \tilde{z} \in B\right]$$

where the function $V_f(\tilde{z})$ is given by formula (29).

Analogously, the function $f_-(z)$ is holomorphic in the Hartogs domain

$$G = [z:\ |z_1| > r(\tilde{z}),\quad \tilde{z} \in B].$$

From this it follows that the function

$$\varphi(z) = f_-\left(\frac{1}{z_1},\ \tilde{z}\right) = \sum_{\alpha=1}^{\infty} z_1^\alpha f_{-\alpha}(\tilde{z})$$

is holomorphic in the complete Hartogs domain

$$\tilde{G} = \left[z:\quad |z_1| < \frac{1}{r(\tilde{z})},\ \tilde{z} \in B\right].$$

Therefore (see section 15.4), the function $\varphi(z)$ is holomorphic in the larger domain

$$\tilde{G}_\infty = [z: \ |z_1| < e^{-v_f(\tilde{z})}, \ \tilde{z} \in B],$$

where the function $v_f(\tilde{z})$ is given by formula (30). From this it follows that the function $f_-(z) = \varphi\left(\frac{1}{z_1}, z\right)$ is holomorphic in the Hartogs domain

$$G_f^- = [z: \ e^{v_f(\tilde{z})} < |z_1|, \ \tilde{z} \in B].$$

It follows from what we have said that the function $f(z)$ is holomorphic in the domain $G_f = G_f^+ \cap G_f^-$. The remaining assertions of the theorem follow easily from the results of section 15.2, which completes the proof.

7. *Logarithmically plurisuperharmonic and plurisubharmonic envelopes*

We shall now construct a holomorphic extension of the Hartogs domain G of the form

$$G = [z: \ r(\tilde{z}) < |z_1| < R(\tilde{z}), \ \tilde{z} \in B].$$

The logarithmically plurisuperharmonic (plurisubharmonic) envelope G^* of the domain G is a holomorphic extension of G.

The envelope G^* is defined as in section 15.5 :

$$G^* = [z: \ e^{v(\tilde{z})} < |z_1| < e^{V(\tilde{z})}, \ \tilde{z} \in B],$$

where $V(\tilde{z})$ (resp. $v(\tilde{z})$) is the smallest (resp. largest) plurisuperharmonic majorant (resp. plurisubharmonic minorant) of the function $\ln R(\tilde{z})$ (resp. $\ln r(\tilde{z})$) in the domain B. Clearly, $G \subset G^*$.

Let $f \in H_G$. From the preceding theorem, the function $f(z)$ is holomorphic in the domain G_f, where the functions $v_f(\tilde{z})$ and $V_f(\tilde{z})$ are plurisubharmonic and plurisuperharmonic, respectively, in B and satisfy the inequalities

$$v_f(\tilde{z}) \leqslant \ln r(\tilde{z}), \quad V_f(\tilde{z}) \geqslant \ln R(\tilde{z}). \tag{31}$$

The inequalities

$$v_f(\tilde{z}) \leqslant v(z), \quad V_f(\tilde{z}) \geqslant V(\tilde{z}), \quad \tilde{z} \in B$$

follow easily from inequalities (31) (see section 10.4). These last inequalities mean that $G^* \subset G_f$. From this we conclude that $H_G = H_{G^*}$, which completes the proof.

8. *Proof of Hartogs' fundamental theorem*

As an application of the results that we have obtained concerning Hartogs' series, let us prove the fundamental theorem of Hartogs that was stated without proof in section 4.2:

If a function $f(z)$ is such that, at an arbitrary point z^0 in a domain G, the functions $f(z_1^0, \ldots, z_{j-1}^0, z_j, z_{j+1}^0, \ldots, z_n^0)$, for $j = 1, 2, \ldots, n$, are holomorphic respectively with respect to z_j at the point z_j^0, then $f(z)$ is holomorphic in the domain G.

As a preliminary, let us prove the

LEMMA (**Osgood**). *Suppose that A is a closed and O an open bounded set and that a function $\lambda(x, y) \geq 0$, defined in $A \times O$, is continuous with respect to x in A for arbitrary $y \in O$ and with respect to y in O for arbitrary $x \in A$. Then, there exists an open set $O' \subset O$ such that $\lambda(x, y)$ is bounded on $A \times O'$.*

Proof: The function

$$\mu(y) = \sup_{x \in A} \lambda(x, y) \tag{32}$$

assumes finite values for every $y \in O$. Furthermore, the sets $A_M = [y:\ \mu(y) \leqslant M,\ y \in O]$ are closed in O. To see this, let $y^{(k)} \to y^0 \in O$ where $y^{(k)} \in O$, and suppose that $\mu(y^{(k)}) \leqslant M$. Then, it follows from (32) that $\lambda(x, y^{(k)}) \leqslant M$ for all $x \in A$ and k. Because of the continuity with respect to y, we conclude that $\lambda(x, y^0) \leqslant M$ for $x \in A$; that is, $\mu(y^0) \leqslant M$. Now suppose that the conclusion of the lemma is untrue. Then, each of the sets A_M, for $M = 1, 2, \ldots$ is nowhere dense in O. But this contradicts the relation $O = \bigcup_{M \geqslant 1} A_M$ and completes the proof.

Proof of Hartogs' theorem: It will be sufficient to show that $f(z)$ is holomorphic at an arbitrary point $z^0 \in G$. Let us suppose that $z^0 = 0$. Suppose that $S(0, r_1) \Subset G$. Let us set

$$A = \overline{S}(0, r_1), \quad O = S\left(0, \frac{r_2}{3}\right) \times \ldots \times S\left(0, \frac{r_n}{3}\right).$$

From Osgood's lemma, there exists a closed polydisk (see Fig. 29)

$$\overline{S}(0, r_1) \times \overline{S}(z_2', \rho_2) \times \ldots \times \overline{S}(z_n', \rho_n),$$
$$S(z_j', \rho_j) \subset S\left(0, \frac{r_j}{3}\right), \quad j = 2, \ldots, n,$$

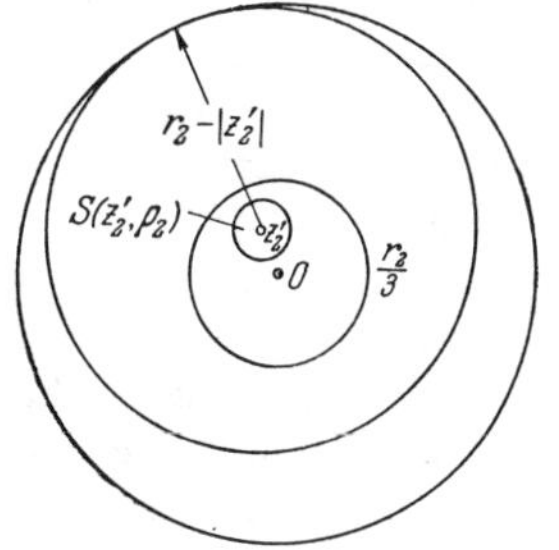

Fig. 29

in which the function $f(z)$ is bounded. Since $f(z)$ is holomorphic with respect to each variable individually, it follows on the basis of section 4.2 that it is holomorphic in the polydisk

$$S(0, r_1) \times S(z_2', \rho_2) \times \ldots \\ \ldots \times S(z_n', \rho_n).$$

Furthermore, since $|z_2'| < r_2/3$, the circle $S(z_2', r_2 - |z_2'|)$ in the z_2-plane is contained in the circle $S(0, r_2)$ and contains the circle $S(0, r_2/3)$ (see Fig. 29). Therefore, for every $(z_1^0, z_3^0, \ldots, z_n^0)$ in $S(0, r_1) \times S(z_3', \rho_3) \times \ldots \times S(z_n', \rho_n)$, the function $f(z_1^0, z_2, z_3^0, \ldots, z_n^0)$ is holomorphic with respect to z_2 in the circle $S(z_2', r_2 - |z_2'|)$. Therefore, from Hartogs' theorem of section 15.3 (with z_1 replaced by z_2), we conclude that the function $f(z)$ is holomorphic in the polydisk

$$S(0, r_1) \times S(z_2', r_2 - |z_2'|) \times S(z_3', \rho_3) \times \ldots \times (z_n', \rho_n).$$

Since $S\left(0, \frac{r_2}{3}\right) \subset S(z_2', r_2 - |z_2'|)$, the function $f(z)$ is holomorphic in the polydisk

$$S(0, r_1) \times S\left(0, \frac{r_2}{3}\right) \times S(z_3', \rho_3) \times \ldots \times S(z_n', \rho_n).$$

If we now repeat our reasoning for the variable z_3, we see that $f(z)$ is holomorphic in the polydisk

$$S(0, r_1) \times S\left(0, \frac{r_2}{3}\right) \times S\left(0, \frac{r_3}{3}\right) \times S(z_4', \rho_4) \times \ldots \times S(z_n', \rho_n)$$

etc. Finally, we conclude that $f(z)$ is holomorphic in the polydisk $S(0, r_1) \times S\left(0, \frac{\tilde{r}}{3}\right)$, which completes the proof.

16. HOLOMORPHIC CONVEXITY

1. *Simultaneous continuation of functions*

We shall say that a set K_G of functions that are holomorphic in a domain G constitutes a *class* if the relation $f \in K_G$ implies that all the functions $D^\alpha f$ and af^p, where p is an arbitrary integer $\geqslant 1$ and a is an arbitrary complex number, also belong to K_G.

For example, the set H_G of all functions that are holomorphic in G constitutes a class. The set P of all polynomials constitutes a class. Obviously, H_G contains every class K_G.

THEOREM (Cartan-Thullen [7]). *Suppose that the maximum principle holds for a point $z^0 \in G$ and a set $A \Subset G$ with respect to the moduli of functions of a class K_G, that is, that*

$$|f(z^0)| \leqslant \sup_{z \in A} |f(z)|, \quad f \in K_G. \tag{33}$$

Then, every function $f \in K_G$ *can be holomorphically continued into the polycircle* $S(z^0, rI)$, *where* $r = \delta_G(A)$, *and for all* $\rho < r$, *the maximum principle*

$$\sup_{z \in S(z^0, \rho I)} |f(z)| \leqslant \sup_{z \in A_\rho} |f(z)| \tag{34}$$

where $A_\rho = \bigcup_{z' \in A} S(z', \rho I)$ (see Fig. 30), *holds.*

Remark: If the open set $S(z^0, rI) \cap G$ is a domain, then, on the basis of the holomorphic continuation theorem (see section 6.1), the continued function $f(z)$ is single-valued in $S(z^0, rI) \cup G$.

Proof: Since $A_\rho \Subset G$ and $\rho < r$, if we define

$$\sup_{z \in A_\rho} |f(z)| = M_{f, \rho},$$

we obtain

$$|f(z)| \leqslant M_{f, \rho}, \quad z \in S(z', \rho I), \quad z' \in A.$$

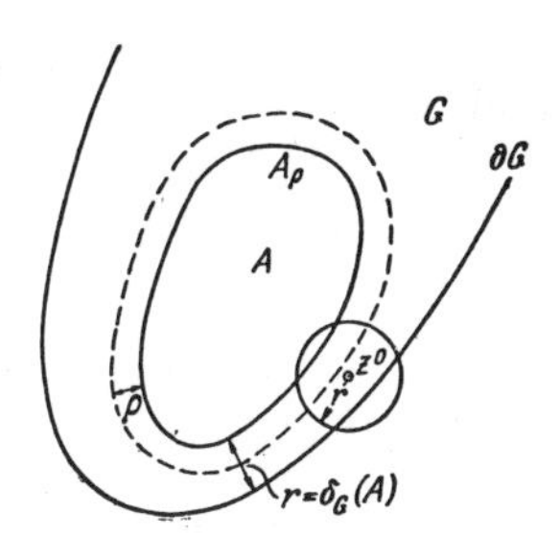

Fig. 30

From this and from Cauchy's inequality (see section 4.4), we deduce

$$\left|\frac{1}{\alpha!} D^\alpha f(z')\right| \leqslant M_{f, \rho} \rho^{-|\alpha|}, \quad z' \in A. \tag{35}$$

Since $D^\alpha f \in K_G$, by using inequality (33), we obtain, on the basis of the above inequality,

$$\left|\frac{1}{\alpha!} D^\alpha f(z^0)\right| \leqslant M_{f, \rho} \rho^{-|\alpha|}. \tag{36}$$

It follows from this inequality that the series

$$f(z) = \sum_{|\alpha| \geqslant 0} \frac{1}{\alpha!} D^\alpha f(z^0)(z - z^0)^\alpha$$

converges uniformly in the polydisk $S(z^0, \rho I)$ for arbitrary $\rho < r$. This means that the function $f(z)$ is holomorphic in $S(z^0, rI)$.

To prove inequality (34), it will be sufficient to show that

$$|f(z)| \leqslant M_{f, \rho}, \quad z \in S(z^0, \rho_1 I) \tag{37}$$

for arbitrary $\rho_1 < \rho$. Suppose that inequality (37) does not hold for some $\rho_1 < \rho$, that is, that

$$\sup_{z \in S(z^0, \rho_1 I)} \frac{|f(z)|}{M_{f, \rho}} = \alpha > 1.$$

It would then follow that the function

$$\varphi_P(z) = \left[\frac{f(z)}{M_{f,\rho}}\right]^P,$$

where P is a positive integer, satisfies the equation

$$\sup_{z \in S(z^0, \rho_1 l)} |\varphi_P(z)| = \alpha^P. \tag{38}$$

On the other hand, $\varphi_P \in K_G$, $M_{\varphi_P, \rho} = 1$, and on the basis of inequality (36), which we have proven, we have

$$\left|\frac{1}{\alpha!} D^\alpha \varphi_P(z^0)\right| \leqslant \rho^{-|\alpha|}.$$

Therefore, the function $\varphi_P(z)$ satisfies the following inequality in the polydisk $S(z^0, \rho_1 l)$:

$$|\varphi_P(z)| \leqslant \sum_{|\alpha| \geqslant 0} \frac{1}{\alpha!} |D^\alpha \varphi_P(z^0)| \, |z - z^0|^\alpha \leqslant$$

$$\leqslant \sum_{|\alpha| \geqslant 0} \frac{1}{\alpha!} \left(\frac{\rho_1}{\rho}\right)^{|\alpha|} = \left(1 - \frac{\rho_1}{\rho}\right)^{-n}.$$

This inequality, together with Eq. (38), yields the inequality

$$\alpha^P \leqslant \left(1 - \frac{\rho_1}{\rho}\right)^{-n}$$

for arbitrary integral $P > 0$. But this inequality is impossible for sufficiently large P since, by hypothesis, $\alpha > 1$. This contradiction proves inequality (36) for arbitrary $\rho_1 < \rho$ and this completes the proof.

This theorem has the

COROLLARY. *Suppose that the maximum principle holds for sets* $S \Subset G$ *and* $T \Subset G$ *with respect to the moduli of functions of a class* K_G. *Then, every function in* K_G *can be holomorphically continued in* S_r *(possibly nonsingle-valuedly* $S_r \cup G$*), where* $r = \delta_G(T)$. *Also, for arbitrary* $\rho < r$, *the maximum principle is valid for the sets* S_ρ *and* T_ρ *with respect to the moduli of functions of the class* K_G.

2. *K-convex domains*

In connection with the results obtained in the preceding section, we may introduce the concept of a K-convex domain.

A domain G is said to be K-convex (convex with respect to the class of functions K_G) if, for an arbitrary set $A \Subset G$, the set

$$F_A \equiv \bigcap_{f \in K_G} \left[z: \quad |f(z)| \leqslant \sup_{z \in A} |f(z)|, \ z \in G \right]$$

is compact in G.

In other words, a domain G is said to be K-convex if, for every set $A \Subset G$, there exists a set F_A such that $A \subset F_A \Subset G$ (see Fig. 31) and for every point $z^0 \in G \setminus F_A$, there exists a function $f \in K_G$ such that

$$\sup_{z \in A} |f(z)| < |f(z^0)|. \tag{39}$$

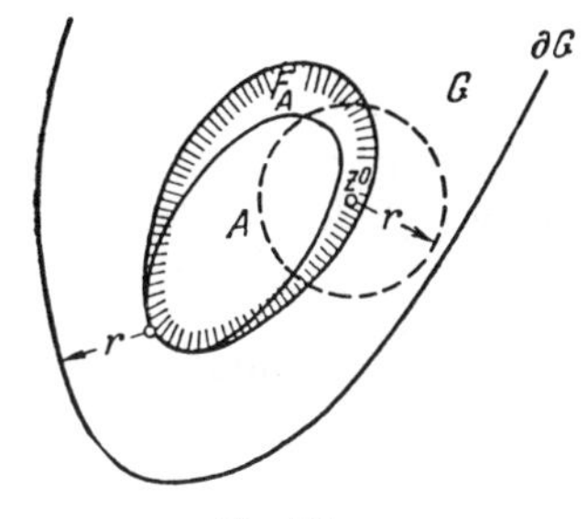

Fig. 31

If a domain G is H-convex ($K_G = H_G$), it is said to be holomorphically convex. We shall say that P-convex domains ($K_G = P$) are polynomially convex. Obviously, every K-convex domain is holomorphically convex.

We note that if G is a K-convex domain, the set F_A is compact and

$$\delta_G(F_A) = \delta_G(A). \tag{40}$$

Proof: Since $F_A \Subset G$, F_A is bounded and (being the intersection of closed sets) closed. Therefore, F_A is compact. Let us prove Eq. (40). Since $A \subset F_A$, we have $\delta_G(F_A) \leqslant \delta_G(A) = r$. It remains to prove the opposite inequality $\delta_G(F_A) \geqslant r$ (see Fig. 31).

Consider a point $z^0 \in F_A$. Since inequalities (33) hold for all functions $f \in K_G$, we conclude from the theorem on simultaneous continuation of functions of class K_G that $f(z)$ is holomorphic in the polydisk $S(z^0, rI)$ and satisfies inequality (34) for all $\rho < r$. But the domain G is K-convex. Therefore, $S(z^0, \rho I) \Subset G$ for all $\rho < r$. From this we see that $\delta_G(z^0) \geqslant r$ for arbitrary $z^0 \in F_A$. This means that $\delta_G(F_A) \geqslant r$ which completes the proof.

It follows from this that *every component of the interior of the intersection K-convex domains is a K-convex domain.*

To see this, let G be a component of int $\bigcap_\alpha G_\alpha$ and suppose that $A \Subset G$. We need to show that $F_A \Subset G$. Since $A \Subset G_\alpha$ and the G_α are K-convex domains, we have

$$F_{A,\alpha} = \bigcap_{f \in K_{G_\alpha}} \left[z: \quad |f(z)| \leqslant \sup_{z \in A} |f(z)|, \ z \in G_\alpha \right] \Subset G_\alpha.$$

We define the class K_G as the smallest class of holomorphic functions in G that contains $\bigcup_\alpha K_{G_\alpha}$. Since $G \subset G_\alpha$ and $K_G \supset K_{G_\alpha}$, the set F_A (for the domain G) is contained in all the sets $F_{A,\alpha}$ and hence is bounded. Therefore, it remains to show that $\delta_G(F_A) > 0$. But this follows from (3) [see section 1.3] and (40):

$$\delta_G(F_A) = \inf_\alpha \delta_{G_\alpha}(F_A) \geqslant \inf_\alpha \delta_{G_\alpha}(F_{A,\alpha}) = \inf_\alpha \delta_{G_\alpha}(A) = \delta_G(A) > 0.$$

Suppose that K and K_1 are sets of functions that are holomorphic in a domain G. We shall say that K_1 is dense in K if, for arbitrary $f \in K$, $G' \Subset G$, and $\varepsilon > 0$, there exists a function $f_1 \in K_1$ such that

$$|f_1(z) - f(z)| < \varepsilon, \quad z \in G'.$$

For K_1 to be dense in K, it is necessary and sufficient that, for an arbitrary function $f \in K$, there exists a sequence of functions $f_\alpha \in K_1$, for $a = 1, 2, \ldots$, that converges uniformly to f in the domain G (in the sense of section 1.6).

The sufficiency is obvious. Let us prove the necessity. Suppose that $G_\alpha \Subset G_{\alpha+1}$ that $\bigcup_\alpha G_\alpha = G$ and that $\varepsilon_\alpha \to 0$ as $\alpha \to \infty$. The functions $f_\alpha \in K_1$ such that

$$|f(z) - f_\alpha(z)| < \varepsilon_\alpha, \quad z \in G_\alpha, \quad \alpha = 1, 2, \ldots,$$

form a sequence of the type required.

LEMMA. *If a domain G is K-convex and if some class K'_G is dense in K_G, the domain G is K'-convex.*

Proof: Suppose that $A \Subset G$. Then, there exists a set F_A such that $A \subset F_A \Subset G$ and, for every point $z^0 \in G \setminus F_A$, there exists a function $f \in K_G$ satisfying inequality (39). Therefore, there exists a $\delta > 0$, such that

$$\delta + \sup_{z \in A} |f(z)| < |f(z^0)|. \tag{41}$$

Since K'_G is dense in K_G, let us take a function $f_1 \in K'_G$ such that

$$|f(z) - f_1(z)| < \frac{\delta}{2}, \quad z \in A \cup [z^0],$$

that is, such that

$$|f_1(z)| - \frac{\delta}{2} < |f(z)| < |f_1(z)| + \frac{\delta}{2}, \quad z \in A \cup [z^0].$$

Inequality (39) follows from the above inequality and inequality (41):

$$|f_1(z^0)| > |f(z^0)| - \frac{\delta}{2} > \frac{\delta}{2} + \sup_{z \in A} |f(z)| > \sup_{z \in A} |f_1(z)|,$$

which means that the domain G is K'-convex, which completes the proof.

If G is a K-convex domain, then, for arbitrary $\rho > 0$, every component G of the open set

$$G(\rho) = [z: \ \delta_G(z) > \rho, \ z \in G]$$

is a K-convex domain.

Proof: Consider an arbitrary $A \Subset G'$. Then, $\delta_G(A) > \rho$. Furthermore, since G is a K-convex domain and $K_{G'} \supset K_G$, it follows that

$$F'_A = \bigcap_{f \in K_{G'}} \left[z: \quad |f(z)| \leqslant \sup_{z \in A} |f(z)|, \quad z \in G'\right] \subset F_A \Subset G.$$

Therefore, taking Eq. (40) into account, we have

$$\delta_G(F'_A) \geqslant \delta_G(F_A) = \delta_G(A) > \rho,$$

which, in view of the boundedness of F'_A, implies that $F'_A \Subset G'$, which completes the proof.

3. *A condition for K-convergence, I*

In what follows, we shall assume that the class K_G contains the functions $z_1, z_2, \ldots, z_n$ if G is an unbounded domain. Let us prove the

LEMMA. *If a domain G is not K-convex, there exists a set $A \Subset G$ and a sequence of points $z^{(k)} \in G$, for $k = 1, 2, \ldots$, such that $z^{(k)} \to z^0 \in \partial G$, and for an arbitrary function $f \in K_G$,*

$$|f(z^{(k)})| \leqslant \sup_{z \in A} |f(z)|, \quad k = 1, 2, \ldots. \tag{42}$$

Proof: If the domain G is not K-convex, it follows from the definition (see section 16.2) that there exist a set $A \Subset G$ and a sequence of points $z^{(k)} \in G$, for $k = 1, 2, \ldots$, possessing no points of accumulation in G such that, for an arbitrary function $f \in K_G$, inequality (42) holds for all points $z^{(k)}$. If we apply inequality (42) to the functions z_j, for $j = 1, 2, \ldots, n$, we see that the sequence $\{z^{(k)}\}$ is bounded. Therefore, on the basis of the Bolzano-Weierstrass theorem, this sequence has a subsequence $\{z^{(k)}\}$ converging to the point z^0. Obviously, $z^0 \in \partial G$, which completes the proof.

For a domain to be K-convex, it is sufficient that, for an arbitrary point $z^0 \in \partial G$, there exists a sequence of points $z^{(k)}$, for $k = 1, 2, \ldots$, such that $z^{(k)} \to z^0$, where $z^{(k)} \notin G$ and, for every point $z^{(k)}$, there exists a function in the class K_G that is not holomorphic at $z^{(k)}$.

Proof: Suppose that the domain G is not K-convex. From the preceding lemma, there exist a set $A \Subset G$ and a sequence $\{z^{(l)}\}$, where $z^{(l)} \in G$ and $z^{(l)} \to z^0 \in \partial G$, such that inequalities (42) hold for arbitrary $f \in K_G$. From this, we conclude on the basis of the theorem in section 15.1 that $f(z)$ can be holomorphically continued into the polydisks $S(z^{(l)}, rI)$, for $l = 1, 2, \ldots$, where $r = \delta_G(A) > 0$. Since $z^{(l)} \to z^0$, it then follows that there exists a number N such that $z^0 \in S(z^{(N)}, rI)$. This means that every function $f \in K_G$ is holomorphic in some fixed neighborhood of the point z^0. This, however, is impossible because of the assumption. Therefore, the domain G is K-convex, which completes the proof.

COROLLARY. *Every domain of holomorphy G of a function $f \in K_G$ is a K-convex domain.*

4. A condition for K-convex, II

For a domain G to be K-convex, it is necessary and sufficient that, for arbitrary sets S and T such that $S \cup T \Subset G$, and the maximum principle holds with respect to the moduli of functions of class K_G, the inequality

$$\delta_G(T) \leqslant \delta_G(S). \tag{43}$$

holds.

Proof of sufficiency: Suppose that inequality (43) holds. Let us show that G is a K-convex domain. Assume, to the contrary, that G is not a K-convex domain. Then, according to the lemma in section 16.3, there exist a set $A \Subset G$ and a sequence of points $z^{(k)} \in G$, for $k = 1, 2, \ldots$, that converges to some point $z^0 \in \partial G$ such that, for an arbitrary function $f \in K_G$, the inequalities (42) hold; that is, the maximum principle holds for the sets $S = z^{(k)}$ and $T = A$ with respect to the moduli of functions of class K_G. If we apply inequality (43) we obtain

$$\delta_G(z^{(k)}) \geqslant \delta_G(A) > 0,$$

which contradicts the limit relation $z^{(k)} \to z^0 \in \partial G$. This contradiction proves that the domain G is K-convex.

Proof of necessity: Suppose that G is a K-convex domain. Suppose also that the maximum principle with respect to the moduli functions of class K_G holds for sets S and T, where $S \Subset G$ and $T \Subset G$; that is, suppose that

$$|f(z)| \leqslant \max_{z \in T} |f(z)|, \quad z \in S, \quad f \in K_G.$$

It follows from this that $S \subset F_T$ (see section 16.2). Keeping Eq. (40) in mind, we obtain from the last relation

$$\delta_G(S) \geqslant \delta_G(F_T) = \delta_G(T), \text{ q. e. d.}$$

5. Approximation of K-convex domains

If a domain G is K-convex, it is the union of an increasing sequence of bounded analytic polyhedra (see section 12.15)

$$|f(z)| \leqslant \max_{z \in T} |f(z)|, \quad z \in S, \quad f \in K_G$$

such that $\mathfrak{P}_k \Subset \mathfrak{P}_{k+1} \Subset G$ and the functions $\chi_\alpha^{(k)} \in K_G$ (see Oka [72]).

To prove this, it will be sufficient to exhibit an analytic polyhedron $\mathfrak{P}$ for an arbitrary subdomain $G' \Subset G$ such that $G' \subset \mathfrak{P} \Subset G$ and the functions $\chi_\alpha(z)$ defining $\mathfrak{P}$, belong to the class K_G.

Since G is a K-convex domain and $G' \Subset G$ there exists a closed set F such that $G' \subset F \Subset G$ and, for an arbitrary point $z^0 \in G \setminus F$, there exists a function $f_{z^0} \in K_G$ satisfying the inequalities

$$|f_{z^0}(z)| < 1, \quad z \in G', \quad |f_{z^0}(z^0)| > 1 \tag{44}$$

(see section 16.2). Because of the continuity, for every point $z^0 \in G \setminus F$, there exists a neighborhood $U(z^0, r)$, contained in the open set $G \setminus F$ such that $|f_{z^0}(z)| > 1$ for $z \in U(z^0, r)$.

Since $F \Subset G$, $\Delta_G(F) = 2r > 0$. Therefore, the open set $F^r = [z: |z - z'| < r, z' \in F]$ (see Fig. 32) is such that $G' \Subset F^r \Subset G$. Then, its boundary ∂F^r is compact: $\partial F^r \Subset G$. From the Heine-Borel theorem, the compact set ∂F^r can be covered by a finite number of the neighborhoods $U(z^{(\alpha)}, r_\alpha)$, for $\alpha = 1, 2, \ldots, N$, that we constructed above. The corresponding functions $\chi_\alpha \equiv f_{z^{(\alpha)}} \in K_G$ satisfy inequalities (44):

$$|\chi_\alpha(z)| < 1, \quad z \in G';$$
$$|\chi_\alpha(z)| > 1, \quad z \in U(z^{(\alpha)}, r_\alpha).$$

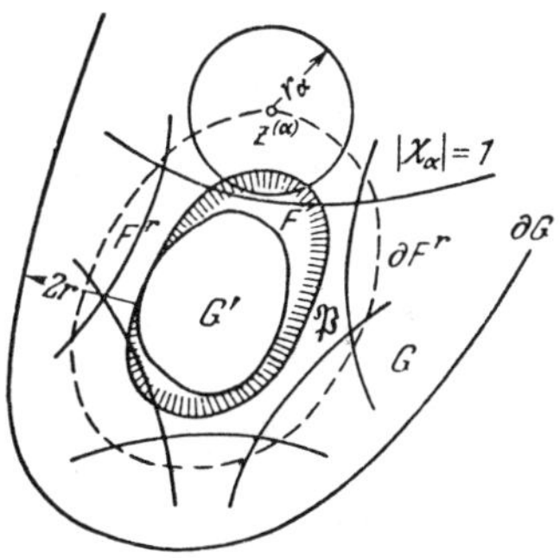

Fig. 32

Thus, the analytic polyhedron

$$\mathfrak{P} = [z: \ |\chi_\alpha(z)| < 1, \quad \alpha = 1, 2, \ldots, N; \ z \in F^r]$$

possesses all the required properties, which completes the proof.

Remark: The above theorem is analogous to the theorem on the approximation of a convex domain by an increasing sequence of convex bounded polyhedra.

Corollary. *Suppose that the conditions of the preceding theorem are satisfied. Then, if $K_G = H_D$ (or if $K_G = P$), where $G \subset D$, we may take Weil domains as the* $\mathfrak{P}_k$ (see section 12.15).

Proof: From what we have proven, for an arbitrary subdomain $G' \Subset G$ there exists a domain

$$\sigma = [z: \ |\chi_\alpha(z)| < 1, \ \alpha = 1, 2, \ldots, N; \ z \in U(\bar{\sigma})], \qquad N \geqslant n,$$

such that $G' \Subset G'' \subset \sigma \Subset G$, where the functions $\chi_\alpha \in H_D$ (or $\chi_\alpha \in P$). By varying the functions $\chi_\alpha(z)$, we can arrange for the intersection of any k hypersurfaces $|\chi_{\alpha_i}(z)| = 1$, where $(1 \leqslant k \leqslant n)$ and $i = 1, 2, \ldots$, in $U(\bar{\sigma})$ to be of dimension not exceeding $2n - k$ and for $G' \subset \sigma \Subset G$, which completes the proof.

Every connected analytic polyhedron

$$\sigma=[z:\ |\chi_\alpha(z)|<1,\ \alpha=1,2,\dots,N;\ z\in U(\bar\sigma)],\ \chi_\alpha\in H_{U(\bar\sigma)}$$

is a K-convex domain, where $F \Subset G$ *is the smallest class containing the functions* z_j *and* $\Delta_G(F) = 2r > 0$, *where* $j = 1, 2, \dots, n$ *and* $\alpha = 1, 2, \dots, N$.

Proof: If $\sigma' \Subset \sigma$,

$$\max_{1\leqslant\alpha\leqslant N}\ \sup_{z\in\sigma'}\ |\chi_\alpha(z)|=\rho<1,\qquad \max_{1\leqslant j\leqslant n}\ \sup_{z\in\sigma'}|z_j|=L.$$

Since $z_j\in K_\sigma$, we then have (see section 16.2)

$$F_{\sigma'}\subset\bigcap_{j,\ \alpha}[z:\ |z_j|\leqslant L,\ |\chi_\alpha(z)|\leqslant\rho,$$

$$j=1,2,\dots,n,\ \alpha=1,2,\dots,N,\ z\in\sigma]\Subset\sigma,\ \text{q. e. d.}$$

which completes the proof.

6. *Holomorphically convex domains*

For a given domain to be holomorphically convex, it is necessary and sufficient that it be a domain of holomorphy.

The sufficiency was proven in section 16.3.

To prove the necessity of the condition, it will be sufficient to prove the following

LEMMA. *If G is a holomorphically convex domain, there exists a function* $f(z)$ *that is holomorphic in G and unbounded at every point* $z^0\in\partial G$ (see Cartan and Thullen [7] and Bochner and Martin [3]).

Let us subtract from the set ∂G a countable everywhere-dense set E. To prove the lemma, it will be sufficient to construct a function $f(z)$, that is holomorphic in G and unbounded at every point of the set E. Let $z^{(k)}$, for $k=1,2,\dots$, be a sequence possessing the property that every point of the set E occurs infinitely many times in that sequence. Suppose also that A_α, for $\alpha=1,2,\dots$, is an increasing sequence of sets such that

$$A_\alpha\Subset A_{\alpha+1}\Subset G\quad\text{and}\quad\bigcup_\alpha A_\alpha=G.$$

Since G is a holomorphically convex domain for the set A_1, there exist a set F_1, a point $a^{(1)}\in G\setminus F_1$ and a function $f_1\in H_G$ such that $A_1\subset F_1\Subset G$, $|a^{(1)}-z^{(1)}|\leqslant 1$, and

$$\sup_{z\in A_1}|f_1(z)|=1,\quad |f_1(a^{(1)})|>1.$$

Furthermore, since $\bigcup A_\alpha = G$, there exists a set A_{α_2}, $\alpha_2 > \alpha_1 = 1$, such that $F_1 \cup \{a^{(1)}\} \subset A_{\alpha_2}$. For the set A_{α_2}, there exist a set F_2, a point $a^{(2)} \in G \setminus F_2$ and a function $f_2 \in H_G$ such that $A_{\alpha_2} \subset F_2 \Subset G$, $|a^{(2)} - z^{(2)}| \leqslant \leqslant 1/2$, and

$$\sup_{z \in A_{\alpha_2}} |f_2(z)| = 1, \quad |f_2(a^{(2)})| > 1$$

and so on. Consider a number $\alpha_k > \alpha_{k-1}$ such that $F_{k-1} \cup \{a^{(k-1)}\} \subset A_{\alpha_k}$. For the set A_{α_k}, there exist a set F_k, a point $a^{(k)} \in G \setminus F_k$ and a function $f_k \in H_G$ such that $A_{\alpha_k} \subset F_k \Subset G$, $|a^{(k)} - z^{(k)}| \leqslant \frac{1}{k}$, and

$$\sup_{z \in A_{\alpha_k}} |f_k(z)| = 1, \quad |f_k(a^{(k)})| > 1 \tag{45}$$

and so on. Since, by construction, $|f_k(a^{(k)})| > 1$, there exists a sequence of integers l_k, for $k = 1, 2, \ldots$, where $(l_1 = 1)$, that satisfies the inequalities

$$\frac{|f_k(a^{(k)})|^{l_k}}{k^2} - \sum_{1 \leqslant i \leqslant k-1} \frac{|f_i(a^{(k)})|^{l_i}}{i^2} \geqslant k, \quad k \geqslant 2. \tag{46}$$

Let us define a function $f(z)$ in the form of the series

$$f(z) = \sum_{1 \leqslant i \leqslant \infty} \frac{[f_i(z)]^{l_i}}{i^2}. \tag{47}$$

Since, on the basis of (45), $|f_i(z)| \leqslant 1$ for $z \in A_{\alpha_i}$ and $A_{\alpha_i} \Subset G$, where $\bigcup_i A_{\alpha_i} = G$, it follows that the series (47) converges absolutely and uniformly in the domain G, thus defining the function $f(z)$, which is holomorphic in G. Also, since $a^{(k)} \in A_{\alpha_{k+1}} \subset A_{\alpha_i}$, for $i > k$, it follows on the basis of (45) that $|f_i(a^{(k)})| \leqslant 1$ for $i > k$. Therefore, we conclude from (46) and (47) that

$$\begin{aligned} |f(a^{(k)})| \geqslant \frac{|f_k(a^{(k)})|^{l_k}}{k^2} - \sum_{1 \leqslant i \leqslant k-1} \frac{|f_i(a^{(k)})|^{l_i}}{i^2} - \\ - \sum_{k+1 \leqslant i} \frac{1}{i^2} \geqslant k - \frac{\pi^2}{6} + 1. \end{aligned} \tag{48}$$

Let us choose an arbitrary point $z^0 \in E$. It follows from the construction that the point z^0 occurs infinitely many times in the sequence $z^{(k)}$, for $k = 1, 2, \ldots$. Suppose that $z^{(k_i)} = z^0$ for $1 \leqslant k_1 < k_2 < \ldots$. Then, we conclude from the inequality $|a^{(k_i)} - z^0| \leqslant 1/k_i \leqslant 1/i$ that the sequence $a^{(k_i)}$, for $i = 1, 2, \ldots$ of points of the domain G converges to the point $z^0 \in E \subset \partial G$. On the other hand, it follows from (48) that $|f(a^{(k_i)})| \to +\infty$ as $i \to \infty$. This means that the function $f(z)$ is infinite at the point $z^0 \in E$ (in the sense of the definition in section 1.6), which completes the proof.

7. *Characterizations of domains of holomorphy*

It follows from the results of the preceding section that every K-convex domain is a domain of holomorphy. In particular, every analytic polyhedron is an open set of holomorphy.

Using the first condition for K-convexity (see section 16.3), we obtain the following theorem, which is useful for applications.

For a domain G to be a domain of holomorphy, it is necessary that there exist a function $f(z)$ that is holomorphic in G and infinite at every point $z^0 \in \partial G$; it is sufficient that, for an arbitrary point $z^0 \in \partial G$, there exist a sequence of points $\{z^{(k)}\}$ such that $z^{(k)} \to z^0$, where $z^{(k)} \bar{\in} G$, and that, for every $z^{(k)}$, there exist a function that is holomorphic in G but not holomorphic at $z^{(k)}$. (In particular, it is sufficient that, for an arbitrary point $z^0 \in \partial G$, there exist a barrier function $f_{z^0} \in H_G$, that is not holomorphic at the point z^0.)

Let us look at an example illustrating an application of this theorem.

Every convex domain is a domain of holomorphy.

To see this, we may take the functions $[(a-ib)(z^0-z)]^{-1}$ where $a(x^0-x)+b(y^0-y)=0$ is the supporting plane for the domain, as the barrier function $f_{z^0}(z)$, $z^0 \in \partial G$.

By using the second condition for K-convexity (see section 16.4), we obtain the following result:

For a domain G to be a domain of holomorphy, it is necessary and sufficient that inequality (43) hold for arbitrary sets S and T, where $S \cup T \Subset G$, for which the maximum principle holds with respect to the moduli of functions that are holomorphic in G.

Finally, we note that the interior of the intersection of domains of holomorphy is an open set of holomorphy (see section 16.2).

8. *Invariance under biholomorphic mappings*

A biholomorphic mapping $\zeta = \zeta(z)$ maps a domain of holomorphy G onto a domain of holomorphy G_1.

From the theorem in section 16.6, it will be sufficient to show that the domain G_1 is holomorphically convex. Let $A_1 \Subset G_1$ and denote by A the preimage of A_1 under our mapping. Then, $A \Subset G$ (cf. section 12.11). But the domain G is holomorphically convex. Therefore, there exists a set F_A such that $A \subset F_A \Subset G$ and, for an arbitrary point $z^0 \in G \setminus F_A$, there exists a function $f \in H_G$ satisfying the inequality

$$\sup_{z \in A} |f(z)| < |f(z^0)| \tag{49}$$

(see section 16.2). Let us denote by F_{A_1} the image of the set F_A. Consider a point $\zeta^0 \in G_1 \setminus F_{A_1}$ and the point $z^0 = z(\zeta^0)$, where $z = z(\zeta)$ is the inverse mapping. Clearly, $z^0 \in G \setminus F_A$. For the point z^0, there exists a function $f \in H_G$ that satisfies inequality (49). Therefore, the function $f[z(\zeta)]$, which is holomorphic in G_1, satisfies the inequality

$$\sup_{z \in A} |f(z)| = \sup_{\zeta \in A_1} |f[z(\zeta)]| < |f[z(\zeta^0)]| = |f(z^0)|.$$

This means that the domain G_1 is holomorphically convex, which completes the proof.

By using Riemann's theorem, we obtain from this the

COROLLARY. *Every domain in* C^1 is a *domain of holomorphy.*

9. *Increasing sequences of domains of holomorphy*

LEMMA. *If an increasing sequence of domains* $G_\alpha \subset G_{\alpha+1}$, *for* $a = 1, 2, \ldots$, *is such that every domain* G_α *is K-convex with respect to some class* K_G, *where* $G = \bigcup_\alpha G_\alpha$, *then G is a K-convex domain.*

Proof: Let S and T, where $S \cup T \Subset G$, be sets for which the maximum principle holds with respect to the moduli of functions of class K_G. There exists a number N such that $S \cup T \Subset G_\alpha$ for $\alpha \geqslant N$. Since the G_α are K-convex domains, it follows on the basis of the second condition for K-convexity (see section 16.4) that the inequalities

$$\delta_{G_\alpha}(T) \leqslant \delta_{G_\alpha}(S), \quad \alpha \geqslant N \tag{50}$$

hold. In view of formula (2) of section 1.3, we obtain from (50)

$$\delta_G(T) = \lim_{\alpha \to \infty} \delta_{G_\alpha}(T) \leqslant \lim_{\alpha \to \infty} \delta_{G_\alpha}(S) = \delta_G(S).$$

Therefore, if we again apply the second condition for K-convexity, we conclude that G is a K-convex domain.

COROLLARY. *If an increasing sequence of domains* $\{G_\alpha\}$ *is such that each domain* G_α *is convex with respect to the class of functions that are holomorphic in* $G = \bigcup_\alpha G_\alpha$, *then G is a domain of holomorphy.*

This follows from the lemma just proven and the theorem in section 16.6.

THEOREM. *If an increasing sequence of domains of holomorphy* $G_\alpha \Subset G_{\alpha+1}$, *for* $a = 1, 2, \ldots$, *is such that each class* $H_{G_{\alpha+1}}$ *is dense in the class* H_{G_α}, *then* $G = \bigcup_\alpha G_\alpha$ *is a domain of holomorphy* (see Behnke and Stein [48]).

To prove this theorem, it will, on the basis of the corollary of the preceding lemma, be sufficient to show that each domain G_α is convex with respect to the class H_G. To prove this last assertion, it will, on the basis of the lemma in section 16.2, be sufficient to show that the class H_G is dense in each class H_{G_α}.

Suppose that $f_0 \in H_{G_\alpha}$, $G' \Subset G_\alpha$, and $\varepsilon > 0$. By hypothesis, there exists a function $f_1 \in H_{G_{\alpha+1}}$ such that

$$|f_0(z) - f_1(z)| < \varepsilon 2^{-1}, \quad z \in G'. \tag{51}$$

Let us now choose a function $f_2 \in H_{G_{\alpha+2}}$ such that

$$|f_2(z) - f_1(z)| < \varepsilon 2^{-2}, \quad z \in G_\alpha.$$

And, in general, we choose a function $f_k \in H_{G_{\alpha+k}}$ such that

$$|f_k(z) - f_{k-1}(z)| < \varepsilon 2^{-k}, \quad z \in G_{\alpha+k-2}. \tag{52}$$

In every fixed subdomain $G_\nu \Subset G$, where $\nu \geqslant \alpha$, this sequence of functions $f_k(z)$ for $k = 0, 1 \ldots$ converges uniformly in itself since, on the basis of (52), the inequalities

$$|f_{k+p} - f_k| \leqslant \sum_{l=1}^{p} |f_{k+l} - f_{k+l-1}| < \varepsilon \sum_{l=1}^{p} 2^{-l-k} < \varepsilon 2^{-k}, \quad z \in G_\nu$$

hold for arbitrary $p \geqslant 1$ and $k \geqslant \nu + 1 - \alpha$. Consequently, this sequence converges uniformly in G_ν and thus (since G_ν is arbitrary) defines a holomorphic function $f(z)$ in the domain G (see section 4.4). Obviously, the function $f(z)$ can be written as a series of the form

$$f = f_0 + (f_1 - f_0) + (f_2 - f_1) + \ldots.$$

Therefore, by virtue of (51) and (52), we obtain

$$|f(z) - f_0(z)| \leqslant \sum_{k=1}^{\infty} |f_k(z) - f_{k-1}(z)| < \varepsilon \sum_{k=1}^{\infty} 2^{-\nu} = \varepsilon$$

for $z \in G'$, which completes the proof.

10. The Behnke-Stein theorem

The union of an increasing sequence $G_\alpha \subset G_{\alpha+1}$, for $\alpha = 1, 2, \ldots$, of domains of holomorphy is a domain of holomorphy (see Behnke and Stein [48, 120]).

We shall prove this theorem for a bounded domain $G = \bigcup_\alpha G_\alpha$. For unbounded domains, the proof can be found in the article [120].

As a preliminary, we note that it will be sufficient to prove this theorem for the case in which the G_α are compact in $G_{\alpha+1}$, where $G_\alpha \Subset G_{\alpha+1}$, for $\alpha = 1, 2, \ldots$. This is true because the G_α are bounded domains of holomorphy and hence the $G_\alpha\left(\frac{1}{\alpha}\right)$ are open sets of holomorphy (see sections 16.2 and 16.6). Clearly,

$$G_\alpha\left(\frac{1}{\alpha}\right) \Subset G_{\alpha+1}\left(\frac{1}{\alpha+1}\right) \text{ and } \bigcup_\alpha G_\alpha\left(\frac{1}{\alpha}\right) = G.$$

If, in each open set $G_\alpha\left(\frac{1}{\alpha}\right)$, we take that component that contains the fixed point of the domain G, we obtain the required sequence of domains.

We adopt the notations (see Fig. 33)

$$m_{\alpha\beta} = \delta_{G_\beta}(G_\alpha), \quad M_{\alpha\beta} = \sup_{z \in \partial G_\alpha} \delta_{G_\beta}(z), \quad \beta \geqslant \alpha;$$

$$m_\alpha = \delta_G(G_\alpha), \quad M_\alpha = \sup_{z \in \partial G_\alpha} \delta_G(z), \quad \alpha = 1, 2, \ldots .$$

Since $G_\alpha \Subset G_{\alpha+1}$ and since $\bigcup_\alpha G_\alpha = G$ is a bounded domain, these numerical sequences possess the properties (see section 1.3): $0 < m_{\alpha\beta} \leq M_{\alpha\beta}$, $m_{\alpha\beta} \to m_\alpha$, and $M_{\alpha\beta} \to M_\alpha$ as $\beta \to +\infty$ (monotonically increasing), and $0 < m_\alpha \leq M_\alpha \to 0$ as $\alpha \to +\infty$ (monotonically decreasing).

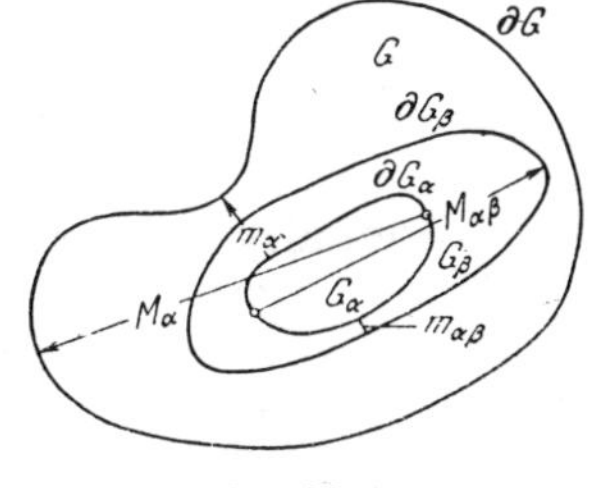

Fig. 33

From the sequence of domains G_α, where $\alpha = 1, 2, \ldots$, let us extract a subsequence G_{α_k} as follows: (1) We take $G_{\alpha_1} = G_1$. (2) We take G_{α_2} in such a way that $M_{\alpha_2} < m_{\alpha_1}$. (3) We take G_{α_3} in such a way that $M_{\alpha_2\alpha_3} < m_{\alpha_1\alpha_3}$ and $M_{\alpha_3} < m_{\alpha_2}$. Because of the preceding inequality $M_{\alpha_2} < m_{\alpha_1}$, such a choice is always possible. (4) In general, we take G_{α_k} in such a way that $M_{\alpha_{k-1}\alpha_k} < m_{\alpha_{k-2}\alpha_k}$ and $M_{\alpha_k} < m_{\alpha_{k-1}}$. Thus for an arbitrary $z^0 \in \partial G_{\alpha_k}$,

$$\delta_{G_{\alpha_{k+1}}}(z^0) < \delta_{G_{\alpha_{k+1}}}(G_{\alpha_{k-1}}). \tag{53}$$

Since the domain $G_{\alpha_{k+1}}$ is holomorphically convex and $G_{\alpha_{k-1}} \Subset G_{\alpha_{k+1}}$, it follows from inequality (53) that, for an arbitrary point $z^0 \in \partial G_{\alpha_k}$, there exist a function $f_{z^0}(z)$ that is holomorphic in $G_{\alpha_{k+1}}$, and a neighborhood $U(z^0, r)$ such that

$$|f_{z^0}(z)| < 1, \quad z \in G_{\alpha_{k-1}}; \quad |f_{z^0}(z)| > 1, \quad z \in U(z^0, r)$$

(see sections 16.2 and 16.5). If we cover the compact set ∂G_{α_k} with a finite number of these neighborhoods $U(z^{(l)}, r_l)$, where $l = 1, 2, \ldots, N_k$, and reason as in section 16.5, we can exhibit a Weil domain

$$\sigma_k = [z: |\chi_l^{(k)}(z)| < 1, \quad l = 1, 2, \ldots, N_k; \quad z \in U(\bar{\sigma}_k)]$$

such that $G_{\alpha_{k-1}} \Subset \sigma_k \Subset G_{\alpha_k}$ and the functions $\chi_l^{(k)}$ are holomorphic in $G_{\alpha_{k+1}}$.

Thus, we have constructed a sequence of Weil domains σ_k, for $k=1, 2, \ldots$, with the following properties:

$$\sigma_k \Subset \sigma_{k+1}, \quad \bigcup_k \sigma_k = G, \quad \chi_l^{(k)} \in H_{\sigma_{k+1}}.$$

Therefore, in accordance with the theorem in section 24.7, the class $H_{\sigma_{k+1}}$ is dense in H_{σ_k}.

According to the theorem in section 16.9, G is a domain of holomorphy, which completes the proof.

11. *Polynomially convex domains*

A domain

$$\sigma = [z: |P_\alpha(z)| < 1, \quad \alpha = 1, 2, \ldots, N; \quad z \in U(\bar{\sigma})],$$

where the P_α are polynomials, is called a *polynomial domain.*

Every polynomial domain is a polynomially convex domain (see section 16.5).

The union of an increasing of polynomially convex domains is a polynomially convex domain.

This follows from the lemma of section 16.9 with $K_G = P$.

For a domain G to be polynomially convex, it is necessary and sufficient that it be the union of an increasing sequence $\sigma_k \Subset \sigma_{k+1}$, *for* $k = 1, 2, \ldots$, *of polynomial Weil domains.*

The sufficiency follows from the preceding assertions. The necessity was proven in section 16.5.

Every convex domain is convex with respect to the class of polynomials of the form $\{cP(az)\}$, *where a is an arbitrary complex vector,* $P(\xi)$ *is an arbitrary polynomial in a single variable with real coefficients, and c is an arbitrary complex number.*

Proof: Every convex domain is the union of an increasing sequence of bounded convex polyhedra. On the basis of the lemma in section 16.9, it will be sufficient to show that each such polyhedron

$$\sigma = [z: \quad d^{(k)}x + b^{(k)}y < e^{(k)}, \ k = 1, 2, \ldots, N_k] \tag{54}$$

is a convex domain with respect to the class $\{cP(az)\}$. But since, by virtue of (54),

$$\sigma = \left[z: \quad \left|e^{a^{(k)}z - e^{(k)}}\right| < 1, \ k=1, 2, \ldots, N_k\right], \ a^{(k)} = d^{(k)} - ib^{(k)},$$

it follows that the domain σ is convex with respect to the smallest class containing the functions $z_1, z_2, \ldots, z_n$, and ce^{az}. But the class

$\{cP(az)\}$ is obviously dense in this class. From the lemma in section 16.2, the domain σ is convex with respect to the class $\{cP(az)\}$, which completes the proof.

17. CONTINUITY PRINCIPLES

1. *The continuity theorem*

We shall say that the continuity principle holds for a domain G if the following assertion is true: Let S_α and T_α, for $\alpha = 1, 2, \ldots$ be sequences of sets for which the maximum principle holds with respect to the moduli of functions that are holomorphic in G and suppose that $S_\alpha \cup T_\alpha \Subset G$, $\lim_{\alpha\to\infty} S_\alpha = S_0$ is bounded, and $\lim_{\alpha\to\infty} T_\alpha = T_0 \Subset G$. Then, $S_0 \Subset G$ (see Fig. 15).

We note that the continuity principle implies the weak continuity principle (see section 12.3) since we may take domains for the sets S_α and their boundaries for the sets T_α (where the domains and their boundaries lie on analytic surfaces). Specifically, on the basis of section 10.14, the maximum principle holds for such sets $S_\alpha \Subset G$ and $\partial S_\alpha \Subset G$ with respect to the absolute values of functions that are holomorphic in G.

We have the following theorem on the location of singular points of a holomorphic function (see Bremermann [13]), which dates back to Hartogs [119]. For brevity, we shall call this theorem the continuity theorem (Kontinuitatssatz).

If G is a domain of holomorphy, the continuity principle holds for it.

Proof: Suppose that $\{S_\alpha\}$ and $\{T_\alpha\}$ are sequences of sets for which the maximum principle holds for the moduli of functions in H_G and suppose that $S_\alpha \cup T_\alpha \Subset G$ that $\lim S_\alpha = S_0$ is bounded, and that $\lim T_\alpha = T_0 \Subset G$. Since G is a holomorphically convex domain, inequality (43) holds for the sets S_α and T_α (see section 16.4):

$$\delta_G(T_\alpha) \leqslant \delta_G(S_\alpha). \tag{55}$$

Since the function $\delta_G(z)$ is continuous in G (see section 1.3) and the set $S_0 \cup T_0$ is bounded and is contained in G, inequality (55) remains valid in the limit. Therefore,

$$\delta_G(T_0) = \lim_{\alpha\to\infty} \delta_G(T_\alpha) \leqslant \lim_{\alpha\to\infty} \delta_G(S_\alpha) = \delta_G(S_0).$$

Since $T_0 \Subset G$, we have $\delta_G(T_0) > 0$ and, hence, $\delta_G(S_0) > 0$; that is, because of the boundedness of the set S_0, we have $S_0 \Subset G$, which completes the proof.

COROLLARY. *Every domain of holomorphy is pseudoconvex.*

This follows from the first condition for pseudoconvexity (see section 12.4).

For a particular function, the continuity theorem takes the following form:

Suppose that a function $f(z)$ is holomorphic in a domain G and that S_α and T_α, for $\alpha = 1, 2, \ldots$, are sequences of sets for which the maximum principle with respect to the moduli of functions of class H_G holds and suppose that $S_\alpha \cup T_\alpha \Subset G$, $\lim S_\alpha = S_0$ is bounded, and $\lim T_\alpha = T_0 \Subset G$. Then, $f(z)$ can be holomorphically continued into points of the set S_0.

Proof: On the basis of the corollary to the Cartan-Thullen theorem (see section 16.1), the function $f(z)$ can be holomorphically continued into the points of the set $\bigcup_\alpha (S_\alpha)_{r_\alpha}$, where $r_\alpha = \delta_G(T_\alpha)$. Since $\delta_G(z)$ is a continuous function and $\lim T_\alpha = T_0 \Subset G$, we have

$$\lim_{\alpha\to\infty} r_\alpha = \lim_{\alpha\to\infty} \delta_G(T_\alpha) = \lim_{\alpha\to\infty} \delta_G(T_0) = 2\varepsilon > 0.$$

Consequently, from some α on, we have $r_\alpha \geqslant \varepsilon$. Therefore, since $S_\alpha \to S_0$, we obtain the inclusion

$$S_0 \subset \bigcup_\alpha (S_\alpha)_\varepsilon \subset \bigcup_\alpha (S_\alpha)_{r_\alpha}, \quad \text{q.e.d.}$$

Remark: The basic difference in the theory of domains of holomorphy of functions of one and of several complex variables consists in the following: In the case of a single variable, bounded sets S and T for which the maximum principle holds with respect to the moduli of functions that are holomorphic in G are subdomains of $S \Subset G$ and their boundary $T = \partial S$. The condition $\delta_G(S) \geqslant \delta_G(\partial S)$ imposes no restrictions at all on the domain G. On the other hand, in the case of several variables, it is possible to exhibit domains $S \Subset G$ lying, for example, on two-dimensional analytic surfaces for which $\delta_G(S) < \delta_G(\partial S)$ (see Fig. 34) if G is not a domain of holomorphy. Therefore, the condition $\delta_G(S) \geqslant \delta_G(\partial S)$ is valid only for a certain class of domains in C^n (for $n \geqslant 2$) (in every case for domains of holomorphy [see section 16.7]).

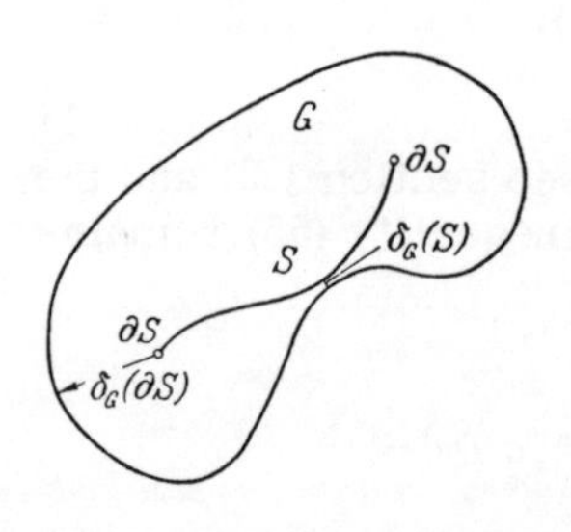

Fig. 34

2. "Disk" theorems

We note the following special cases of the continuity theorem. They are referred to as "disk" theorems. These theorems are very convenient for applications.

Suppose that $(z, w) \in C^{n+m}$ and $z = z(t)$, where $0 \leq t \leq 1$, is a continuous curve in C^n. Suppose also that $D(t)$, for $0 \leq t \leq 1$, is a family of domains in C^m possessing the property that, for an arbitrary compact set $K' \subset D(0)$, there exists a number $\eta = \eta(K')$, in $(0, 1]$ such that $K' \subset D(t)$ for all $0 \leq t < \eta$, there.

Suppose that the domain of holomorphy G contains the "disks" $[(z, w): z = z(t), w \in D(t)]$, *where* $0 < t \leq 1$, *and the points* $[(z, w): z = z(0), w = D(0)\setminus K]$, *where K is a compact set contained in* $D(0)$. *Then, G contains also the limit "disk"* $[(z, w): z = z(0), w \in D(0)]$.

Suppose that $G(t)$, *for* $0 \leq t \leq 1$, *is a family of bounded domains in* C^1 *possessing the property that, for an arbitrary compact set* $K \subset g(0)$, *there exists a number* $\eta = \eta(K)$, *in* $(0,1]$ *such that* $K \subset g(t)$ *for all t in* $[0, \eta)$. *Suppose also that the family of "disks"*

$$D(t) = [z: \; z = z(\lambda, \; t), \;\; \lambda \in \bar{g}(t)], \;\; 0 \leqslant t \leqslant 1,$$

is such that (1) *the vector-valued function* $z(\lambda, t)$ *is holomorphic with respect to* λ *in the domain* $\bar{g}(t)$ *for every* $t \in [0,1]$, *it is continuous, and* $\dfrac{\partial z}{\partial \lambda} \neq 0$ *in* $\bar{g}(r) \times [0,1]$; (2) *the domain of holomorphy G contains "disks"* $D(t)$ *for* $0 < t \leq 1$ *and the boundary of the limit "disk"* $\partial D(0) = [z: z = z(\partial, 0), \lambda \in \partial g(0)]$. *Then,* $D(0) \Subset G$.

Analogous "disk" theorems hold for a specific function $f(z)$. For example, corresponding to Theorem II is the

THEOREM. *Suppose that a function* $f(z)$ *is holomorphic at points of the set*

$$\bigcup_{0 < t \leqslant 1} D(t) \cup \partial D(0).$$

Then, $f(z)$ *can be holomorphically continued into points of the "disk"* $D(0)$.

3. *A strong "disk" theorem*

In the "disk" theorems that we have given, it was required that all points of the boundary of the limit "disk" be contained in a domain of holomorphy. In the following theorem, it will be possible to weaken this requirement, replacing it with the requirement that at least one point in the limit "disk" belong to a domain of holomorphy. We shall call this theorem the strong "disk" theorem (Bremermann [18]: see also Bros [63]).

Suppose that a Jordan curve of the form

$$\tilde{z} = \tilde{z}(t) \equiv \tilde{z}^0 + b\lambda(t), \;\; 0 \leqslant t \leqslant 1,$$

where b is a vector, is given in the space C^{n-1}. *Suppose also that the family of domains* $D(t)$, *where* $\leq t \leq$, *lying in the* z_1*-plane possesses the property that, for every compact set* $K \subset D(0)$, *there exists a number* $\eta = \eta(K)$ *in* $(0,1]$ *such that* $K \subset D(t)$ *for all t in* $[0, \eta)$. *If the function* $f(z)$ *is holomorphic at points of the "disks"*

$$[z: \;\; z_1 \in D(t), \;\; \tilde{z} = \tilde{z}(t)], \;\; 0 < t \leqslant 1,$$

and at least one point of the limit "disk"

$$[z: \;\; z_1 \in D(0), \;\; \tilde{z} = \tilde{z}(0)],$$

then, this function is holomorphic at all points of the limit "disk".

Proof: Suppose that the assertion of the theorem is not true. Then, there exists a point $(z_1', \tilde{z}(0))$, where $z_1' \in D(0)$, at which $f(z)$ is not holomorphic. Then, from the condition of the theorem, there exists a circle $|z_1 - z_1^0| \leqslant r$, contained in $D(0)$ such that the function $f(z)$ is holomorphic at the points $[z: |z_1 - z_1^0| < r\ \tilde{z} = \tilde{z}(0)]$; and at some point $(z_1'', \tilde{z}(0))$ for which $|z_1'' - z_1^0| = r$, this function is not holomorphic (see Fig. 35). Furthermore, for some $r' > r$, the circle $|z_1 - z_1^0| \leqslant r'$ is also contained in $D(0)$. Therefore, there exists a $\eta > 0$, such that

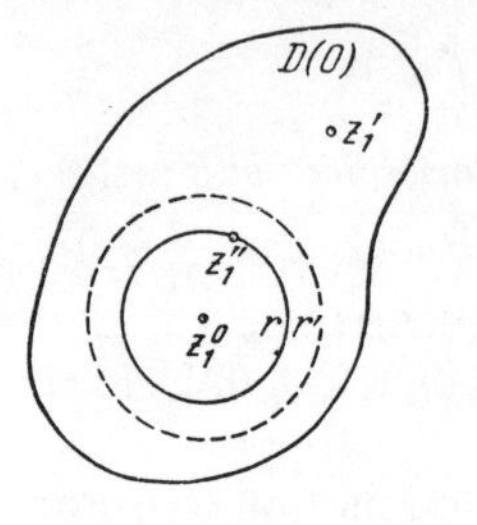

Fig. 35

$$[|z_1 - z_1^0| \leqslant r'] \subset D(t) \text{ for all } 0 \leqslant t < \eta.$$

On the basis of what has been said, the function $f(z)$ is holomorphic in a complete Hartogs domain of the form

$$G' = [z: \ |z_1 - z_1^0| < R(\tilde{z}), \ \tilde{z} \in U],$$

where U is some neighborhood of the curve $\tilde{z} = \tilde{z}(t)$, where $0 \leqslant t \leqslant \eta_1 \leqslant \eta$, and the function $R(\tilde{z})$ is lower-semicontinuous and satisfies the conditions

$$R[\tilde{z}(0)] = r, \quad R[\tilde{z}(t)] \geqslant r' > r, \quad 0 < t \leqslant \eta_1. \tag{56}$$

According to the theorem in section 15.4, the function $-\ln R(\tilde{z})$ is plurisubharmonic in U. But the curve $\tilde{z} = \tilde{z}(t)$ lies on the two-dimensional analytic plane $\tilde{z} = \tilde{z}^0 + b\lambda$. Therefore, the function $-\ln R(\tilde{z}^0 + b\lambda)$ is subharmonic with respect to λ in

$$U_{\tilde{z}^0, b} = [\lambda: \ \tilde{z}^0 + \lambda b \in U].$$

But then, from the theorem in section 9.16,

$$\overline{\lim_{t \to 0,\, t \neq 0}} \{-\ln R[\tilde{z}^0 + b\lambda(t)]\} = \\ = -\ln R[\tilde{z}(0)],$$

which contradicts conditions (56). This contradiction proves the strong "disk" theorem.

Remark: It is still an open question as to whether the theorem in section 9.16 is valid for plurisubharmonic functions in domains of (complex) dimension $n \geqslant 2$. An affirmative answer to this question would enable us to dispense with the requirement in the strong "disk" theorem that the curve $z = z(t)$ lie on a two-dimensional analytic plane.

Let us now look at some examples illustrating the application of the "disk" theorems.

4. *Holomorphic extension of the boundary of a domain*

If a function $f(z)$ is holomorphic at all points of the boundary ∂G of a bounded simply connected domain $G \subset C^n$ (for $n \geq 2$), where ∂G is a connected set, then $f(z)$ is holomorphic in the domain G.

Proof: The function $f(z)$ is holomorphic in some neighborhood $U(\partial G)$ of the connected set ∂G. Since the domain G is bounded, we have $G \setminus U(\partial G) \Subset G$ and $\partial G \Subset U(\partial G)$. Let us take an arbitrary point $z^0 \in G \cap U(\partial G)$ and pass through it a two-dimensional analytic plane $z = a\lambda + z^0$, where $|a| = 1$. Consider a point $z' \in G$. Since G is a domain, there exists a continuous curve $z = b(t)$, for $0 \leqslant t \leqslant 1$, that lies entirely in G and that connects the points $z^0 = b(0)$ and $z' = b(1)$ (see Fig. 36). We denote by $g(t)$ that component (in the λ-plane) of the intersection of the domain G and the plane $z = a\lambda + b(t)$, that contains the point $\lambda = 0$. Consider the family of "disks":

$$D(t) = [z: \quad z = a\lambda + b(t), \quad \lambda \in \bar{g}(t)], \quad 0 \leqslant t \leqslant 1.$$

According to the hypothesis of the theorem $D(0) \Subset U(\partial G)$. Therefore, there exists a number $t_0 \leqslant 1$, such that $D(t) \Subset U(\partial G)$ for $0 \leqslant t < t_0$. If we note now that

$$\partial D(t) = [z: \quad z = a\lambda + b(t), \quad \lambda \in \partial g(t)] \subset \partial G \Subset U(\partial G),$$

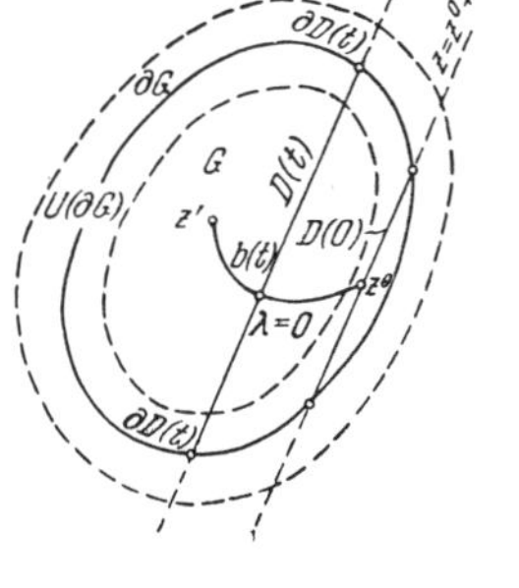

Fig. 36

we conclude from the "disk" theorem (see section 17.2) that the function $f(z)$ can be holomorphically continued into all points of the "disk" $D(1)$ and, in particular, to the point $z' \in G$.

Thus, the function $f(z)$ can be holomorphically continued to an arbitrary point $z' \in G$ from an arbitrary point $z^0 \in G \cap U(\partial G)$ along any path connecting them that lies in the domain G. But G is assumed to be a simply connected domain. From the monodromy theorem (see section 6.6), the function $f(z)$ is single-valued in G, which completes the proof.

In other words, we may say that the domain $U(\partial G) \cup G$ is a holomorphic extension of the neighborhood $U(\partial G)$.

We note that this theorem remains valid without the assumption that the domain G is simply connected. This assertion constitutes the Osgood-Brown theorem (see [1], p. 324).

COROLLARY. *A holomorphic function cannot have isolated singularities if $n \geq 2$.*

5. *Holomorphic extension of tubular domains*

Let us use the strong "disk" theorem to prove Bochner's theorem on tubular domains (see [3], Chapter V).

Every function $f(z)$ that is holomorphic in a tubular domain $T_B = B + iR^n$ is holomorphic in its convex envelope $O(T_B) = O(B) + iR^n$; that is, $O(T_B)$ is a holomorphic extension of T_B.

Proof: It will be sufficient to show that $f(z)$ can be continued holomorphically into points of the straight line segment $[z', z'']$, connecting arbitrary points z' and z'' in T_B. That the holomorphic continuation of the function $f(z)$ into the domain $O(T_B)$ is single-valued follows from the monodromy theorem (see section 6.6) since the domain $O(T_B)$ is convex and hence simply connected.

Since T_B is a domain, the points z' and z'' can be connected by a broken line with a finite number of links that lies entirely in T_B. The problem can thus be reduced, by induction, to the simpler problem: Suppose that segments $[a_2, a_1]$ and $[a_1, a_3]$ not lying on a single straight line are contained in the domain T_B. Show that $f(z)$ can be holomorphically continued into points of the segment $[a_2, a_3]$.

Now let us solve this problem. Consider a point $a \in [a_2, a_3]$. Since the domain T_B is a tubular domain, we may assume that $\operatorname{Im} a_1 = \operatorname{Im} a_2 = \operatorname{Im} a_3 = \operatorname{Im} a$. By a suitable real linear transformation of the form $z = Az' + a_1$, we may arrange for the points $a_k' = A^{-1}(a_k - a_1)$, for $k = 1, 2, 3$, to lie in the plane $x_1' x_2'$, where $a_1' = 0$, $a_2' = (\alpha, 1, 0, \ldots, 0)$ and $a_3' = (\beta, 1, 0, \ldots, 0)$ (see Fig. 37). This transformation maps the domain T_B into $T_{B'}$.

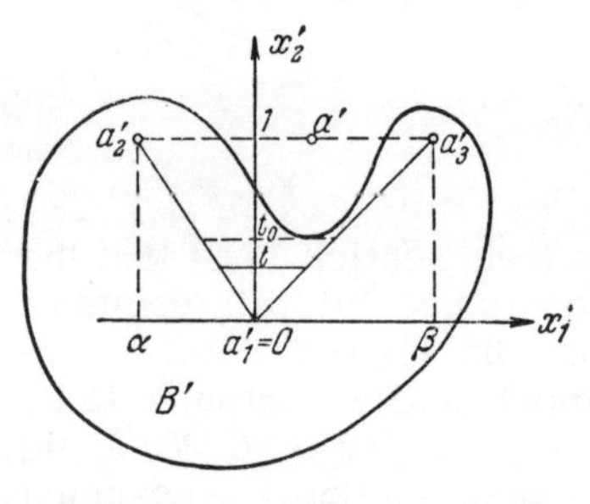

Fig. 37

In the plane z_2', let us take the curve $z_2' = t$ for $0 \leqslant t \leqslant 1$. We choose a corresponding sequence of domains $D(t)$ in the plane z_1' in the form of infinite strips (assuming $\alpha < \beta$):

$$D(t) = [z_1': \; t\alpha < x_1' < t\beta, \; |y_1'| < \infty], \quad 0 < t \leqslant 1.$$

Clearly, the sequence $D(t)$, for $0 < t \leqslant 1$ depends continuously on t. Since the broken line $[a_2', a_1', a_3']$ lies entirely in $T_{B'}$, there exists a neighborhood U of this broken line that also lies entirely in $T_{B'}$. Therefore, there exists a number t_0 in $(0, 1]$ such that the "disks"

$$[z': \; z_1' \in D(t), \; z_2' = t, \; z_3' = \ldots = z_n' = 0], \quad 0 \leqslant t < t_0,$$

are contained in $T_{B'}$ and certain points of the limit "disk" corresponding to $t = t_0$ are also contained in $T_{B'}$.

By applying the strong "disk" theorem (see section 17.3), we see that function $f(Az' + a_1)$ is holomorphic at all points of this limit "disk." From this we conclude that it is also holomorphic at all points of the "disk" for $t = 1$

$$[z': \ z_1' \in D(1), \ z_2' = 1, \ z_3' = \ldots = z_n' = 0]$$

and, in particular, at the point $a' = A^{-1}\operatorname{Re}(a - a_1)$. This means that $f(z)$ is holomorphic at the point $z = a$. This completes the proof of the theorem.

18. LOCAL PSEUDOCONVEXITY

There are many different definitions of local pseudoconvexity (see Hitotsumatsu [52]). Here, we shall consider two of these: pseudoconvexity in the sense of Cartan and pseudoconvexity in the sense of Levi.

1. *Definition of local pseudoconvexity of domains*

A domain G is said to be pseudoconvex in the sense of Cartan at a point $z^0 \in \partial G$ if there exists a hypersphere $U(z^0, r)$ such that $G \cap U(z^0, r)$ is an open set of holomorphy.

A domain G is said to be pseudoconvex in the sense of Levi at a point $z^0 \in \partial G$ if every two-dimensional analytic surface passing through this point in every neighborhood of the point z^0 contains points not belonging to $G \cup \{z^0\}$ (see Fig. 38).

We note that the latter definition is analogous to the usual definition of convexity of a domain G at a point $z^0 \in \partial G$ if, instead of the set of two-dimensional analytic surfaces passing through the point z^0, we have the pencil of straight lines which meet G at the point z^0.

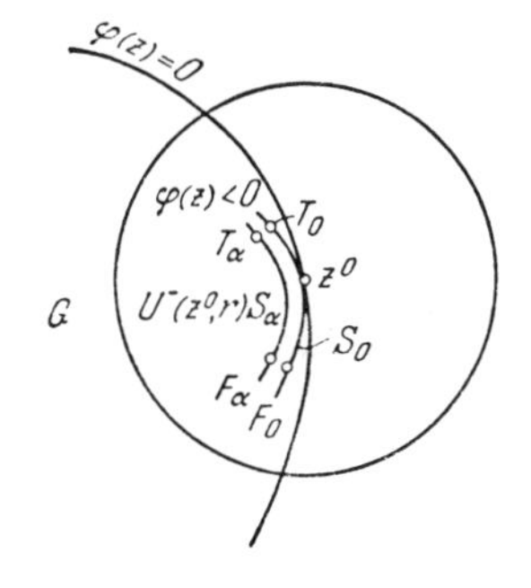

Fig. 38

If a domain is pseudoconvex in the sense of Cartan or Levi at every point of its boundary, it is said to be locally pseudoconvex in the sense of Cartan or Levi, respectively.

In connection with these definitions, the problem arises as to the relationship of these two concepts of local pseudoconvexity with each other and with the concepts of (global) pseudoconvexity and holomorphic convexity.

We point out that *every domain of holomorphy is locally pseudoconvex in the sense of Cartan.*

This is true because the intersection of a domain of holomorphy with every hypersphere is an open set of holomorphy (see section 16.7).

2. *Definition of local pseudoconvexity of surfaces*

Suppose that the boundary ∂G of a domain G in a neighborhood $U(z^0, r)$ of a point $z^0 \in \partial G$ coincides with a set $S = [z\colon \varphi(z) = 0]$, where $\varphi(z)$ is a sufficiently smooth function in $U(z^0, r)$. In what follows, we shall assume that G lies on that side of S corresponding to $\varphi(z) < 0$, that is, that $G \cap U(z^0, r) = U^-(z^0, r)$, where $U^-(z^0, r) = [z\colon \varphi(z) < 0 \text{ and } |z - z^0| < r]$ (see Fig. 38).

We shall say that a sufficiently smooth hypersurface $S = [z\colon \varphi(z) = 0]$ (see section 1.4) is pseudoconvex in the sense of Cartan at a point $z^0 \in S$ on the $(\varphi(z) < 0)$ side if there exists an r such that $U^-(z^0, r)$ is an open set of holomorphy. Analogously, we shall say that a hypersurface S is pseudoconvex in the sense of Levi at a point z^0 on the $(\varphi(z) < 0)$ side if every two-dimensional analytic surface passing through z^0 contains points (other than the point z^0 itself) in every neighborhood of the point z^0 at which $\varphi(z) \geqslant 0$.

If any (hyper)surface is pseudoconvex (in either sense of the word) at all its points, we shall say that this hypersurface is locally pseudoconvex (in the same sense of the word).

Our problem consists in finding the conditions under which hypersurfaces will be locally pseudoconvex. These conditions are quite significant in applications.

3. *Conditions for domains to be locally pseudoconvex*

If a domain G is locally pseudoconvex in the sense of Cartan, G is a pseudoconvex domain.

Proof: Consider a point $z^0 \in \partial G$. Then, there exists a positive number r such that each component of the open set $G' = G \cap U(z^0, r)$ is a domain of holomorphy and hence a pseudoconvex domain (see section 17.1). Therefore, the function $-\ln \Delta_{G'}(z)$ is plurisubharmonic in G'. But

$$-\ln \Delta_{G'}(z) = -\ln \Delta_G(z), \quad z \in G \cap U\left(z^0, \frac{r}{3}\right)$$

and hence the function $-\ln \Delta_G(z)$ is plurisubharmonic in some boundary strip of the domain G. According to the theorem in section 12.6, the domain G is pseudoconvex, which completes the proof.

Suppose that a domain G is pseudoconvex and that, in some neighborhood $U(z^0, r)$ of a boundary point z^0, the boundary $\partial G = [z : \phi(z) = 0]$ belongs to the class $C^{(1)}$. Then, the domain G is pseudoconvex in the sense of Levi at the point z^0.

Proof: Let us suppose that the domain G is not pseudoconvex in the sense of Levi at the point $z^0 \in \partial G$. Then, there exists a two-dimensional analytic surface $F_0 = [z\colon z = z(\lambda), |\lambda| < \rho]$ passing through the point z^0, where $z^0 = z(0)$, and lying, except for the point

z^0 in the domain G (see Fig. 38); that is,

$$\varphi[z(\lambda)] < 0, \quad |\lambda| < \rho, \quad \lambda \neq 0. \tag{57}$$

Let us set $c = \overline{\text{grad}\, \varphi(z^0)}$. Since grad $\varphi \neq 0$ in $U(z^0, r)$, it follows that $c \neq 0$ and, because of the continuity, we have

$$\text{Re}\,(c, \text{grad}\, \varphi(z)) \geqslant \varkappa > 0, \qquad z \in U(z^0, r_1) \tag{58}$$

for some $r_1 \leqslant r$ and $\varkappa > 0$. Let us take a sequence of analytic surfaces

$$F_\alpha = [z: \quad z = z(\lambda) - \alpha c, \quad |\lambda| < \rho], \qquad \alpha \geqslant 0.$$

If we apply Taylor's formula to the function $\varphi[z(\lambda) - \alpha c]$ for small values of $\alpha > 0$ and use inequalities (57) and (58), we obtain, for all sufficiently small $|\lambda| \leqslant \rho_1 < \rho$,

$$\begin{aligned} &\varphi[z(\lambda) - \alpha c] = \\ &\qquad = \varphi[z(\lambda)] - 2\alpha\, \text{Re}\,(c, \text{grad}\, \varphi[z(\lambda)]) + o(\alpha) < 0. \end{aligned} \tag{59}$$

Thus, by virtue of (59), we see that the sequence of domains

$$S_\alpha = [z: \quad z = z(\lambda) - \alpha c, \quad |\lambda| < \rho_1],$$

lying on the two-dimensional analytic surfaces F_α is such that $S_\alpha \cup \partial S_\alpha \Subset G$, $S_\alpha \to S_0 = [z: \quad z = z(\lambda), \ |\lambda| < \rho_1]$, is bounded, and $\partial S_\alpha \to T_0 = [z: \quad z = z(\lambda), \ |\lambda| = \rho_1] \Subset G$. But the domain G is assumed to be pseudoconvex. Therefore, from the weak continuity principle (see section 12.4), $S_0 \Subset G$, which contradicts the fact that $z^0 = z(0) \in \partial G$. This contradiction proves that the domain G is pseudoconvex in the sense of Levi at the point z^0, which completes the proof.

From this theorem and the corollary in section 17.1, we have the

COROLLARY. *If a domain G is pseudoconvex in the sense of Cartan at a point $z^0 \in \partial G$ and if $\partial G \in C^{(1)}$ in some neighborhood of that point, the domain G is pseudoconvex in the sense of Levi at the point z^0.*

4. *The Levi-Krzoska theorem* (see [115, 116])

If a class $C^{(1)}$ hypersurface $S = [z : \phi(z) = 0]$ is pseudoconvex in the sense of Levi at a point $z^0 \in S$ on the $(\phi(z) < 0)$ side, then

$$(H(z^0; \varphi)\, a, \bar{a}) \geqslant 0 \tag{60}$$

for all a satisfying the condition

$$(a, \text{grad}\, \varphi(z^0)) = 0; \tag{61}$$

the form H being defined in section 10.10.

Conversely, if

$$(H(z^0;\ \varphi)\,a,\ \bar{a}) > 0 \tag{62}$$

at a point z^0 of a class $C^{(2)}$ hypersurface $S = [z : \phi(z) = 0]$ for all $a \neq 0$ satisfying condition (61), *then S is pseudoconvex in the senses of Cartan and Levi at the point z^0 on the $(\phi(z) < 0)$ side. Furthermore, it is possible to pass through point z^0 a $(2n - 2)$-dimensional analytic surface $P(z) = 0$, where $P(z)$ is a second-degree polynomial lying entirely on the $(\phi(z) > 0)$ side in some neighborhood of the point z^0, excluding, of course, the point z^0.*

*Proof:** Since $\varphi \in C^{(2)}$, it follows that, on the manifold $z = z^0 + a\lambda + b\lambda^2$ passing through z^0, this function may, for small λ, be expressed in accordance with Taylor's formula in the form

$$\begin{aligned}\varphi(z^0 + a\lambda + b\lambda^2) &= 2\,\mathrm{Re}\,(a\lambda + b\lambda^2,\ \mathrm{grad}\,\varphi(z^0)) + \\ &+ \mathrm{Re}\left[\lambda^2 \sum_{j,k} \frac{\partial^2 \varphi(z^0)}{\partial z_j\, \partial z_k} a_j a_k\right] + |\lambda|^2 (H(z^0;\ \varphi)\, a,\ \bar{a}) + o(|\lambda|^2).\end{aligned} \tag{63}$$

To prove the first part of the theorem, let us assume that, contrary to the assertion, the inequality

$$(H(z^0;\ \varphi)\, a^0,\ \bar{a}^0) < 0$$

is satisfied for some vector $a^0 \neq 0$, that satisfies Eq. (61). Since $\mathrm{grad}\,\varphi(z^0) \neq 0$, there exists a vector b^0 such that

$$2\,(b^0,\ \mathrm{grad}\,\varphi(z^0)) = \sum_{j,k} \frac{\partial^2 \varphi(z^0)}{\partial z_j\, \partial z_k}\, a_j^0 a_k^0.$$

Therefore, formula (63) for the vectors a^0 and b^0 takes the form

$$\varphi(z^0 + a^0\lambda + b^0\lambda^2) = |\lambda|^2\, [(H(z^0;\ \varphi)\, a^0,\ \bar{a}^0) + o(1)],$$

from which we conclude because of continuity that

$$\varphi(z^0 + a^0\lambda + b^0\lambda^2) < 0, \qquad |\lambda| \leqslant \rho,\quad \lambda \neq 0$$

if ρ is sufficiently small. Furthermore, for small ρ, the manifold $F = [z:\quad z = z^0 + a^0\lambda + b^0\lambda^2,\ \ |\lambda| < \rho]$ is a two-dimensional analytic surface since $\partial z / \partial\lambda = a^0 + 2b^0\lambda \neq 0$. From what has been said, it follows that the two-dimensional analytic surface F in some neighborhood of the point z^0 lies entirely on the $\varphi(z) < 0$ side of S (except for the point z^0 [see Fig. 38]). But this means that the hypersurface S cannot be pseudoconvex in the sense of Levi at the point z^0. This contradiction proves inequality (60) under the condition (61).

*See Behnke and Sommer [51].

Now let us prove the second part of the theorem. Since $S = [z\colon \varphi(z) = 0] \in C^{(2)}$ and $z^0 \in S$, there exists an r such that $\varphi(z) \in C^{(2)}$ and $\operatorname{grad} \varphi(z) \neq 0$ in $U(z^0, r)$. Suppose that $\rho < r/3$. Then, in the neighborhood $U(z', 2\rho)$ of each point $z' \in S \cap U(z^0, \rho)$ (see Fig. 39), the Taylor expansion

$$\varphi(z) = 2\operatorname{Re}(z - z', \operatorname{grad}\varphi(z')) + \\ + \operatorname{Re}\sum_{j,k} \frac{\partial^2 \varphi(z')}{\partial z_j \partial z_k}(z_j - z'_j)(z_k - z'_k) + \\ + (H(z'; \varphi)(z - z'), \bar{z} - \bar{z}') + o(|z - z'|^2) \tag{64}$$

is valid. Let us consider the function φ on the $(2n-2)$-dimensional analytic surface

$$P_{z'}(z) \equiv 2(z - z', \operatorname{grad}\varphi(z')) + \\ + \sum_{j,k} \frac{\partial^2 \varphi(z')}{\partial z_j \partial z_k}(z_j - z'_j)(z_k - z'_k) = 0$$

which passes through the point z'. On the basis of (64), on this surface in the neighborhood $U(z', 2\rho)$, we have

$$\varphi(z) = (H(z; \varphi)(z - z'), \\ \bar{z} - \bar{z}') + o(|z - z'|^2) = \\ = \left[\frac{(H(z; \varphi)(z - z'), \bar{z} - \bar{z}')}{|z - z'|^2} + \right. \\ \left. + o(1)\right]|z - z'|^2. \tag{65}$$

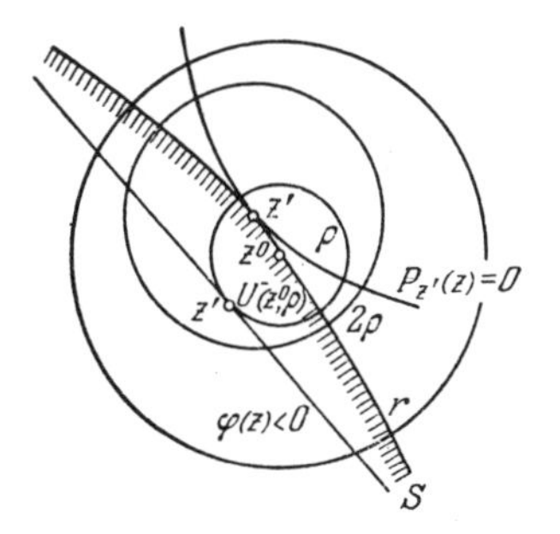

Fig. 39

Furthermore, we conclude from condition (62) that $(H(z^0; \varphi)a, \bar{a}) \geqslant \varkappa > 0$ for some a and for all a such that $|a| = 1$ which satisfy condition (61). But on the surface $P_{z'}(z) = 0$ in the hypersphere $U(z', 2\rho)$ condition (61) is approximately satisfied for the vectors $a = \frac{z - z'}{|z - z'|}$ up to a quantity of the order ρ. Specifically,

$$\left|\left(\frac{z - z'}{|z - z'|}, \operatorname{grad}\varphi(z^0)\right)\right| \leqslant \left|\left(\frac{z - z'}{|z - z'|}, \operatorname{grad}\varphi(z')\right)\right| + \\ + \left|\left(\frac{z - z'}{|z - z'|}, \operatorname{grad}\varphi(z^0) - \operatorname{grad}\varphi(z')\right)\right| \leqslant \\ \leqslant \frac{1}{2|z - z'|}\left|\sum_{j,k} \frac{\partial^2 \varphi(z')}{\partial z_j \partial z_k}(z_j - z'_j)(z_k - z'_k)\right| + \\ + |\operatorname{grad}\varphi(z^0) - \operatorname{grad}\varphi(z')| \leqslant c\rho.$$

On the basis of what has been said, we conclude from (65) that

$$\varphi(z) \geqslant \frac{\varkappa}{2}|z - z'|^2 \quad \text{for} \quad z' \in U(z^0, \rho) \cap S, \\ z \in U(z', 2\rho) \cap [z\colon \; P_{z'}(z) = 0],$$

if we choose the number ρ sufficiently small. This inequality means that through every point $z' \in S$ of the hypersphere $U(z^0, \rho)$ it is possible to pass a $(2n-2)$-dimensional analytic surface $P_{z'}(z)=0$ where $P_{z'}(z)$ is a second-degree polynomial in z that lies entirely on the $(\varphi(z) \geqslant 0)$ side in the neighborhood $U(z, 2\rho)$. From this, it follows that the domain $U^-(z^0, \rho)=[z: \varphi(z)<0, |z-z^0|<\rho]$ is a domain of holomorphy since, for every boundary point z' of this domain, there exists a barrier, that is, a function $f_{z'}(z)$ that is holomorphic in $U^-(z^0, \rho)$ and that has a singularity at the point z' (see section 16.7). Specifically, we may take

$$f_{z'}(z)=\begin{cases} [P_{z'}(z)]^{-1}, & \text{if } \varphi(z')=0; \\ [\rho^2-(z-z^0)(\bar{z}'-\bar{z}^0)]^{-1}, & \text{if } |z'-z^0|=\rho. \end{cases}$$

Therefore, S is pseudoconvex in the sense of Cartan at the point z^0 on the $\varphi(z)<0$ side. From this and from the corollary of the theorem in section 18.3, we conclude that the hypersurface S is also pseudoconvex in the sense of Levi at the point z^0 on the $(\varphi(z)<0)$ side. This completes the proof of the theorem.

5. *Examples of locally pseudoconvex surfaces*

For $n=2$, conditions (60)-(61) for local pseudoconvexity are equivalent to the inequality

$$L(\varphi) \equiv -\begin{vmatrix} 0 & \dfrac{\partial\varphi}{\partial z_1} & \dfrac{\partial\varphi}{\partial z_2} \\ \dfrac{\partial\varphi}{\partial \bar{z}_1} & \dfrac{\partial^2\varphi}{\partial z_1 \partial \bar{z}_1} & \dfrac{\partial^2\varphi}{\partial \bar{z}_1 \partial z_2} \\ \dfrac{\partial\varphi}{\partial \bar{z}_2} & \dfrac{\partial^2\varphi}{\partial z_1 \partial \bar{z}_2} & \dfrac{\partial^2\varphi}{\partial z_2 \partial \bar{z}_2} \end{vmatrix} \geqslant 0.$$

The determinant $L(\varphi)$ is known as Levi's determinant [115].

Let us evaluate Levi's determinant for certain surfaces.

(1) The boundary of a complete multiple-circular domain is given by the equation (see section 13.6)

$$\varphi \equiv |z_1| - R(|z_2|) = 0, \quad |z_2| < R_0.$$

Therefore,

$$L(\varphi) = -\begin{vmatrix} 0 & \dfrac{1}{2}\sqrt{\dfrac{\bar{z}_1}{z_1}} & -\dfrac{1}{2}R'\sqrt{\dfrac{\bar{z}_2}{z_2}} \\ \dfrac{1}{2}\sqrt{\dfrac{z_1}{\bar{z}_1}} & \dfrac{1}{4|z_1|} & 0 \\ -\dfrac{1}{2}R'\sqrt{\dfrac{z_2}{\bar{z}_2}} & 0 & -\dfrac{1}{4}R''-\dfrac{1}{4|z_2|}R' \end{vmatrix} =$$

$$= -\frac{R^2}{16|z_2|^2}\frac{d^2 \ln R}{d(\ln|z_2|)^2}.$$

In this case, the condition $L(\varphi) \geqslant 0$ means that the function $\ln R(|z_2|)$ is convex with respect to $\ln|z_2|$ for $0 < |z_2| < R_0$; that is, the domain

$$G = [z: \quad |z_1| < R(|z_2|), \ |z_2| < R_0]$$

is pseudoconvex (see section 13.6). (The monotonic decrease of the function $R(\rho)$ in $[0, R_0)$ follows from the completeness of the multiple-circular domain G.)

(2) For the complete Hartogs domain

$$G = [z: \quad |z_1| < R(z_2), \ z_2 \in B]$$

(see section 12.13), let us consider that portion of the boundary of the form $\varphi \equiv |z_1| - R(z_2) = 0$, where $z_2 \in B$. We have

$$L(\varphi) = -\begin{vmatrix} 0 & \frac{1}{2}\sqrt{\frac{\bar{z}_1}{z_1}} & -\frac{\partial R}{\partial z_2} \\ \sqrt{\frac{z_1}{\bar{z}_1}} & \frac{1}{4|z_1|} & 0 \\ -\frac{\partial R}{\partial \bar{z}_2} & 0 & -\Delta R \end{vmatrix} = -\frac{R}{4}\Delta \ln R.$$

The condition $L(\varphi) \geqslant 0$ means that the function $-\ln R(z_2)$ is subharmonic in B; that is, the domain G is pseudoconvex (see section 12.3).

(3) Suppose that a portion of the boundary of the tubular domain $T_B = B + iR^2$ is $\varphi \equiv x_1 - \psi(x_2) = 0$ (see section 7.5). Then,

$$L(\varphi) = -\begin{vmatrix} 0 & \frac{1}{2} & -\frac{1}{2}\psi' \\ \frac{1}{2} & 0 & 0 \\ -\frac{1}{2}\psi' & 0 & -\frac{1}{4}\psi'' \end{vmatrix} = -\frac{1}{16}\psi''.$$

The condition $L(\varphi) \geqslant 0$ means that $\psi'' \leqslant 0$; that is, the curve $x_1 = \psi(x_2)$ is convex (cf. section 13.5).

(4) For the semitubular domain

$$G = [z: \quad v(z_2) < x_1 < V(z_2), \ z_2 \in B, \ |y_1| < \infty]$$

let us consider a portion of the boundary $\varphi_1 \equiv x_1 - V(z_2)$ or $\varphi_2 \equiv v(z_2) - x_1$, where $z_2 \in B$. We have

$$L(\varphi_1) = -\begin{vmatrix} 0 & \frac{1}{2} & \frac{\partial V}{\partial z_2} \\ \frac{1}{2} & 0 & 0 \\ \frac{\partial V}{\partial \bar{z}_2} & 0 & -\Delta V \end{vmatrix} = -\frac{1}{4}\Delta V, \quad L(\varphi_2) = \frac{1}{4}\Delta v.$$

Therefore, the conditions $L(\varphi_i) \geqslant 0$ (for $i=1, 2$) mean that the functions v and V are respectively subharmonic and superharmonic. This means that the domain G is pseudoconvex (see section 12.14).

6. *Supporting analytic sets*

If through a boundary point z^0 of a domain G it is possible to pass a $2k$-dimensional (for $1 \leqslant k \leqslant n-1$) analytic set (see section 10.12) lying outside G in a sufficiently small neighborhood of that point, we shall call this set a locally supporting $2k$-dimensional analytic set (surface) for the domain G at the point z^0.

The second part of the Levi-Krzoska theorem asserts that if in a neighborhood of a boundary point z^0 of the domain G, the boundary $\partial G = [z: \varphi(z) = 0]$ belongs to the class $C^{(2)}$ and if inequality (62) is satisfied at the point z^0 itself when condition (61) is satisfied, then there exists a locally supporting $(2n-2)$-dimensional analytic surface for the domain G at the point z^0 (see Fig. 39). In connection with this, the question arises as to the existence of locally supporting analytic sets. On this subject, we have the following

THEOREM. *If, in a neighborhood $U(z^0, r)$ of a boundary point z^0 of a domain G, the boundary $\partial G = [z: \phi(z) = 0]$ is such that $\phi \in C^{(2)}$ in $U(z^0, r)$ and*

$$(H(z^0; \varphi)a, \bar{a}) \geqslant \sigma |a|^2, \qquad \sigma > 0, \tag{66}$$

then the domain G is pseudoconvex in the sense of Cartan at the point z^0 and there exists a locally supporting $(2n-2)$-dimensional analytic set $P(z) = 0$ for the domain G at the point z^0, where $P(z)$ is a second-degree polynomial.

Proof: The assertion of the theorem has already been proven (see section 18.4) for the case in which $\operatorname{grad} \varphi(z^0) \neq 0$. Suppose that $\operatorname{grad} \varphi(z^0) = 0$. In this case, formula (64) yields

$$\varphi(z) = \operatorname{Re} \sum_{j,k} \frac{\partial^2 \varphi(z')}{\partial z_j \partial z_k}(z_j - z'_j)(z_k - z'_k) + \\ + (H(z^0; \varphi)(z - z'), \bar{z} - \bar{z}') + o(|z - z'|^2)$$

for all $z' \in \partial G \cap U(z^0, \rho)$ and $z \in U(z', 2\rho)$ if $\rho < r/3$. From this and from (66), it follows that

$$\varphi(z) = (H(z'; \varphi)(z - z'), \bar{z} - \bar{z}') + \\ + o(|z - z'|^2) \geqslant \frac{\sigma}{2} |z - z'|^2.$$

at points of the $(2n-2)$-dimensional analytic sets

$$P_{z'}(z) \equiv \sum_{j,k} \frac{\partial^2 \varphi(z')}{\partial z_j \partial z_k}(z_j - z'_j)(z_k - z'_k) = 0, \quad z' \in \partial G \cap U(z^0, \rho).$$

lying in the hyperspheres $U(z', 2\rho)$ if the number ρ is chosen sufficiently small. Now this inequality means that each analytic set $P_{z'}(z)=0$ for $z' \in \partial G \cap U(z^0, \rho)$ has no points in common with the set $G \cap U(z^0, \rho)$ (see Fig. 39). Just as in the proof of the second portion of the theorem in section 18.4, it follows from this that $G \cap U(z^0, \rho) = U^-(z^0, \rho)$ is an open set of holomorphy, that is, that the domain G is pseudoconvex in the sense of Cartan at the point $z^0 \in \partial G$, which completes the proof.

From this theorem and the theorem in section 12.10, we have the

COROLLARY. *Every pseudoconvex domain is the union of an increasing sequence of bounded domains that are locally pseudoconvex in the sense of Cartan.*

7. *Analytic hypersurfaces*

An analytic hypersurface is a $(2n-1)$-dimensional set S defined in a neighborhood $U(z^0, r)$ of each point $z^0 \in S$ by an equation of the form

$$f(z^0; z, t)=0, \qquad |t|<\varepsilon,$$

where the function $f(z^0; z, t)$ is holomorphic with respect to z in $U(z^0, r)$ for each $t \in (-\varepsilon, \varepsilon)$ and continuous with respect to t in $(-\varepsilon, \varepsilon)$ for each z in $U(z^0, r)$ and where $\sum_j \left|\frac{\partial f}{\partial z_j}\right| \neq 0$ for all $(z, t) \in U(z^0, r) \times (-\varepsilon, \varepsilon)$. For definiteness, we assume that $f(z^0; z^0, 0)=0$ and $\partial f / \partial z_1 \neq 0$. We note that ε depends on z^0.

Examples of analytic hvpersurfaces:

(a) The cut $C=[z: \ f(z)=t, a<t<b]$ where $f(z)$ is a holomorphic function in a neighborhood of C and $\operatorname{grad} \varphi \neq 0$ on C.

(b) The level surface of the absolute value of a holomorphic function $S_0=[z: \ |f(z)|=1]$, where $f(z)$ is a holomorphic function in a neighborhood of S_0 and $\operatorname{grad} f \neq 0$ on S_0. This is an example since $S_0=[z: \ f(z)=e^{it}, \ |t|<\infty]$.

It follows from the definition of an analytic hypersurface S that, through every point $z' \in S \cap U(z^0, \rho)$, where $\rho < r/3$, it is possible to pass the supporting $(2n-2)$-dimensional analytic surface $f(z^0; z, \rho')=0$, where $f(z^0; z', \rho')=0$, that lies on S in the sphere $U(z', 2\rho)$.

From this it follows that $U(z^0, \rho) \setminus S$ is an open set of holomorphy. To see this, note that, for every point z' of the boundary of this set, there exists a barrier function that is holomorphic in $U(z^0, \rho) \setminus S$ and that has a singularity at the point z': for points $z' \in S$, we may choose as a barrier the function $[f(z^0; z, \rho')]^{-1}$, and, for the remaining points of the boundary, we may choose the function

$$[\rho^2-(z-z')(\bar{z}^0-\bar{z}')]^{-1}$$

Then, we only need to use the theorem in section 16.7.

Thus, *an analytic hypersurface is locally convex in the sense of Cartan on both sides* (see section 18.2).

8. *Levi's theorem* [115]

For a hypersurface $S = [z : \phi(z) = 0]$ *of a class* $C^{(2)}$ *to be analytic, it is necessary and sufficient that*

$$(H(z; \varphi)\, a, \bar{a}) = 0 \tag{67}$$

on S *for all vectors* a *satisfying the condition*

$$(a, \operatorname{grad} \varphi(z)) = 0. \tag{68}$$

Proof of the necessity: Suppose that S is an analytic hypersurface class $C^{(2)}$. From what we have proven (see section 18.7), S is locally pseudoconvex in the sense of Cartan on both sides. Therefore, on the basis of the corollary to the theorem in section 18.3, the hypersurface S is locally pseudoconvex in the sense of Levi on both sides. From the Levi-Krzoska theorem (see section 18.4), we then conclude that $(H(z; \pm\varphi)\, a, \bar{a}) \geqslant 0$ on S when condition (68) is satisfied. It follows from this that Eq. (67) holds on S when condition (68) is satisfied.

Proof of the sufficiency: Suppose that Eq. (67) is satisfied on a hypersurface S of class $C^{(2)}$ for all vectors a satisfying condition (68). Suppose that $\partial\varphi/\partial z_1 \neq 0$ in a neighborhood of the point $z^0 \in S$. We shall show that in another (possibly smaller) neighborhood of this point, the hypersurface S is analytic, that is, that there exists a family $z_1 = z_1(\tilde{z}, \rho)$ of analytic surfaces that depend continuously on the real parameter ρ, that lie on S, and that exhaust all points of S. This means that the function $z_1(\tilde{z}, \rho)$ must, in a neighborhood of the point $\tilde{z}^0$, satisfy the equation for the surface S,

$$\varphi[z_1(\tilde{z}, \rho), \bar{z}_1(\tilde{z}, \rho), \tilde{z}, \bar{\tilde{z}}] = 0, \tag{69}$$

and the system of first-order partial differential equations

$$\frac{\partial z_1}{\partial z_j} = -\frac{\partial \varphi}{\partial z_j} \Big/ \frac{\partial \varphi}{\partial z_1}, \qquad \frac{\partial z_1}{\partial \bar{z}_j} = 0, \quad j = 2, 3, \ldots, n. \tag{70}$$

Here, the right-hand members of the first set of equations (70) are once continuously differentiable with respect to all arguments in some neighborhood of the point z^0. The second set of equations (70) constitute the Cauchy-Riemann conditions for the function z_1.

Let us now use the general theory of canonical systems of differential equations of the form

$$\frac{\partial u}{\partial x_j} = L_j(x, u), \qquad j = 1, 2, \ldots, n, \tag{71}$$

with the hypothesis that the functions L_j are once continuously differentiable with respect to all arguments (see, for example, Stepanov [53], p. 355). For the system (71) to have a solution, it is necessary and sufficient that

$$\frac{\partial L_j}{\partial x_k} + \frac{\partial L_j}{\partial u} L_k = \frac{\partial L_k}{\partial x_j} + \frac{\partial L_k}{\partial u} L_j, \qquad 1 \leqslant j, \ k \leqslant n. \tag{72}$$

Here, the solution u passing through the given point (u^0, x^0) is unique, and, thus, the manifold of solutions of the system (71) with conditions (72) depends on a single parameter. This dependence is continuous.

Let us apply this result to our system of differential equations (70) when the relationship (69) holds for two independent functions x_1 and y_1, where $z_1 = x_1 + iy_1$. If we use Eq. (69) to eleiminate one of the functions, let us say, y_1, and use the Cauchy-Riemann conditions, we obtain a canonical system of $2n-2$ equations in the function x_1. However, this procedure would lead to extremely tedious calculations. We can perform essentially the same calculations by a simpler method, namely, the technique of formal derivatives (see section 4.5). In these terms, the integrability conditions (72) take the form

$$\left.\begin{aligned} &\frac{d}{dz_k}\left(\frac{\partial \varphi}{\partial z_j} \Big/ \frac{\partial \varphi}{\partial z_1}\right) = \frac{d}{dz_j}\left(\frac{\partial \varphi}{\partial z_k} \Big/ \frac{\partial \varphi}{\partial z_1}\right), \\ &\frac{d}{d\bar{z}_k}\left(\frac{\partial \varphi}{\partial z_j} \Big/ \frac{\partial \varphi}{\partial z_1}\right) = 0, \qquad 2 \leqslant j, \ k \leqslant n, \end{aligned}\right\} \tag{73}$$

where the symbol $\frac{d}{dz}$ denotes the total derivative. The first set of equations (73) can be verified by direct calculation. Let us prove the validity of the second set of equations (73). From Eqs. (69) and (70), we have

$$-\frac{d}{d\bar{z}_k}\left(\frac{\partial \varphi}{\partial z_j} \Big/ \frac{\partial \varphi}{\partial z_1}\right) = \\ = \frac{1}{\left|\frac{\partial \varphi}{\partial z_1}\right|^2 \frac{\partial \varphi}{\partial z_1}} \begin{vmatrix} 0 & \frac{\partial \varphi}{\partial z_j} & \frac{\partial \varphi}{\partial z_1} \\ \frac{\partial \varphi}{\partial \bar{z}_k} & \frac{\partial^2 \varphi}{\partial z_j \partial \bar{z}_k} & \frac{\partial^2 \varphi}{\partial z_1 \partial \bar{z}_k} \\ \frac{\partial \varphi}{\partial \bar{z}_1} & \frac{\partial^2 \varphi}{\partial z_j \partial \bar{z}_1} & \frac{\partial^2 \varphi}{\partial z_1 \partial \bar{z}_1} \end{vmatrix}. \tag{74}$$

Furthermore, by eliminating a_1 from Eqs. (67) and (68) and using formula (74), we obtain the equation

$$\frac{\partial \varphi}{\partial z_1} \sum_{2 \leqslant j,\, k \leqslant n} \frac{d}{d\bar{z}_k} \left(\frac{\partial \varphi}{\partial z_j} \Big/ \frac{\partial \varphi}{\partial z_1} \right) a_j \bar{a}_k = 0,$$

which is valid for all $(a_2, a_3, \ldots, a_n)$. The validity of the second set of equations (73) then follows.

Thus, the conditions for integrability of the system (69)-(70) are satisfied. Consequently, this system has a one-parameter family of solutions $z_1 = z_1(\tilde{z}, \rho)$, where $|\rho| < \varepsilon$, which depends continuously on the real parameter ρ. On the basis of the second set of equations (70), the function $z_1(\tilde{z}, \rho)$ is holomorphic with respect to $\tilde{z}$ in some neighborhood of the point $\tilde{z}^0$. (On the basis of the Heine-Borel theorem, this neighborhood may be chosen independently of ρ if ε is sufficiently small.) Then through every point $z' \in S$ in some neighborhood of the point z^0 there passes a unique integral surface $z_1 = z_1(\tilde{z}, \rho')$ if $z_1' = z_1(\tilde{z}', \rho')$. This means that the hypersurface S coincides with the analytic surface $z_1 = z_1(\tilde{z}, \rho)$, where $|\rho| < \varepsilon$, in some neighborhood of the point $z^0 \in S$. This completes the proof of the theorem.

9. *Local pseudoconvexity of analytic hypersurfaces*

For a hypersurface $S = [z : \phi(z) = \sigma]$ of a class $C^{(2)}$ to be analytic, it is necessary and sufficient that it be locally pseudoconvex (in the sense of Cartan or Levi) on both sides (see Levi [116]).

Proof: If a surface S of class $C^{(2)}$ is locally pseudoconvex in the sense of Cartan on both sides, it is also locally pseudoconvex in the sense of Levi on both sides (see section 18.3). But then, from the Levi-Krzoska theorem (see section 18.4), Eq. (67) is satisfied on S when condition (68) holds. On the basis of Levi's theorem (see section 18.8), the hypersurface S is analytic. Conversely, if the class $C^{(2)}$ hypersurface S is analytic, it is locally pseudoconvex in th sense of Cartan on both sides (see section 18.7), which completes the proof.

10. *Local pseudoconvexity of piecewise-smooth hypersurfaces*

Suppose that two class $C^{(2)}$ hypersurfaces $S_i = [z\colon \varphi_i(z) = 0]$, for $i = 1, 2$, are locally pseudoconvex on the $\varphi_i(z) < 0$ side respectively and that they intersect each other along an edge F, which is assumed to be a $(2n-2)$-dimensional surface of class $C^{(2)}$. The question arises as to whether the piecewise-smooth hypersurface $S = [z\colon \varphi(z) = 0]$, where $\varphi(z) = \min[\varphi_1(z), \varphi_2(z)]$ (see Fig. 40), is locally pseudoconvex on the $\varphi(z) < 0$ side. Clearly, we can expect a violation of local pseudoconvexity only at points of the edge F since the union of two domains of holomorphy is not always a domain of holomorphy. Therefore, the answer to the question just raised is not always affirmative.

The theorem on the "embedded edge" that we shall prove in the next section asserts that if the hypersurface S is pseudoconvex at points of the edge F, it is necessary that F be an analytic surface. Whether this is a sufficient condition or not remains an open question. We note in passing that the question of local pseudoconvexity (in the sense of Cartan or Levi) of the hypersurface $\varphi^*(z)=0$, where $\varphi^*(z)=\max[\varphi_1(z), \varphi_2(z)]$, on the $\varphi^*(z)<0$ side is trivially answered in the affirmative since the intersection of two domains of holomorphy is always a domain of holomorphy (see section 16.7 and Fig. 40).

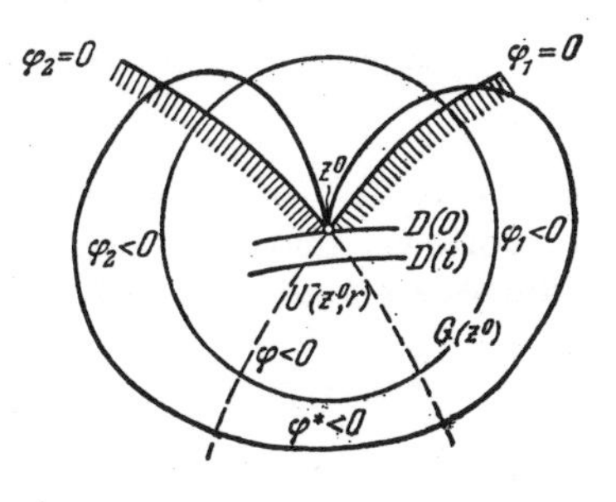

Fig. 40

LEMMA 1. *Suppose that two class* $C^{(2)}$ *hypersurfaces* $\phi_i(z) = 0$, *(for* $i = 1, 2$*) intersect at a point* z^0 *and that at that point the rank of the matrix*

$$\left\| \begin{matrix} \operatorname{grad} \varphi_1 \\ \operatorname{grad} \varphi_2 \end{matrix} \right\| \tag{75}$$

is 2. *Then, no domain* $G(z^0)$ *containing the domain* $U^-(z^0, r)$ *for some* r *and having* z^0 *on its boundary can be pseudoconvex in the sense of either Cartan or Levi at the point* z^0. *Here* (see Fig. 40),

$$U^-(z^0, r)=[z:\varphi(z)<0, \ |z-z^0|<r], \quad \varphi(z)= \\ =\min[\varphi_1(z), \varphi_2(z)].$$

Proof: Since the rank of the matrix (75) at the point z^0 is 2, there exist vectors p, a, and b such that

$$(p, \operatorname{grad}\varphi_i(z^0))=-1, \ i=1, 2, \quad (a, \operatorname{grad}\varphi_1(z^0))=1, \\ (a, \operatorname{grad}\varphi_2(z^0))=-1;$$

$$(b, \operatorname{grad}\varphi_i(z^0))+\frac{1}{2}\sum_{j,k}\frac{\partial^2\varphi_i(z^0)}{\partial z_j\,\partial z_k}a_j a_k- \\ -\frac{1}{2}(H(z^0; \varphi_i)a, \bar{a})=1, \quad i=1, 2. \tag{76}$$

Since $a\neq 0$, then, for sufficiently small ρ, we have $a+2b\lambda\neq 0$ for all $|\lambda|\leqslant\rho$. Therefore, the sets

$$D(t)=[z: \ z=z^0+pt+a\lambda+b\lambda^2, \ |\lambda|\leqslant\rho], \quad 0\leqslant t<\delta,$$

are "disks" (in the sense of section 17.2) that depend continuously on the parameter t. By virtue of formulas (63) and (76), at points of the "disks" $D(t)$,

$$\varphi_1(z^0+pt+a\lambda+b\lambda^2)=2u-v^2+2u^2(H(z^0;\ \varphi_1)a,\ \overline{a})+$$
$$+2t\,\mathrm{Re}(p,\ \mathrm{grad}\,\varphi_1(z^0+a\lambda+b\lambda^2))+o(t)+o(|\lambda|^2),$$
$$\varphi_2(z^0+pt+a\lambda+b\lambda^2)=-2u-v^2+2u^2(H(z^0;\ \varphi_2)a,\ \overline{a})+$$
$$+2t\,\mathrm{Re}(p,\ \mathrm{grad}\,\varphi_2(z^0+a\lambda+b\lambda^2))+o(t)+o(|\lambda|^2),$$

where $\lambda=u+iv$. From these relations and continuity considerations, it follows that, for given $r_1\leqslant r$, there exist sufficiently small numbers ρ and δ that $|pt+a\lambda+b\lambda^2|<r_1$ and

$$\varphi(z^0+pt+a\lambda+b\lambda^2)<0,\quad |\lambda|\leqslant\rho,\quad 0\leqslant t<\delta,\quad |\lambda|+t\neq 0.$$

This conclusion means that all the "disks" $D(t)$ for $0\leqslant t<\delta$, are contained in the domain $U^-(z^0,\ r_1)$ except the point z^0 corresponding to $\lambda=0$, which lies in the limit "disk" $D(0)$ (see Fig. 40). From this it follows that no domain $G(z^0)$ containing the domain $U^-(z^0,\ r)$ and having the point z^0 on its boundary can be pseudoconvex in the sense of Levi at the point z^0 (see section 18.1).

Let us show that $G(z^0)$ cannot be pseudoconvex in the sense of Cartan at the point z^0. Suppose that this were not the case. Then, each component of the open set $G(z^0)\cap U(z^0,\ r_1)$ for some $r_1\leqslant r$ would be a domain of holomorphy (see section 18.1). In this case, the "disk" theorem (see section 17.2) would be valid for $G(z^0)\cap U(z^0,\ r_1)$. But the sequence of "disks" $D(t)$ that we constructed above contradicts this theorem. This contradiction proves that the domain $G(z^0)$ cannot be pseudoconvex in the sense of Cartan at the point z^0. This completes the proof of the lemma.

LEMMA 2. *Suppose that two hypersurfaces $\phi_i(z)=0$ (for $i=1,2$) of class $C^{(2)}$ intersect along a $(2n-2)$-dimensional surface F of class $C^{(2)}$. If the rank of the matrix (75) is 1 at points on F, then F is an analytic surface.*

Proof: Since the rank of the matrix (75) is 1 on F, we have

$$\mathrm{grad}\,\varphi_2(z)=\eta(z)\,\mathrm{grad}\,\varphi_1(z),\qquad z\in F, \tag{77}$$

where $\mathrm{Im}\,\eta(z)\neq 0$. This is true because $F\in C^{(2)}$ and hence the rank of the matrix

$$\left\|\begin{matrix}\mathrm{grad}\,\varphi_1, & \overline{\mathrm{grad}\,\varphi_1}\\ \mathrm{grad}\,\varphi_2, & \overline{\mathrm{grad}\,\varphi_2}\end{matrix}\right\| \tag{78}$$

is F on 2.

Let $(dz_1,\ \dots,\ dz_n)=dz$ denote an arbitrary displacement vector lying in the plane tangent to the surface F at the point z and suppose that

$$\mathrm{Re}(dz,\ \mathrm{grad}\,\varphi_1(z))=0,\qquad \mathrm{Re}(dz,\ \mathrm{grad}\,\varphi_2(z))=0.$$

In view of (77), we easily obtain from this

$$(dz, \operatorname{grad} \varphi_i(z)) = 0, \quad i = 1, 2, \qquad z \in F.$$

These equations indicate that F is an analytic surface. This completes the proof of the lemma.

11. The "embedded edge" theorem

Suppose that two class $C^{(2)}$ hypersurfaces $\phi_i(z) = 0$ (for $i = 1, 2$) intersect along a $(2n - 2)$-dimensional class $C^{(2)}$ surface F. If the hypersurface $\phi(z) \equiv \min[\phi_1(z), \phi_2(z)] = 0$ is pseudoconvex in the sense of either Cartan or Levi at every point F on the $(\phi(z) < 0)$ side, then F is an analytic surface (see Kneser (54).

Proof: Since the hypersurface $\varphi(z) = 0$ is pseudoconvex in the sense of Cartan or Levi at points F on the $\varphi(z) < 0$ side, it follows that, for every $z^0 \in F$, there exists an r such that the domain $U^-(z^0, r)$ is pseudoconvex in the sense of Cartan or Levi at the point z^0. By applying Lemma 1 (see section 18.10), we conclude from this that the rank of the matrix (75) is equal to 1 or 0 on F. But the hypersurface $\varphi_1(z) = 0$ belongs to the class $C^{(2)}$, and hence $\operatorname{grad} \varphi_1(z) \neq 0$ on F. Therefore, the rank of the matrix (75) cannot be zero on F. Consequently, this rank is 1. From Lemma 2 (see section 18.10), the surface F is analytic, which completes the proof.

12. Local pseudoconvexity of smooth surfaces

For a $(2n - 2)$-dimensional class $C^{(2)}$ surface F to be analytic, it is necessary and sufficient that it be locally pseudoconvex.

Proof: Suppose that F is a locally pseudoconvex class $C^{(2)}$ surface. This means that, for every point $z^0 \in F$, there exists a hypersphere $U(z^0, r)$ in which F is defined by the equations $\varphi_i(z) = 0$ (for $i = 1, 2$, where $\varphi_i \in C^{(2)}$ and the rank of the matrix (78) is 2, and which has the property that $U(z^0, r) \setminus F$ is a domain of holomorphy). Since z^0 belongs to the boundary of the domain $U(z^0, r) \setminus F$, this domain is convex in the sense of Cartan at the point z^0. But then, if we set $G(z^0) = U(z^0, r) \setminus F$ in Lemma 1 of section 18.10, we conclude that the rank of the matrix (75) is 1 on F. From Lemma 2 of section 18.10, the surface F is analytic.

Conversely, suppose that F is an analytic surface. Then, in a neighborhood $U(z^0, r)$ of each point $z^0 \in F$, this surface is defined by the equation $f(z) = 0$, where $f(z)$ is holomorphic in $U(z^0, r)$ and $\operatorname{grad} f \neq 0$. From this it follows that $U(z^0, r) \setminus F$ is a domain of holomorphy since, for every boundary point z' of this domain, there exists a barrier function $f_{z'}(z)$, that is holomorphic in $U(z^0, r)$ and that has a singularity at the point z' (see section 16.7); specifically, we can take

$$f_{z'}(z) = \begin{cases} [f(z)]^{-1}, & \text{if} \quad z' \in F, \\ [r^2 - (z - z^0)(\overline{z'} - \overline{z^0})]^{-1}, & \text{if} \quad |z' - z^0| = r. \end{cases}$$

Thus, the surface F is pseudoconvex at the point z^0, which completes the proof.

13. *Removal of removable singularities of holomorphic functions*

Suppose that F is a connected (2n − 2)-dimensional surface of class $C^{(2)}$ and let U(F) denote a neighborhood of F. Suppose that $U(F) \setminus F$ is contained in a domain of holomorphy G. Then, either $F \subset G$ or $F \subset \partial G$. In the latter case, F is an analytic surface (see Hartogs [119]).

Proof: Suppose that some point of the surface F belongs to a domain of holomorphy G. Let us show that then $F \subset G$. Suppose that this is not the case. Since the surface F is assumed to be connected, there exist a hypersphere $U(z', r') \subset G$ (where $z' \in F$ and a point $z^0 \in F \cap \partial G \cap \partial U(z', r')$ (see Fig. 41). But the $(2n-2)$-dimensional surface F belongs to the class $C^{(2)}$. Therefore, in some neighborhood $U(z^0, r) \Subset U(F)$, it is given by equations $\varphi_i(z) = 0$ (for $i = 1, 2$, where $\varphi \in C^{(2)}$ and the rank of the matrix (78) is 2 in $U(z^0, r)$. But, by assumption, G is a domain of holomorphy and hence is pseudoconvex in the sense of Cartan at the point $z^0 \in \partial G$. If we set $G(z^0) = G$, in Lemma 1 (see section 18.10), we conclude that the rank of the matrix (75) is 1 at the point z^0, that is, that

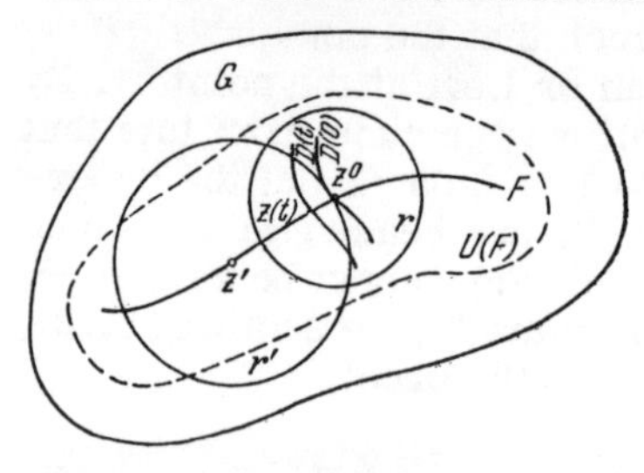

Fig. 41

$$\operatorname{grad} \varphi_2(z^0) = \eta \operatorname{grad} \varphi_1(z^0), \qquad \operatorname{Im} \eta \neq 0.$$

From this, by setting $c = \overline{\operatorname{grad} \varphi_1(z^0)} \neq 0$, we conclude that

$$\begin{vmatrix} (c, \operatorname{grad} \varphi_1(z^0)), & (\bar{c}, \operatorname{grad} \varphi_1(z^0)) \\ (c, \operatorname{grad} \varphi_2(z^0)), & (\bar{c}, \operatorname{grad} \varphi_2(z^0)) \end{vmatrix} = \tag{79}$$
$$= -2i \operatorname{Im} \eta |(c, \operatorname{grad} \varphi_1(z^0))|^2 \neq 0.$$

Because of the continuity, inequality (79) will be retained in $U(z^0, r)$ provided we take r sufficiently small.

Let us take a smooth curve $z = z(t)$ (where $0 < t \leqslant \eta$) that lies on F in $U(z', r') \cap U(z^0, r)$ [see Fig. 41] with end at the point $z^0 = z(0)$ and let us construct the family of "disks"

$$D(t) = [z: \quad z = z(t) + c\lambda, \ |\lambda| \leqslant \rho], \ 0 \leqslant t \leqslant \eta,$$

that are contained in $U(F)$ and that depend continuously on the parameter t. Let us show that for η and ρ sufficiently small, each "disk" $D(t)$ (for $0 \leqslant t \leqslant \eta$) has only one point $z(t)$ in common with the surface F. This assertion follows from the fact that the system of equations $\varphi_i[z(t)+c\lambda]=0$ (for $i=1, 2$) has a unique solution $\lambda=\bar{\lambda}=0$ in the circle $|\lambda| \leqslant \rho$ for each $t \in [0, \eta]$ since (by virtue of (79)) the Jacobian determinant of the functions $\varphi_i[z(t)+c\lambda]$ with respect to λ and $\bar{\lambda}$ does not vanish in $U(z^0, r)$. Furthermore, by construction, the curve $z=z(t)$ (for $0 < t \leqslant \eta$ lies in $U(z', r') \subset G$. Therefore, all the "disks" $D(t)$ (for $0 < t \leqslant \eta$) lie in the domain of holomorphy G. For the same reason, the limit "disk" $D(0)$ with the exception of the point z^0 corresponding to $\lambda=0$ also lies in G. From the "disk" theorem (see section 17.2), $z^0 \in G$. This contradiction proves that $F \subset G$.

Thus, either F lies entirely in G or F lies entirely on ∂G. In the latter case, F is a locally pseudoconvex surface and, hence, by virtue of the theorem in section 18.12, is an analytic surface. This completes the proof of the theorem.

COROLLARY. *Suppose that a k-dimensional* (*for* $k = 0, 1, \ldots, 2n-3$) *surface of class* $C^{(2)}$ F *is such that* $U(F) \backslash F$ *is contained in a domain of holomorphy* G. *Then*, $F \subset G$.

19. GLOBAL PSEUDOCONVEXITY

1. *Oka's theorem*

As was shown in section 17.1, every domain of holomorphy is a pseudoconvex domain. The converse assertion that every pseudoconvex domain is a domain of holomorphy constitutes the famous problem of Levi [116], which he posed as far back as 1911. The Levi problem was first solved by Oka for $n=2$ in 1942 (see [19]). For $n \geqslant 2$, this problem was solved in 1953-1954 by Oka [20], Norquet [21], and Bremermann [15]. The solution to the Levi problem for $n=2$ is also given in detail in the book by Fuks [2].

Solution of the Levi problem reduces to proving the following preparation theorem, discovered by Oka.

If G is a bounded domain such that, for certain numbers a_1 *and* a_2, *where* $a_1 < a_2$, *the sets*

$$G_1 = [z: \quad x_1 > a_1, \quad z \in G],$$

$$G_2 = [z: \quad x_1 < a_2, \quad z \in G], \quad a_1 < a_2,$$

are open sets of holomorphy, then G is a domain of holomorphy.

The proof of this theorem is rather complicated and requires supplementary information on the theory of functions of several complex variables that we are not treating in the present book (for example, solvability of Cousin's problem). Therefore, we shall not go into the proof of this theorem. Several variations of the proof occur in the original references cited. We note that this theorem has a trivial analog for convex domains: If the domains G_1 and G_2 are convex, the domain is also convex.

Solution of the Levi problem reduces to proving the preparation theorem of Oka by using the following considerations: Since every pseudoconvex domain is the union of an increasing sequence of bounded domains that are locally convex in the sense of Cartan (see section 18.6), it follows on the basis of the Behnke-Stein theorem (see section 16.10) that it will be sufficient to solve the Levi problem for bounded domains that are locally pseudoconvex in the sense of Cartan.

Suppose that G is a bounded domain that is locally pseudoconvex in the sense of Cartan. According to the Heine-Borel theorem, there exists a positive number r such that all the components of every open set $U(z^0, r) \cap G$ for $z^0 \in \partial G$ are domains of holomorphy. Since every component of the intersection of a finite number of domains of holomorphy is a domain of holomorphy, there exists a positive number δ such that all the components of each open set $K(z^0, \delta) \cap G$, for $z^0 \in C^n$, where $K(z^0, \delta)$ is a cube of side 2δ with center at the point z^0, are domains of holomorphy. In the space R^{2n}, let us draw planes parallel to the coordinate axes at a distance δ from each other. This partitions the domain G into a finite number of open sets of holomorphy $G_k = K\left(z^{(k)}, \frac{\delta}{2}\right) \cap G$ for $k = 1, 2, \ldots N$. Suppose that the cubes $K\left(z^{(k)}, \frac{\delta}{2}\right)$ and $K\left(z^{(l)}, \frac{\delta}{2}\right)$ are adjacent, let us say, along the face $x_1 = a_1$; that is, $x_1^{(k)} = a_1 - \delta/2$, $x_1^{(l)} = a_1 + \delta/2$ $y_1^{(k)} = y_1^{(l)}$, $z_j^{(k)} = z_j^{(l)}$, for $j \geqslant 2$ (see Fig. 42).

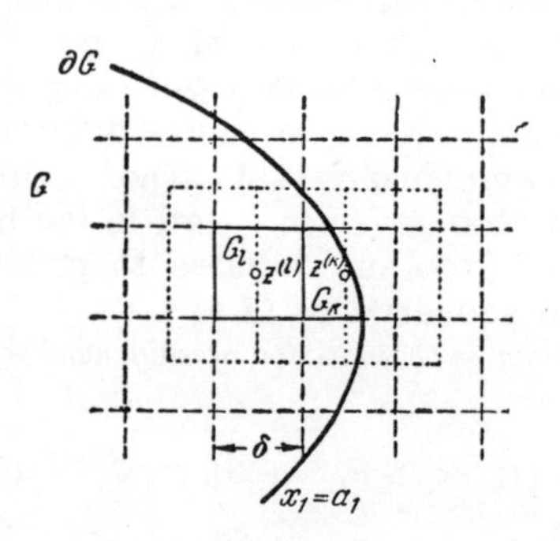

Fig. 42

On the basis of Oka's preparation theorem, we conclude that the set

$$G_k \cup G_l \cup [z: \ z \in G, \ x_1 = a_1] = \Big[z: \ |x_1 - a_1| < \delta,$$
$$|y_1 - y_1^{(k)}| < \frac{\delta}{2}, \ |x_j - x_j^{(k)}| < \frac{\delta}{2}, \ |y_j - y_j^{(l)}| < \frac{\delta}{2},$$
$$j = 2, 3, \ldots, n, \ z \in G\Big]$$

is an open set of holomorphy. If we apply this theorem to all adjacent faces of all the cubes $K\left(z^{(k)}, \delta\right)/2$, for $k=1, 2, \ldots, N$, we see that G is a domain of holomorphy.

Thus, we have the following theorem (see Oka [19, 20]). *For a given domain to be a domain of holomorphy, it is necessary and sufficient that it be a pseudoconvex domain.*

2. *Some properties of domains of holomorphy*

Thus, (on the basis of Oka's theorem), the class of (single-sheeted) domains of holomorphy coincides with the class of pseudoconvex domains, the geometric properties of which were studied in section 12. This makes it possible to reformulate for domains of holomorphy the properties proven for pseudoconvex domains, and vice versa.

Let us state some of these properties.

Every domain in C^1 is a domain of holomorphy (see section 12.4; cf. section 8.2).

For a domain to be a domain of holomorphy, it is necessary that the continuity principle apply to it and sufficient that the weak continuity principle apply to it (see sections 12.4 and 17.1).

For a domain G to be a domain of holomorphy, it is necessary and sufficient that the function $\ln \Delta_{a,G}(z)$ be plurisubharmonic for all a such that $|a| = 1$ (see section 12.5).

For a domain G to be a domain of holomorphy, it is necessary and sufficient that there be a function that is plurisubharmonic in G and that approaches $+\infty$ everywhere on ∂G (see section 12.8).

If a domain G can be represented in the form

$$G = [z: \ V(z) < 0, \ z \in U(\overline{G})],$$

where $V(z)$ is a plurisubharmonic function in a domain $U(\overline{G})$, then G is a domain of holomorphy (see section 12.7).

Suppose that G is a domain of holomorphy and that a function $V(z)$ is plurisubharmonic in G. Then, $[z: \ V(z) < 0, \ z \in G]$ is an open set of holomorphy (see section 12.9).

For a domain G to be a domain of holomorphy, it is necessary and sufficient that it be the union of a sequence of domains of holomorphy G_α of the form

$$G_\alpha = [z: \ V_\alpha(z) < 0, \ z \in U(\overline{G}_\alpha)], \quad \alpha = 1, 2, \ldots,$$

such that $G_\alpha \Subset G_{\alpha+1} \Subset G$ and $V_\alpha(z) \in C^{(\infty)}$ in $U(\overline{G}_\alpha)$ and

$$(H(z, V_\alpha)a, \bar{a}) \geqslant \frac{1}{\alpha}|a|^2, \quad z \in U(\overline{G}_\alpha) \quad \text{(see section 12.10).}$$

For a domain G to be a domain of holomorphy, it is necessary and sufficient that it be locally pseudoconvex in the sense of Cartan (see section 18.3).

Let us now consider a number of important examples of domains of holomorphy.

3. *Multiple-circular and tubular domains of holomorphy*

If a multiple-circular domain is a domain of holomorphy, it is a logarithmically convex domain. If a multiple-circular domain with center at zero is complete (or if it does not contain points of the manifold $z_1 z_2 \ldots z_n = 0$) and is logarithmically convex, then it is a domain of holomorphy.

It follows from this that a complete logarithmically convex domain of convergence of a power series.

In section 14.4, this assertion was proven for bounded domains. We have now proven it for the general case with the aid of Oka's theorem. Let us now prove it without using Oka's theorem.

Suppose that G is a complete logarithmically convex domain: $G = G^*$. Then, G is the union of an increasing sequence of complete bounded logarithmically convex domains $G_R = G \cap S(0, RI)$. From what we have proven, G_R is a region of absolute convergence of some power series. Let us denote by $f_R(z)$ the sum of that power series. Suppose that $z^0 \in \partial G_R$. Then, there exists a point $z' \in \partial G_R$, where $|z_j^0| = |z_j'|$ for $j = 1, 2, \ldots, n$, at which the function $f_R(z)$ is not holomorphic (because otherwise, from the Heine-Borel theorem, the function f_R would be holomorphic in a larger domain than G_R and, from the theorem in section 14.1, G_R would not be the domain of absolute convergence of the power series for f_R). Therefore, the function $f_R\left(\ldots, z_j e^{i\arg z_j' - i\arg z_j^0}, \ldots\right)$ is a barrier at the point $z^0 \in \partial G_R$. According to the theorem in section 16.7, G_R is a domain of holomorphy. According to the Behnke-Stein theorem (see section 16.10), G is a domain of holomorphy and, hence, by virtue of the theorem in section 14.1, is the domain of convergence of a power series.

The following assertion follows from Oka's theorem and the theorem in section 13.5:

For a tubular domain of holomorphy, it is necessary and sufficient that it be a convex domain (see section 13.5).

4. *Hartogs domains of holomorphy*

For a complete Hartogs domain

$$[z: \quad |z_1| < e^{V(\tilde{z})}, \quad \tilde{z} \in B]$$

to be a domain of holomorphy, it is necessary and sufficient that B be a domain of holomorphy and that V(z) be a plurisuperharmonic function in B (see section 12.13).

For a Hartogs domain of the form

$$G = [z: \quad e^{v(\tilde{z})} < |z_1| < e^{V(\tilde{z})}, \ \tilde{z} \in B],$$

where B is a domain of holomorphy in C^{n-1}, to be a domain of holomorphy, it is necessary and sufficient that the function $v(\tilde{z})$ (resp. $V(\tilde{z})$) be a plurisubharmonic (resp. plurisuperharmonic) function in B.

Proof: The sufficiency of the conditions follows from the theorem in section 12.13 and from Oka's theorem (see section 19.1). To prove the necessity, suppose that G is a domain of holomorphy. Then, there exists a function $f(z)$ that is holomorphic in G but not holomorphic in any larger domain. According to the theorem in section 15.6, this function is holomorphic in the domain

$$G_f = [z: \quad e^{v_f(\tilde{z})} < |z_1| < e^{V_f(\tilde{z})}, \ \tilde{z} \in B],$$

where $v_f(\tilde{z})$ (resp. $(V_f(\tilde{z}))$) is plurisubharmonic (resp. plurisuperharmonic) in B and satisfies the inequality $v_f(\tilde{z}) \leqslant v(\tilde{z})$ (resp. $(V_f(\tilde{z}) \geqslant V(\tilde{z}))$). From this it follows that $G \subset G_f$. But G is a domain of holomorphy of the function f. Therefore, $G = G_f$. It remains to show that

$$v_f(\tilde{z}) = v(\tilde{z}), \ V_f(\tilde{z}) = V(\tilde{z}), \quad \tilde{z} \in B. \tag{80}$$

Let us suppose that $v_f(\tilde{z}^0) < v(\tilde{z}^0)$ at some point $\tilde{z}^0 \in B$. Since the function $v_f(\tilde{z})$ is upper-semicontinuous in B, there exists a neighborhood $U(\tilde{z}^0, r)$ such that $v_f(\tilde{z}) < v(\tilde{z}^0)$ for $\tilde{z} \in U(\tilde{z}^0, r)$. But this contradicts the equation $G_f = G$. Therefore, $v_f(\tilde{z}) = v(\tilde{z})$ for $\tilde{z} \in B$. The second of Eqs. (80) is proven in an analogous manner. This completes the proof.

5. *Semitubular domains of holomorphy*

Let G be a semitubular domain of the form

$$G = [z: \quad v(\tilde{z}) < x_1 < V(\tilde{z}), \ \tilde{z} \in B, \ |y_1| < \infty],$$

where B is a domain of holomorphy in C^{n-1}. For G to be a domain of holomorphy, it is necessary and sufficient that the functions $v(\tilde{z})$ and $-V(\tilde{z})$ be plurisubharmonic in B (see Bremermann [18]).

Proof: The sufficiency of the conditions follows from the theorem in section 12.14 and Oka's theorem (see section 19.1). Let us prove the necessity. Since G is a domain of holomorphy, it follows from Oka's theorem that the function $\ln \Delta_G(z)$ is plurisubharmonic in G. Since G is a semitubular domain, this function is independent of y_1 because

$$-\ln \Delta_G(z) = -\ln \Delta_D(x_1, \tilde{z}),$$

where

$$D = [(x_1, \tilde{z}): v(\tilde{z}) < x_1 < V(\tilde{z}); \tilde{z} \in B] \subset R^{2n-1}$$

is the base of G. Therefore, the function

$$\varphi(x_1, \tilde{z}) = \max[-\ln \Delta_D(x_1, \tilde{z}), x_1^2 + |\tilde{z}|^2]$$

is plurisubharmonic in G and approaches $+\infty$ everywhere on ∂D.

The pseudoconformal transformation (see section 7.1) $\zeta_1 = e^{z_1}$, $\tilde{\zeta} = \tilde{z}$ maps the domain G onto the Hartogs domain

$$G_1 = [\zeta: e^{v(\tilde{\zeta})} < |\zeta_1| < e^{V(\tilde{\zeta})}, \tilde{\zeta} \in B].$$

The inverse transformation

$$z_1 = \ln \zeta_1 \equiv \ln |\zeta_1| + i \arg \zeta_1, \quad \tilde{z} = \tilde{\zeta}$$

is not single-valued. However, since the function $\varphi(x_1 \tilde{z})$ is independent of y_1, this inverse transformation maps it into the single-valued function $\varphi(\ln|\zeta_1|, \tilde{\zeta})$, which is plurisubharmonic in the domain G_1 (see remark in section 10.11). Furthermore, since the function $\varphi(x_1, \tilde{z})$ approaches $+\infty$ everywhere on ∂D (in the sense of the definition in section 1.6), the function $\varphi(\ln|\zeta_1|, \tilde{\zeta})$ approaches $+\infty$ everywhere on ∂G_1. Therefore, G_1 is a domain of holomorphy (see section 19.2). But then, the functions $v(\tilde{z})$ and $-V(\tilde{z})$ are plurisubharmonic in B (see section 19.4), which completes the proof.

20. ENVELOPES OF HOLOMORPHY

1. *Definition of an envelope of holomorphy*

We have seen from numerous examples that the space C^n, for $n \geqslant 2$, contains domains G for which there exist nontrivial holomorphic extensions $\tilde{G}$ (see section 8.4). Obviously, this fact is connected with the geometric nature of the domain G itself and is independent of any particular function that is holomorphic in G. The question naturally arises as to how we may construct a domain that will be the *largest* of all the holomorphic extensions $\tilde{G}$ of the domain G. It turns out that this problem cannot always be solved in the class of single-sheeted domains (see section 20.2). However, in the broader class of nonsingle-sheeted domains (see section 8.6), this problem is always solvable even for an arbitrary nonsingle-sheeted domain G (see Fuks [1], section 13). Let us suppose now that the domain G is such that it does have a largest (single-sheeted) holomorphic extension. We shall call this extension the *envelope of holomorphy* of the domain G and we shall denote it by $H(G)$. Clearly, $G \subset H(G)$.

Thus, we shall call a domain $H(G)$ possessing the following properties an *envelope of holomorphy*: (1) Every function that is holomorphic in G is holomorphic (and single-valued) in $H(G)$; (2) $H(G)$ contains every holomorphic extension of the domain G (see Cartan and Thullen [7]).

It follows from this definition that every domain in the space C^1 coincides with its envelope of holomorphy (see section 19.2). Therefore, the concept of an envelope of holomorphy in the space C^1 has no meaning. It plays an important role in the spaces C^n, for $n \geqslant 2$, since, as we recall, these spaces contain domains G for which $G \neq H(G)$. Clearly, if G is a domain of holomorphy, then $H(G) = G$.

In what follows, unless the contrary is explicitly stated, we shall consider only (bounded and single-sheeted) domains possessing single-sheeted envelopes of holomorphy.*

2. *An example of a single-sheeted domain whose envelope of holomorphy is nonsingle-sheeted* (see Thullen [65]).

Consider a Hartogs domain of the form

$$G = \Big[(z_1, z_2):\ -\frac{1}{2} + \arg z_2 < |z_1| < \frac{1}{2} + \arg z_2,\quad \frac{1}{2} < |z_2| < 1,\ 0 < \arg z_2 < 2\pi\nu\Big],$$

where ν is an integer greater than unity (see Fig. 43). The Hartogs domain G contains points of its plane of symmetry $z_1 = 0$. (In the space of the variables $|z_1|$, x_2, and y_2, this domain is a spiral that winds about the $|z_1|$-axis.) From the theorem in section 15.2, every function $f(z)$ that is holomorphic in the domain G can be holomorphically continued to every point of the smallest complete Hartogs domain $\pi(G)$ containing the given domain G:

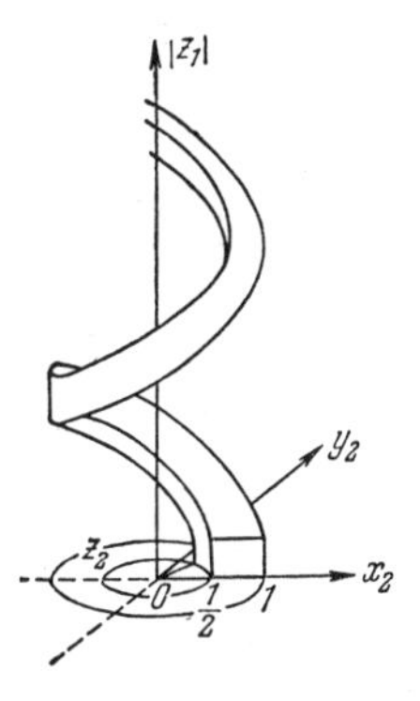

Fig. 43

$$\pi(G) = [z:\ z_1 = \lambda z_1',\ z_2 = z_2',\ z' \in G,\ |\lambda| \leqslant 1].$$

*It would be extremely interesting to find a geometric characteristic of the class of single-sheeted domains possessing single-sheeted envelopes of holomorphy.

However, not every such function is single-valued in $\pi(G)$. For example, the function $\ln z_2 = \ln|z_2| + i \arg z_2$ is holomorphic in G but nonsingle-valued in $\pi(G)$. Specifically, at the point $(0, z_2) \in \pi(G)$, where $1/2 < |z_2| < 1$ and $0 < \arg z_2 < 2\pi\nu$, this function assumes the $\nu - 1$ distinct values $\mathrm{Ln}\, z_2 + 2\pi i k$, for $k = 1, 2, \ldots, \nu - 1$. It follows immediately from this that the domain G cannot be a single-sheeted envelope of holomorphy: its envelope of holomorphy is at least $(\nu - 1)$-sheeted.

Proof: On the other hand, there exist examples of nonsingle-sheeted domains whose envelopes of holomorphy are single-sheeted (see Fuks [1], p. 227). Such domains include all nonsingle-sheeted multiple-circular domains that contain their centers and all nonsingle-sheeted tubular domains (see section 21.2).

3. *Properties of envelopes of holomorphy*

An envelope of holomorphy $H(G)$ of a domain G is a domain of holomorphy.

Proof: If $H(G)$ were not a domain of holomorphy, it would also not be a holomorphically convex domain (see section 16.6). But then, reasoning as in the proof of the theorem in section 16.3, we see that there exists a point $z^0 \in \partial H(G)$, such that every function $f \in H_G$ is holomorphic in some neighborhood $S(z^0, \rho I)$, where ρ is independent of f. But this contradicts the definition of an envelope of holomorphy $H(G)$. Therefore, $H(G)$ is a holomorphically convex domain, that is, a domain of holomorphy (see section 16.6), which completes the proof.

From this it follows on the basis of Oka's theorem (see section (19.1) that *$H(G)$ is a pseudoconvex envelope of the domain G (and is the smallest one); that is,*

$$H(G) = \operatorname{int} \bigcup_{G' \supset G} G', \tag{81}$$

where G' is an arbitrary pseudoconvex domain containing G.

It follows from (81) that *if $G_1 \subset G_2$, then $H(G_1) \subset H(G_2)$.*

Every function $f(z)$ that is holomorphic in G assumes in $H(G)$ those values that it assumes in G and only those values. In particular,

$$\sup_{z \in G} |f(z)| = \sup_{z \in H(G)} |f(z)|. \tag{82}$$

Proof: If $f(z) \neq a$ in G, then $f(z) \neq a$ in $H(G)$ since, otherwise, the function $[f(z) - a]^{-1}$ would be holomorphic in G but not holomorphic in $H(G)$, which is impossible.

From this follows the corollary: *If a domain G is bounded, its envelope of holomorphy $H(G)$ is also bounded.*

Proof: On the basis of Eq. (82), we have

$$\sup_{z \in H(G)} |z|^2 \leqslant \sum_{j=1}^{n} \sup_{z \in H(G)} |z_j^2| = \sum_{j=1}^{n} \sup_{z \in G} |z_j^2| < +\infty.$$

If $G_\alpha \subset G_{\alpha+1}$, and $\bigcup_\alpha G_\alpha = G$, then $\bigcup_\alpha H(G_\alpha) = H(G)$.

To see this, let set $\bigcup_\alpha H(G_\alpha) = B$. From the Behnke-Stein theorem (see section 16.10), we conclude that B is a domain of holomorphy. Since every function that is holomorphic in G is holomorphic in B, is holomorphic in G_α, in $H(G_\alpha)$, and, consequently, in B, it follows that $B \subset H(G)$. But $B \supset G$. Therefore, $H(G) = B$, which completes the proof.

4. *Other properties of envelopes of holomorphy*

Suppose that $G' \Subset G$ *and* $\phi_G(G') = r > 0$. *Then,*

$$\delta_{H(G)}[H(G')] \geqslant r.$$

Proof: Suppose that $z^0 \in H(G')$. From Eq. (82) with G placed by G', we derive tne inequality

$$|f(z^0)| \leqslant \sup_{z \in H(G')} |f(z)| = \sup_{z \in G'} |f(z)| \tag{83}$$

which holds for all functions f that are holomorphic in $H(G')$. Since $G' \subset G$, we have $H(G') \subset H(G)$. Thus, it follows from (83) that the maximum principle holds for the sets $\{z^0\}$ and G' with respect to the absolute values of the functions that are holomorphic in $H(G)$. Since $H(G)$ is a domain of holomorphy, we see, by applying inequality (43) (see Sections 16.4 and 16.7), that

$$\delta_{H(G)}(z^0) \geqslant \delta_{H(G)}(G') \geqslant \delta_G(G') = r$$

for all $z^0 \in H(G')$, which completes the proof.

COROLLARY. *If* $G' \Subset G$, *then* $H(G') \Subset H(G)$.

THEOREM. *Suppose that a function* $f(z)$ *is holomorphic in a bounded domain* G *and possesses a singularity at a point* $z^0 \in \partial H(G)$. *Then, this function has a singularity at some point of the boundary* ∂G.

Proof: Suppose that this theorem is not true. Then, the function $f(z)$ is holomorphic in some neighborhood U of the domain $\bar{G}$. Consequently, $f(z)$ is also holomorphic in the envelope of holomorphic $H(U)$. Since G is a bounded domain, we have $G \Subset U$. From this we conclude that $H(G) \Subset H(U)$, which contradicts the fact that $f(z)$ has a singularity at the point $z^0 \in \partial H(G)$. This contradiction proves the theorem.

It follows from this theorem that if, to every boundary point z^0 of a bounded domain of holomorphy G there corresponds a function $f_{z^0}(z)$ that is holomorphic in $\bar{G} \setminus \{z^0\}$, then there is no subdomain $G' \subset G$ such that $H(G') = G$ and $\partial G \setminus \partial G' \neq \varnothing$.

For example, for the hypersphere $|z| < 1$, the function $(1 - z\bar{z}^0)^{-1}$ would be such a function $f_{z^0}(z)$. Therefore, the hypersphere is not a domain of holomorphy of any of its subdomains G' for which $[z : |z| = 1] \setminus \partial G' \neq \varnothing$.

5. The Cartan-Thullen theorem

As a preliminary we prove:

Suppose that a biholomorphic mapping $\xi = \xi(z)$ maps a domain G onto a domain G_1. Suppose also that it maps $H(G)$ one-to-one onto a domain D_1. Then, $H(G)$ is single-sheeted and $H(G_1) = D_1$.

Proof: The functions $\zeta_j(z)$, (for $j=1, 2, \ldots, n$) and $\partial(\zeta)/\partial(z)$, which are holomorphic in G, are holomorphic (and single-valued) in $H(G)$. Furthermore, since $\partial(\zeta)/\partial(z) \neq 0$ in G, it follows that $\partial(\zeta)/\partial(z) \neq 0$ in $H(G)$ (see section 20.3). Therefore, the transformation $\zeta = \zeta(z)$ maps $H(G)$ holomorphically (and, by hypothesis, one-to-one) onto the domain $D_1 \supset G_1$ (see section 7.1). But then, D_1 is a domain of holomorphy (see section 16.8). Let $f(\zeta)$ be an arbitrary function that is holomorphic in G_1. Then, the function $f[\zeta(z)]$ is holomorphic in G (see section 4.6) and, consequently, in $H(G)$. By performing the inverse transformation $z = z(\zeta)$ (which is assumed to be single-valued in D_1), we conclude that the function $f(\zeta)$ is holomorphic (and single-valued) in D_1. Since D_1 is a domain of holomorphy and contains G_1, this means that $H(G_1) = D_1$, which completes the proof.

A consequence of this theorem is the following

THEOREM (Cartan-Thullen [7]). *The set of automorphisms of the envelope of holomorphy $H(G)$ of a domain G contains the set of automorphisms of any domain G' for which $H(G') = H(G)$.*

In particular, the group of automorphisms of the domain $H(G)$ contains the group of automorphisms of the domain G as a subgroup.

6. The envelope of holomorphy of the product of two domains

As a preliminary, we give the following theorem, which is a generalization of Hartogs theorem (see section 15.3).

If a function $f(z, w)$ is holomorphic in the domain $G \times D_1$, where $G \subset C^n$ and $D_1 \subset C^m$, and is holomorphic with respect to w in a domain $D \subset D_1$ for every fixed $z \in G$, then f is holomorphic in $G \times D$.

The proof of this theorem is analogous to the proof of Hartogs' theorem in section 15.3 and is done by expanding the function $f(z,w)$ in a generalized Hartogs series

$$f(z, w) = \sum_{|\alpha| \geqslant 0} f_\alpha(z) w^\alpha.$$

Therefore, we shall not stop to write out the proof. It appears in the book by Bochner and Martin [3], Chapter VII.

We have the formula

$$H(G \times D) = H(G) \times H(D). \tag{84}$$

Proof: By twice applying the theorem just given, we see that every function $f(z, w)$ that is holomorphic in $G \times D$, is holomorphic in $H(G) \times H(D)$. Since the domains $H(G)$ and $H(D)$ are domains of holomorphy, it follows that $H(G) \times H(D)$ is a domain of holomorphy. The validity of Eq. (84) follows from this.

21. CONSTRUCTION OF ENVELOPES OF HOLOMORPHY

In this section, we shall look at a number of examples of the construction of envelopes of holomorphy.

1. *Envelopes of holomorphy of tubular domains*

The envelope of holomorphy $H(T_B)$ of a tubular domain T_B coincides with the convex envelope $O(T_B)$ of the domain T_B; that is,

$$H(T_B) = O(T_B) = T_{O(B)}.$$

Proof: It follows from Bochner's theorem (see section 17.5) that $O(T_B) \subset H(T_B)$. But $O(T_B) = T_{O(B)}$, which completes the proof.

If we assume in advance that $H(T_B)$ is single-sheeted, this theorem can be proven in a different manner without using Bochner's theorem (see Bremermann [18]).

Specifically, since $O(T_B)$ is a domain of holomorphy and $O(T_B) \supset T_B$, we have $H(T_B) \subset O(T_B)$. Let T_{B_1} be the largest tubular domain contained in $H(T_B)$. (Since $H(T_B)$ is assumed to be single-sheeted, the domain T_{B_1} exists.) From the construction, we have $T_B \subset T_{B_1} \subset H(T_B)$. Let us show that $T_{B_1} \neq H(T_B)$. Let us suppose, to the contrary, that $T_{B_1} = H(T_B)$. Then, on ∂T_{B_1}, there exists a point z^0 in the interior of $H(T_B)$ such that $\Delta_{H(T_B)}(z^0) > 0$. From the Cartan-Thullen theorem, $H(T_B)$ is a tubular domain. Therefore,

$$\Delta_{H(T_B)}(x^0 + iy) = \Delta_{H(T_B)}(z^0) > 0, \qquad |y| < \infty.$$

From this it follows that the domain T_{B_1} can be enlarged somewhat in such a way that the enlarged domain will be tubular and will be contained in $H(T_B)$. But this contradicts the maximality of T_{B_1}. Therefore, $H(T_B) = T_{B_1}$. This means that T_{B_1} is a domain of holomorphy and hence B_1 is a convex domain (see section 19.3). From this and from the inclusions $T_B \subset T_{B_1} = H(T_B) \subset O(T_B) = T_{O(B)}$, it follows that $B_1 = O(B)$, which completes the proof.

EXAMPLE (Ruelle's lemma [50]). *The envelope of holomorphy of a domain $T^C \setminus \epsilon$, where $T^C = R^3 + iC$ is a tubular domain with base (cone) $C = [y : y_j > 0, j = 1, 2, 3]$ and $\epsilon = [z : 0 \leq \arg z_1 \leq \arg z_2 \leq \arg z_3 \leq \pi]$, coincides with T^C.*

Proof: Consider the tubular domain $T^B = R^3 + iB$ in the space C^3 of the variables $\zeta_j = \xi_j + i\eta_j$, for $j = 1, 2, 3$ where

$$B = [\eta\colon\ 0 < \eta_j < \pi,\ j = 1,\ 2,\ 3] \setminus [\eta\colon\ 0 \leqslant \eta_1 \leqslant \eta_2 \leqslant \eta_3 \leqslant \pi].$$

From the theorem that we have just proven, $H(T^B) = T^{O(B)}$, where $O(B) = [\eta\colon\ 0 < \eta_j < \pi,\ j = 1,\ 2,\ 3]$ (see Fig. 44). The mapping

$$z_j = e^{\zeta_j},\quad \zeta_j = \ln z_j \equiv \ln |z_j| + i \arg z_j,\quad j = 1,\ 2,\ 3$$

biholomorphically maps the domain T^B onto $T^C \setminus \varepsilon$ and the domain $T^{O(B)}$ onto T^C. From the theorem in section 20.5, $H(T^C \setminus \varepsilon) = T^C$, which completes the proof.

2. *Envelopes of holomorphy of multiple-circular domains*

The envelope of holomorphy of a multiple-circular domain G that contains its center coincides with the logarithmically convex envelope G^* to that domain.

Proof: Every function $f(z)$ that is holomorphic in G is holomorphic in the smallest complete multiple-circular domain $\pi(G)$ containing G (see section 14.1). But then, this function is holomorphic in the logarithmically convex envelope $[\pi(G)]$ (see section 14.4). Since the logarithmically convex envelope of a complete multiple-circular domain is a complete logarithmically convex envelope (see section 14.3), it follows that $[\pi(G)]^*$ is a domain of holomorphy (see section 19.3).

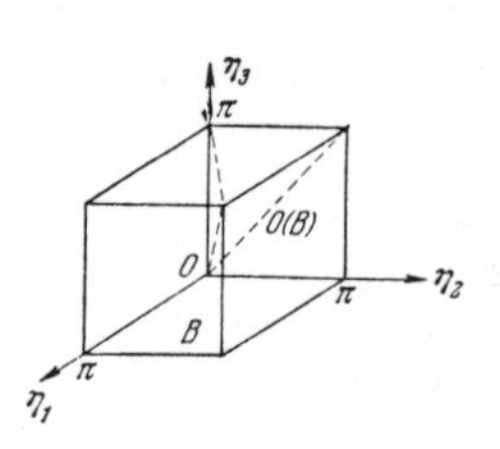

Fig. 44

Therefore, $H(G) = [\pi(G)]^*$. But it is easy to see that $[\pi(G)]^* = G^*$ (see section 14.3). Consequently, $H(G) = G^*$, which completes the proof.

Remark: From the very beginning, we could have assumed the multiple-circular domain G containing its center to be nonsingle-sheeted. Reasoning as in section 14.1, we would have shown that every function that is holomorphic in G can be expanded in $\pi(G)$ in an absolutely convergent power series. From this it follows that every function that is holomorphic in G is single-valued in it, that is, it assumes the same values at analytic points with the same projections. Here, we have examples of nonsingle-sheeted domains whose envelopes of holomorphy are single-sheeted (see section 20.2). Analogous remarks are valid for tubular domains.

The envelope of holomorphy of a multiple-circular domain G with center at zero that does not contain points of the manifold $z_1 z_2 \ldots z_n = 0$ coincides with the logarithmically convex envelope G^* of that domain.

Proof: Since the logarithmically convex envelope G^* of the domain G has zero for its center and since it does not contain points of the manifold $z_1 z_2 \ldots z_n = 0$, it follows that G^* is a domain of holomorphy (see section 19.3). Therefore, we need only show that every function $f(z)$ that is holomorphic in G is holomorphic in G^*.

The holomorphic transformation

$$z_j = e^{\zeta_j}; \ \zeta_j = \ln z_j \equiv \ln |z_j| + i \arg z_j, \ j = 1, 2, \ldots, n, \tag{85}$$

maps the multiple-circular domain G onto the tubular domain $T_Q = Q + iR^n$, where Q is the image of G in the space of the logarithms of the absolute values:

$$Q = [\zeta: \ \zeta_j = \ln |z_j|, \ \ j = 1, 2, \ldots, n; \ \ \ z \in G].$$

This mapping maps the function $f(z)$ into the function $\varphi(\zeta) = f(e^{\zeta_j})$, which is holomorphic in T_Q and which is periodic with respect to each variable ζ_j individually with period $2\pi i$. According to the theorem in section 21.1, the function $\varphi(\zeta)$ is holomorphic in the convex envelope $O(T_B) = T_{O(B)}$. From the holomorphic continuation theorem (see section 6.1), the equations $\varphi(\zeta + 2\pi i \alpha) = \varphi(\zeta)$ where $\alpha = (\alpha_1, \alpha_2, \ldots, \alpha_n)$ (where in turn the α_j are integers), which are valid in T_B, remain valid in $T_{O(B)}$. Thus, the function $\varphi(\zeta)$ is periodic with respect to each variable ζ_j individually (with period $2\pi i$) in the domain $T_{O(B)}$. Therefore, on the basis of (85), the function $\varphi(\ln |z_j| + i \arg z_j)$ is holomorphic (and single-valued) in

$$G^* = [z: \ (\ln |z_1|, \ \ldots, \ \ln |z_n|) \in O(Q)].$$

Obviously, the function $\varphi(z)$ coincides with the function $f(z)$ in the domain G. This means that $f(z)$ is holomorphic in G^*, which completes the proof.

3. *Envelopes of holomorphy of Hartogs domain*

As we saw in section 20.2, the envelope of holomorphy of a (single-sheeted) Hartogs domain is not always single-sheeted (cf. section 15.2). However, for Hartogs domains of the form considered above

$$[z: \ \ |z_1| < R(\tilde{z}), \ \ \tilde{z} \in B] \text{ and } [z: \ \ r(\tilde{z}) < |z_1| < R(\tilde{z}), \ \ \tilde{z} \in B],$$

where B is a domain of holomorphy in C^{n-1}, the envelopes of holomorphy are single-sheeted and they coincide with the logarithmically plurisuperharmonic and plurisubharmonic envelopes respectively.

These assertions follow from the results of sections 15.5, 15.7, and 19.4.

In the article by Vladimirov and Shirinbekov [67], the envelope of holomorphy of the complete Hartogs domain

$$G = [z: \ \ |z_1| < R(\tilde{z}), \ \ \tilde{z} \in B]$$

is constructed under the assumption that $H(B)$ is single-sheeted. This envelope of holomorphy is of the form

$$H(G)=[z:\ |z_1|<e^{V(\tilde{z})},\quad \tilde{z}\in H(B)],$$

where $V(\tilde{z})$ is the smallest plurisuperharmonic majorant in $H(B)$ of the function

$$M(\tilde{z})=\begin{cases}\ln R(\tilde{z}), & \tilde{z}\in B,\\ -\infty, & \tilde{z}\in H(B)\setminus B.\end{cases}$$

We note that the smallest plurisuperharmonic majorant $V(\tilde{z})$ of the function $M(\tilde{z})$ in $H(B)$ exists and is given by the formula

$$V(\tilde{z})=\varliminf_{\tilde{z}'\to\tilde{z}}\ \inf v(\tilde{z}'),$$

where the infimum is over all plurisuperharmonic functions v in the domain $H(B)$ that satisfy the inequality $M(\tilde{z})\leqslant v(\tilde{z})$ for $\tilde{z}\in B$.

4. *Functions that are holomorphic in semitubular domains*

Consider a semitubular domain of the form

$$G=[z:\ v(\tilde{z})<x_1<V(\tilde{z}),\ \tilde{z}\in B,\ |y_1|<\infty],$$

where B is a domain of holomorphy in C^{n-1} and the functions $V(\tilde{z})$ and $v(\tilde{z})$ are lower and upper-semicontinuous in B respectively (see section 2.2). Let us prove the following

THEOREM. *Every function $f(z)$ that is holomorphic in a semitubular domain G is also holomorphic in the larger domain*

$$G_f=[z:\ v_f(\tilde{z})<x_1<V_f(\tilde{z}),\ \tilde{z}\in B,\ |y_1|<\infty],$$

where $v_f(\tilde{z})$ (resp. $V_f(\tilde{z})$) is some plurisubharmonic (resp. plurisuperharmonic) function in B that satisfies the inequality $v_f\leq v$ (resp. the inequality $V_f\leq V$).

Proof: Let us fix the point $\tilde{z}\in B$ arbitrarily and consider the function $f(z_1,\tilde{z})$ of the single complex variable z_1. This function is holomorphic in the strip $v(\tilde{z})<x_1<V(\tilde{z})$ and it may possibly be holomorphic in a wider strip. Let us denote by $v_1(\tilde{z})<x_1<V_1(\tilde{z})$ the largest strip in which the function $f(z_1,\tilde{z})$ is holomorphic. Since a strip is a convex domain, it follows from the monodromy theorem (see section 6.6) that the function $f(z_1,\tilde{z})$ continues to be single-valued in the strip $v_1(\tilde{z})<x_1<V_1(\tilde{z})$. By construction, $v_1(\tilde{z})\leqslant v(\tilde{z})$ and $V_1(\tilde{z})\geqslant V(\tilde{z})$ in the domain B. We set

$$v_f(\tilde{z})=\varlimsup_{\tilde{z}'\to\tilde{z}} v_1(\tilde{z}'),\quad V_f(\tilde{z})=\varliminf_{\tilde{z}'\to\tilde{z}} V_1(\tilde{z}').$$

The functions $v_f(\tilde{z})$ and $V_f(\tilde{z})$ are upper- and lower-semicontinuous in B respectively. Since the functions $v(\tilde{z})$ and $V(\tilde{z})$ are assumed to be upper- and lower-semicontinuous, respectively, it follows that the functions constructed satisfy the inequalities $v_f(\tilde{z}) \leqslant v(\tilde{z})$ and $V_f(\tilde{z}) \geqslant V(\tilde{z})$ for $\tilde{z} \in B$.

Let us show that the function $f(z)$ is holomorphic in the domain G_f that we have constructed. Consider a set $G' \Subset G_f$. According to the Heine-Borel theorem, the compact set $\overline{G'}$ can be covered by a finite number of semitubular domains of the form

$$g = [z:\ a < x_1 < A,\ \tilde{z} \in U(\tilde{z}^0,\ r),\ |y_1| < \infty],\ U(\tilde{z}^0,\ r) \Subset B,$$

that are contained in G_f (cf. section 15.4). For a and A, we may take the numbers

$$a = \max_{\tilde{z} \in \overline{U}(\tilde{z}^0,\ r)} v_f(\tilde{z}), \quad A = \min_{\tilde{z} \in \overline{U}(\tilde{z}^0,\ r)} V_f(\tilde{z}).$$

Here, the radii $r = r(\tilde{z}^0)$ are chosen sufficiently small that

$$a' = \max_{\tilde{z} \in \overline{U}(\tilde{z}^0,\ r)} v(\tilde{z}) < A' = \min_{\tilde{z} \in \overline{U}(\tilde{z}^0,\ r)} V(\tilde{z}).$$

This is always possible since the functions $v(\tilde{z})$ and $V(\tilde{z})$ are upper- and lower-semicontinuous, respectively (see section 2.2). Clearly, $a \leqslant a'$ and $A' \leqslant A$ (see Fig. 45).

From the hypothesis, $f(z)$ is holomorphic in the semitubular domain

$$g' = [z:\ a' < x_1 < A',\ \tilde{z} \in U(\tilde{z}^0,\ r),\ |y_1| < \infty]$$

and is holomorphic with respect to z_1 in the strip $a < x_1 < A$ for every fixed $\tilde{z} \in U(\tilde{z}^0,\ r)$. On the basis of the generalized Hartogs theorem (see section 20.6), $f(z)$ is holomorphic in the domain

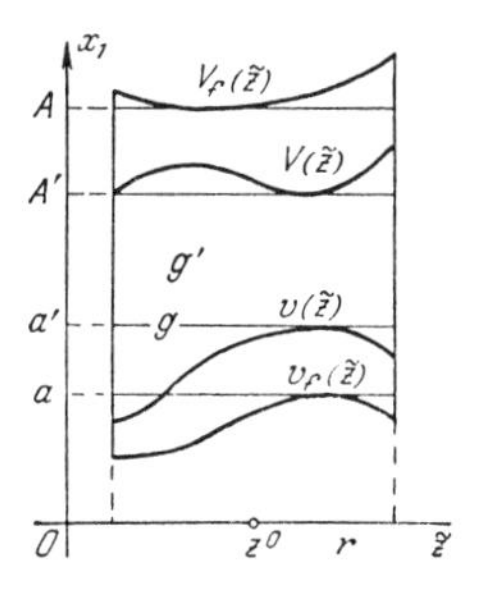

Fig. 45

$$g = [z_1:\ a < x_1 < A] \times U(\tilde{z}^0,\ r).$$

Actually, it is easy to prove that $f(z)$ is holomorphic in the domain g without using the generalized Hartogs theorem. For this, it is sufficient to map the strip a $a < x_1 < A$ single-sheetedly and conformally onto the unit circle in such a way that the point $a' + A'/2$ is mapped into the point 0 and to use Hartogs' theorem of section 15.3.

Since any subdomain $G' \Subset G_f$ can be covered by a finite number of domains of the type g, the function $f(z)$ is holomorphic in G_f.

Let us show that the functions $v_f(\tilde{z})$ and $-V_f(\tilde{z})$ are plurisubharmonic in B. On the basis of the theorem in section 19.5, it will be sufficient to show that the domain G_f is a domain of holomorphy. Consider a point $z^0 \in \partial G_f$. Let us assume first that $\tilde{z}^0 \in \partial B$. Since B is assumed to be a domain of holomorphy, there exists a function $f_0(\tilde{z})$ that is holomorphic in B and that has a singularity at every point of ∂B. Thus, in this case, there exists a barrier function $f_0(\tilde{z})$ that is holomorphic in G_f and that has a singularity at the point z^0. (This function is independent of z_1.)

Let us suppose now that $\tilde{z}^0 \in B$. Then, either $x_1^0 = v_f(\tilde{z}^0)$ or $x_1^0 = V_f(\tilde{z}^0)$. Let us consider the first case (the second being analogous to it). Since $v_f(\tilde{z}^0) = \varlimsup_{\tilde{z} \to \tilde{z}^0} v_1(\tilde{z})$, there exists a sequence $\tilde{z}^{(k)}$, for $k = 1, 2, \ldots$, of points in the domain B such that $\tilde{z}^{(k)} \to \tilde{z}^0$ and $v_1(\tilde{z}^{(k)}) \to v_f(\tilde{z}^0)$ as $k \to \infty$. From this it follows that, for arbitrary $\varepsilon > 0$, there exists a point $\tilde{z}'$ such that

$$|\tilde{z}' - \tilde{z}^0| < \frac{\varepsilon}{\sqrt{2}}, \qquad |v_1(\tilde{z}') - v_f(\tilde{z}^0)| < \frac{\varepsilon}{2\sqrt{2}}. \tag{86}$$

Furthermore, since the strip $v_1(\tilde{z}') < x_1 < V_1(\tilde{z}')$, throughout which the function $f(z_1, \tilde{z}')$ is holomorphic, is of maximum width, this function has a (finite) singular point z_1'' such that

$$0 \leqslant v_1(\tilde{z}') - x_1'' < \frac{\varepsilon}{2\sqrt{2}}.$$

From this and from (86), it follows that the point $z' = (x_1'' + iy_1^0, \tilde{z}')$ (see Fig. 46) does not belong to G and that it satisfies the inequality

$$|z' - z^0| = \sqrt{|x_1'' - x_1^0|^2 + |\tilde{z}' - \tilde{z}^0|^2} \leqslant$$
$$\leqslant \sqrt{[|x_1'' - v_1(\tilde{z}')| + |v_1(\tilde{z}') - v_f(\tilde{z}^0)|]^2 + \frac{\varepsilon^2}{2}} < \varepsilon.$$

Also, the function $f[z_1 + i(y_1'' - y_1^0), \tilde{z}]$ is holomorphic in G but is not holomorphic at the point z' since, by construction, the point z_1'' is a singular point of the function $f(z_1, \tilde{z}')$; that is, it lies on the boundary of the domain of the domain of holomorphy of that function.

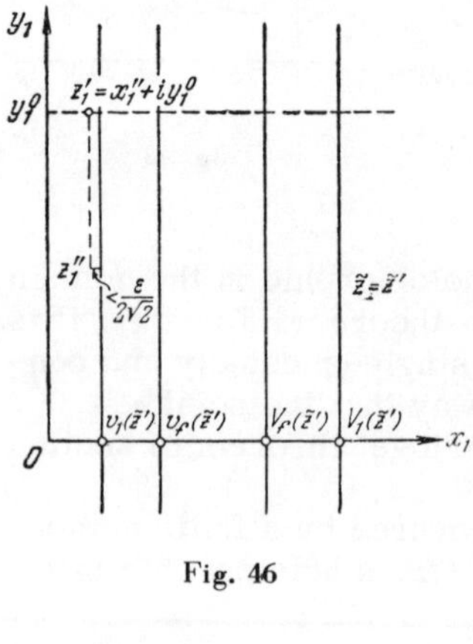

Fig. 46

Thus, since ε is arbitrary, this point $z^0 \in \partial G_f$ satisfies the conditions of the theorem in section 16.7. Since there are no boundary points of the domain G_f other than those that we have been examining, it follows that G_f satisfies the conditions of the theorem in section 16.7 and hence is a domain of holomorphy. This completes the proof.

5. *Envelopes of holomorphy of semitubular domains*

Suppose that G is a domain. We shall say that a semitubular domain G^* of the form

$$G^* = [z: \quad u(\tilde{z}) < x_1 < U(\tilde{z}), \ \tilde{z} \in B, \ |y_1| < \infty],$$

where $U(\tilde{z})$ (resp. $u(\tilde{z})$) is the smallest plurisuperharmonic majorant (resp. largest plurisubharmonic minorant) of the function $V(\tilde{z})$ (resp. $v(\tilde{z})$) in the domain B, is the plurisuperharmonic (resp. plurisubharmonic) envelope of G.

The envelope of holomorphy of a semitubular domain G for which B is the domain of holomorphy coincides with the plurisuperharmonic (plurisubharmonic) envelope G^*.

Proof: Since B is a domain of holomorphy, G^* is also a domain of holomorphy (see section 19.5). Also, $G \subset G^*$. It remains only to show that G^* is a holomorphic extension of G. Suppose that $f \in H_G$. From the preceding theorem, $f(z)$ is holomorphic in the semitubular domain

$$G_f = [z: \quad v_f(\tilde{z}) < x_1 < V_f(\tilde{z}), \quad \tilde{z} \in B, \quad |y_1| < \infty],$$

where v_f and V_f are respectively plurisubharmonic and plurisuperharmonic functions in B such that $v_f \leqslant v$ and $V_f \geqslant V$. This implies the inequalities $v_f \leqslant u$ and $V_f \geqslant U$ in the domain B (see sections 10.4 and 9.7), from which we conclude that $G^* \subset G_f$. This inclusion means that every function f in H_G is holomorphic in G^*, which completes the proof.

Remark: Bremermann [18] constructed an envelope of holomorphy for an arbitrary semitubular domain in the space C^2 of the form

$$G = [z: \quad (x_1, z_2) \in D, \quad |y_1| < \infty].$$

As in the case of Hartogs domains (see sections 15.2 and 20.2), this envelope of holomorphy is not always single-sheeted. (For appropriate examples, see [18].)

We note in passing that if a semitubular domain G is single-sheeted, its projection B onto the plane $z_1 =$ const does not have to be single-sheeted. For example, the domain

$$G = \left[z: \quad \arg z_2 - \frac{1}{2} < x_1 < \arg z_2 + \frac{1}{2}, \ z_2 \in B, \ |y_1| < \infty\right],$$

where B is the Riemann surface of $\ln z_2$ for $1 < |z_2| < 2$, is single-sheeted (cf. Fig. 43).

6. *The method of successive approximations for constructing domains of holomorphy*

We saw above that, for a number of domains of relatively simple geometric structure (tubular, multiple-circular, semicircular, semitubular, etc.), the domain of holomorphy can be actually constructed: Either it is completely characterized in geometric terms or an explicit analytic expression is obtained for its boundary. However, for more complicated domains, such visually comprehensible characteristics of domains of holomorphy cannot always be obtained because of the exceeding complexity of the problem. Therefore, it is of interest to develop approximative (in particular, numerical) methods of constructing envelopes of holomorphy. We shall present one of these methods, that of successive approximations, in the form suggested by Taylor [47]. A variant of the method of successive approximations, developed by Bremermann [150], is expounded in the book by Fuks [2], section 17.

As always, we assume that the domain $G \neq C^n$ and its envelope of holomorphy $H(G)$ are single-sheeted. Then, $H(G)$ is a pseudoconvex envelop of G (see section 20.3). We have the following theorems:

The sequence of domains

$$G_\alpha = [z: \ |z - z'| < e^{-V_{\alpha-1}(z')}, \ z' \in G_{\alpha-1}], \quad G_0 = G,$$
$$\alpha = 1, 2, \ldots,$$

where $V_\alpha(z)$ is the largest plurisubharmonic majorant of the function $-\ln \Delta_{G_\alpha}(z)$ in the domain G_α, is such that

$$G_\alpha \subset G_{\alpha+1} \subset H(G), \qquad \bigcup_\alpha G_\alpha = H(G).$$

Proof: Let us show first that $G_1 \subset H(G)$. The function $-\ln \Delta_{H(G)}(z)$ is plurisubharmonic in $H(G)$ (see section 19.1) and, consequently, in $G_0 = G \subset H(G)$. Also,

$$-\ln \Delta_{H(G)}(z) \leqslant -\ln \Delta_{G_0}(z), \qquad z \in G_0.$$

From this and from the definition of greatest plurisubharmonic minorant (see section 10.4) $V_0(z)$ of the continuous function $-\ln \Delta_{G_0}(z)$ in the domain G_0, it follows that

$$-\ln \Delta_{H(G)}(z) \leqslant V_0(z), \qquad z \in G_0,$$

that is,

$$e^{-V_0(z)} \leqslant \Delta_{H(G)}(z).$$

But this inequality means that $G_1 \subset H(G)$.

Continuing this reasoning, we see that $G_\alpha \subset H(G)$. The inclusions $G_\alpha \subset G_{\alpha+1}$, for $\alpha = 1, 2, \ldots$ are obvious. Let us define

$$D = \bigcup_\alpha G_\alpha.$$

Obviously, $D \subset H(G)$. It remains to show that $D = H(G)$. For this, it will be sufficient to show that D is a pseudoconvex domain, that is, that the function $-\ln \Delta_D(z)$ is plurisubharmonic in D.

Consider a point $z^0 \in D$. Then, there exists an r such that $U(z^0, r) \Subset D$. Therefore, there exists an N such that $U(z^0, r) \Subset G_\alpha$ for all $\alpha \geqslant N$. Furthermore, from the construction, we have the inequalities

$$-\ln \Delta_{G_{\alpha+1}}(z) \leqslant V_\alpha(z) \leqslant -\ln \Delta_{G_\alpha}(z), \quad z \in U(z^0, r), \quad \alpha \geqslant N. \tag{87}$$

On the basis of (2) of section 1.3,

$$\lim_{\alpha \to \infty} [-\ln \Delta_{G_\alpha}(z)] = -\ln \Delta_D(z), \qquad z \in U(z^0, r)$$

so that it follows from inequalities (87) that

$$V_{\alpha+1}(z) \leqslant V_\alpha(z), \quad \lim_{\alpha \to \infty} V_\alpha(z) = -\ln \Delta_D(z), \qquad z \in U(z^0, r).$$

Consequently, the function $-\ln \Delta_D(z)$ is plurisubharmonic in $U(z^0, r)$ (see section 10.3). Since $U(z^0, r)$ is a neighborhood of an arbitrary point $z^0 \in D$, it follows that the function $-\ln \Delta_D(z)$ is plurisubharmonic in D (see section 10.3), which completes the proof.

Suppose that $V(\lambda)$ is the largest plurisubharmonic majorant of the function $-\ln \Delta_G(z^0 + \lambda a)$, where $|a| = 1$, in the open set $G_{z^0,\alpha} = [\lambda: z^0 + \lambda a \in G]$. Then, $G \cup \pi_{z^0,a} \subset H(G)$, where

$$\pi_{z^0, a} = [z: \ |z - (z^0 + \lambda a)| < e^{-V(\lambda)}, \qquad \lambda \in G_{z^0, a}].$$

Proof: Let $V_0(z)$ be the greatest plurisubharmonic minorant of the function $-\ln \Delta_G(z)$ in the domain G. Then, the function $V_0(z^0 + \lambda a)$ is subharmonic in $G_{z^0, a}$ (see section 9.1) and satisfies the inequality

$$V_0(z^0 + \lambda a) \leqslant -\ln \Delta_G(z^0 + \lambda a).$$

Consequently,

$$V_0(z^0 + \lambda a) \leqslant V(\lambda), \ \lambda \in G_{z^0, a},$$

from which it follows that $\pi_{z^0, a} \subset G_1$, where

$$G_1 = [z: \ |z - z'| < e^{-V_0(z')}, \ z' \in G].$$

From the preceding theorem, $G_1 \subset H(G)$, which completes the proof.

In conclusion, let us look at an example illustrating the application of this last theorem. Suppose that $n = 2$ and $G = T^\Gamma \cup D$, where $T^\Gamma = [z: \ |y_1| > |y_2| \ |x| < \infty]$ and D is some complex neighborhood of the set of real points $[x: \ |x_1| < f(x_2)]$. The function $f(x_2)$

is assumed to be positive and lower-semicontinuous. Suppose that $a=(1, 0)$ and $z^0=(0, x_2^0)$. Then, $G_{z^0, a}\equiv G(x_2^0)$ is the λ-plane deleted by the cuts $|\mathrm{Re}\,\lambda|\geqslant f(x_2^0)$, $\mathrm{Im}\,\lambda=0$. It is easy to show (see Fig. 47) that

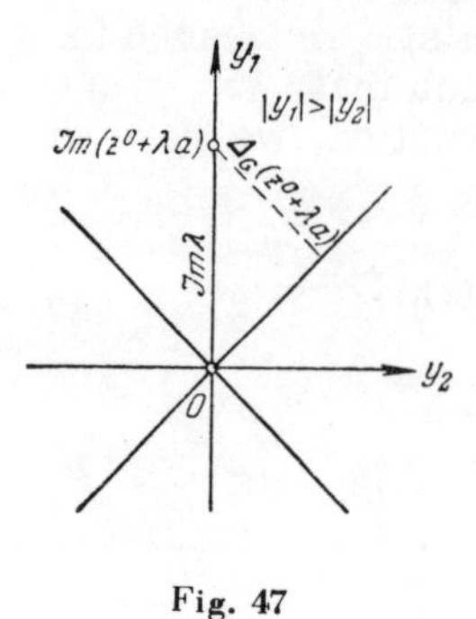

Fig. 47

$$-\ln\Delta_G(z^0+\lambda a)\leqslant-\ln\frac{|\mathrm{Im}\,\lambda|}{\sqrt{2}}. \qquad (88)$$

The smallest superharmonic majorant of of the function $|\mathrm{Im}\,\lambda|/\sqrt{2}$ in the domain $G(z^0)$ is the function $|\mathrm{Im}\ \sqrt{\lambda^2-f^2(x_2^0)/2}\ |$ (see section 27.5). Therefore, if we denote by $V(\lambda)$ the greatest subharmonic minorant of the function $-\ln\Delta_G(z^0+\lambda a)$ in $G(z^0)$, we conclude from (88) that

$$V(\lambda)\leqslant-\ln\left|\mathrm{Im}\sqrt{\frac{\lambda^2-f^2(x_2^0)}{2}}\right|, \qquad \lambda\in G(z^0).$$

On the basis of our last theorem, it follows from this that $G\cup\pi\subset H(G)$, where

$$\pi=\bigcup_{x_2^0}\left[z:\quad |z-(z^0+\lambda a)|<\left|\mathrm{Im}\sqrt{\frac{\lambda^2-f^2(x_2^0)}{2}}\right|,\right.$$

$$\left.\lambda\in G(x_2^0)\right].$$

In particular, we obtain from this

$$[z:\quad \sqrt{2}\,|y_2|<|\mathrm{Im}\sqrt{z_1^2-f^2(x_2)}|\,]\subset H(G).$$

CHAPTER IV

Integral Representations

Integral represenations constitute one of the more powerful tools in the theory of holomorphic functions. An integral representation expresses the values of an arbitrary function that is holomorphic in a domain G in terms of its values on the boundary ∂G or even on the portion of ∂G of the smallest number of dimensions (the hull of the domain G). The simplest example of an integral representation is Cauchy's formula (see section 4.3), which is valid for domains G of the form $G_1 \times G_2 \times \ldots \times G_n$ where G_j is a bounded domain in the plane of the complex variable z_j that has a piecewise-smooth boundary ∂G_j for $j = 1, 2, \ldots, n$. Cauchy's formula expresses the values in a domain G of an arbitrary function that is holomorphic in G and continuous in $\bar{G}$ in terms of its values on the n-dimensional manifold $\partial G_1 \times \partial G_2 \times \ldots \times \partial G_n$, that is, the hull of ∂G. Here, the kernel in Cauchy's formula is independent of the particular form of the domains G_j.

There are other integral representations generalizing the Cauchy representation that are suitable for domains with a more complicated geometrical structure. In this chapter, we shall treat three types of integral representations: the Martinelli-Bochner representation for bounded domains with piecewise-smooth boundary, the Bergmann-Weil representation for Weil domains, and the Bochner representation for tubular radial domains. In the following chapter, we shall look at the Jost-Lehmann-Dyson representation for the special form of domains that arise in quantum field theory.

22. FACTS FROM THE THEORY OF DIFFERENTIAL FORMS

The information that we are about to present from the theory of differential forms is contained in the book by de Rham [58]. In what follows, we shall confine ourselves to consideration of piecewise-smooth surfaces. Therefore, our exposition will be accompanied with corresponding simplifications.

1. Definition of differential forms and the simplest operations on them

A *differential form* of kth order (for $0 \leqslant k \leqslant n$) is an expression of the form

$$\alpha = \sum_{j_1, j_2, \ldots, j_k} \alpha_{j_1, j_2, \ldots, j_k}(x)\, dx_{j_1} dx_{j_2} \ldots dx_{j_k}, \tag{1}$$

where $x = (x_1, x_2, \ldots, x_n)$ and the coefficients $a_{j_1 \ldots j_k}(x)$ are continuous functions in some domain of the space R^n. Two forms are considered equal if each can be transformed into the other by a transformation of the products of the differentials according to the formula

$$dx_l\, dx_j = -dx_j\, dx_l. \tag{2}$$

With the use of rule (2), every kth-order form can be reduced to the canonical form

$$\sum_{j_1 < \ldots < j_k} a_{j_1 \ldots j_k}(x)\, dx_{j_1} \ldots dx_{j_k}. \tag{3}$$

The *exterior product* $\alpha\beta$ of two differential forms α and β of orders k and m, respectively, is the form of order $k+m$ obtained by formal algebraic multiplication of the forms α and β by use of rule (2). Clearly, $\alpha\beta = (-1)^{km}\beta\alpha$; $\alpha\beta = 0$ if $k+m > n$.

Under a continuously differentiable coordinate transformation $x_j = x_j(x'_1, \ldots, x'_n)$, for $j = 1, 2, \ldots, n$, the form (3) is mapped into the form

$$\sum_{j_1 < \ldots < j_k} a'_{j_1 \ldots j_k}(x')\, dx'_{j_1} \ldots dx'_{j_k}, \tag{4}$$

where

$$a'_{j_1 \ldots j_k}(x') = \sum_{i_1 < \ldots < i_k} D\begin{pmatrix} x_{i_1}, & \ldots, & x_{i_k} \\ x'_{j_1}, & \ldots, & x'_{j_k} \end{pmatrix} a_{i_1 \ldots i_k}[x(x')]. \tag{5}$$

In particular, for an nth-order form, formulas (4)–(5) take the form

$$a(x)\, dx_1 \ldots dx_n = a[x(x')]\, D\begin{pmatrix} x_1, & \ldots, & x_n \\ x'_1, & \ldots, & x'_n \end{pmatrix} dx'_1 \ldots dx'_n. \tag{6}$$

The *exterior differential* $d\alpha$ of a kth-order differential form α (where $0 \leqslant k \leqslant n-1$) with continuously differentiable coefficients is defined as the $(k+1)$st-order form

$$d\alpha = \sum_{j_1 < \dots < j_k} \left(\sum_{j=1}^{n} \frac{\partial a_{j_1 \dots j_k}(x)}{\partial x_j} dx_j \right) dx_{j_1} \dots dx_{j_k}$$

where, of course, Eqs. (2) may be used. It follows from this definition that if the coefficients of the form α are twice continuously differentiable, then

$$dd\alpha = 0.$$

For example, suppose that $n = 3$. Then, for $k = 0$, we have $\alpha = a_0(x)$ and

$$d\alpha = \frac{\partial a_0}{\partial x_1} dx_1 + \frac{\partial a_0}{\partial x_2} dx_2 + \frac{\partial a_0}{\partial x_3} dx_3;$$

for $k = 1$, $\alpha = a_1(x)\,dx_1 + a_2(x)\,dx_2 + a_3(x)\,dx_3$ and

$$d\alpha = \left(\frac{\partial a_2}{\partial x_1} - \frac{\partial a_1}{\partial x_2} \right) dx_1\,dx_2 + \left(\frac{\partial a_3}{\partial x_1} - \frac{\partial a_1}{\partial x_3} \right) dx_1\,dx_3 + \\ + \left(\frac{\partial a_3}{\partial x_2} - \frac{\partial a_2}{\partial x_3} \right) dx_2\,dx_3;$$

for $k = 2$, $\alpha = a_1(x)\,dx_2\,dx_3 + a_2(x)\,dx_3\,dx_1 + a_3(x)\,dx_1\,dx_2$ and

$$d\alpha = \left(\frac{\partial a_1}{\partial x_1} + \frac{\partial a_2}{\partial x_2} + \frac{\partial a_3}{\partial x_3} \right) dx_1\,dx_2\,dx_3.$$

2. *The orientation of surfaces*

The concept of orientation of a surface will be of great importance in what follows. (For definitions pertaining to surfaces, see section 1.4.) A connected k-dimensional class $C^{(1)}$ surface S is said to be *orientable* if in a neighborhood of each point $x^0 \in S$, it is possible to choose a local coordinate system $(t_1, t_2, \dots, t_k)$ in a consistent manner, that is, in the intersection of arbitrary neighborhoods, a transformation from one local coordinate system to another can be achieved by means of functions whose Jacobians are positive. Thus, on an orientable surface S, all systems of local coordinates fall into two classes. If the Jacobian of one coordinate system with respect to the other is positive in the intersection of their domains of definition, both these systems belong to the same class; in the opposite case, they belong to different classes. We shall refer to these classes of coordinate systems as right-hand and left-hand systems. Therefore, on an orientable surface S it is possible to define two continuous quantities ε and $-\varepsilon$, which are respectively equal to +1 and -1 in any right-handed system and equal respectively to -1 and +1 in any left-handed system.* Each of these quantities is called the *orientation* of the surface S. For definiteness, let us agree henceforth to choose the quantity ε for the orientation of S. Thus, the choice of a consistent system of local coordinates on S determines the orientation of the surface S. In this case, we shall say that S is *oriented*.

Clearly, this definition of orientation remains applicable for the case in which $k=n$, that is, for the orientation of domains in R^n. This orientation is defined by the choice (right-hand or left-hand) of Cartesian coordinate system in R^n. Thus, any domain in R^n is orientable. For the other extreme case, that of $k=0$, let us agree that all isolated points in R^n are orientable. There exist, however, nonorientable surfaces of intermediate dimensions, for example, the Möbius band. In what follows, we shall assume that all surfaces under discussion are orientable.

Let V denote a connected k-dimensional surface with class $C^{(1)}$ boundary and suppose that its boundary ∂V is a $(k-1)$-dimensional piecewise-smooth surface. This means that V lies on some connected k-dimensional surface of class $C^{(1)}$ and that ∂V consists of a finite number of connected $(k-1)$-dimensional surfaces v_i each with a boundary of class $C^{(1)}$. $\partial V = \bigcup_{1\leqslant i\leqslant N} v_i$ (see Fig. 48). Let us assume that the surface S is orientable. Let us orient S, for example, by means of a right-handed coordinate system (in which, as we have agreed, $\varepsilon = +1$). This determines the orientation of the surface V in a natural manner. Let us show that the orientation of the surface V induces a definite orientation of each subsurface v_i as follows: In a k-dimensional neighborhood (on the surface S) of each point $x^0 \in v_i$, let us choose a local coordinate system $(t_1, t_2, \ldots, t_k)$ satisfying the following two conditions: (1) we direct the coordinate curve t_1 along the outer normal on S to v_i at the point x^0; the remaining coordinate curves $t_2, \ldots, t_k$ are located on the surface v_i in such a way that the coordinate system $(t_1, t_2, \ldots, t_k)$ will be a right-handed system and will thus determine the orientation $\varepsilon = +1$ of the surface S and hence of V. Since the surfaces S and v_i are assumed to be orientable, a local coordinate system $(t_2, \ldots, t_k)$ chosen in this manner in a neighborhood of each point x^0 of the surface v_i will be consistent and will determine the orientation ε of the surface v_i. (In accordance with our convention, $\varepsilon = +1$ if this coordinate system is a right-handed system and $\varepsilon = -1$ if it is a left-handed system.)

Now that we have thus oriented each subsurface v_i, for $i = 1, 2, \ldots, N$, of the boundary ∂V, we can speak of the *orientation of the entire piecewise-smooth boundary* ∂V of the surface V.

---

*Such quantities are called pseudoscalars.

This orientation corresponds to the *positive direction of the outer normal vector*.

Thus, if the boundary ∂B of the domain $B \subset R^n$ is a piecewise-smooth surface, that is, if it consists of a finite number of connected $(n-1)$-dimensional surfaces V_i with boundary of class $C^{(1)}$ $(\partial B = \bigcup_i V_i)$, the orientation of B induces an orientation of each subsurface V_i. If the boundary ∂V_i of each portion V_i is a piecewise-smooth surface, the orientation of V_i in turn induces an orientation of ∂V_i, etc. Finally, the orientation of a smooth curve induces an orientation of both its ends if we take as the "outer normal" vector to the curve at an end point the tangent at that point directed away from the curve. Therefore, if the boundary of the domain B possesses the properties listed, to orient all its k-dimensional surfaces (for $0 \leqslant k \leqslant n-1$) with boundary of class $C^{(1)}$, it is sufficient to give the orientation of the domain B itself, that is, to choose a definite coordinate system in R^n.

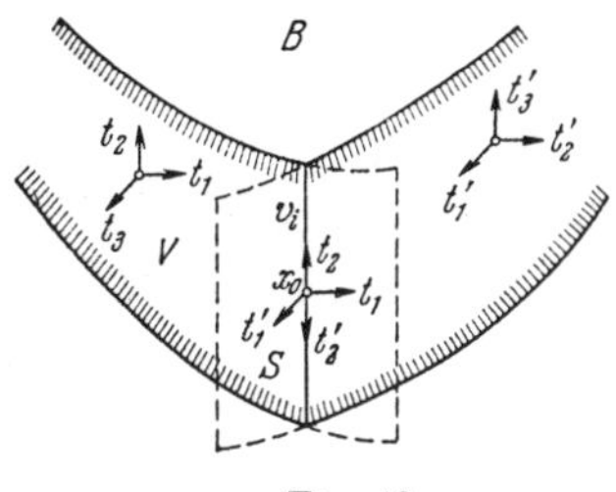

Fig. 48

We note that if a k-dimensional (for $0 \leqslant k \leqslant n-2$) surface $v \subset \partial B$ lies on the boundary of several $(k+1)$-dimensional surfaces $V_i \subset \partial B$, v may have different orientations induced by the orientations of the different surfaces V_i (see Fig. 48).

3. *Integration of differential forms*

We now define the integral of a form α of order k over a k-dimensional oriented bounded surface V (for $0 \leqslant k \leqslant n$) with boundary of class $C^{(1)}$. The coefficients in the form α are assumed to be continuous in a neighborhood of V.

Let us partition V into subsurfaces $V_1, \ldots, V_N$ with boundary of class $C^{(1)}$. In each of these subsurfaces, we introduce local coordinates $(t_1, t_2, \ldots, t_k)$ in a consistent manner. As a result of this, one of the two possible orientations is defined on V. By using rules (2) and (5), we reduce the form α on each subsurface V_i to the form

$$\alpha = a_i(t_1, \ldots, t_k)\, dt_1 \ldots dt_k$$

in the chosen local coordinates. Let us integrate the function $a_i(t_1, \ldots, t_k)$ according to the usual rules of integration over the corresponding closed domain D_i of definition of the variables $t_1, \ldots, t_k$ and let us add the results. We shall call the number just obtained the integral of the form α over the oriented bounded surface V including its boundary and we adopt the notation

$$\int_V \alpha = \sum_{1 \leqslant i \leqslant N} \int_{D_i} a_i(t_1, \ldots, t_k)\, dt_1 \ldots dt_k. \tag{7}$$

Clearly, the value of the integral defined in this way is independent of the partition of the surface V into subsurfaces V_i. Furthermore, on the basis of (4)-(5), this integral is independent of the choice of local coordinates of each subsurface V_i which correspond to a fixed orientation. Thus, the integral (7) is defined except for sign and the sign in turn is determined by the choice of orientation of the surface V.

For example, for $k = n$, we have $\alpha = x(a)dx_1 \ldots dx_n$ and the integral (7) is reduced to an ordinary n-dimensional integral

$$\int_V \alpha = \int_V a(x)\, dx$$

except that the formula for change of variables in this integral contains not the absolute value of the Jacobian but the Jacobian itself:

$$\int_V \alpha = \int_{V'} \alpha', \quad \alpha' = a[x(x')]\, D\begin{pmatrix} x_1, & \ldots, & x_n \\ x'_1, & \ldots, & x'_n \end{pmatrix} dx'_1 \ldots dx'_n.$$

For $k = 0, V = x^0$ and $\alpha = a_0(x)$. Obviously, we need to set $\int_V \alpha = \pm a_0(x^0)$, where the sign + or - is chosen according to the orientation of the point x^0.

Finally, the integral of a form α over an oriented bounded piecewise-smooth surface V is defined as the sum of the integrals of α over the oriented bounded surfaces with boundary of class $C^{(1)}$ that constitute the surface V.

4. *Stokes' formula*

Let V be a k-dimensional bounded surface (for $1 \leq k \leq n$) with boundary of class $C^{(1)}$. Suppose ∂V is a $(k-1)$-dimensional piecewise-smooth surface whose orientation is induced by the orientation of V. Suppose that a form a of order $k-1$ is once continuously differentiable in a neighborhood of the surface V. Then, Stokes' formula

$$\int_V d\alpha = \int_{\partial V} \alpha. \tag{8}$$

Formula (8) contains as a special case the classical Gauss-Ostrogradskiy formula (for $k = n$). For example,

$$\int_{\partial V} x_1 dx_2 dx_3 \dots dx_n - x_2 dx_1 dx_3 \dots dx_n + \dots \\ \dots + (-1)^n x_n dx_1 dx_2 \dots dx_{n-1} = n \operatorname{mes} V. \tag{9}$$

Remark: Our definition of a k-dimensional surface of class $C^{(1)}$ requires that the rank of the matrix composed of the first derivatives of the functions defining this surface with respect to the local coordinates be equal to k everywhere on the surface (see section 1.4). However, differential forms can be integrated over surfaces for which this condition is not satisfied, and Stokes' formula remains in force for them (see de Rham [58]).

5. *Differential forms in the space C^n*

Differential forms in the space C^n are defined as differential forms in the space $R^{2n} = R^n \times R^n$ of the variables (x, y). However, to shorten our calculations, it is convenient to consider differential forms $dz_j = dx_j + i\,dy_j$ and $d\bar{z}_j = dx_j - i\,dy_j$ in the variables $z_j = x_j + iy_j$ and $\bar{z}_j = x_j - iy_j$, for $j = 1, 2, \dots, n$, in $C^n = R^n + iR^n$ and to use the technique of formal derivatives (see section 4.5). Thus, a kth order form in C^n is of the form

$$\sum_{j_1, j_2, \dots, j_k} a_{j_1 j_2 \dots j_k}(z, \bar{z})\, dz_{j_1} \dots dz_{j_l} d\bar{z}_{j_{l+1}} \dots dz_{j_k}.$$

Here, in accordance with rule (2), we need to assume that

$$d\bar{z}_j dz_j = 2i\, dx_j dy_j, \quad dz_j dz_k = -dz_k dz_j \quad \text{, etc.} \tag{10}$$

Finally, let us agree to orient the space C^n (and hence each domain in C^n) in such a way that, for some polycircle $S(z^0, r)$,

$$\int_{S(z^0, r)} d\bar{z}_1 dz_1 \dots d\bar{z}_n dz_n = (2\pi i)^n r_1^2 \dots r_n^2.$$

Obviously, this orientation is independent of the choice of polycircle $S(z^0, r)$.

6. *The Cauchy-Poincaré theorem*

It is easy to obtain by means of Stokes' formula a number of relations generalizing Cauchy's theorem for a single complex variable:

Let $V \subset C^n$ be an $(n+k)$-dimensional bounded surface (for $k = 1, 2, \ldots, n$) with boundary of class $C^{(1)}$; let ∂V be an $(n+k-1)$-dimensional piecewise-smooth surface, and let $f(z)$ be a holomorphic function in a neighborhood of V. Then,

$$\int_{\partial V} f(z)\, dz_1 \ldots dz_n\, d\bar{z}_{\alpha_1} \ldots d\bar{z}_{\alpha_{k-1}} = 0,$$

$$\alpha_j = 1, 2, \ldots, n, \quad j = 1, 2, \ldots, k-1; \quad k = 1, 2, \ldots, n. \tag{11}$$

To prove Eqs. (11), we need only note that the exterior differential of the form in the integrand is, by virtue of the Cauchy-Riemann conditions, equal to zero and to use Stokes' formula. Specifically,

$$d\left[f(z)\, dz_1 \ldots dz_n\, d\bar{z}_{\alpha_1} \ldots d\bar{z}_{\alpha_{k-1}}\right] = \\ = \left(\sum_{1 \leqslant j \leqslant n} \frac{\partial f}{\partial z_j} dz_j + \frac{\partial f}{\partial \bar{z}_j} d\bar{z}_j\right) dz_1 \ldots dz_n\, d\bar{z}_{\alpha_1} \ldots d\bar{z}_{\alpha_{k-1}} = 0.$$

The following theorem (Cauchy-Poincaré) follows from this theorem for $k = 1$: *If a function $f(z)$ is holomorphic in a domain G, if $V \subset G$ is an $(n+1)$-dimensional bounded surface with class $C^{(1)}$ boundary, and if ∂V is an n-dimensional piecewise-smooth surface, then*

$$\int_{\partial V} f(z)\, dz_1 \ldots dz_n = 0. \tag{12}$$

7. A generalization of Morera's theorem

The converse of the Cauchy-Poincaré theorem is also valid. It is a generalization of Morera's theorem for a single complex variable.

Let $f(z)$ denote a function that is continuous in a domain G and suppose that, for an arbitrary surface $V = l_1 \times \ldots \times l_{k-1} \times g_k \times l_{k+1} \times \ldots \times l_n$ (for $k = 1, 2, \ldots n$) contained in G, Eq. (12) is satisfied. Then, $f(z)$ is holomorphic in the domain G. Here, the l_j, for $j \neq k$, are class $C^{(1)}$ curves with ends z'_j and z''_j in the z_j-plane and g_k is a closed bounded simply connected domain in the z_k-plane with a piecewise-smooth boundary.

Proof: Let z^0 denote an arbitrary point in the domain G and suppose that $S(z^0, r) \Subset G$. Obviously, we need only show that $f(z)$ is holomorphic in $S(z^0, r)$. Suppose that an arbitrary surface V of the type in question is contained in $S(z^0, r)$. Clearly, V is an $(n+1)$-dimensional orientable bounded surface with boundary of class $C^{(1)}$. Its boundary ∂V is an n-dimensional piecewise-smooth surface. It consists of the following surfaces (see Fig. 49):

$$v_k = l_1 \times \ldots \times l_{k-1} \times \partial g_k \times l_{k+1} \times \ldots \times l_n,$$
$$v'_s = [z\colon\; z_s = z'_s,\; z_k \in g_k,\; z_j \in l_j,\; j \neq s,\; j \neq k], \qquad s \neq k,$$
$$v''_s = [z\colon\; z_s = z''_s,\; z_k \in g_k,\; z_j \in l_j,\; j \neq s,\; j \neq k], \qquad s \neq k.$$

From the hypothesis of the theorem, we have

$$\int\limits_{\partial V} f(z)\,dz_1 \ldots dz_n \equiv \int\limits_{v_k} f(z)\,dz_1 \ldots dz_n +$$

$$+ \sum_{\substack{1 \leqslant s \leqslant n \\ s \neq k}} \left[\int\limits_{v_s'} f(z)\,dz_1 \ldots dz_n + \int\limits_{v_s''} f(z)\,dz_1 \ldots dz_n \right] = 0.$$

Each term under the summation sign is equal to zero. Therefore,

$$\int\limits_{z_1'}^{z_1''} dz_1 \ldots \int\limits_{z_{j-1}'}^{z_{j-1}''} dz_{j-1} \oint\limits_{\partial g_k} dz_k \int\limits_{z_{j+1}'}^{z_{j+1}''} dz_{j+1} \ldots \int\limits_{z_n'}^{z_n''} f(z)\,dz_n = 0, \qquad k = 1, 2, \ldots, n, \tag{13}$$

where the integrations are taken over the curves l_j, for $j \neq k$, lying in the circles $S(z_j^0, r_j)$ and connecting the points z_j' and z_j''. Since the paths l_j, for $j \neq k$, and the closed curves ∂g_k are arbitrary, it follows from Eq. (13) that the function

$$F(z) = \int\limits_{z_1^0}^{z_1} dz_1' \ldots \int\limits_{z_n^0}^{z_n} f(z')\,dz_n' \tag{14}$$

is single-valued in the polycircle $S(z^0, r)$ (that is, it is independent of the path connecting z_j^0 and z_j and lying in the circle $S(z_j^0, r_j)$).

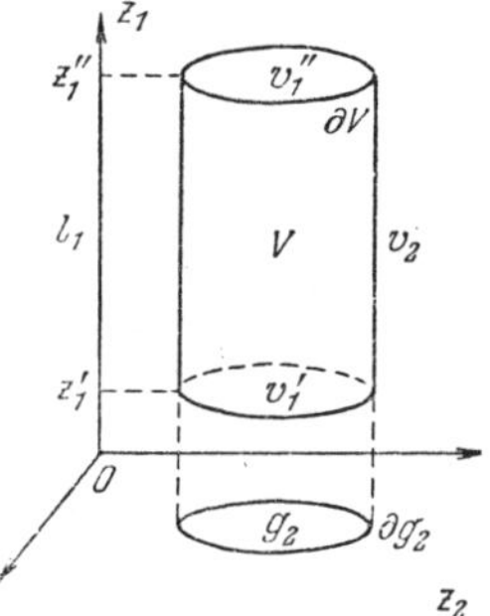

Fig. 49

Let us fix a point $(a_1, \ldots, a_{k-1}, a_{k+1}, \ldots, a_n)$ in the polycircle

$$S(z_1^0, r_1) \times \ldots \times S(z_{k-1}^0, r_{k-1}) \times \times S(z_{k+1}^0, r_{k+1}) \times \ldots \times S(z_n^0, r_n).$$

Then, it follows from (13) and (14) that the function

$$F(a_1, \ldots, a_{k-1}, z_k, a_{k+1}, \ldots, a_n) = \int\limits_{z_k^0}^{z_k} \left[\int\limits_{z_1^0}^{a_1} dz_1' \ldots \int\limits_{z_{k-1}^0}^{a_{k-1}} dz_{k-1}' \int\limits_{z_{k+1}^0}^{a_{k+1}} dz_{k+1}' \ldots \int\limits_{z_n^0}^{a_n} f(z')\,dz_n' \right] dz_k'$$

satisfies the conditions of Morera's theorem for a single complex variable z_k in the circle $S(z_k^0, r_k)$. Therefore, this function is holomorphic with respect to z_k in $S(z_k^0, r_k)$. Thus, the function $F(z)$ constructed in accordance with formula (14) is holomorphic in the polycircle $S(z^0, r)$ with respect to each variable z_k individually. This means that $F(z)$ is holomorphic in $S(z^0, r)$ [see section 4.2]. But, it follows from (14) that

$$f(z) = \frac{\partial^n F(z)}{\partial z_1 \partial z_2 \ldots \partial z_n}.$$

Therefore, $f(z)$ is holomorphic in $S(z^0, r)$, which completes the proof.

23. THE MARTINELLI-BOCHNER INTEGRAL REPRESENTATION

The Martinelli-Bochner integral representation (see Bochner [55], Martinelli [151]) is a generalization of Cauchy's integral formula for a single complex variable to the case of several variables. This formula is derived by the methods of potential theory by use of Stokes' formula. However, in contrast with the case in which $n = 1$, the Dirichlet problem in the class of pluriharmonic functions (see section 4.1) is not in general solvable (see Fuks [1], p. 316). Therefore, not every continuous function defined on a sufficiently "well-behaved" boundary ∂G of a domain G can be regarded as the limiting value of the real or imaginary part of some function that is holomorphic in G. This complicates the application of potential-theory methods to the theory of functions of several complex variables.

1. *The Martinelli-Bochner formula*

Suppose that a function $f(z)$ is holomorphic in a bounded domain G and is continuous $\bar{G}$. Suppose also that ∂G is a piecewise-smooth surface oriented by the orientation of G. Then, the Martinelli-Bochner formula is valid:

$$\frac{(n-1)!}{(2\pi i)^n} \int\limits_{\partial G} \frac{f(\zeta)}{|z-\zeta|^{2n}} \sum_{j=1}^{n} (\bar{\zeta}_j - \bar{z}_j)\, d\bar{\zeta}_1\, d\zeta_1 \ldots [d\bar{\zeta}_j]\, d\zeta_j \ldots$$

$$\ldots d\bar{\zeta}_n\, d\zeta_n = \begin{cases} f(z), & \text{if} \quad z \in G; \\ 0, & \text{if} \quad z \bar{\in} \bar{G}. \end{cases} \qquad (15)$$

Here, the square brackets indicate that the enclosed term is omitted.

Let us suppose that the space C^n is oriented in accordance with section 22.5.

To prove formula (15), we take a sequence of domains G_α with piecewise-smooth boundaries ∂G_α that satisfies the condition that

$$G_\alpha \subset G_{\alpha+1} \Subset G, \quad \bigcup_\alpha G_\alpha = G.$$

As a preliminary, let us prove formula (15) for every domain G_α. The exterior differential of the form in the integrand vanishes in the domain $G_\alpha \setminus \{z\}$. This assertion can be verified directly by using rules (10) and the Cauchy-Riemann conditions $\partial f / \partial \bar{z}_j = 0$, for $j = 1, 2, \ldots, n$. Therefore, from Stokes' formula (see section 22.4), the integral (15) [with ∂G replaced by ∂G_α] vanishes for $z \bar{\in} \bar{G}_\alpha$.

Suppose now $z \in G_\alpha$. Let us remove a hypersphere $U(z, \varepsilon)$ of sufficiently small radius ε from the domain G_α (see Fig. 50) and apply Stokes' formula to the domain $\bar{G}_\alpha \setminus U(z, \varepsilon)$. We then obtain

$$\left(\int_{\partial G_\alpha} + \int_{|z-\zeta|=\varepsilon} \right) \frac{f(\zeta)}{|z-\zeta|^{2n}} \sum_{j=1}^{n} (\bar{\zeta}_j - \bar{z}_j)\, d\bar{\zeta}_1\, d\zeta_1 \ldots \\ \ldots [d\bar{\zeta}_j]\, d\zeta_j \ldots d\bar{\zeta}_n d\zeta_n = 0.$$

To obtain formula (15) it thus remains to show that

$$\lim_{\varepsilon \to 0} \int_{|z-\zeta|=\varepsilon} \frac{f(\zeta)}{|z-\zeta|^{2n}} \sum_{j=1}^{n} (\bar{\zeta}_j - \bar{z}_j)\, d\bar{\zeta}_1\, d\zeta_1 \ldots [d\bar{\zeta}_j]\, d\zeta_j \ldots \\ \ldots d\bar{\zeta}_n\, d\zeta_n = \frac{(2\pi i)^n}{(n-1)!} f(z), \tag{16}$$

where the orientation of the sphere $|z-\zeta| = \varepsilon$ corresponds to the positive direction of the outer normal to it.

Also as a preliminary, we evaluate the integral

$$J = \int_{|\zeta|=\varepsilon} \frac{1}{|\zeta|^{2n}} \sum_{j=1}^{n} \bar{\zeta}_j\, d\bar{\zeta}_1\, d\zeta_1 \ldots \\ \ldots [d\bar{\zeta}_j]\, d\zeta_j \ldots d\bar{\zeta}_n\, d\zeta_n.$$

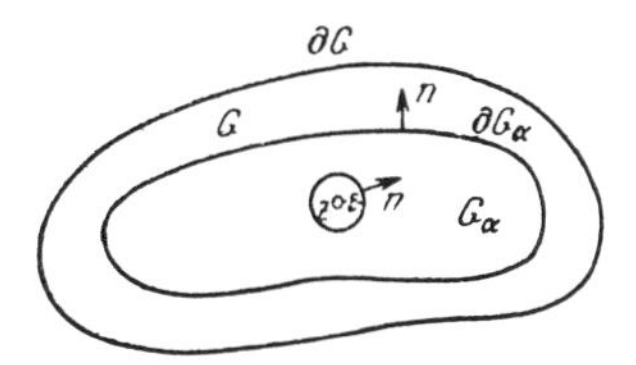

Fig. 50

If we shift to the real coordinates $\zeta_j = \xi_j + i\eta_j$, $\bar{\zeta}_j = \xi_j - i\eta_j$, for $j = 1, 2, \ldots, n$, and use rule (10), we obtain

$$J = \frac{(2i)^{n-1}}{\varepsilon^{2n}} \int_{|\zeta|=\varepsilon} \sum_{j=1}^{n} d\xi_1\, d\eta_1 \ldots (\xi_j\, d\xi_j + \eta_j\, d\eta_j + \\ + i\xi_j\, d\eta_j - i\eta_j\, d\xi_j) \ldots d\xi_n\, d\eta_n. \tag{17}$$

Since

$$d\left[\sum_{j=1}^{n} d\xi_1\, d\eta_1 \ldots (\xi_j\, d\xi_j + \eta_j\, d\eta_j) \ldots d\xi_n\, d\eta_n\right] = 0,$$

it follows on the basis of Stokes' formula that the terms in (17) containing the expressions $\xi_j\, d\xi_j + \eta_j\, d\eta_j$ are equal to zero. Therefore, we deduce from (17) that

$$J = \frac{i\,(2i)^{n-1}}{\varepsilon^{2n}} \int\limits_{|\zeta|=\varepsilon} \sum_{j=1}^{n} d\xi_1\, d\eta_1 \ldots (\xi_j\, d\eta_j - \eta_j\, d\xi_j) \ldots d\xi_n\, d\eta_n\, .$$

By applying formula (9), we obtain

$$J = \frac{i\,(2i)^{n-1}}{\varepsilon^{2n}}\, 2n \operatorname{mes} U(0,\, \varepsilon) = \frac{i\,(2i)^{n-1}}{\varepsilon^{2n}}\, 2n\, \frac{2\pi^n}{\Gamma(n)}\, \frac{\varepsilon^{2n}}{2n} = \frac{(2\pi i)^n}{(n-1)!}.$$

Formula (16) follows from this because of the continuity of the function $f(\zeta)$ in a neighborhood of the point z.

Thus, formula (15) is established for an arbitrary subdomain $G_\alpha \Subset G$. Since the function $f(z)$ is assumed to be continuous in G, formula (15) remains valid even for the domain $G = \bigcup\limits_{\alpha} G_\alpha$ itself, which completes the proof.

Remark: The variables $\bar{z}_j$ that appear in the integral representation (15) should be dropped in the evaluation of the integral since the function $f(z)$ is independent of $\bar{z}$. Therefore, if we replace $\bar{z}_j$ by w_j in the integral representation (15), where w is a point lying in a sufficiently small neighborhood of the point $\bar{z}^0$, for $z^0 \in G$, we obtain a function $F(z, w)$ that is holomorphic with respect to the set of $2n$ complex variables (z, w) in a neighborhood of the point $(z^0, \bar{z}^0)$ and that satisfies the equation $F(z, w) \equiv f(z)$ on the surface $w = \bar{z}$. It follows from this that $F(z, w) = f(z)$ in the neighborhood referred to (see section 6.2). Therefore, in a sufficiently small neighborhood of the point $z^0 \in G$, the function $f(z)$ is represented by the integral (15), where $\bar{z}$ should be replaced with $\bar{z}^0$:

$$f(z) = \frac{(n-1)!}{(2\pi i)^n} \int\limits_{\partial G} \frac{f(\zeta)}{[(z-\zeta)(\bar{z}^0 - \bar{\zeta})]^n} \sum_{j=1}^{n} (\bar{\zeta}_j - \bar{z}_j^0)\, d\bar{\zeta}_1\, d\zeta_1 \ldots \ldots [d\bar{\zeta}_j]\, d\zeta_j \ldots d\bar{\zeta}_n\, d\zeta_n.$$

2. *Generalization of the principle of holomorphic continuation*

Suppose that a function $f(z)$ is holomorphic in a domain G except possibly at points of a piecewise-smooth hypersurface S and that f is continuous in G. Then, $f(z)$ is holomorphic in G.

Proof: Let z^0 denote a point in $S \cap G$. Let us show that $f(z)$ is holomorphic at the point z^0. Since S consists of a finite number of hypersurfaces S_i with boundary of class $C^{(1)}$, there exists a hypersphere $U(z^0, r) \Subset G$ such that S partitions it into a finite number of domains $U_1, U_2, \ldots, U_N$ with piecewise-smooth boundaries ∂U_l (see Fig. 51). Suppose that $z \in U_l$. Then, formula (15) yields the following equations:

$$\frac{(n-1)!}{(2\pi i)^n} \int\limits_{\partial U_k} \frac{f(\zeta)}{|z-\zeta|^{2n}} \sum_{j=1}^{n} (\bar{\zeta}_j - \bar{z}_j)\, d\bar{\zeta}_1\, d\zeta_1 \ldots [d\bar{\zeta}_j]\, d\zeta_j \ldots$$

$$\ldots d\bar{\zeta}_n\, d\zeta_n = \begin{cases} f(z), & \text{if} \quad k = l, \\ 0, & \text{if} \quad k \neq l, \end{cases} \qquad k = 1, 2, \ldots, N.$$

Adding these equations, we obtain, for $z \in U_l$,

$$f(z) = \frac{(n-1)!}{(2\pi i)^n} \int\limits_{\bigcup_k \partial U_k} \frac{f(\zeta)}{|z-\zeta|^{2n}} \sum_{j=1}^{n} (\bar{\zeta}_j - \bar{z}_j)\, d\bar{\zeta}_1\, d\zeta_1 \ldots [d\bar{\zeta}_j]\, d\zeta_j \ldots d\bar{\zeta}_n\, d\zeta_n. \tag{18}$$

The piecewise-smooth hypersurface $S \cap \bar{U}(z^0, r)$ can be partitioned into a finite number of surfaces σ_s with boundary of class $C^{(1)}$. Each surface σ_s belongs to two domains, let us say, U_m and U_n, which are abutting along σ_s. Therefore, the surface σ_s is oriented in one way if it belongs to ∂U_m and another way if it belongs to ∂U_n (see section 22.2). Thus, the surface σ_s appears twice, opposite orientations, in the expression $\bigcup_k \partial U_k$ and we may write

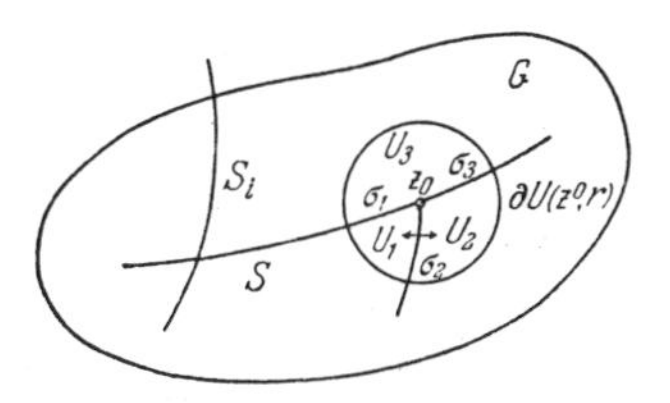

Fig. 51

$$\bigcup_k \partial U_k = \bigcup_s [\sigma_s \cup (-\sigma_s)] \cup \partial U(z^0, r). \tag{19}$$

By hypothesis, $f(z)$ is continuous in G. Because of this and Eq. (19), the integrals over the surfaces σ_s and $-\sigma_s$ in Eq. (18) cancel and there remains only the integral over the surface $\partial U(z^0, r)$. Thus, we obtain the representation

$$f(z) = \frac{(n-1)!}{(2\pi i)^n} \int\limits_{|\zeta - z^0| = r} \frac{f(\zeta)}{|z-\zeta|^{2n}} \sum_{j=1}^{n} (\bar{\zeta}_j - \bar{z}_j)\, d\bar{\zeta}_1\, d\zeta_1 \ldots [d\bar{\zeta}_j]\, d\zeta_j \ldots d\bar{\zeta}_n\, d\zeta_n, \tag{20}$$

which is valid for all $z \in U_i$, where $i = 1, 2, \ldots, N$.

Let us show that the function $f(z)$ is holomorphic in the polysphere $U(z^0, \theta r)$, where $0 < \theta < \sqrt{2} - 1$. For this, we introduce the function $F(z, w)$ of the two complex n-tuples (z, w) obtained by replacing $\bar{z}$ by w in the right member of (20):

$$F(z, w) = \frac{(n-1)!}{(2\pi i)^n} \int\limits_{|\zeta - z^0| = r} \frac{f(\zeta)}{[(z-\zeta)(w-\bar{\zeta})]^n} \sum_{j=1}^{n} (\bar{\zeta}_j - w_j) d\bar{\zeta}_1 d\zeta_1 \ldots \ldots [d\bar{\zeta}_j] d\zeta_j \ldots d\bar{\zeta}_n d\zeta_n.$$

This function is defined and holomorphic in $U(z^0, \theta r) \times U(\bar{z}^0, \theta r)$ because, in that domain, the denominator in the integrand does not vanish in the region of integration. This is true because, on the basis of the Bunyakovskiy-Schwarz inequality, we have for such z, w, and ζ,

$$|(z-\zeta)(w-\bar{\zeta})| = = ||\zeta - z^0|^2 + (z - z^0)(w - \bar{z}^0) - (z - z^0)(\bar{\zeta} - \bar{z}^0) - - (\zeta - z^0)(w - \bar{z}^0)| \geqslant r^2 - \theta^2 r^2 - \theta r^2 - \theta r^2 > 0.$$

Furthermore, the function $F(z, w)$ coincides with $f(z)$ on the surface $w = \bar{z}$ if $z \in U_i \cap U(z^0, \theta r)$, where $i = 1, 2, \ldots, N$. Therefore (see section 6.2),

$$F(z, w) \equiv f(z), \quad (z, w) = U(z^0, \theta r) \times U(z^0, \theta r).$$

From this it follows* that the function $f(z)$ is holomorphic in $U(z^0, \theta r)$, which completes the proof.

We note in passing the following closely associated result, which also generalizes the principle of holomorphic continuation (see Fuks [1], p. 118). *Suppose that a function $f(z)$ is bounded in $U(z^0, r)$ and holomorphic in $U(z^0, r) \setminus F$, where F is a $(2n - 2)$-dimensional analytic surface passing through the point z^0. Then, $f(z)$ is holomorphic and $U(z^0, r)$.*

24. THE BERGMAN-WEIL INTEGRAL REPRESENTATION

The Bergman-Weil integral representation (see Bergman [68] and Weil [69]) has to do with functions that are holomorphic in a neighborhood of a Weil domain

$$\sigma = [z: \ |\chi_j(z)| < 1, \ j = 1, 2, \ldots, N; \ z \in U(\bar{\sigma})].$$

This representation is derived from the Martinelli-Bochner formula. The derivation that we give is due to Sommer [56].

*The function $F(z, w)$ is actually independent of w (see section 23.1).

1. *The hull of a Weil domain*

For simplicity of exposition, we shall consider Weil domains possessing the property that

$$\operatorname{rank}\left\| \begin{matrix} \operatorname{grad} \chi_{\alpha_1}(z) \\ \cdots \\ \operatorname{grad} \chi_{\alpha_k}(z) \end{matrix} \right\| = k, \qquad 1 \leqslant \alpha_1, \ldots, \alpha_k \leqslant N, \; k = 1, 2, \ldots, n \tag{21}$$

on a set

$$S_{\alpha_1 \ldots \alpha_k} = [z: \; |\chi_{\alpha_1}(z)| = \ldots = |\chi_{\alpha_k}(z)| = 1, \; z \in U(\bar{\sigma})].$$

It follows from this hypothesis that each set $S_{\alpha_1 \ldots \alpha_k}$ either is empty or is a $(2n - k)$-dimensional surface of class $C^{(\infty)}$ (see section 1.4). This is true because the functions $\varphi_\alpha(z) \equiv \chi_\alpha(z)\bar{\chi}_\alpha(z) - 1$ are infinitely differentiable in $U(\bar{\sigma})$ and, by virtue of the Cauchy-Riemann conditions,

$$\frac{\partial \varphi_\alpha}{\partial z_j} = \chi_\alpha \frac{\partial \bar{\chi}_\alpha}{\partial z_j} + \bar{\chi}_\alpha \frac{\partial \chi_\alpha}{\partial z_j} = \bar{\chi}_\alpha \frac{\partial \chi_\alpha}{\partial z_j}, \quad j = 1, 2, \ldots, n.$$

Therefore, by virtue of (21), in $S_{\alpha_1 \ldots \alpha_k}$, we have

$$\operatorname{rank}\left\| \begin{matrix} \operatorname{grad} \varphi_{\alpha_1}, \; \operatorname{grad} \bar{\varphi}_{\alpha_1} \\ \cdots\cdots \\ \operatorname{grad} \varphi_{\alpha_k}, \; \operatorname{grad} \bar{\varphi}_{\alpha_k} \end{matrix} \right\| = \operatorname{rank}\left\| \begin{matrix} \bar{\chi}_{\alpha_1} \operatorname{grad} \chi_{\alpha_1}, \; \chi_{\alpha_1} \operatorname{grad} \bar{\chi}_{\alpha_1} \\ \cdots\cdots\cdots \\ \bar{\chi}_{\alpha_k} \operatorname{grad} \chi_{\alpha_k}, \; \chi_{\alpha_k} \operatorname{grad} \bar{\chi}_{\alpha_k} \end{matrix} \right\| = k.$$

In particular, all the S_α are analytic hypersurfaces (see section 18.7).

Let us assume that all the functions χ_α are essential for the definition of the polyhedron σ. Let us orient C^n and, consequently, σ in accordance with section 22.5.

The boundary of the domain σ consists of a finite number of $(2n - 1)$-dimensional bounded orientable hypersurfaces σ_{α_1} with class $C^{(\infty)}$ boundary $\partial\sigma = \bigcup_{\alpha_1} \sigma_{\alpha_1}$, that lie on the analytic surfaces S_{α_1}. The number of hypersurfaces σ_{α_1} may actually be less than N. In this case, we assume that the corresponding σ_{α_1} are empty: $\sigma_{\alpha_1} = \varnothing$. The orientation of σ induces an orientation of the hypersurfaces σ_{α_1} (in accordance with section 22.2). The boundary $\partial\sigma_{\alpha_1}$ of each hypersurface σ_{α_1} consists of a finite number of $(2n - 2)$-dimensional bounded orientable surfaces $\sigma_{\alpha_1\alpha_2}$ with boundary of class $C^{(\infty)}$ $\partial\sigma_{\alpha_1} = \bigcup_{\alpha_2} \sigma_{\alpha_1\alpha_2}$ that lie on $S_{\alpha_1\alpha_2}$ (see Fig. 52). Let us assume that $\sigma_{\alpha_1\alpha_2} = \varnothing$ if $\alpha_1 = \alpha_2$ or if $\partial\sigma_{\alpha_1}$ and $\partial\sigma_{\alpha_2}$ do not have a common $(2n - 2)$-dimensional subsurface. The orientation of the hypersurface σ_{α_1} induces an orientation of the surfaces $\sigma_{\alpha_1\alpha_2}$ such that $\sigma_{\alpha_1\alpha_2} = -\sigma_{\alpha_2\alpha_1}$, etc. The boundary $\partial\sigma_{\alpha_1 \ldots \alpha_k}$ (for $1 \leqslant k \leqslant n - 1$) of the $(2n - k)$-dimensional surface $\sigma_{\alpha_1 \ldots \alpha_k} \subset \partial\sigma_{\alpha_1 \ldots \alpha_{k-1}}$ consists of a finite number of $(2n - k - 1)$-dimensional bounded orientable surfaces $\sigma_{\alpha_1 \ldots \alpha_k \alpha_{k+1}}$ with boundary of class $C^{(\infty)}$:

$$\partial\sigma_{\alpha_1 \ldots \alpha_k} = \bigcup_{\alpha_{k+1}} \sigma_{\alpha_1 \ldots \alpha_k \alpha_{k+1}}, \quad \sigma_{\alpha_1 \ldots \alpha_k \alpha_{k+1}} \subset S_{\alpha_1 \ldots \alpha_{k+1}}. \tag{22}$$

Here, we take $\sigma_{\alpha_1 \ldots \alpha_k \alpha_{k+1}} = \varnothing$ if $\alpha_{k+1} = \alpha_l$ for any $l = 1, 2, \ldots, k$ or if the surfaces $\partial\sigma_{\alpha_1 \ldots \alpha_k}$ and $\partial\sigma_{\alpha_{k+1}}$ have no common $(2n - k - 1)$-dimensional subsurface. The orientation of the surface $\sigma_{\alpha_1 \ldots \alpha_k}$ induces an orientation of the surfaces $\sigma_{\alpha_1 \ldots \alpha_k \alpha_{k+1}}$ such that

$$\sigma_{\alpha_1 \ldots \alpha_k \alpha_{k+1}} = \pm \sigma_{\alpha_{\nu_1} \ldots \alpha_{\nu_k} \alpha_{\nu_{k+1}}}. \tag{23}$$

Here, the sign + or − is taken according to whether the permutation $(\nu_1, \ldots, \nu_k, \nu_{k+1})$ is even or odd.

The oriented n-dimensional set

$$\Delta(\sigma) = (-1)^{\frac{n(n-1)}{2}} \bigcup_{\alpha_1 < \ldots < \alpha_n} \sigma_{\alpha_1 \ldots \alpha_n}$$

is called the *hull* of the Weil domain σ.

A special case of a Weil domain is the polycircle

$$S(0, r) = [z: |z_j| < r_j, \ j = 1, 2, \ldots, n] = S(0, r_1) \times \ldots \times S(0, r_n).$$

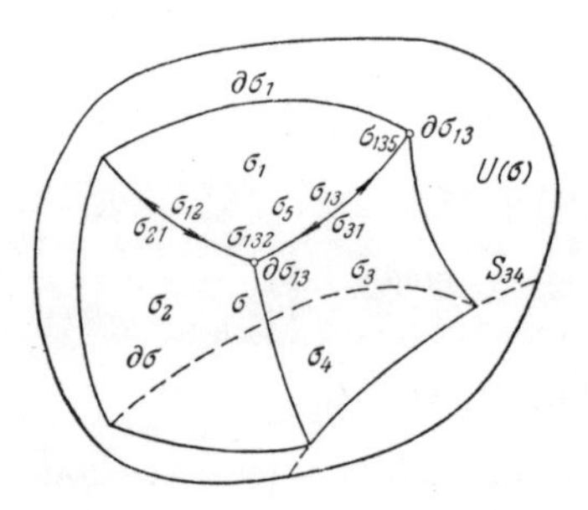

Fig. 52

In section 4.3, we defined the hull of a polycircle by formula $\partial S(0, r_1) \times \ldots \times \partial S(0, r_n)$. In this case, the positive direction on $\partial S(0, r_j)$ is taken counterclockwise. On the other hand, the hull of the Weil domain $S(0, r)$ is the oriented surface

$$\Delta[S(0, r)] = (-1)^{\frac{n(n-1)}{2}} \sigma_{12 \ldots n}.$$

For these two definitions of the hull of the domain $S(0, r)$ to be consistent, we need to show that

$$\partial S(0, r_1) \times \ldots \times \partial S(0, r_n) = (-1)^{\frac{n(n-1)}{2}} \sigma_{12 \ldots n}. \tag{24}$$

To prove Eq. (24), we note that

$$\int_{\partial S(0,r_1)\times\ldots\times\partial S(0,r_n)} \bar{z}_1\bar{z}_2\ldots\bar{z}_n\,dz_1\,dz_2\ldots dz_n = \prod_j \int_{\partial S(0,r_j)} \bar{z}_j\,dz_j = (2\pi i)^n r_1 r_2\ldots r_n. \tag{25}$$

On the other hand, making use of the convention adopted regarding orientation (see section 22.5), we obtain by use of Stokes' formula

$$(2\pi i)^n r_1\ldots r_n = \int_{S(0,r)} d\bar{z}_1\,dz_1\ldots d\bar{z}_n\,dz_n = \int_{\sigma_1} \bar{z}_1\,dz_1\,d\bar{z}_2\,dz_2\ldots d\bar{z}_n\,dz_n =$$

$$= \int_{\sigma_{12}} \bar{z}_1\bar{z}_2\,dz_1\,dz_2\,d\bar{z}_3\,dz_3\ldots d\bar{z}_n\,dz_n = \ldots$$

$$\ldots = (-1)^{\frac{n(n-1)}{2}} \int_{\sigma_{12\ldots n}} \bar{z}_1\bar{z}_2\ldots\bar{z}_n\,dz_1\,dz_2\ldots dz_n.$$

From this equation and Eq. (25), we derive the relation

$$\int_{\partial S(0,r_1)\times\ldots\times\partial S(0,r_n)} \bar{z}_1\ldots\bar{z}_n\,dz_1\ldots dz_n = (-1)^{\frac{n(n-1)}{2}} \int_{\sigma_{12\ldots n}} \bar{z}_1\ldots\bar{z}_n\,dz_1\ldots dz_n.$$

Since the surfaces $\partial S(0,r_1)\times\ldots\times\partial S(0,r_n)$ and $\sigma_{12\ldots n}$ consist of the same geometric points, the required relation (24) follows from this.

It can be shown in an analogous manner that if a Weil domain σ is of the form $G_1\times G_2\times\ldots\times G_n$, then

$$\Delta(\sigma) = \partial G_1\times\partial G_2\times\ldots\times\partial G_n.$$

2. *Hefer's theorem*

Suppose that a function $\chi(z)$ is holomorphic in a domain of holomorphy $D\subset C^n$. Then, there exist functions $P_j(\zeta,z)$ that are holomorphic in $D\times D$ such that

$$\chi(\zeta)-\chi(z) = \sum_{j=1}^{n}(\zeta_j - z_j)P_j(\zeta,z). \tag{26}$$

Let us accept this theorem without proof. Its proof can be found in the article by Hefer [59] or the book by Fuks [2], p. 137. We note that if $\chi(z)$ is a polynomial, this theorem is trivial (in which case $P_j(\zeta, z)$ are also polynomials).

For $n \geqslant 2$, the functions P_j are nonsingle-valuedly defined by Eqs. (26).

3. *Certain differential forms*

Suppose that we are given the kn functions $q_{lj}(\zeta, \bar{\zeta})$, for $l = 1, 2, \ldots, k$ and $j = 1, 2, \ldots, n$, and the $(n-k)n$ differential first-order forms

$$\theta_{lj} = \sum_{\nu=1}^{n} [a_{lj\nu}(\zeta, \bar{\zeta})\, d\zeta_\nu + b_{lj\nu}(\zeta, \bar{\zeta})\, d\bar{\zeta}_\nu], \quad l = k+1, \ldots, n,$$

where $j = 1, 2, \ldots, n$ (and $k = 0, 1, 2, \ldots, n$). Let us form the vectors

$$q_l = (q_{l1}, q_{l2}, \ldots, q_{ln}), \; l = 1, 2, \ldots, k;$$
$$\theta_l = (\theta_{l1}, \theta_{l2}, \ldots, \theta_{ln}), \; l = k+1, \ldots, n$$

and the new differential form of order $n - k$

$$D(q_1, \ldots, q_k, \theta_{k+1}, \ldots, \theta_n) = \begin{vmatrix} q_{11}, & \ldots, & q_{k1}, & \theta_{k+1\,1}, & \ldots, & \theta_{n1} \\ q_{12}, & \ldots, & q_{k2}, & \theta_{k+1\,2}, & \ldots, & \theta_{n2} \\ \cdot & \cdot & \cdot & \cdot & \cdot & \cdot \\ q_{1n}, & \ldots, & q_{kn}, & \theta_{k+1\,n}, & \ldots, & \theta_{nn} \end{vmatrix}$$

The nth-order determinant in the right member of this equation can be expanded according to the usual rules except that the differentials must be multiplied according to rules (10). The position of each factor in the terms of the sum is determined by the number of the column to which this factor belongs. Therefore, this determinant vanishes if one of its rows is a linear combination of the remaining ones. This property does not apply to the columns. For example,

$$D(d\zeta, \ldots, d\zeta) = \begin{vmatrix} d\zeta_1, & \ldots, & d\zeta_n \\ \cdot & \cdot & \cdot \\ d\zeta_1, & \ldots, & d\zeta_n \end{vmatrix} = n!\, d\zeta_1\, d\zeta_2 \ldots d\zeta_n, \tag{27}$$

where we set $d\zeta = (d\zeta_1, d\zeta_2, \ldots, d\zeta_n)$.

Let us note other, easily proven, properties of the determinant D:

$$D(\ldots, q_l, \ldots, q_m, \ldots, \theta_{k+1}, \ldots, \theta_n) = \\ = -D(\ldots, q_m, \ldots, q_l, \ldots, \theta_{k+1}, \ldots, \theta_n), \tag{28}$$

$$D(\ldots, qq_l, \ldots, \theta_{k+1}, \ldots, \theta_n) = \\ = qD(\ldots, q_l, \ldots, \theta_{k+1}, \ldots, \theta_n), \tag{29}$$

$$D(\ldots, q_l, \ldots, \theta_{k+1}, \ldots, \theta_n) = \\ = D\Big(\ldots, q_l + \sum_{j \neq l} \alpha_j q_j, \ldots, \theta_{k+1}, \ldots, \theta_n\Big), \tag{30}$$

$$D(q_1, \ldots, q_k, \ldots, \theta_l, \ldots, \theta_m, \ldots) = \\ = D(q_1, \ldots, q_k, \ldots, \theta_m, \ldots, \theta_l, \ldots), \tag{31}$$

$$D(q_1, \ldots, q_k, \ldots, \theta\theta_l, \ldots) = \\ = \theta D(q_1, \ldots, q_k, \ldots, \theta_l, \ldots), \tag{32}$$

$$D(q_1, \ldots, q_k, \ldots, \theta_l, \ldots) = \\ = D\Big(q_1, \ldots, q_k, \ldots, \theta_l + \sum_{j \neq l} q_j \omega_j, \ldots\Big), \tag{33}$$

where $\omega_1, \ldots, \omega_k$ are first-order differential forms.

Let us suppose now that the vectors q_l satisfy the equations

$$(\zeta - z) q_l = 1, \tag{34}$$

where z is some fixed point. For $\zeta \neq z$, let us define

$$\theta_k = \frac{d\bar{\zeta}}{N^{n-k}} - (n-k)\frac{(\zeta - z)\, d\bar{\zeta}}{N^{n-k+1}}(\bar{\zeta} - \bar{z}), \quad k = 1, 2, \ldots, n \tag{35}$$

where

$$N = |z - \zeta|^2 = (z - \zeta)(\bar{z} - \bar{\zeta}). \tag{36}$$

It is easy to verify that

$$\begin{vmatrix} 1, & 1, \ldots, 1, & \frac{1+k-n}{N^{n-k}}(\zeta - z)\, d\bar{\zeta}, & (\zeta - z)\, d\bar{\zeta}, \ldots, (\zeta - z)\, d\bar{\zeta} \\ \frac{\bar{\zeta}_1 - \bar{z}_1}{N}, & q_{11}, \ldots, q_{k1}, & \theta_{k1}, & d\bar{\zeta}_1, \ldots, d\bar{\zeta}_1 \\ \cdots & \cdots & \cdots & \cdots \\ \frac{\bar{\zeta}_n - \bar{z}_n}{N}, & q_{1n}, \ldots, q_{kn}, & \theta_{kn}, & d\bar{\zeta}_1, \ldots, d\bar{\zeta}_1 \end{vmatrix} = 0, \tag{37}$$

since, on the basis of (34)-(36), the first row of this determinant is a linear combination of the remaining rows. If we expand this determinant according to elements of the first row, we obtain from Eqs. (37) the following equations for $k = 1, 2, \ldots, n$:

$$\begin{aligned} D(q_1, \ldots, q_k, \theta_k, d\bar{\zeta}, \ldots, d\bar{\zeta}) = \\ = \sum_{l=1}^{k} (-1)^{l-1} D\left(\frac{\bar{\zeta}-\bar{z}}{N}, q_1, \ldots, [q_l], \ldots, q_k, \theta_k, d\bar{\zeta}, \ldots, d\bar{\zeta}\right) + \\ + (-1)^k \frac{1-n+k}{N^{n-k}} (\zeta - z)\, d\bar{\zeta} \times \\ \times D\left(\frac{\bar{\zeta}-\bar{z}}{N}, q_1, \ldots, q_k, d\bar{\zeta}, \ldots, d\bar{\zeta}\right) + \\ + (-1)^k (n-k-1)(\zeta - z)\, d\bar{\zeta} \times \\ \times D\left(\frac{\bar{\zeta}-\bar{z}}{N}, q_1, \ldots, q_k, \theta_k, d\bar{\zeta}, \ldots, d\bar{\zeta}\right). \end{aligned} \tag{38}$$

We take the factor N^{-n+k} in the $(k+2)$nd column of the last determinant outside the determinant. Also, we multiply the first column of this determinant by $(n-k)(\zeta - z)\, d\bar{\zeta}$ and add the result to the $(k+2)$nd column. As a result of this, we see by virtue of Eqs. (35) and (36) that the last two terms in Eqs. (38) cancel each other. We proceed in a similar manner with the individual terms of the remaining sum. If we now transfer the first column in each of these terms to the position of the kth column, we obtain from (38) the following equations:

$$\begin{aligned} D(q_1, \ldots, q_k, \theta_k, d\bar{\zeta}, \ldots, d\bar{\zeta}) = \\ = \sum_{l=1}^{k} (-1)^{k+l} D\left(q_1, \ldots, [q_l], \ldots, q_k, \frac{\bar{\zeta}-\bar{z}}{N^{n-k+1}}, d\bar{\zeta}, \ldots, d\bar{\zeta}\right), \\ k = 1, 2, \ldots, n. \end{aligned} \tag{39}$$

We note that we used properties (28)-(33) in our operations with the determinants.

4. *The Bergman-Weil formula*

Suppose that $f(z)$ is a function that is holomorphic in a neighborhood of a Weil domain

$$\sigma = [z\colon |\chi_\alpha(z)| < 1,\ \alpha = 1, 2, \ldots, N;\ z \in U(\bar{\sigma})],$$

where the functions χ_α are holomorphic in the domain of holomorphy $D \supset U(\bar{\sigma})$. Then, the Bergman-Weil formula

$$\frac{(-1)^{\frac{n(n-1)}{2}}}{(2\pi i)^n} \sum_{\alpha_1 < \ldots < \alpha_n} \int_{\sigma_{\alpha_1 \ldots \alpha_n}} f(\zeta) D(q_{\alpha_1}, \ldots, q_{\alpha_n})\, d\zeta_1, \ldots, d\zeta_n = \\ = \begin{cases} f(z), & z \in \sigma, \\ 0, & z \in D \setminus (\sigma \cup \delta), \end{cases} \tag{40}$$

where $\sigma_{\alpha_1 \ldots \alpha_n}$ *are n-dimensional surfaces constituting the hull* $\Delta(\sigma)$, *holds* (see section 24.1). The coordinates of the vectors q_α, for $\alpha = 1, 2, \ldots, N$, are given by the formula

$$q_{\alpha j} = \frac{P_{\alpha j}(\zeta, z)}{\chi_\alpha(\zeta) - \chi_\alpha(z)}, \quad j = 1, 2, \ldots, n, \tag{41}$$

where the functions $P_{\alpha j}(\zeta, z)$ are holomorphic in $D \times D$ and are determined in accordance with Hefer's theorem (see section 24.2):

$$\chi_\alpha(\zeta) - \chi_\alpha(z) = \sum_{j=1}^{n} (\zeta_j - z_j) P_{\alpha j}(\zeta, z). \tag{42}$$

The determinant D is of the form (see section 24.3)

$$D(q_{\alpha_1}, \ldots, q_{\alpha_n}) = \begin{vmatrix} q_{\alpha_1 1}, & \ldots, & q_{\alpha_n 1} \\ \cdot & \cdot\ \cdot\ \cdot & \cdot \\ q_{\alpha_1 n}, & \ldots, & q_{\alpha_n n} \end{vmatrix}. \tag{43}$$

Finally (see Fig. 53),

$$\delta = \bigcup_{1 \leqslant \alpha \leqslant N} [z: \ |\chi_\alpha(z)| = 1, \ z \in D]. \tag{44}$$

5. *Proof of the Bergman-Weil formula*

Let us prove the Bergman-Weil formula for Weil domains satisfying condition (21). The proof that we shall give is valid also for arbitrary Weil domains (see Sommer [56]). Here, it is necessary to extend the class of surfaces over which it is possible to integrate the differential forms (see section 22.4, remark).

Let us apply the Martinelli-Bochner formula (15) to the function $f(z)$, remembering that $\partial\sigma = \bigcup_{\alpha_1} \sigma_{\alpha_1}$ (see section 24.1). Then we obtain

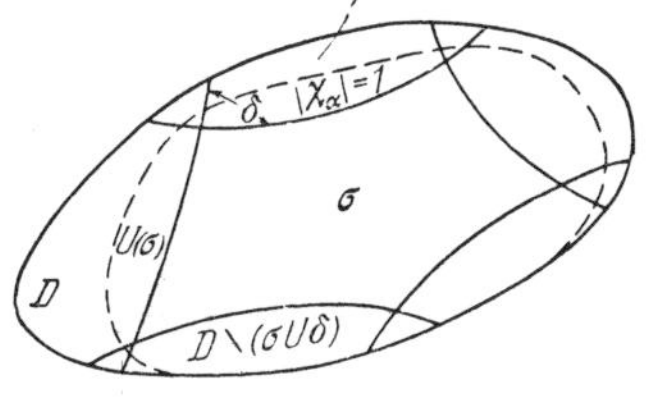

Fig. 53

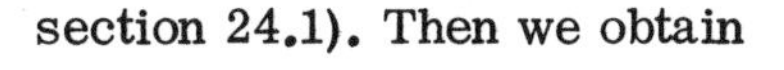

$$\frac{(-1)^{\frac{n(n-1)}{2}}}{(2\pi i)^n} \sum_{\alpha_1} \int_{\sigma_{\alpha_1}} f(\zeta) D\left(\frac{\bar{\zeta} - \bar{z}}{N^n}, d\bar{\zeta}, \ldots, d\bar{\zeta}\right) d\zeta_1\, d\zeta_2 \ldots d\zeta_n = \tag{45}$$

$$= \begin{cases} f(z), & z \in \sigma, \\ 0, & z \bar{\in} \bar{\sigma}, \end{cases}$$

Since, on the basis of formula (27),

$$\frac{(n-1)!}{N^n}\sum_{j=1}^{n}(\bar{\zeta}_j-\bar{z}_j)\,d\bar{\zeta}_1\,d\zeta_1\ldots[d\bar{\zeta}_j]\,d\zeta_j\ldots d\bar{\zeta}_n\,d\zeta_n=$$

$$=(-1)^{\frac{n(n-1)}{2}}D\left(\frac{\bar{\zeta}-\bar{z}}{N^n},\,d\bar{\zeta},\,\ldots,\,d\bar{\zeta}\right)d\zeta_1\,d\zeta_2\ldots d\zeta_n.$$

If we apply Stokes' formula $k-1$ times to the integral in the left member of Eq. (45), we obtain

$$\begin{gathered}\frac{(-1)^{\frac{n(n-1)}{2}}}{(2\pi i)^n}\sum_{\alpha_1<\ldots<\alpha_k}\int_{\sigma_{\alpha_1\ldots\alpha_k}}f(\zeta)\sum_{l=1}^{k}(-1)^{k+l}\times\\ \times D\left(q_{\alpha_1},\,\ldots,\,[q_{\alpha_l}],\,\ldots,\,q_{\alpha_k},\,\frac{\bar{\zeta}-\bar{z}}{N^{n-k+1}},\,d\bar{\zeta},\,\ldots,\,d\bar{\zeta}\right)\times\\ \times d\zeta_1\,d\zeta_2\ldots d\zeta_n=\begin{cases}f(z), & z\in\sigma,\\ 0, & z\in D\setminus(\sigma\cup\delta),\end{cases}\quad k=1,\,2,\,\ldots,\,n.\end{gathered}\tag{46}$$

For $k = 2, 3, \ldots, n$, the integrals in (46) are, by virtue of (41), meaningful only for $z\in D\setminus\delta$, where the set δ is defined in (44). Here, the integrands are holomorphic functions of the variables ζ and $\bar{\zeta}$ in a neighborhood of $\partial\sigma$.

We prove formula (46) by induction on k. For $k = 1$, formula (46) coincides with formula (45), which we have already proven. It follows from Eqs. (41) and (42) that the vectors q_α satisfy Eqs. (34). Therefore, the identities (39) are valid. Keeping these identities in mind, we rewrite formulas (46) in the form

$$\begin{gathered}\frac{(-1)^{\frac{n(n-1)}{2}}}{(2\pi i)^n}\sum_{\alpha_1<\ldots<\alpha_k}\int_{\sigma_{\alpha_1\ldots\alpha_k}}f(\zeta)\times\\ \times D\left(q_{\alpha_1},\,\ldots,\,q_{\alpha_k},\,\theta_k,\,d\bar{\zeta},\,\ldots,\,d\bar{\zeta}\right)d\zeta_1\ldots d\zeta_n=\\ =\begin{cases}f(z), & z\in\sigma,\\ 0, & z\in D\setminus(\sigma\cup\delta),\end{cases}\end{gathered}\tag{47}$$

where the functions θ_k are determined by Eqs. (35) and (36).

Suppose that $k = 1, 2, \ldots, n-1$. It will now be more convenient for us to treat the indices $\alpha_1, \ldots, \alpha_k$ in the summation in formulas (47) as independent. Since the determinant D changes sign (see section 24.3) when the two columns q_{α_ν} and q_{α_μ} are switched and since the surfaces $\sigma_{\ldots\alpha_\nu\ldots\alpha_\mu\ldots}$ and $\sigma_{\ldots\alpha_\mu\ldots\alpha_\nu\ldots}$ are oppositely oriented (see section 24.1), Eq. (47) leads to the following equations for $k = 1, 2, \ldots, n-1$:

$$\frac{(-1)^{\frac{n(n-1)}{2}}}{(2\pi i)^n (k+1)!} \sum_{l=1}^{k+1} \sum_{\alpha_1 \ldots [\alpha_l] \ldots \alpha_{k+1}} \int_{\sigma_{\alpha_1 \ldots [\alpha_l] \ldots \alpha_{k+1}}} f(\zeta) \times \\ \times D(q_{\alpha_1}, \ldots, [q_{\alpha_l}], \ldots, q_{\alpha_{k+1}}, \theta_k, d\bar{\zeta}, \ldots, d\bar{\zeta}) \times \\ \times d\zeta_1 d\zeta_2 \ldots d\zeta_n = \begin{cases} f(z), & z \in \sigma, \\ 0, & z \in D \setminus (\sigma \cup \delta). \end{cases} \tag{48}$$

If we now use the Cauchy-Riemann conditions, the properties of the determinant D, and formulas (35)-(36), we can verify directly that the equations

$$f(\zeta) D(q_{\alpha_1}, \ldots, q_{\alpha_k}, \theta_k, d\bar{\zeta}, \ldots, d\bar{\zeta}) d\zeta_1 d\zeta_2 \ldots d\zeta_n = \\ = d\left[f(\zeta) D\left(q_{\alpha_1}, \ldots, q_{\alpha_k}, \frac{\bar{\zeta} - \bar{z}}{N^{n-k}}, d\bar{\zeta}, \ldots, d\bar{\zeta}\right)\right] \times \\ \times d\zeta_1 d\zeta_2 \ldots d\zeta_n, \qquad k = 1, 2, \ldots, n-1$$

are valid. Keeping these equations in mind, we conclude from (48) that

$$\frac{(-1)^{\frac{n(n-1)}{2}}}{(2\pi i)^n (k+1)!} \sum_{l=1}^{k+1} \sum_{\alpha_1 \ldots [\alpha_l] \ldots \alpha_{k+1}} \int_{\sigma_{\alpha_1 \ldots [\alpha_l] \ldots \alpha_{k+1}}} \times \\ \times d\left[f(\zeta) D\left(q_{\alpha_1}, \ldots, [q_{\alpha_l}], \ldots, q_{\alpha_{k+1}}, \frac{\bar{\zeta} - \bar{z}}{N^{n-k}}, d\bar{\zeta}, \ldots, d\bar{\zeta}\right) \times \right. \\ \left. \times d\zeta_1 d\zeta_2 \ldots d\zeta_n\right] = \begin{cases} f(z), & z \in \sigma, \\ 0, & z \in D \setminus (\sigma \cup \delta). \end{cases} \tag{49}$$

Noting, on the basis of (22) and (23), that

$$\partial \sigma_{\alpha_1 \ldots [\alpha_l] \ldots \alpha_{k+1}} = \bigcup_{\alpha_l} \sigma_{\alpha_1 \ldots [\alpha_l] \ldots \alpha_{k+1} \alpha_l} = \\ = \bigcup_{\alpha_l} (-1)^{k+1-l} \sigma_{\alpha_1 \ldots \alpha_l \ldots \alpha_{k+1}},$$

we apply Stokes' formula to each integral in (49) and obtain

$$\frac{(-1)^{\frac{n(n-1)}{2}}}{(2\pi i)^n (k+1)!} \sum_{l=1}^{k+1} \times$$
$$\times \sum_{\alpha_1 \ldots [\alpha_l] \ldots \alpha_{k+1}} \sum_{\alpha_l} (-1)^{k+1-l} \int_{\sigma\alpha_1 \ldots \alpha_{k+1}} f(\zeta) \times$$
$$\times D\left(q_{\alpha_1}, \ldots, [q_{\alpha_l}], \ldots, q_{\alpha_{k+1}}, \frac{\bar{\zeta}-\bar{z}}{N^{n-k}}, d\bar{\zeta}, \ldots, d\bar{\zeta}\right) \times$$
$$\times d\zeta_1 d\zeta_2 \ldots d\zeta_n = \begin{cases} f(z), & z \in \sigma, \\ 0, & z \in D \setminus (\sigma \cup \delta), \end{cases} \quad k = 1, \ldots, n-1.$$

These formulas coincide with formulas (46) (with k replaced by $k+1$). Thus, formulas (46), and hence formulas (47), are proven. If we set $k = n$ (in 47), we obtain the Bergman-Weil formula (40).

6. *Remarks*

(1) By means of the Bochner-Martinelli (for $k = 1$) and Bergman-Weil (for $k = n$) formulas, formula (47) gives yet another set of intermediate representations corresponding to $k = 2, \ldots, n-1$.

If we replace $\bar{z}$ by $\bar{z}^0$, where z^0 is a fixed point in a domain σ, just as in section 23.1, we obtain new integral representations, which are valid in a sufficiently small neighborhood of the point z^0.

(2) Integration in the Bergman-Weil formula is over the hull $\Delta(\sigma)$ of the Weil domain σ. Therefore, every function $f(z)$ that is holomorphic in the closure of a Weil domain σ is completely determined by its values on the hull $\Delta(\sigma)$.

We note in passing that if the Weil domain σ is a polycircle $S(0, r)$, the Bergman-Weil formula reduces to Cauchy's formula (see section 4.3) since, in this case, we may take $P_{\alpha j} = \delta_{\alpha j}$ for $1 \leqslant \alpha$ and $j \leqslant n$, where $\delta_{\alpha j}$ is the Kronecker delta, and then use formula (24).

(3) For $n = 1$, the Bergman-Weil formula reduces to Cauchy's formula

$$\frac{1}{2\pi i} \sum_{\alpha} \int_{\sigma_\alpha} \frac{f(\zeta)\, d\zeta}{\zeta - z} = \begin{cases} f(z), & z \in \sigma, \\ 0, & z \,\bar{\in}\, \bar{\sigma}. \end{cases}$$

For $n = 2$, the Bergman-Weil formula takes the form

$$\frac{1}{4\pi^2} \sum_{\alpha<\beta} \int_{\sigma_{\alpha\beta}} f(\zeta) \frac{[P_{\alpha 1} P_{\beta 2} - P_{\beta 1} P_{\alpha 2}](\zeta, z)\, d\zeta_1\, d\zeta_2}{[\chi_\alpha(\zeta) - \chi_\alpha(z)][\chi_\beta(\zeta) - \chi_\beta(z)]} =$$
$$= \begin{cases} f(z), & z \in \sigma, \\ 0, & z \in D \setminus (\sigma \cup \delta). \end{cases}$$

(4) Every domain of holomorphy G is the union of an increasing sequence σ_k, for $k = 1, 2, \ldots,$ of Weil domains that are compact in G. Here, the corresponding functions $\chi_\alpha^{(k)}(z)$, for $\alpha = 1, 2, \ldots, N_k$, are holomorphic in G (see section 16.5). Therefore, every function $f(z)$ that is holomorphic in a domain G can be represented in an arbitrary subdomain $G' \Subset G$ in the form of Bergman-Weil integrals (40) over every Weil domain σ_k such that $G' \subset \sigma_k \Subset G$.

7. *The Weil expansion*

Let σ be a Weil domain and suppose that $\sigma' \Subset \sigma$. Then, there exists a number $\rho = \rho(\sigma') < 1$ such that $|\chi_\alpha(z)| < \rho$, where $\alpha = 1, 2, \ldots, N$, for all $z \in \sigma'$. Therefore,

$$\frac{1}{[\chi_{\alpha_1}(\zeta) - \chi_{\alpha_1}(z)] \ldots [\chi_{\alpha_n}(\zeta) - \chi_{\alpha_n}(z)]} = \sum_{|s| \geqslant 0} \frac{[\chi_\alpha(z)]^s}{[\chi_\alpha(\zeta)]^{s+I}}.$$

These expansions cover absolutely and uniformly for $(z, \zeta) \in \sigma' \times \sigma_{\alpha_1 \ldots \alpha_n}$. Here, $\chi_\alpha = (\chi_{\alpha_1}, \ldots, \chi_{\alpha_n})$. If we substitute these series into the determinants (43) and make use of formulas (41), we obtain

$$D(q_{\alpha_1}, \ldots, q_{\alpha_n}) = D(P_{\alpha_1}, \ldots, P_{\alpha_n}) \sum_{|s| \geqslant 0} \frac{[\chi_\alpha(z)]^s}{[\chi_\alpha(\zeta)]^{s+I}}.$$

If we now substitute these expressions into the Bergman-Weil formula (40) and integrate termwise, we obtain the following Weil expansion [69] for an arbitrary function $f(z)$ that is holomorphic in $\bar{\sigma}$:

$$f(z) = \sum_{\alpha_1 < \ldots < \alpha_n} \sum_{|s| \geqslant 0} Q_{\alpha, s}(z) [\chi_\alpha(z)]^s, \tag{50}$$

where

$$Q_{\alpha, s}(z) = \\ = \frac{(-1)^{\frac{n(n-1)}{2}}}{(2\pi i)^n} \int_{\sigma_{\alpha_1 \ldots \alpha_n}} f(\zeta) D(P_{\alpha}, \ldots, P_{\alpha_n}) [\chi_\alpha(\zeta)]^{-s-I} d\zeta. \tag{51}$$

If the functions $\chi_\alpha(z)$ are holomorphic in a domain of holomorphy $D \Supset \sigma$, it follows on the basis of Hefer's theorem (see section 24.2) that the functions $Q_{\alpha, s}(z)$ are also holomorphic in D. Therefore, all the terms of the series (50) are holomorphic functions in the domain D. This series converges absolutely and uniformly in an arbitrary relatively compact subdomain $\sigma' \Subset \sigma$. Thus, we have proven that

Every function $f(z)$ that is holomorphic in the closure of a Weil domain σ can be represented in this domain by the series (50), *which converges absolutely and uniformly in σ.*

If we represent the Weil domain σ in the form of the sum of an increasing sequence of Weil domains $\sigma_k \Subset \sigma_{k+1}$ (see section 16.5) and use the result we have just obtained, we conclude that *every function $f(z)$ that is holomorphic in σ is the limit of a sequence of holomorphic functions in D that converges uniformly in σ.*

Thus, if a Weil domain σ is defined by functions χ_α that are holomorphic in a domain of holomorphy $D \Supset \sigma$, the class H_D is dense in the class H_{σ}.

8. *Runge domains*

A domain G is called a Runge domain if the class of polynomials is dense in H_G. A Runge domain is always single-sheeted.

The following theorem, due to Runge (see Markushevich [70], p. 285), is always applicable in the space C^1. *For a domain G to be a Runge domain, it is necessary and sufficient that it be simply connected.*

In the space C^n for $n \geqslant 2$, we have no such simple geometric characteristic of Runge domains: not every simply connected domain is a Runge domain, and not every Runge domain is simply connected (see Fuks [2], p. 68).

First of all, we shall show that *a domain G is a Runge domain if its envelope of holomorphy $H(G)$ is a Runge domain.*

The assertion that if $H(G)$ is a Runge domain, so will G be a Runge domain is trivial and follows from the more general fact that if any holomorphic extension G^* of a domain G is a Runge domain, then G is a Runge domain.

Conversely, suppose that G is a Runge domain. Let us show that $H(G)$ is also a Runge domain. Let $f(z)$ be a function that is holomorphic in G and hence in $H(G)$. Then, there exists a sequence of polynomials $P_\alpha(z)$ for $\alpha = 1, 2, \ldots$, that converges uniformly to a function f in a domain G (see section 16.2). We denote by G_f the (largest) domain of uniform convergence of the sequence $\{P_\alpha\}$. Let us show that G_f is a polynomially convex domain. Suppose that this is not the case. According to the lemma of section 16.3, there exists a set $A \Subset G_f$, where $\delta_{G_f}(A) = r > 0$, and a point $z^0 \in G_f$, where $\delta_{G_f}(z^0) < r$, such that the inequality

$$|P(z^0)| \leqslant \sup_{z \in A} |P(z)|$$

is valid for an arbitrary polynomial. Then, we conclude from the Cartan-Thullen theorem (see section 16.1) that

$$|P(z)| \leqslant \sup_{z \in A_\rho} |P(z)|, \qquad z \in S(z^0, \rho I) \tag{52}$$

for all $\rho < r$. Since $A_\rho \Subset G_f$, it follows from inequality (52) that the sequence $\{P_\alpha\}$ converges uniformly in $S(z^0, \rho l)$ and, consequently, that $S(z^0, \rho l) \subset G_f$ for all $\rho < r$, which is impossible. Therefore, G_f is a polynomially convex domain.

Thus, we have shown that G_f is a domain of holomorphy (see section 16.7). Since $G_f \supset G$, it follows that $H(G) \subset G_f$. Therefore, the sequence $\{P_\alpha\}$ converges uniformly in $H(G)$, defining a function that is holomorphic in $H(G)$ (see section 4.4). On the basis of the holomorphic continuation theorem (see section 6.1), this function coincides with the original function f. This establishes the fact that an arbitrary function $f \in H_G$ can be represented as the limit of a sequence of polynomials that converges uniformly in $H(G)$. This means that $H(G)$ is a Runge domain, which completes the proof.

Remark: The reasoning followed in the above proof can easily be generalized to the case of functions belonging to any class K_G (in the present case, $K_G = P$). In this way, we obtain a solution of Julia's problem, namely, that every domain of uniform convergence of a sequence of holomorphic functions is a domain of holomorphy (see Cartan and Thullen [7]).

Let us give some examples of Runge domains.

(1) Multiple-circular domains that contain their centers (this follows from section 14.1).

(2) Complete Hartogs domains $[z : |z_1| < R(\tilde{z}), \ \tilde{z} \in B]$ if the base B is a Runge domain in C^{n-1} (this follows from section 15.3).

(3) Semitubular domains $[z : v(\tilde{z}) < x_1 < V(\tilde{z}), \ \tilde{z} \in B]$ if B is a Runge domain (see Shirinbekov [74]).

9. *Weil's theorem*

For a domain G to be a Runge domain, it is necessary and sufficient that its envelope of holomorphy H(G) be polynomially convex (see Weil [71]).

Proof of the necessity: Suppose that G is a Runge domain. Then, from the preceding theorem, $H(G)$ is a Runge domain. Since $H(G)$ is a holomorphically convex domain, on applying the lemma of section 16.2 where we take $K_{H(G)} = H_{H(G)}$ and $K'_{H(G)} = P$, we conclude that $H(G)$ is a polynomially convex domain.

Proof of the sufficiency: Suppose that $H(G)$ is a polynomially convex domain. Then, it is the union of an increasing sequence of polynomial Weil domains (see section 16.11). Therefore, for an arbitrary subdomain D that is compact in $H(G)$, there exists a polynomial Weil domain

$$\sigma = [z: \ |P_\alpha(z)| < 1, \ \alpha = 1, 2, \ldots, N; \ z \in U(\bar{\sigma})]$$

such that $D \Subset \sigma \Subset H(G)$.

Suppose that a function $f(z)$ is holomorphic in $H(G)$. Then, in the domain σ, this function can be represented in the form of an absolutely and uniformly convergent series (50)

$$f(z) = \sum_{\alpha_1 < \cdots < \alpha_n} \sum_{|s| \geqslant 0} Q_{\alpha, s}(z) [P_\alpha(z)]^s,$$

where the functions $Q_{\alpha, s}(z)$ defined by formula (51) are polynomials [since in (42) we may take polynomials for the functions $P_{\alpha j}(z)$]. It follows from this that, for every $\varepsilon > 0$, there exists a polynomial $P(z)$ such that

$$|f(z) - P(z)| < \varepsilon, \qquad z \in D.$$

Thus, $H(G)$, and hence G, must be a Runge domain (see section 24.8). This completes the proof of the theorem.

COROLLARIES. 1. *Every polynomial domain is a Runge domain; every convex domain is a Runge domain* (see Bremermann [73]).

This is true because every convex domain is a polynomially convex domain (see section 16.11).

2. *For a K-convex domain G to be a Runge domain, it is necessary and sufficient that the class of polynomials be dense in the class K_G.*

This follows from Weil's theorem and the lemma in section 16.2.

25. BOCHNER'S INTEGRAL REPRESENTATION

Bochner's representation has to do with unbounded tubular radial domains. Therefore, in contrast with the representations that we have considered above, this representation is valid not for all functions that are holomorphic in the domains in question but only for those that decrease sufficiently rapidly at infinity.

1. Cones

A cone $C \subset R^n$ (with vertex at zero) is defined as a set of points with the property that $y \in C$ implies $\lambda y \in C$ for all positive λ.

The intersection of the cone C with the unit sphere $|y| = 1$ is called its *projection*, which we shall denote by pr C. A cone C' such that pr $\overline{C'} \subset$ pr C will be called a *compact subcone* of the cone C. For any cone C, we define the *dual cone* (see Fig. 54)

$$C^* = [\xi: \ \xi y \geqslant 0, \ y \in C].$$

The cone C^* is closed and convex. It is easy to verify that $C^* = \overline{C}^* = O(C)^*$ and $C^{**} = \overline{O(C)}$, where $O(C)$ is the convex envelope of the cone C.

From this it follows that, *for* $\overline{O(C)}=\overline{O(C_1)}$, *it is necessary and sufficient that* $C^*=C_1^*$.

If $C^*=\overline{C}$, we shall call the cone C self-dual. For example, the light cones $\Gamma^+=[y: y_1>|\tilde{y}|]$ and $\Gamma^-=[y: y_1<-|\tilde{y}|]$ (future and past respectively) (see Fig. 55) and also the octant $\Gamma^0=[y: y_j>0,\ j=1, 2, \ldots, n]$ are self-dual cones. Here, $\tilde{y}=(y_2, y_3, \ldots, y_n)$.

We shall call the function

$$\mu_C(\xi)=\sup_{y\in \mathrm{pr}\, C}(-\xi y)$$

the indicatrix of a cone C. It is a convex (and hence continuous) homogeneous function of degree 1 and satisfies the inequalities

$$|\mu_C(\xi)|\leqslant|\xi|, \quad \mu_C(\xi_1+\xi_2)\leqslant\mu_C(\xi_1)+\mu_C(\xi_2).$$

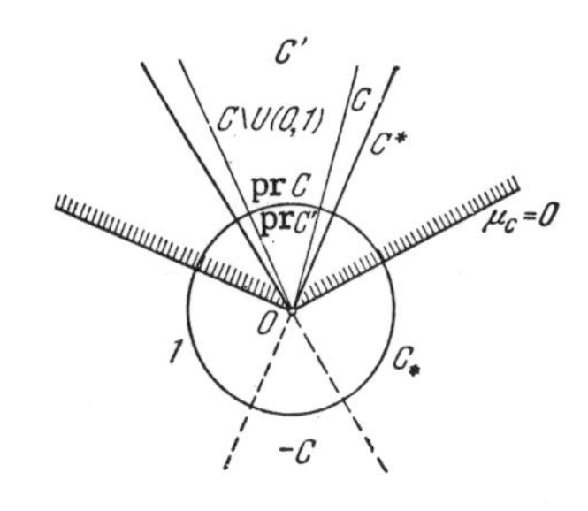

Fig. 54

LEMMA 1. *For* $\overline{C}$ *to be equal to* $\overline{C}_1$, *it is necessary and sufficient that* $\mu_C(\xi)\equiv\mu_{C_1}(\xi)$.

The necessity of the condition is obvious since $\mu_C(\xi)\equiv\mu_{\overline{C}}(\xi)$. Let us prove its sufficiency. Suppose that $\overline{C}\neq\overline{C}_1$. Then, there exists a point $\xi^0\in\mathrm{pr}\,\overline{C}$ such that $\xi^0\,\overline{\in}\,\mathrm{pr}\,\overline{C}_1$ (see Fig. 56). Therefore, $\mu_C(-\xi^0)=|\xi^0|=1$ and

$$\mu_{C_1}(-\xi^0)=\sup_{y\in \mathrm{pr}\, C_1}\xi^0 y<1=\mu_C(-\xi^0),$$

which, by hypothesis, is impossible.

If $\xi\in C^*$, then $\mu_C(\xi)>0$; if $\xi\in C^*$, then $\mu_C(\xi)\leqslant 0$. Thus (see Fig. 54),

$$C^*=[\xi: \mu_C(\xi)\leqslant 0]; \quad C_*=R^n\setminus C^*=[\xi: \mu_C(\xi)>0].$$

We note that, for all ξ,

$$\mu_C(\xi)\leqslant\mu_{O(C)}(\xi). \tag{53}$$

Let us show that if $\xi\in C^*$, then

$$\mu_C(\xi)=\mu_{O(C)}(\xi). \tag{54}$$

Since $\mu_{O(C)}(\xi)\leqslant 0$ for $\xi\in C^*$, then $\xi y\geqslant 0$ for all $y\in O(C)$. Furthermore, since the function ξy is linear with respect to y, its extreme values in an arbitrary set and in the convex envelope of that set coincide. Therefore (see Fig. 54),

$$\mu_C(\xi) = -\inf_{y \in \mathrm{pr}\, C} \xi y = -\inf_{y \in C \setminus U(0,\,1)} \xi y =$$
$$= -\inf_{y \in O[C \setminus U(0,\,1)]} \xi y \geqslant -\inf_{y \in O(C) \setminus U(0,\,1)} \xi y =$$
$$= -\inf_{y \in O(C)} \xi y = \mu_{O(C)}(\xi), \quad \xi \in C^*$$

(since $O(C) \setminus U(0,\ 1) \subset O[C \setminus U(0,\ 1)]$), which, together with inequality (53), proves Eq. (51).

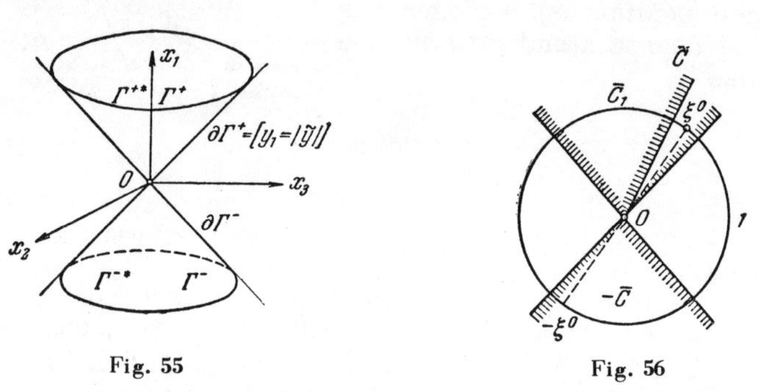

Fig. 55 Fig. 56

We now introduce the number

$$\rho_C = \sup_{\xi \in C_*} \frac{\mu_{O(C)}(\xi)}{\mu_C(\xi)}.$$

Since $\mu_C(\xi) > 0$ for $\xi \in C_*$, it follows from inequality (53) that $\rho_C \geqslant 1$. The number ρ_C characterizes the nonconvexity of the cone C. Specifically, we have

LEMMA 2. *For a cone C to be convex, it is necessary and sufficient that* $\rho_C = 1$.

The necessity is obvious. To prove the sufficiency, suppose that $\rho_C = 1$. Then, on the basis of (53) and (54), $\mu_C(\xi) = \mu_{O(C)}(\xi)$ for all ξ. From Lemma 1, $\bar{C} = \overline{O(C)}$, which completes the proof.

LEMMA 3. *If a cone C is open and consists of a finite number of components,* $\rho_C < +\infty$.

Proof: Let us first show that

$$\mu_{O(C)}(\xi) = 1 \text{ and } \mu_C(\xi) \geqslant \sigma > 0 \text{ for all } \xi \in \mathrm{pr}\, \overline{O(-C)}. \quad (55)$$

Since $\xi \in \mathrm{pr}\ \overline{O(-C)}$, there exists a $y = y(\xi) \in \mathrm{pr}\ \bar{C}$ at which $-\xi y > 0$) since otherwise, we should have $\xi y \geqslant 0$ for all $y \in \mathrm{pr}\ \bar{C}$ and hence $\xi y \geqslant 0$ for all $y \in \mathrm{pr}\ \overline{O(C)}$, which is impossible for $y = -\xi \in \mathrm{pr}\ \overline{O(C)}$. From this, we have the equation $\mu_{O(C)}(\xi) = 1$

and the inequality $\mu_C(\xi) > 0$ for all $\xi \in$ pr $\overline{O(-C)}$. Thus, the continuous function $\mu_C(\xi)$ is positive on the compact set pr $\overline{O(-C)}$ and therefore

$$\inf_{\xi \in \mathrm{pr}\, \overline{O(-C)}} \mu_C(\xi) = \sigma > 0.$$

Suppose that $O(C) = R^n$. In this case, pr $C_* = [\xi\colon |\xi| = 1]$, and it follows from (55) that $\rho_C = 1/\sigma + \infty$, which proves the lemma.

Suppose now that $O(C) \neq R^n$. In this case, we may assume that $O(C)$ and C^* lie in the semispace $y_1 \geqslant 0$ (this is true because there exists a point $y'' \in O(C)$ such that $y''y \geqslant 0$ for all $y \in O(C)$, and since $y'' \in O(C)$, it follows that $y''\xi \geqslant 0$ for all $\xi \in C^*$ [see Fig. 57]).

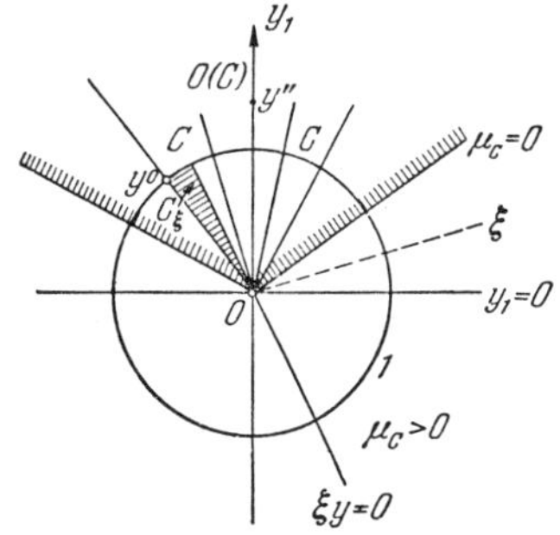

Fig. 57

Let us prove that

$$\mu_C(\xi) = \mu_{O(C)}(\xi) \text{ for } \xi_1 > 0. \tag{56}$$

On the basis of Eq. (54), it will be sufficient to prove this equation for $\mu_C(\xi) > 0$, where $\xi_1 > 0$. Suppose that

$$\begin{aligned} \mu_C(\xi) &= -\xi y^0 > 0, \quad y^0 \in \mathrm{pr}\, \overline{C}; \\ \mu_{O(C)}(\xi) &= -\xi y' > 0, \quad y' \in \mathrm{pr}\, \overline{O(C)}. \end{aligned} \tag{57}$$

We introduce the closed set (see Fig. 57)

$$C_\xi = [y\colon\ y \in \overline{C},\ \xi y \leqslant 0,\ |y| \leqslant 1], \tag{58}$$

which contains the point y^0. Since the cone $\overline{O(C)}$ lies in the semispace $y_1 \geqslant 0$, it follows that

$$O(C_\xi) = [y\colon\ y \in \overline{O(C)},\ \xi y \leqslant 0,\ |y| \leqslant 1];\ y' \in O(C_\xi). \tag{59}$$

By consideration of (57)-(59), we obtain Eq. (56):

$$\begin{aligned} \mu_C(\xi) &= -\xi y^0 = \sup_{y \in \mathrm{pr}\, \overline{C}} (-\xi y) = \sup_{y \in C_\xi,\ |y|=1} (-\xi y) = \\ &= \sup_{y \in C_\xi} (-\xi y) = \sup_{y \in O(C_\xi)} (-\xi y) = \sup_{y \in O(C_\xi),\ |y|=1} (-\xi y) = \\ &\qquad = -\xi y' = \mu_{O(C)}(\xi). \end{aligned}$$

Since the cone C is open, the cone C^* is compact in the cone $[\xi\colon \xi_1 > 0]$ (see Fig. 57). Therefore, the function $\mu_C(\xi)$ which is continuous on the compact set $[\xi\colon \xi_1 \leqslant 0,\ |\xi| = 1]$, is positive, so that

$$\inf_{\xi_1 \leqslant 0,\ |\xi|=1} \mu_C(\xi) = \sigma > 0.$$

From this and from Eq. (56), we conclude that $\rho_C = \frac{1}{\sigma} < +\infty$, which completes the proof.

EXAMPLES

(1) $C = R^n$. Then $R^{n*} = \{0\}$, $\mu_{R^n}(\xi) = |\xi|$, and $\rho_{R^n} = 1$.
(2) $C = \Gamma^+$ is a future light cone. Then, $\Gamma^{+*} = \overline{\Gamma}$, $\rho_{\Gamma^+} = 1$, and

$$\mu_{\Gamma^+}(\xi) = \begin{cases} |\xi|, & \xi \in -\Gamma^+, \\ \frac{1}{\sqrt{2}}(|\tilde{\xi}| - \xi_1), & \xi \overline{\in} -\Gamma^+. \end{cases}$$

(3) $C = \Gamma = \Gamma^+ \cup \Gamma^- = [y: y_1^2 > |\tilde{y}|^2]$ is a light cone. Then, $O(\Gamma) = R^n$, $\Gamma^* = \{0\}$, $\rho_\Gamma = \sqrt{2}$, $\mu_{O(\Gamma)}(\xi) = |\xi|$ and

$$\mu_\Gamma(\xi) = \begin{cases} |\xi|, & \xi \in \Gamma, \\ \frac{1}{\sqrt{2}}(|\tilde{\xi}| + |\xi_1|), & \xi \overline{\in} \Gamma. \end{cases}$$

2. *Tubular radial domains*

We shall refer to a cone $T^C = R^n + iC$, where C is an open cone in R^n, as a tubular cone. If the cone C is connected, we shall call T^C a *tubular radial domain.*

Consider a connected open cone C such that $\overline{O(C)}$ contains no complete straight line.* Then, int $C^* \neq \varnothing$. This follows from the following lemma:

LEMMA 1. *For a cone $\overline{O(C)}$ to contain an entire straight line, it is necessary and sufficient that the cone C^* lie in some $(n-1)$-dimensional plane.*

Proof: If the cone C^* lies in an $(n-1)$-dimensional plane $\xi e = 0$, where $|e| = 1$, then $\pm e \in (C^*)^* = O(C)$. In this case, an arbitrary straight line $y = y^0 + te$ for $-\infty < t < \infty$, where $y^0 \in \overline{O(C)}$, is contained in $\overline{O(C)}$ (see section 58).

Conversely, suppose that $\overline{O(C)}$ contains a straight line $y = y^0 + te$ for $-\infty < t < \infty$, where $y^0 \in \overline{O(C)}$. For an arbitrary point $\xi \in C^*$, we then have $y^0\xi + t\xi e \geqslant 0$ for all t. From this it follows that $\xi e = 0$; that is, the cone C^* lies in the plane $\xi e = 0$, which completes the proof.

---

*In this case, a tubular radial domain T^C is equivalent (see section 7.5) to a bounded domain. For this reason, T^C is sometimes called an essentially bounded domain (see Gindikin [76]).

LEMMA 2. *Let C be an open cone and C' a cone that is compact in $O(C)$. Then, there exist a number $\sigma = \sigma(C') > 0$ and an open cone $C'' = C''(C')$* (see Fig. 59) *containing the cone C^* such that*

$$\xi y \geqslant \sigma |y| |\xi|, \qquad y \in C', \qquad \xi \in C''. \tag{60}$$

Proof: Since the cone C' consists of interior points of the cone $C^{**} = \overline{O(C)}$, we have $y\xi > 0$ for all $\xi \in \text{pr } C^*$ and all $y \in \text{pr } C'$ (see section 25.1). Inequality (60) for some $\sigma > 0$ and C'' follows from this and from the homogeneity and continuity of the form $y\xi$. In this case, the cone C^* is compact in C'', which completes the proof.

For tubular radial domain T^C, we introduce the kernel

$$K(z) = \int_{C^*} e^{iz\xi}\, d\xi. \tag{61}$$

The function $K(z)$ is holomorphic in $T^{O(C)}$. For all y in the cone C' that is compact in $O(C)$,

$$\|K(x+iy)\|^2 \leqslant M(C')|y|^{-n}. \tag{62}$$

Proof: If the cone $\overline{O(C)}$ contains an entire straight line, it follows from Lemma 1 that mes $C^* = 0$ and the assertion is trivial. Therefore, let us assume that $\overline{O(C)}$ does not contain any entire line. Suppose that $U(z^0, \delta) \Subset T^{O(C)}$. Then, it follows from inequality (60) (see Lemma 2) that $\xi y \geqslant |\xi| \sigma$, where $\sigma > 0$, for all $\xi \in C^*$ and $z \in U(z^0, \delta)$. From this we conclude that the integral (61) and all its derivatives

$$D^\alpha K(z) = \int_{C^*} e^{-y\xi + ix\xi} (i\xi)^\alpha\, d\xi$$

converge uniformly with respect to $z \in U(z^0, \delta)$ since

$$\left| \int_{C^*} e^{-y\xi + ix\xi} (i\xi)^\alpha\, d\xi \right| \leqslant \sigma_n \int_0^\infty e^{-\sigma r} r^{|\alpha|+n-1} dr < +\infty,$$

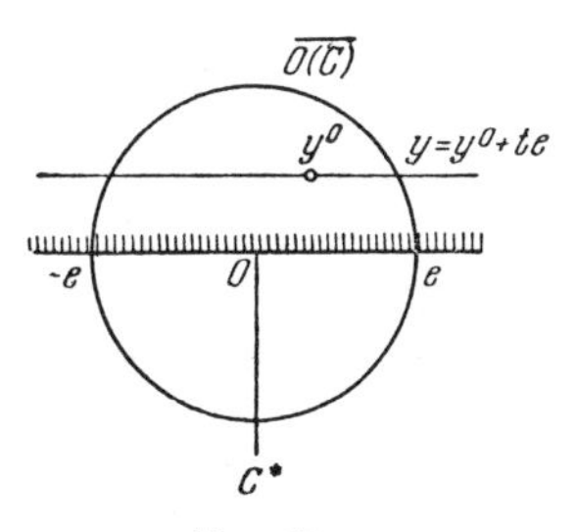

Fig. 58

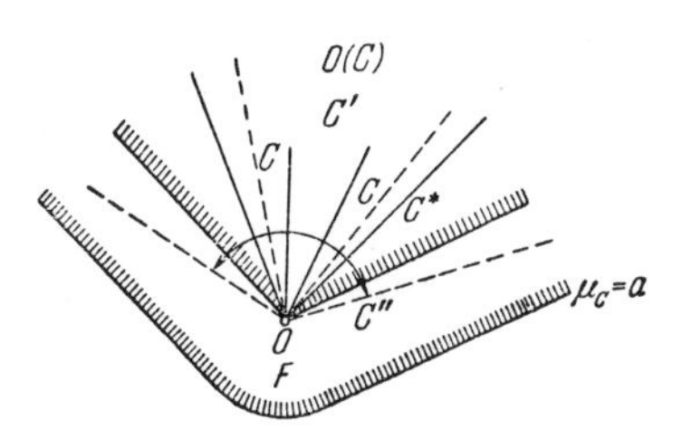

Fig. 59

where σ_n is the area of the surface of a unit sphere in R^n. From this and the arbitrariness of the neighborhood $U(z^0, \delta)$, it follows that the function $K(z)$ is holomorphic in $T^{O(C)}$ (see section 4.2).

Furthermore, it follows from (61) that the function $K(x+iy)$ is, for each fixed $y \in O(C)$ the Fourier transform of the function $e^{-y\xi}e_{C^*}(\xi)$, where $e_{C^*}(\xi)$ is the characteristic function of the cone C^* (see section 1.2). Therefore, Parseval's inequality (see section 3.9) holds:

$$\int |K(x+iy)|^2\, dx = (2\pi)^n \int\limits_{C^*} e^{-2\xi y}\, d\xi, \tag{63}$$

Inequality (62) follows from this equality and from inequality (60) for all $y \in C'$:

$$\|K(x+iy)\|^2 = (2\pi)^n \sigma_n \int\limits_0^\infty e^{-2\sigma r|y|} r^{n-1}\, dr \leqslant M(C')\,|y|^{-n}.$$

EXAMPLES

(1) Suppose that $C = \Gamma^0 = [y\colon\; y_j > 0,\; j = 1, 2, \ldots, n]$. Then, $\Gamma^{0*} = \Gamma^0$ and

$$K(z) = \int\limits_0^\infty \ldots \int\limits_0^\infty e^{i \sum\limits_j z_j \xi_j}\, d\xi_1 \ldots d\xi_n = \frac{i^n}{z_1 z_2 \ldots z_n}.$$

(2) Suppose that $C = \Gamma^+$, then $\Gamma^{+*} = \bar{\Gamma}^+$ and (see section 30)

$$K(z) = \int\limits_{\bar{\Gamma}^+} e^{iz\xi}\, d\xi = \frac{\Gamma\left(\frac{n}{2}\right) 2^{n-1} \pi^{\frac{n-2}{2}}}{(-z_1^2 + z_2^2 + \ldots + z_n^2)^{n/2}}.$$

For $n = 1$, these kernels can be transformed into the Cauchy kernel.

3. *Functions that are holomorphic in tubular radial domains*

Suppose that a function $f(z)$ is holomorphic in a tubular radial domain T^C and that, for all y in a cone C' that is compact in C,

$$\|f(x+iy)\| \leqslant M_{\varepsilon, f}(C')\, e^{\varepsilon|y|} \tag{64}$$

for any $\varepsilon > 0$.

Let us show that the function

$$g_y(\xi) = (2\pi)^{-n} e^{\xi y} \int f(x+iy)\, e^{-i\xi x}\, dx = \\ = (2\pi)^{-n} \int\limits_{z=x+iy} f(z)\, e^{-i\xi z} dz_1 \ldots dz_n \tag{65}$$

is independent of $y \in C$. It follows from condition (64) that, for an arbitrary subdomain $B \Subset C$,

$$\int\limits_B |f(x+iy)|^2\, dy \to 0 \qquad \text{as } |x| \to \infty.$$

From this and from the Cauchy-Poincaré theorem (see section 22.6), we conclude that the integral (65) is independent of the plane of integration $z = x + iy$, $|x| < +\infty$ no matter how fast y varies in the domain B. This means that the function $g_y(\xi)$ is independent of $y \in B$. Since B is an arbitrary compact subdomain of the cone C, $g_y(\xi)$ is independent of y for all y in C.

We now denote the function $g_y(\xi)$ by $g(\xi)$. This function is measurable and, on the basis of (65), it possesses the properties

$$g(\xi)\, e^{-\xi y} \in L_2 \qquad \text{for all } y \in C; \tag{66}$$

$$f(z) = \int g(\xi)\, e^{iz\xi}\, d\xi \qquad \text{for all } z \in T^C. \tag{67}$$

Here, Parseval's equation holds:

$$(2\pi)^n \int |g(\xi)|^2\, e^{-2\xi y}\, d\xi = \|f(x+iy)\|^2, \qquad y \in C. \tag{68}$$

Let us show that $g(\xi) \in L_2$. On the basis of inequality (64), we obtain from (68)

$$(2\pi)^n \int |g(\xi)|^2\, e^{-2\xi y}\, d\xi \leqslant M^2_{\varepsilon, f}(C')\, e^{2\varepsilon |y|}, \qquad y \in C'. \tag{69}$$

From this and from Fatou's lemma (see section 2.8), we conclude that

$$\int\limits_{|\xi| \leqslant R} |g(\xi)|^2\, d\xi \leqslant \varliminf_{\substack{y \to 0 \\ y \in C'}} \int\limits_{|\xi| \leqslant R} |g(\xi)|^2\, e^{-2\xi y}\, d\xi \leqslant M_1(C')$$

for arbitrary $R > 0$. This inequality means that $g(\xi) \in L_2$.

Let us show that $g(\xi) = 0$ almost everywhere in the cone C_*. In particular, if the cone $O(C)$ contains an entire straight line, then $g(\xi) = 0$ almost everywhere in R^n. (In this case, mes $C^* = 0$ [see section 25.2].)

Suppose on the contrary that $g(\xi) \neq 0$ on a set of positive measure in the cone $C_* = R^n \setminus C^*$. Then, there exists a point $\xi^0 \in C_*$ such that $g(\xi) \neq 0$ on a set of positive measure in the sphere $|\xi - \xi^0| < \delta$ for arbitrary $\delta > 0$. Since $\xi^0 \in C_*$, there exists a point $y^0 \in$ pr C such that $\xi^0 y^0 < -\sigma < 0$(see Fig. 60). From continuity considerations, this inequality must hold in a sufficiently small sphere $|\xi - \xi^0| < \delta' \leqslant \delta$; that is, $\xi y^0 < -\sigma$ for all $|\xi - \xi^0| < \delta'$. Therefore, if we set $C' = [y\colon y = ty^0,\ t > 0]$ in inequality (69), we obtain

$$(2\pi)^{-n} M^2_{\varepsilon, f}(C') e^{2\varepsilon t} \geqslant \int |g(\xi)|^2 e^{-2ty^0\xi} d\xi \geqslant$$

$$\geqslant \int\limits_{|\xi - \xi^0| < \delta'} |g(\xi)|^2 e^{-2ty^0\xi} d\xi \geqslant e^{2t\sigma} \int\limits_{|\xi - \xi^0| < \delta'} |g(\xi)|^2 d\xi,$$

which is impossible for $\varepsilon < \sigma$ and $t \to \infty$. Therefore, $g(\xi) = 0$ almost everywhere in C_*.

Thus, the function

$$f(x + i0) = \int\limits_{C^*} g(\xi) e^{ix\xi} d\xi \qquad \in L_2 \tag{70}$$

and, on the basis of (67), Parseval's equality holds:

$$\|f(x + iy) - f(x + i0)\|^2 = (2\pi)^n \int\limits_{C^*} |g(\xi)|^2 (1 - e^{-\xi y})^2 d\xi. \tag{71}$$

Since $\xi y \geqslant 0$ for $\xi \in C^*$ and $y \in C$, by using Lebesgue's theorem on taking the limit under the Lebesgue integral sign (see section 2.5), we obtain from (71)

$$\int |f(x + iy) - f(x + i0)|^2 dx \to 0$$

$$\text{for } y \to 0, \quad y \in C.$$

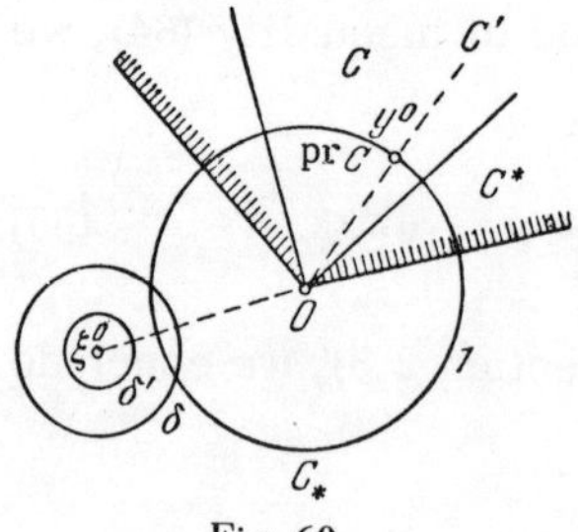

Fig. 60

Thus, we have obtained the following result (see Bochner [57]): *Every function $f(z)$ that is holomorphic in a tubular radial domain T^C and satisfies conditions* (64) *has the limiting value*

$$f(x + i0) = \lim_{y \to 0,\ y \in C} f(x + iy),$$

which is independent of the sequence $y \to 0$ for $y \in C$. Convergence here is understood in the L_2 sense. If the cone $O(C)$ contains an entire straight line, $f(z) \equiv 0$.

Remark: An investigation will be made in section 26.3 of boundary values of functions that are holomorphic in T^C and that satisfy a weaker condition than (64).

In Bochner's work [57], a study is made of the class of functions satisfying a more restrictive condition than (64): $\|f(x+iy)\| \leqslant M_f$. It is assumed in this connection that the cone C is convex and that $\bar{C}$ does not contain any complete straight line.

4. *Bochner's formula*

Every function $f(z)$ that is holomorphic in a tubular radial domain T^C and that satisfies inequality (64) *can be represented in the form of an integral*

$$f(z) = \frac{1}{(2\pi)^n} \int K(z - x') f(x' + i0)\, dx', \tag{72}$$

where $K(z)$ is the kernel of T^C and $f(x + i0)$ is the boundary value of the function $f(z)$ as $y \to 0$ for $y \in C$.

Proof: Since $g(\xi) = 0$ almost everywhere in $R^n \setminus C^*$, formula (67) can be represented in the form

$$f(z) = \int g(\xi)\, e_{C^*}(\xi)\, e^{-\xi y + i x \xi}\, d\xi. \tag{73}$$

If we apply the convolution theorem to the integral (73) and make use of formulas (61) and (70), we obtain formula (72).

Corollary. *Every function $f(z)$ that is holomorphic in a tubular radial domain T^C and that satisfies equality* (63) *is holormophic in the convex envelope $O(T^C) = T^{O(C)}$. For all y in a cone C' that is compact in $O(C)$, it satisfies an inequality of the form*

$$\|f(x+iy)\| \leqslant N(C')\, |y|^{-\frac{n}{2}} \|f(x+i0)\|, \tag{74}$$

where the number $N(C')$ is independent of $f(z)$. Here, if the cone $\overline{O(C)}$ contains an entire straight line, then $f(z) \equiv 0$.

Remark: That $f(z)$ is holomorphic follows from the representation (73) and from the properties of the kernel $K(z)$ and also from Bochner's theorem [see section 17.5]). Inequality (74) follows from the representation (72), from inequality (62), and from the Bunyakovskiy-Schwarz inequality:

$$\begin{aligned} \|f(x+iy)\| &\leqslant (2\pi)^{-n} \|K(x+iy-x')\|\, \|f(x'+i0)\| = \\ &= (2\pi)^{-n} \|K(x+iy)\|\, \|f(x+i0)\| \leqslant \\ &\leqslant (2\pi)^{-n} \sqrt{M(C')}\, |y|^{-\frac{n}{2}} \|f(x+i0)\|. \end{aligned}$$

Remark: The integration in formula (72) is over the "edge" R^n of the "wedge" T^C, that is, the hull of the domain T^C. The orientations of the hull R^n and the cone T^C must be consistent (see section 22.2).

More general integral representations have been found by Gindikin [76] for tubular domains, by Ayzenberg [122] for convex domains, and by Temlyakov [152] and Ayzenberg [153] for multiple-circular domains.

CHAPTER V

Some Applications of the Theory of Functions of Several Complex Variables

Wishing to illustrate the application of the methods of the theory of functions of several complex variables in quantum field theory with a number of examples, we shall, in the present chapter, expound several characteristic mathematical results and statements of problems associated with one direction taken by this theory, namely, with what are known as dispersion relations. The exposition is presented under more general conditions than is required for quantum field theory. Therefore, the results presented, apart from their immediate purpose, have an independent, purely mathematical, interest. They find applications in connection with certain questions concerning the theory of functions and the theory of partial differential equations with constant coefficients.

The mathematical results discussed are sufficient for proving the dispersion relations. Use of them reduces the mathematical portion of this proof to basically algebraic calculations. Therefore, we shall not prove here the dispersion relations in general but shall confine ourselves to an exposition of the more difficult and interesting of them, those that are connected with nontrivial use of the theory of functions of several complex variables.

Complete proofs of the dispersion relations (from a standpoint of both physical and mathematical aspects) are given in the works by Bogolyubov, Medvedev, and Polivanov [22], Bogolyubov, and Vladimirov [39], Logunov and Stepanov [100], Bremermann, Oehme, and Taylor [27, 34], Lehmann [28], Vladimirov and Logunov [29], Todorov [101], and Omnes [31].* A mathematically correct proof of the dispersion relations was first given by Bogolyubov in 1956 in an address in Seattle, Washington (see also [22], Appendix A and [39]).

We shall barely touch on another interesting branch of the theory associated with the axiomatization of Wightman, namely, the study of the analytic properties of the vacuum expectation values of the products of field operators. The reader who is interested in this

*We also mention a new approach to the proof of the dispersion relations that has been developed in recent works by Bros, Epstein, and Glaser [124], and Bros, Messiah, and Stora [26] and that does not use the Jost-Lehmann-Dyson representation.

topic is referred to the original works by Wightman [24, 32, 61], Källén and Wightman [25], Källén [125], Dyson [35], Jost [113], Ruelle [50, 127], Brown [64], Streater [102–104, 114], Steinmann [128, 129], Araki [126], and Bros, Epstein, and Glaser [124].

It should be noted that along with functions of several complex variables, we shall use generalized functions extensively in this chapter, especially, generalized functions of slow increase (the space S^*, see section 3). In section 26, we shall discuss the properties of functions that are holomorphic in tubular cones. Here, we shall pay especial attention to the establishment of a connection between the growth of these functions and the properties of their spectral functions. In section 27, we shall prove the "edge of a wedge" theorem of Bogolyubov, which we shall strengthen and generalize somewhat. Section 28 contains the theorem on a C-convex envelope. In section 29, an exposition is made of various applications of the results obtained. We shall also consider generalized functions connected with a light cone (see section 30), and various representations of generalized solutions of the wave equation (see section 31). In section 32, we shall derive the Jost-Lehmann-Dyson integral representation. We shall use this representation in section 33 to construct an envelope of holomorphy $K(T \cup \tilde{G})$ and shall apply the results obtained to questions on the verification of the dispersion relations.

26. FUNCTIONS THAT ARE HOLOMORPHIC IN TUBULAR CONES

Functions that are holomorphic in tubular cones (see section 25.2) are of great significance in various fields of mathematics and theoretical physics, for example, in the theory of the (Fourier-) Laplace transformations, the theory of generalized functions, and quantum field theory. One of the basic problems connected with such functions consists in finding the relations between the increase in these functions and the properties of their spectral functions.

1. *Spectral functions*

Suppose that a function $f(z)$ is holomorphic in a tubular domain $T^B = R^n + iB$.. We shall use the term *spectral function* of the function $f(z)$ to denote the generalized function $g \in D^*$ possessing the following properties:

(a) $g(\xi) e^{-\xi y} \in S^*$ for all $y \in B$;

(b) $f(z) = F[g(\xi) e^{-y\xi}](x) = \int g(\xi) e^{iz\xi} d\xi$ for all $z \in T^B$.

Here, we shall call the function $f(z)$ the *Fourier-Laplace transform* of the spectral function $g(\xi)$.

We shall refer to the carrier of the spectral function g as the *spectrum* of the function $f(z)$. As is clear from the definition, if a given function has a spectral function, it is unique.

If B is an open set and consists of components $B_1, B_2, \ldots$, we can consider $f(z)$ on each component T^{B_k} separately and speak of the set of spectral functions g_k corresponding to the components B_k (for $k = 1, 2, \ldots$).

The question of the existence of spectral functions is answered by the following theorem, which follows from the results of Schwartz [45].

THEOREM.* *For a function $f(z)$ that is holomorphic in a tubular domain T^B to be the Fourier-Laplace transform of some spectral function, it is* (a) *sufficient that, for an arbitrary compact set $K \subset B$, there exist numbers $M = M(K)$ and $m = m(K)$ such that*

$$|f(z)| \leqslant M(1+|x|)^m, \qquad z \in R^n + iK; \tag{1}$$

and (b) *necessary that $f(z)$ be holomorphic in the convex envelope $O(T^B) = T^{O(B)}$ and that it and all its derivatives satisfy an inequality of the type* (1).

COROLLARY. *If $f(z)$ is holomorphic in T^B and satisfies inequality* (1), *then $f(z)$ is holomorphic*** *in $T^{O(B)}$ and it and its derivatives satisfy an inequality of the type* (1) *in $T^{O(B)}$.*

Let us prove the sufficiency. Suppose that $f(z)$ is holomorphic in a tubular domain $R^n + iB = T^B$ and satisfies inequality (1) in it. Then, for every $y \in B$, the function $f(x+iy)$ defines a functional on S. Consequently,

$$\varphi_y(\xi) = F^{-1}[f(x+iy)](\xi) \in S^* \tag{2}$$

and, therefore,

$$g_y(\xi) = e^{\xi y}\varphi_y(\xi) \in D^*. \tag{3}$$

Let us show that $g_y(\xi)$ is independent of $y \in B$. Let G be an arbitrary subdomain that is compact in B and let N be an integer $\geqslant 1/2\,(m+n+1)$ where $m = m(\bar{G})$. Let us choose a number $A > 0$ such that, for all $z \in R^n + i\bar{G}$, the inequality

$$|A + zz| \geqslant \frac{1}{2}(1+|x|)^2 \tag{4}$$

is satisfied. By using formulas (20) of Chapter 1, we obtain from (2) and (3)

*For L_2 functions, the corresponding theorem was proven by Bochner [149].

**That $f(z)$ is holomorphic in $T^{O(B)}$ also follows from Bochner's theorem (see section 17.5).

$$g_y(\xi) = \frac{(A-\Delta)^N}{(2\pi)^n} \int \frac{f(x+iy)\, e^{-i\xi(x+iy)}}{[A+(x+iy)(x+iy)]^N}\, dx.$$

It follows from inequalities (1) and (4) that the function

$$f(z)(A+zz)^{-N} e^{-i\xi z}$$

is holomorphic in T^{G} and that its modulus does not exceed $C(\overline{G})\, e^{\xi y}$ $(1+|x|)^{-n-1}$. Therefore, the integral

$$\int \frac{f(x+iy)\, e^{-i\xi(x+iy)}}{[A+(x+iy)(x+iy)]^N}\, dx =$$

$$= \int_{z=x+iy} f(z)(A+zz)^{-N} e^{-i\xi z}\, dz_1 \ldots dz_n$$

converges and, on the basis of the Cauchy-Poincaré theorem (see section 22.6), it is independent of the plane of integration $z = x+iy$, where $|x| < \infty$, no matter how fast y varies in the domain G. Since G is an arbitrary compact subdomain of the domain B, it follows that $g_y(\xi)$ is independent of y for all y in B (cf. section 25.3). If we denote $g_y(\xi)$ by $g(\xi)$, we conclude on the basis of (2) and (3) that $g \in D^*$ possesses properties (a) and (b), so that $g(\xi)$ is the spectral function of $f(z)$.

The necessity of the conditions follows from the results of the preceding section.

2. *Properties of spectral functions*

For a generalized function $g(\xi) \in D^*$ to be the spectral function of a function

$$f(z) = F[e^{-\xi y} g(\xi)](x),$$

that is holomorphic in a domain $T^{O(B)}$, it is necessary and sufficient that $e^{-\xi y} g(\xi) \in S^*$ for $y \in B$.

The necessity is obvious. Let us prove the sufficiency. Consider a compact set $K \Subset O(B)$. As a preliminary, let us show that, for some $\varepsilon = \varepsilon(K) > 0$, the set of generalized functions

$$[T\colon T(\xi) = g(\xi)\, e^{-\xi y + \varepsilon\sqrt{1+|\xi|^2}},\quad y \in K] \tag{5}$$

is bounded S^* (see section 3.7).

Since $K \Subset O(B)$, it follows that $K^{\varepsilon} \Subset O(B)$ for sufficiently small $\varepsilon > 0$. Therefore, the domain B contains a finite number of points $y^{(1)}, y^{(2)}, \ldots, y^{(l)}$ the closed convex envelope of which (i.e., the polyhedron π contains K^{ε} (see Fig. 61). Let us represent functions belonging to the set (5) in the form

$$g(\xi)\, e^{-\xi y + \varepsilon\sqrt{1+|\xi|^2}} = \sum_{1 \leqslant k \leqslant l} e^{\varepsilon\sqrt{1+|\xi|^2}}\, a(\xi, y)\left[g(\xi)\, e^{-\xi y^{(k)}}\right], \tag{6}$$

where

$$a(\xi, y) = e^{-\xi y}\left(\sum_{1 \leqslant k \leqslant l} e^{-\xi y^{(k)}}\right)^{-1}.$$

The function $a(\xi, y)$ is infinitely differentiable and satisfies the inequalities

$$0 \leqslant a(\xi, y) \leqslant 1, \qquad (\xi, y) \in R^n \times \pi; \tag{7}$$

$$e^{\varepsilon\sqrt{1+|\xi|^2}} a(\xi, y) \leqslant e^{\varepsilon}, \qquad (\xi, y) \in R^n \times K. \tag{8}$$

To see this, note that for $y \in \pi$

$$y = \sum_{1 \leqslant k \leqslant l} t_k y^{(k)}, \qquad t_k \geqslant 0, \qquad \sum_{1 \leqslant k \leqslant l} t_k = 1,$$

Keeping the inequality between the arithmetic and geometric means (see [10], p. 29), we obtain the inequality

$$e^{-\xi y} = \prod_{1 \leqslant k \leqslant l} e^{-\xi t_k y^{(k)}} \leqslant \sum_{1 \leqslant k \leqslant l} t_k e^{-\xi y^{(k)}} \leqslant \sum_{1 \leqslant k \leqslant l} e^{-\xi y^{(k)}},$$

from which it follows that $a \leqslant 1$. By using inequality (7) and the fact that $K^{\varepsilon} \subset \pi$, we obtain inequality (8):

$$e^{\varepsilon\sqrt{1+|\xi|^2}} a(\xi, y) \leqslant e^{\varepsilon + \varepsilon|\xi|} a(\xi, y) =$$
$$= e^{\varepsilon} \max_{|x| \leqslant \varepsilon} e^{-\xi x} a(\xi, y) = e^{\varepsilon} \max_{|x| \leqslant \varepsilon} a(\xi, y + x) \leqslant e^{\varepsilon}.$$

It follows from the definition of the function $a(\xi, y)$ that none of the derivatives $D^{\alpha} a(\xi, y)$ with respect to ξ exceeds a quantity of the form $K_{\alpha}(\xi, y)\, a(\xi, y)$ in absolute value, where the $K_{\alpha}(\xi, y)$ are bounded for all $\xi \in R^n$ and $y \in K$. Since, in addition, the derivatives of all orders of the function $\sqrt{1+|\xi|^2}$ are bounded by using inequalities (7) and (8), we obtain

$$\left| D^{\alpha} e^{\varepsilon\sqrt{1+|\xi|^2}} a(\xi, y) \right| \leqslant C_{\alpha}(\overline{\overline{G}}), \qquad (\xi, y) \in R^n \times K \tag{9}$$

for all sufficiently small $\varepsilon > 0$. Furthermore, since $y^{(k)} \in B$, it follows from condition (a) that $g(\xi) e^{-\xi y^{(k)}} \in S^*$, for $1 \leqslant k \leqslant l$. From this and from (6) and (9), we conclude that the set (5) is bounded in S^*.

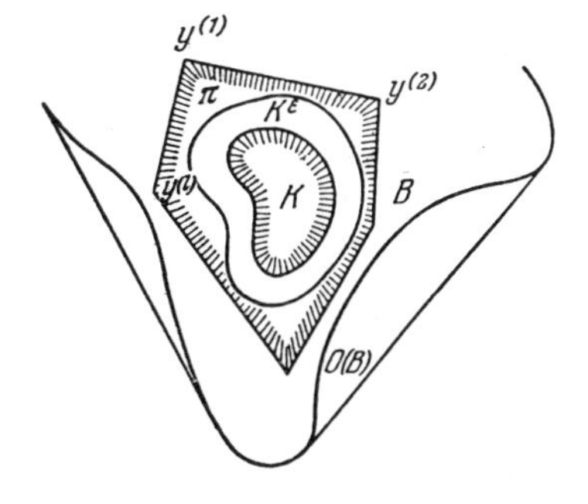

Fig. 61

Since K is an arbitrary compact set contained in $O(B)$, it follows from the boundedness of the set (5) in S^* that $g(\xi)e^{-\xi y} \in S^*$ for all $y \in O(B)$, that is, $g(\xi)$ possesses the property (a) in $O(B)$. Let us show that it also possesses property (b) in $T^{O(B)}$, that is, that $f(z)$ is holomorphic in $T^{O(B)}$.

Let G be an arbitrary domain that is compact in $O(B)$. Then, from what we have proven,

$$g(\xi)e^{-\xi y+\varepsilon\sqrt{1+|\xi|^2}} \in S^*, \qquad y \in G$$

for some $\in R^n_-$. Therefore, we have a chain of equations (see section 3.9):

$$\begin{aligned}(f(x+iy), F[\varphi]) &= (F[g(\xi)e^{-\xi y}], F[\varphi]) = \\ &= (2\pi)^n (g(\xi)e^{-\xi y}, \varphi(\xi)) = \left(g(\xi)e^{-\xi y}, \int e^{-i\xi x} F[\varphi](x)\,dx\right) = \\ &= \int \left(g(\xi)e^{iz\xi+\varepsilon\sqrt{1+|\xi|^2}}, e^{-\varepsilon\sqrt{1+|\xi|^2}}\right) F[\varphi](x)\,dx.\end{aligned}$$

From this follows the representation

$$f(z) = \left(g(\xi)e^{iz\xi+\varepsilon\sqrt{1+|\xi|^2}}, e^{-\varepsilon\sqrt{1+|\xi|^2}}\right), \quad z \in T^G = R^n + iG. \tag{10}$$

We conclude from this representation that the function $f(z)$ defined in the domain T^G possesses all the partial derivatives

$$D^\alpha f(z) = (g(\xi)(i\xi)^\alpha e^{iz\xi+\varepsilon\sqrt{1+|\xi|^2}}, e^{-\sqrt{1+|\xi|^2}}). \tag{11}$$

On the basis of the fundamental theorem of Hartogs (see section 4.2), the function $f(z)$ is holomorphic in R^n+iG. Since G is an arbitrary compact subdomain $O(B)$, it follows that $f(z)$ is holomorphic in the domain $T^{O(B)}$. This proves our assertion.

Let us show that *all the derivatives $D^\alpha f(z)$ in $T^{O(B)}$ satisfy inequality (1).*

Since the set (5) (for $K=\overline{G}$) is bounded in S^*, we obtain inequality (1) from the representation (11) for all $z \in T^G$ (see section 3.7):

$$\begin{aligned}|D^\alpha f(z)| &= \left|\left(g(\xi)e^{-\xi y+\varepsilon\sqrt{1+|\xi|^2}}, (i\xi)^\alpha e^{ix\xi-\varepsilon\sqrt{1+|\xi|^2}}\right)\right| \leqslant \\ &\leqslant \left\|g(\xi)e^{-\xi y+\varepsilon\sqrt{1+|\xi|^2}}\right\|_{-N} \left\|\xi^\alpha e^{ix\xi-\varepsilon\sqrt{1+|\xi|^2}}\right\|_N \leqslant M(1+|x|)^N.\end{aligned}$$

Thus, from what we have proven, $f(x+iy) \in \theta_M$ for $y \in O(B)$ and, consequently,

$$g(\xi)e^{-\xi y} \in \theta_C^*, \qquad y \in O(B),$$

where g is the spectral function of $f(z)$. Thus, for all $y \in O(B)$, the generalized function $g(\xi)e^{-\xi y}$ is a convolutor in S^* (see section 3.6). This makes it possible to define the convolution of any two spectral functions (for a given domain B).

Consider two spectral functions $g_k(\xi)$ where $k=1, 2$. Then, the functions

$$f_k(z) = F[g_k(\xi)e^{-\xi y}](x), \qquad k = 1, 2$$

are holomorphic in $T^{O(B)}$ and they satisfy inequality (1). Their product $f(z) = f_1(z) f_2(z)$ is a holomorphic function in $T^{O(B)}$, and it satisfies inequality (1). Therefore, according to the theorem in section 26.1, $f(z)$ has a (unique) spectral function $g(\xi)$, which we shall call the convolution of $g_1(\xi)$ and $g_2(\xi)$ and we write $g = g_1 * g_2$.

From this definition and from formulas (24) and (25) of Chapter 1, we get the formula

$$(g_1 e^{-\xi y}) * (g_2 e^{-\xi y}) = e^{-\xi y} (g_1 * g_2), \qquad y \in O(B), \tag{12}$$

where the left member is the convolution in S^* defined in section 3.9. Also, $e^{-\xi y}(g_1 * g_2) \in \theta^*_C$.

Thus, we have obtained the following results (see Schwartz [45]). *The set of special functions (for a given domain B) constitutes a commutative ring with unity without zero divisors (with respect to the operation of convolution just defined).*

3. *The existence of boundary values in the space S^**

Suppose that a function $f(z)$ is holomorphic in the tubular domain $T^{C_R} = R^n + iC_R$, where $C_R = C \cap U(0, R)$ and C is a connected cone. Suppose that, for an arbitrary number $R' < R$ and an arbitrary cone C' that is compact in C,

$$\begin{gathered} |f(x+iy)| \leqslant C(R', C') |y|^{-\alpha} (1+|x|)^{\beta}, \\ z \in R^n + i[C' \cap U(0, R')], \end{gathered} \tag{13}$$

where the nonnegative numbers α and β are independent of R' and C'. Then, there exists in S^* a unique boundary value

$$f(x) = \lim_{y \to 0,\ y \to C} f(x+iy) \in S^{(m)*}, \quad m = \alpha + \beta + n + 3, \tag{14}$$

that is independent of the sequence $y \to 0$ for $y \in C$. (Here, we assume that the sequence $y \to 0$ is contained in some cone C' that is compact in C.) (See Vladimirov [42, 30]; for the octant $T^0 = C$, see Tillman [133].)

Proof: We may assume that the nonnegative numbers α and β are integers. Let us choose a compact subcone $C' \subset C$ and a positive number η such that $(\alpha + 2)\eta < R$. Let us show that the set

$$[f:\ f = f(x + iy\tau),\ \ y \in \operatorname{pr} C',\ \ 0 < \tau \leqslant \eta] \tag{15}$$

is bounded in S^*. The function

$$f_0(\lambda;\ x,\ y) = f(x + \lambda y), \qquad \lambda = \sigma + i\tau$$

is holomorphic in the strip $0 < \tau < R$ and, on the basis of (13), satisfies the inequality

$$|f_0(\lambda;\ x,\ y)| \leqslant C^{(0)}\tau^{-\alpha}(1+|x|+|\sigma|)^{\beta},$$
$$y \in \mathrm{pr}\ C',\quad 0<\tau\leqslant\eta(\alpha+2).$$

But then, the function

$$f_1(\lambda;\ x,\ y) = Lf_0 = \int_{i\eta}^{\lambda} f_0(\lambda';\ x,\ y)\,d\lambda',$$

where the integration is taken over the straight line connecting the points $i\eta$ and λ, is holomorphic in the strip $0<\tau<R$ and, for $0<\tau\leqslant\eta(\alpha+1)$ satisfies the inequality

$$|f_1(\lambda;\ x,\ y)| \leqslant \int_0^1 |f_0[t\lambda+(1-t)i\eta;\ x,\ y]|\,|\lambda-i\eta|\,dt \leqslant$$
$$\leqslant C_1(1+|x|+|\sigma|)^{\beta}(1+|\sigma|)\int_0^1 \frac{dt}{\left(t\frac{\tau}{\eta}+1-t\right)^{\alpha}} \leqslant$$
$$\leqslant C^{(1)}\tau^{-\alpha+1}(1+|x|+|\sigma|)^{\beta}(1+|\sigma|).$$

Finally, the function

$$f_{\alpha+1}(\lambda;\ x,\ y) = L^{\alpha+1}f_0$$

is holomorphic in the strip $0<\eta<R$ and satisfies the inequality

$$|f_{\alpha+1}(\lambda;\ x,\ y)| \leqslant C^{(\alpha+1)}(1+|x|+|\sigma|)^{\beta}(1+|\sigma|)^{\alpha+1},$$
$$y\in \mathrm{pr}\ C',\quad 0<\tau\leqslant\eta.$$

From this inequality and the relation

$$f(x+y\sigma+iy\tau) = f_0(\lambda;\ x,\ y) = \frac{\partial^{\alpha+1} f_{\alpha+1}(\sigma+i\tau;\ x,\ y)}{\partial\sigma^{\alpha+1}}$$

we conclude (remembering that the differentiation operator is continuous from S^* into S^*) that the set

$$[f:\ f=f(x+y\sigma+iy\tau),\ y\in\mathrm{pr}\ C',\ 0<\tau\leqslant\eta] \tag{16}$$

is bounded with respect to the norm in $S^{(m)*}_{(x,\sigma)}$, where $m=\beta+\alpha+n+3$.

Suppose that M is a bounded set in $S^{(m)}_{x}$: $\|\varphi\|_m\leqslant A$, where $\varphi\in M$. Let us fix the function $\chi(\sigma)\in S_{\sigma}$ so that $\int\chi(\sigma)\,d\sigma=1$. Then, the set

$$[\psi:\psi(u,\ \sigma)=\varphi(u+y\sigma)\chi(\sigma),\qquad y\in\mathrm{pr}\ C',\ \varphi\in M]$$

is bounded in $S^{(m)}_{(u,\sigma)}$, that is,

$$\|\varphi(u+y\sigma)\chi(\sigma)\|_m \leqslant B\,\|\varphi\|_m \leqslant BA.$$

From this and the boundedness of the set (16), we obtain, for all $\varphi \in M$,

$$\left|\int f(x+iy\tau)\varphi(x)\,dx\right| =$$
$$=\left|\int f(x-y\sigma+y\sigma+iy\tau)\varphi(x)\chi(\sigma)\,dx\,d\sigma\right| =$$
$$=\left|\int f(u+y\sigma+iy\tau)\varphi(u+y\sigma)\chi(\sigma)\,du\,d\sigma\right| \leqslant$$
$$\leqslant \|f(u+y\sigma+iy\tau)\|_{-m}\,\|\varphi(u+y\sigma)\chi(\sigma)\|_m \leqslant LBA.$$

This inequality proves that the set (15) is bounded in $S^{(m)*}$ (see section 3.7).

Note that $g(\xi)$ is the spectral function of $f(z)$ (its existence follows from the theorem in section 26.1). Since $g(\xi)\in D^*$, we have, for arbitrary $\varphi\in D$,

$$\lim_{y\to 0}(g(\xi)e^{-y\xi},\ \varphi(\xi)) = (g,\ \varphi).$$

From this it follows that

$$\lim_{y\to 0,\ y\in C'}(f(x+iy),\ F[\varphi]) = (2\pi)^n(g,\ \varphi),\qquad \varphi\in D \tag{17}$$

for an arbitrary sequence $y\to 0$ in C'. Since D is dense in S (see section 3.6) and the operator F is a homeomorphism of S onto S, the set $Q=[\psi:\psi=F[\varphi],\ \varphi\in D]$ is dense in S. But S is dense in $S^{(m)}$. Therefore, Q is dense in $S^{(m)}$.

Thus, an arbitrary sequence of functionals

$$(f(x+iy),\ \psi(x)),\qquad y\to 0,\qquad y\in C',$$

that is bounded in norm in $S^{(m)*}$ converges to a set Q that is dense in $S^{(m)*}$. From the Banach-Steinhaus theorem (see section 3.7), the function $f\in S^{(m)*}\subset S^*$ has the weak limit:

$$\lim_{y\to 0,\ y\in C'}(f(x+iy),\ \psi(x)) = (f,\ \psi),\qquad \psi\in S^{(m)}.$$

On the basis of (17), the functional $f\in S^{(m)*}$ constructed on the set Q that is dense in $S^{(m)}$ is of the form

$$(f,\ \psi) = (2\pi)^n\left(g,\ F^{-1}[\psi]\right),\qquad \psi\in Q.$$

Consequently, it is independent of the sequence $y\to 0$ contained in the (arbitrary but fixed) subcone C', which is compact in C. Since the space S is perfect (see section 3.6), the sequence $f(x+iy)$, where $y\to 0$, and $y\in C$, also converges to Q in norm in some space $S^{(l)*}$, where $l\geqslant m$ (see section 3.7), which completes the proof.

Thus, condition (13), which is more restrictive than condition (1) ensures that the spectral function $g(\xi)$ will belong to the space

S^*. This is true because the boundary value of $f \in S^*$ and, hence, by virtue of (17), S can be extended to the entire space g according to the formula

$$(g, \varphi) = (2\pi)^{-n}(f, F[\varphi]), \qquad \varphi \in S,$$

from which it follows that

$$g = F^{-1}[f] \in S^*.$$

4. *The class* $H_p(a; C)$

We shall now consider a more restricted class of functions that are holomorphic in the tubular cone T^C. We shall say that a function $f(z)$ belongs to the class $H_p(a; C)$ (where $(p \geqslant 1, \ a \geqslant 0)$ if it is holomorphic in the tubular cone T^C and, for an arbitrary subcone C' that is compact in C the inequality

$$|f(z)| \leqslant M(C')(1+|z|)^{\beta}(1+|y|^{-\alpha})e^{a|y|^p}, \qquad z \in T^{C'} \tag{18}$$

is satisfied for certain nonnegative values of α and β, which do not depend on C'. We define

$$H_p(a+\varepsilon; C) = \bigcap_{a'>a} H_p(a'; C); \quad H_0(C) = H_1(0; C).$$

Suppose that the components of the cone C are the cones C_1, $C_2, \ldots$. It follows from the results of the preceding section that a function $f(z)$ of the class $H_p(a; C)$ has boundary values

$$\lim_{y \to 0,\, y \in C_k} f(x+iy) = f_k(x) \in S^*, \qquad k = 1, 2, \ldots.$$

The corresponding spectral functions $g_k = F^{-1}[f_k]$ also belong to S^*.

Let us review certain definitions associated with an open cone C (see section 25.1). C^* is the dual of the cone C; $C_* = R^n \setminus C^*$; $\mu_C(\xi)$ is the indicatrix of the cone C; ρ_C characterizes the nonconvexity of the cone C.

We have the following theorems for functions of the class $H_p(a; C)$:

THEOREM 1.* *Consider a function* $f(z) \in H_p(a+\epsilon; C)$, *where* C *is a connected cone,* $p > 1$, *and* $a > 0$. *The spectral function* $g(\xi)$ *of* $f(z)$ *can be represented in the form of the sum of a finite number of derivatives of continuous functions* $G_l(\xi)$ *of power increase,*

$$g(\xi) = \sum_l D^l G_l(\xi), \tag{19}$$

---

*See Vladimirov [42].

which, for all $\xi \in C_*'$, *where* C_*' *is an arbitrary cone that is compact in* C_*, *and, for all* $\epsilon > 0$, *satisfy the inequality*

$$|G_l(\xi)| \leqslant M_\varepsilon'(C_*')\, e^{-(a'-\varepsilon)[\mu_C(\xi)]^{p'}}, \tag{20}$$

where the numbers p and a are connected with p' and a' by the relations

$$\frac{1}{p}+\frac{1}{p'}=1, \qquad (p'a')^p(pa)^{p'}=1. \tag{21}$$

Conversely, if $g(\xi)$ satisfies these conditions for certain numbers $a'>0$, $p'>1$ and the cone C^*, then all the derivatives $D^\alpha f(z)$ of its Fourier-Laplace transform $f(z)$ belong to the class $H_p(\rho_C^p a+\varepsilon;\, O(C))$.

COROLLARY 1. *If* $f(z) \in H_p(a+\epsilon;\, C)$, *then* $D^\alpha f(z) \in H_p(\rho_C^p a + \epsilon;\, O(C))$.

COROLLARY 2*. *For an entire function* $f(z)$ *to have a spectral function* $g(\xi)$ *of the form* (19), *where the continuous functions* $G_l(\xi)$ *satisfy the inequality*

$$|G_l(\xi)| \leqslant M_\varepsilon' e^{-(a'-\varepsilon)|\xi|^{p'}} \qquad (p'>1,\ a'>0),$$

it is necessary and sufficient that, for some nonnegative β, *the function* $f(z)$ *satisfy the inequality*

$$|f(z)| \leqslant M_\varepsilon(1+|z|)^\beta e^{(a+\varepsilon)|y|^p}. \tag{22}$$

THEOREM 2.** *Suppose that* $f(z) \in H_1(a+\epsilon;\, C)$, *where* C *is a connected cone and* $a \geq 0$. *Then, its spectral function* $g(\xi) \in S^*$ *and*

$$g(\xi)=0, \qquad \mu_C(\xi)>a.$$

Conversely, if $g \in S^*$ *vanishes in the domain* $\mu_C(\xi) \geq a$ *for some* $a>0$ *and some cone* C, *then all the derivatives* $D^\alpha f(z)$ *of its Fourier-Laplace transform* $f(z)$ *belong to the class* $H_1(\rho_C a;\, O(C))$.

COROLLARY 1. *If* $f(z)$, *then* $D^\alpha f(z) \in H_1(\rho_C a;\, O(C))$. *In particular, if* C *is a convex cone, then* $H_1(a+\epsilon;\, C) = H_1(a;\, C)$.

COROLLARY 2. (Paley-Wiener-Schwartz [11]). *For the spectrum of an entire function* $f(z)$ *to be contained in a sphere* $|\xi| \leq a$, *it is necessary and sufficient that* $f(z)$ *satisfy inequality* (22) *for* $p = 1$.

Remark: It follows from Corollaries 1 to Theorems 1 and 2 that the sets $\bigcup_{a' \geqslant a} H_p(a';\, C)$ and $H_0(C)$ constitute classes in the sense of section 16.1.

*See Eskin [86].

**Schwartz [45], Lions [46], Bogolyubov and Parasyuk [78], and Vladimorov [42]. For functions in L_2, the corresponding results were obtained by Bochner and Martin [3], Chapter VI. See also Paley and Wiener [155] and Plancherel and Pólya [156].

5. *Proof of Theorem 1*

Suppose that $f(z) \in H_p(a+\varepsilon, C)$, where $p > 1$ and $a > 0$, that is, suppose that $f(z)$ satisfies inequality (18) for some nonnegative values of α and β. Then, on the basis of section 26.3, there exists a boundary value

$$f(x) = \lim_{y \to 0,\, y \in C} f(x + iy), \tag{23}$$

which is a weak limit in the space $S^{(m)*}$, where $m = \alpha + \beta + n + 3$.

Let us choose an integer N such that $2N > m$ and let us set

$$\sigma = \left[\frac{p'}{p}\right] + 1, \qquad c = \left(\frac{1}{ap}\right)^{\sigma}, \tag{24}$$

where $[x]$ denotes the greatest integer not exceeding x. Let $g = F^{-1}[f]$ be the spectral function of $f(z)$. We have (see section 26.2)

$$g(\xi) = e^{\xi y} F^{-1}[f(x + iy)](\xi) =$$

$$= F^{-1}\left[\frac{\{2 + c^2 |\xi|^{2\sigma} + (x + iy)(x + iy)\}^N}{\{2 + c^2 |\xi|^{2\sigma} + (x + iy)(x + iy)\}^N} f(x + iy) e^{\xi y}\right](\xi).$$

By using the properties of the Fourier transformation in S^* (see section 3.9), we may represent this last expression in the form of a finite sum (19) with functions G_l of the form

$$G_l(\xi) = P_l(\xi) \int \frac{f(x + iy) e^{-i\xi(x+iy)}\, dx}{[2 + c^2 |\xi|^{2\sigma} + (x + iy)(x + iy)]^{n_l}}, \tag{25}$$

where the P_l are polynomials, the n_l are integers $\geqslant N$, and y is an arbitrary member of C.

Let us show that the integral in (25) is independent of y for all $y \in C$ such that $|y|^2 < c^2 |\xi|^{2\sigma} + 1$.

For such y,

$$\begin{aligned} |2 + c|\xi|^{2\sigma} + (x + iy)(x + iy)| &\geqslant \\ \geqslant |2 + c^2|\xi|^{2\sigma} + |x|^2 - |y|^2| &\geqslant 1 + |x|^2. \end{aligned} \tag{26}$$

Therefore, the integrand in (25) is holomorphic in the domain

$$R^n + i[y : y \in C,\ y^2 < c^2|\xi|^{2\sigma} + 1]$$

and, by virtue of (18), it does not exceed in absolute value the quality $c_k(y)(1 + |x|)^{-n-1}$. Consequently, on the basis of the Cauchy-Poincaré theorem, the integral in (25) does not change if the plane of integration $z = x + iy$ (where $y =$ const) is displaced along y while remaining in the region $y \in C$, where $|y|^2 < c^2|\xi|^{2\sigma} + 1$.

Thus, if $|y|<1$ and $y\in C$, the representation (25) is independent of y for all ξ and defines a continuous function ξ in R^n.

Furthermore, on the basis of the choice of the number N and the inequalities $n_l \geqslant N$, we have

$$y \to 0,\ y\in C$$

$$\frac{e^{-i\xi(x+iy)}}{[2+c^2|\xi|^{2\sigma}+(x+iy)^2]^{n_l}} \to \frac{e^{-i\xi x}}{[2+c^2|\xi|^{2\sigma}+|x|^2]^{n_l}} \equiv \varphi_l(\xi;\ x)$$

as $y\in C\to 0$ in the sense of the norm of the space $S^{(m)}$ for every ξ. It follows from this and from (23) that we can take the limit as $y\in C\to 0$ in the representation (25). As a result, we obtain (see section 26.2)

$$G_l(\xi)=P_l(\xi)(f(x),\ \varphi_l(\xi;\ x)). \tag{27}$$

But the norm of the function $\varphi_l(\xi;\ x)$ in the space $S^{(m)}$ does not increase faster than a polynomial in ξ:

$$\|\varphi_l(\xi;\ x)\|_m \leqslant c_l(1+|\xi|)^m.$$

If we now use the boundedness of the functional f in $S^{(m)}$, we conclude from this inequality and Eq. (27) that the functions $G_l(\xi)$ are of power increase:

$$|G_l(\xi)|=|P_l(\xi)||(f(x),\ \varphi_l(\xi;\ x))| \leqslant \\ \leqslant c_l\|f\|_{-m}|P_l(\xi)|(1+|\xi|)^m.$$

It remains to prove inequality (20). Let ξ range over an arbitrary subcone C_*' that is compact in $C_*=[\xi:\ \mu_C(\xi)>0]$. Since $\mu_C(\xi)$ is a continuous homogeneous function of first degree, there exists a constant $\gamma=\gamma(C_*')$ such that

$$\gamma|\xi| \leqslant \mu_C(\xi) \leqslant |\xi|,\ \xi\in C_*'. \tag{28}$$

Let us now establish the

LEMMA. *Let C_*' be a cone that is compact in the cone $C_*=[\xi:\mu_C(\xi)>0]$. For an arbitrary number η in $(0,1)$, there exists a cone $C'=C'(C_*',\eta)$, that is compact in C such that, for an arbitrary point $\xi\in C_*'$, there exists a point $y_\xi^0\in \mathrm{pr}\, C'$, at which*

$$-\xi y_\xi^0 \geqslant (1-\eta)\,\mu_C(\xi). \tag{29}$$

Proof: Since the function $\mu_C(\xi)$ is homogeneous, it will be sufficient to prove this assertion for $|\xi|=1$. In this case, on the basis of (28), $\mu_C(\xi)\geqslant\gamma>0$ for all $\xi\in \mathrm{pr}\ C_*'$. For the cone C' we take a cone for which (see Fig. 62)

$$\mathrm{pr}\ C'=\bigcap_{y'\in \mathrm{pr}\,\partial C}\left[y:|y-y'|\geqslant\frac{\gamma\eta}{M},\ y\in \mathrm{pr}\ C\right].$$

where the number $M \geqslant 1$ is taken sufficiently large that $\text{pr}\,C' \neq \varnothing$. Suppose that $\mu_C(\xi) = -y_\xi \xi$, $y_\xi \in \text{pr}\,C$, $\xi \in \text{pr}\,C'_*$. Then, there exists a vector $y_\xi^0 \in \text{pr}\,C'$ such that $|y_\xi - y_\xi^0| \leqslant \frac{\gamma\eta}{M}$. Consequently,

$$-\xi y_\xi^0 = -\xi y_\xi + \xi(y_\xi - y_\xi^0) \geqslant \mu_C(\xi) - |y_\xi - y_\xi^0| \geqslant$$

$$\geqslant \mu_C(\xi) - \frac{\gamma\eta}{M} \geqslant \mu_C(\xi)\left(1 - \frac{\eta}{M}\right) \geqslant \mu_C(\xi)(1-\eta), \text{ q.e.d.}$$

Let us continue the proof of the theorem. Suppose that $\xi \in C'_*$. Let η be an arbitrary number in (0.1) and construct the vector

$$y_\xi = y_\xi^0 \left[\frac{\mu_C(\xi)}{ap}\right]^{\frac{p'}{p}}, \tag{30}$$

where y_ξ^0 is a unit vector defined in accordance with the lemma just proven. Then, $[y: y = y_\xi,\ \xi \in C'_*] \subset C'$.

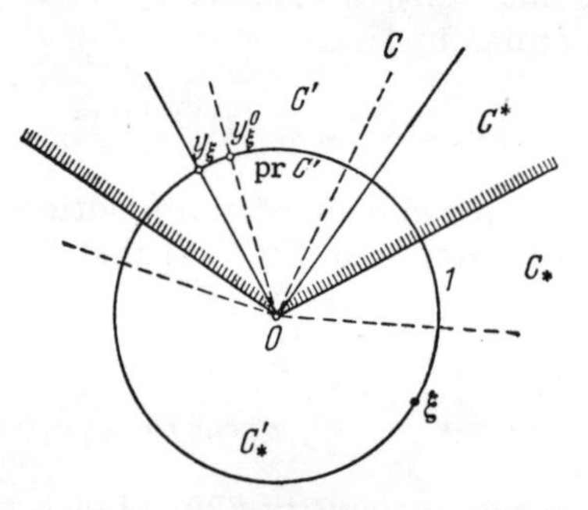

Fig. 62

Suppose that $|\xi| \geqslant ap$. For such values of ξ, we have, in view of (24) and (28),

$$|y_\xi| = \left[\frac{\mu_C(\xi)}{ap}\right]^{\frac{p'}{p}} \leqslant \left(\frac{|\xi|}{ap}\right)^{\frac{p'}{p}} \leqslant \left(\frac{|\xi|}{ap}\right)^{\sigma} = c|\xi|^\sigma < \sqrt{c^2|\xi|^{2\sigma} + 1}.$$

Therefore, we may set $y = y_\xi$ in the representation (25). Finding a bound for its absolute value and taking inequality (26) into consideration, we obtain

$$G_l(\xi)| \leqslant |P_l(\xi)|\, e^{\xi y_\xi} \int \frac{|f(x + iy_\xi)|}{(1+|x|^2)^{n_l}}\, dx,\quad \xi \in C'_*,\ |\xi| \geqslant ap. \tag{31}$$

Let us now observe inequality (18) for $y = y_\xi$ (with a replaced by $a+\eta$) and Eq. (30). As a result, we obtain

$$|f(x + iy_\xi)| \leqslant M^*_\eta(C')(1+|x|)^\beta (1+|\xi|)^\sigma \times$$

$$\times \left(1 + \left[\frac{\mu_C(\xi)}{ap}\right]^{-\frac{p'}{p}\alpha}\right) \exp\left\{(a+\eta)\left[\frac{\mu_C(\xi)}{ap}\right]^{p'}\right\}.$$

If we substitute this inequality into inequality (31) and keep inequalities (28) and (29) and Eqs. (21) in mind, we obtain inequality (20) for all $\xi \in C'_*$ and $|\xi| \geqslant ap$

$$|G_l(\xi)| \leqslant M^*_\eta(C')|P_l(\xi)|(1+|\xi|)^\sigma\left(1+\gamma^{-\frac{p'}{p}\alpha}\right)\times$$
$$\times \exp\left\{(a+\eta)\left[\frac{\mu_C(\xi)}{ap}\right]^{p'} - (1-\eta)\mu_C(\xi)\left[\frac{\mu_C(\xi)}{ap}\right]^{\frac{p'}{p}}\right\}\times$$
$$\times \int \frac{(1+|x|)^\beta}{(1+|x|^2)^{n_l}}\,dx = M''_\eta(C'_*)|P_l(\xi)|(1+|\xi|)^\sigma\times$$
$$\times \exp\left\{-a'\left(1-p'\eta-\frac{p'\eta}{pa}\right)[\mu_C(\xi)]^{p'}\right\} \leqslant$$
$$\leqslant M'_\varepsilon(C'_*)\exp\{-(a'-\varepsilon)[\mu_C(\xi)]^{p'}\}.$$

Obviously, on the basis of inequalities (28), inequality (20) is also valid for $|\xi| < ap$.

Conversely, suppose that $g(\xi)$ is the finite sum (19) with continuous functions $G_l(\xi)$ of power increase that satisfy inequality (20) in the cone C^* for some $a' > 0$ and $p' > 1$.

Let y range over a cone C' that is compact in $O(C)$. From Lemma 2 in section 25.2, there exists a positive number δ and a cone C'' such that

$$\xi y \geqslant \delta|\xi||y|, \quad (\xi, y) \in C'' \times C'. \tag{32}$$

Here, the cone C^* is compact in C^{**} (see Fig. 59).

Denoting the highest power of increase of the functions $G_l(\xi)$ by r, we derive from (32) the following inequality, valid for all $(\xi, y) \in C'' \times C'$:

$$|G_l(\xi)e^{-\xi y}| \leqslant d(1+|\xi|)^r e^{-\delta|\xi||y|}, \tag{33}$$

where d is some positive number. On the other hand, if $\xi \notin C''$, it follows from the inclusion $C'' \Supset C^*$ that ξ ranges over a cone $C'_* = R^n \setminus C''$, which is compact in C^*. In this case, inequality (20) yields

$$|G_l(\xi)e^{-\xi y}| \leqslant M'_\varepsilon(C'_*)\exp\{-(a'-\varepsilon)[\mu_C(\xi)]^{p'} - \xi y\} \tag{34}$$

for arbitrary $\varepsilon > 0$. Let us recall (see section 25.1) that

$$-\xi y \leqslant \mu_{O(C)}(\xi)|y|, \quad \mu_{O(C)}(\xi) \leqslant \rho_C\mu_C(\xi), \quad \xi \in C_*. \tag{35}$$

From inequalities (34) and (35), we conclude that

$$|G_l(\xi)e^{-\xi y}| \leqslant M'_\varepsilon(C'_*)\exp\{-(a'-\varepsilon)[\mu_C(\xi)]^{p'} + \rho_C\mu_C(\xi)|y|\}, \quad \eta > 0 \tag{36}$$

for all $(\xi, y) \in C'_* \times C'$.

In particular, it follows from inequalities (33) and (36) and from the representation (19) that $g(\xi)e^{-\xi y} \in S^*$ for all $y \in O(C)$. Therefore, the function $f(z)$, the Fourier-Laplace transform of $g(\xi)$, is holomorphic in $T^{O(C)}$ (see section 26.2) and can be represented in the form (see section 3.9)

$$f(z) = \sum_l (-iz)^l \int G_l(\xi) e^{iz\xi}\, d\xi, \quad z \in T^{O(C)}. \tag{37}$$

Let us find a bound for each individual integral in the sum (37). By using inequalities (33), (36), and (28) and Eqs. (21), we obtain the following chain of inequalities for $z \in T^{C'}$:

$$\left| \int G_l(\xi) e^{i\xi z}\, d\xi \right| \leqslant \int\limits_{C''} |G_l(\xi)| e^{-\xi y}\, d\xi + \int\limits_{C'_*} |G_l(\xi)| e^{-\xi y}\, d\xi \leqslant$$

$$\leqslant d \int\limits_{C''} (1+|\xi|)^r e^{-\delta |\xi| |y|}\, d\xi +$$

$$+ M'_\varepsilon(C'_*) \int\limits_{C'_*} \exp\{-(a'-\varepsilon)[\mu_C(\xi)]^{p'} + \rho_C \mu_C(\xi)|y|\}\, d\xi \leqslant$$

$$\leqslant d \int (1+|\xi|)^r e^{-\delta |\xi| |y|}\, d\xi +$$

$$+ M''_\varepsilon(C'_*) \int\limits_{C'_*} \exp\{-(a'-2\varepsilon)[\mu_C(\xi)]^{p'} + \rho_C \mu_C(\xi)|y|\} \times$$

$$\times \frac{d\xi}{(1+|\xi|)^{n+1}} \leqslant M^{(1)}_\varepsilon(C') \left[\int_0^\infty (1+\rho^r)\rho^{n-1} e^{-\delta\rho |y|}\, d\rho + \right.$$

$$+ \sup_{\xi \in C'_*} \exp\{-(a'-2\varepsilon)[\mu_C(\xi)]^{p'} +$$

$$\left. + \rho_C \mu_C(\xi)|y|\} \right] \leqslant M^{(2)}_\varepsilon(C') \left[|y|^{-r-n} + \right.$$

$$\left. + |y|^{-n} + \sup_{\rho \geqslant 0} \exp\{-(a'-2\varepsilon)\rho^{p'} + \rho_C \rho |y|\} \right] \leqslant$$

$$\leqslant M^{(2)}_\varepsilon(C') \left[|y|^{-r-n} + |y|^{-n} + \exp\left\{ \frac{1}{p} \left[\frac{1}{p'(a'-2\varepsilon)} \right]^{\frac{p}{p'}} |y|^p \rho_C^p \right\} \right] \leqslant$$

$$\leqslant M_\eta(C')(1+|y|^{-r-n}) \exp\{(a+\eta)|y|^p \rho_C^p\}.$$

From these inequalities and Eq. (37), it follows that

$$f(z) \in H_p(a\rho_C^p + \varepsilon;\ O(C)).$$

In an analogous fashion, it can be proven that all the derivatives

$$D^{\alpha} f(z) = \sum_{l} (-iz)^{l} \int (i\xi)^{\alpha} G_{l}(\xi) e^{iz\xi}\, d\xi$$

also belong to the class $H_p(a\rho_C^p + \varepsilon;\ O(C))$. This proves Theorem 1.

6. *Proof of Theorem 2*

If $f(z) \in H_1(a+\varepsilon; C)$ its spectral function $g(\xi)$ exists (see section 26.1) and belongs to S^* (see section 26.3). It remains to show that $g(\xi) = 0$ for $\mu_C(\xi) > a$. It will be sufficient to prove this equation for a small neighborhood $|\xi - \xi^0| < \eta$ of an arbitrary point ξ^0 in the domain $\mu_C(\xi) > a$ (see section 3.2). Let $\varphi(\xi)$ be an arbitrary function in D such that $S_{\varphi} \subset U(\xi^0, \eta)$. Since $g(\xi)$ is the spectral function of $f(z)$, we have for all $y \in C$, (see section 26.1)

$$\begin{aligned}(g, \varphi) &= \left(e^{\xi y} F^{-1}[f(x+iy)],\ \varphi\right) = \\ &= \left(F^{-1}[f(x+iy)],\ e^{\xi y}\varphi\right) = \left(f(x+iy),\ F^{-1}[e^{\xi y}\varphi(\xi)]\right) = \\ &= (2\pi)^{-n} \int \frac{f(x+iy)}{(1+|x|^2)^N} \int e^{-i\xi x} (1-\Delta)^N [e^{\xi y}\varphi(\xi)]\, d\xi,\end{aligned} \tag{38}$$

where N is an integer such that $2N \geqslant \beta + n + 1$, and Δ is the Laplacian operator.

Since $\xi^0 \in C_*$, it follows from the lemma in section 26.5 that, for an arbitrary sufficiently small number η', there exists a vector $y^0 \in \mathrm{pr}\, C$ (see Fig. 62) such that

$$-\xi^0 y^0 \geqslant \mu_C(\xi^0) - \eta'. \tag{39}$$

Let us set $y = \lambda y^0$, where λ is an arbitrary positive number, in Eqs. (38). Assuming that η and η' are sufficiently small so that

$$-\mu_C(\xi^0) + \eta + \eta' \leqslant -a - \eta,$$

we obtain from (39), for all $\xi \in S_{\varphi}$, the inequality

$$y\xi = \lambda[y^0\xi^0 + y^0(\xi - \xi^0)] \leqslant \lambda[-\mu_C(\xi^0) + \eta + \eta'] \leqslant -\lambda(a+\eta). \tag{40}$$

This inequality makes it possible to give a bound for the integral

$$\begin{aligned}&\left|\int e^{-i\xi x}(1-\Delta)^N [e^{\xi y}\varphi(\xi)]\, d\xi\right| \leqslant \\ &\qquad \leqslant K_{\varphi}(1+\lambda)^{2N} \sup_{\xi \in S_{\varphi}} e^{\xi y} \leqslant K_{\varphi}(1+\lambda)^{2N} e^{-(a+\eta)\lambda}.\end{aligned}$$

When we substitute this inequality into Eqs. (38) and use inequality (18) for $C'=[y: y=\lambda y^0]$ and $p=1$ (with a replaced with $a+\varepsilon$) we obtain, for arbitrary $\varepsilon>0$ and $\lambda>0$,

$$|(g, \varphi)| \leqslant K'_\varepsilon(\varphi, \xi^0)(1+\lambda^{-a})(1+\lambda)^{2N+\beta} e^{-\lambda(\eta-\varepsilon)},$$

where the constant $K'_\varepsilon(\varphi, \xi^0)$ is independent of λ (though dependent on the cone C', that is, on $y^0=y^0(\xi^0)$). If we take $\varepsilon<\eta$ in this inequality and let λ approach infinity, we obtain $(g, \varphi)=0$, that is, $g(\xi)=0$ for $\xi \in U(\xi^0, \eta)$ as asserted.

Conversely, suppose that $g(\xi) \in S^*$ and $g(\xi)=0$ for $\mu_C(\xi)>a$. Therefore, the carrier S_g is contained in the closed set (see Fig. 59)

$$F=[\xi: \mu_C(\xi) \leqslant a].$$

We note that (since $\mu_C(\xi)$ is a convex function (see section 25.1)) *the set F is convex* (see section 13.4). Furthermore, on the basis of (35), the chain of inequalities

$$\xi y \geqslant -\mu_{O(C)}(\xi)|y| \geqslant -\rho_C \mu_C(\xi)|y| \geqslant -\rho_C a|y| \tag{41}$$

is valid for all $(y, \xi) \in O(C) \times F$. Therefore, $e^{-\xi y} g(\xi) \in S^*$ for all $y \in O(C)$. From this it follows that the function $f(z)$ is holomorphic in $T^{O(C)}$ (see section 26.2).

Since the carrier g is contained in the regular set F, it follows from the theorem in section 3.8 that

$$g(\xi)=\sum_{|\alpha| \geqslant m} D^\alpha \mu_\alpha(\xi) \tag{42}$$

where the measures μ_α possess the properties

$$\int (1+|\xi|)^{-m} |d\mu_\alpha(\xi)| < \infty \tag{43}$$

$$S_{\mu_a} \subset F. \tag{44}$$

By using the second of formulas (20) of Chapter 1, we obtain from (42)

$$f(z) = \sum_{|\alpha| \leqslant m} (-iz) F[\mu\, e^{-y\xi}](x) = \sum_{|\alpha| \geqslant m} (-iz) \int e^{i\xi z} d\mu(\xi). \tag{45}$$

Let C' be an arbitrary cone that is compact in $O(C)$. Keeping (43) and (44) in mind, we obtain from (45) the following inequality, which holds for all $z \in T^{C'}$:

$$|f(z)| \leq \sum_{|\alpha| \leqslant m} |z|^m \int (1+|\xi|)^m e^{-y\xi} \frac{|d\mu(\xi)|}{(1+|\xi|)^m}$$

$$\leqslant L(1+|z|)^m \sup_{\xi \in F} (1+|\xi|)^m e^{-y\xi} \leqslant$$

$$\leqslant L(1+|z|)^m \left[\sup_{\xi \in C''} (1+|\xi|)^m e^{-y\xi} + \sup_{\xi \in F \cap C'_*} (1+|\xi|)^m e^{-y\xi}\right], \tag{46}$$

where C'' is a cone in which inequality (32) is verified (see Fig. 59): $\xi y \geqslant \delta|\xi||y|$ for $(\xi, y) \in C'' \times C'$, where $C' = R^n \setminus C''$ is a cone in which inequality (28) is verified: $\mu_C(\xi) \geqslant \gamma|\xi|$ for $\xi \in C'_*$. Keeping these inequalities and inequality (41) in mind, we continue inequality (46):

$$|f(z)| \leqslant L(1+|z|)^m \left[\sup_{\rho \geqslant 0} (1+\rho)^m e^{-\delta\rho|y|} + \right.$$

$$\left. + \sup_{\gamma|\xi| \leqslant \mu_C(\xi) \leqslant a} (1+|\xi|)^m e^{\rho_C a|y|}\right] \leqslant$$

$$\leqslant L_1(1+|z|)^m \left[\sup_{\rho \geqslant 0} \rho^m e^{-\delta\rho|y|} + \left(1+\frac{a}{\gamma}\right)^m e^{\rho_C a|y|}\right] \leqslant$$

$$\leqslant L_2(1+|z|)^m \left(|y|^{-m} + e^{a\rho_C|y|}\right).$$

From this it follows that $f(z) \in H_1(a\rho_C;\ O(C))$.

Analogously, it can be proven that all the derivatives

$$D^\alpha f(z) = F[(i\xi)^\alpha g(\xi) e^{-\xi y}](x)$$

belong to the class $H_1(a\rho_C;\ O(C))$. This completes the proof of Theorem 2.

7. *The case of a disconnected cone C*

In this case, Theorems 1 and 2 of section 26.4 admit the following generalization.

THEOREM.* *Suppose that $f(z) \in H_p(a+\epsilon;\ C)$, where the cone C consists of a finite number of connected components (cones) $C_1, C_2, \ldots, C_t$. Suppose also that its boundary values*

$$f_k(x) = \lim_{y \to 0,\ y \in C_k} f(x+iy),$$

that exist in the space S^* coincide:

$$f_1(x) = f_2(x) = \ldots = f_t(x), \qquad x \in R^n. \tag{47}$$

*See Vladimirov [42].

Then, $f(z) \in H_p\left(a\rho_c^p + \epsilon; O(C)\right)$ *for* $p > 1$ *and* $a > 0$, *and* $f(z) \in H_1\left(a\rho_c; O(C)\right)$ *for* $p = 1$ *and* $a \geq 0$.

COROLLARY. *For* $n = 1$, *this theorem takes the following form. Suppose that a function* $f(z)$ *is holomorphic in the upper and lower half-planes and satisfies the inequality*

$$|f(z)| \leqslant M_\epsilon (1 + |z|)^\beta |y|^{-\alpha} e^{(a+\epsilon)|y|^p}$$

for arbitrary $\epsilon > 0$. *Suppose that its boundary values*

$$f(x \pm i0) = \lim_{y \to \pm 0} f(x + iy),$$

that exist in the space S^* *coincide. Then,* $f(z)$ *is an entire function and satisfies the inequality*

$$|f(z)| \leqslant M'_\epsilon (1 + |z|)^{\beta'} e^{(a+\epsilon)|y|^p}, \quad \beta' \geqslant \beta.$$

(For $p = 1$, we may take $\epsilon = 0$.)

Proof of the theorem: Considering $f(z)$ in each component T^{C_k} separately, we conclude on the basis of (47) that the corresponding spectral functions $g_k = F^{-1}[f_k]$ for $k = 1, 2, \ldots, t$, coincide:

$$g(\xi) = g_1(\xi) = \ldots = g_t(\xi). \tag{48}$$

Suppose that $p = 1$. We conclude from Theorem 2 of section 26.4 and Eqs. (48) that $g(\xi) = 0$ for $\xi \in \bigcup_{1 \leqslant k \leqslant t} [\xi: \mu_{C_k}(\xi) > a]$. But we always have (see section 25.1)

$$\mu_{O(C)}(\xi) \leqslant \rho_C \mu_C(\xi) = \rho_C \max_{1 \leqslant k \leqslant t} \mu_{C_k}(\xi), \quad \rho_C < +\infty. \tag{49}$$

Therefore, $g(\xi)$ vanishes in the domain $\mu_{O(C)}(\xi) > a\rho_C$ (see Fig. 63). But then, we conclude from Theorem 2 of section 26.4 that

$$f(z) = F[e^{-\xi y} g(\xi)](x) \in H_1(a\rho_C; O(C)).$$

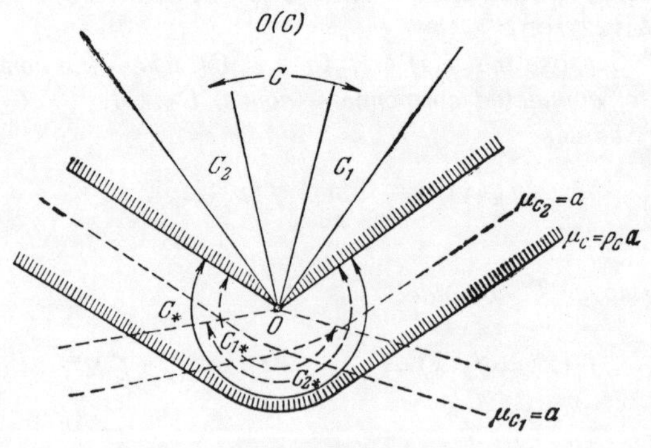

Fig. 63

Suppose now that $p > 1$ and $a > 0$. From Theorem 1 of section 26.4, every spectral function $g_k(\xi)$ can be represented as a sum of the form (19) with continuous functions $G_{k,l}(\xi)$ of power increase that satisfy inequality (20) in the cone C_{k*}. Furthermore, the functions $G_{k,l}(\xi)$ can be expressed in accordance with formula (27)

$$G_{k,l}(\xi) = P_l(\xi)(f_k(x), \varphi_l(\xi, x)).$$

Here, since there are only finitely many components $C_1, C_2, \ldots, C_t$, the polynomials P_l and the functions φ_l can be chosen so as to be independent of k. But then, it follows from Eqs. (47) that $G_{1,l} = G_{2,l} = \ldots = G_{t,l} = G_l$. Consequently, the spectral function $g(\xi)$ can be represented as a sum of the form (19), where the functions $G_l(\xi)$ satisfy inequality (20) in each cone C_{k*}.

$$|G_l(\xi)| \leqslant M'_{k,\varepsilon}(\operatorname{pr}\xi)\, e^{-(a'-\varepsilon)[\mu_C(\xi)]^{p'}}, \quad k = 1, 2, \ldots, t. \tag{50}$$

Here, the numbers a' and p' are connected with a and p by relations (21) and the functions $M'_{k,\varepsilon}(\operatorname{pr}\xi)$ are bounded in each cone C'_{k*} which is compact in C_{k*}.

Now, let us note that (see Fig. 63)

$$C_* = \bigcup_{1 \leqslant k \leqslant t} C_{k*}, \tag{51}$$

which follows from the relation

$$\mu_C(\xi) = \max_{1 \leqslant k \leqslant t} \mu_{C_k}(\xi) \tag{52}$$

and from the definitions of the cones C_* and C_{k*}

Suppose that $\xi \in C_*$. We denote by $n(\xi)$ the set of those indices k for which $\xi \in C_{k*}$, in accordance with (51). Here, on the basis of Eq. (52) and the fact that $\mu_{C_k}(\xi) \leqslant 0$ for all ξ belonging to C_k^* but not to $\xi \in C_{k*}$, we have

$$\max_{k \in n(\xi)} \mu_{C_k}(\xi) = \max_{1 \leqslant k \leqslant t} \mu_{C_k}(\xi) = \mu_C(\xi) \geqslant \frac{1}{\rho_C} \mu_{O(C)}(\xi).$$

If we take the minimum over all $k \in n(\xi)$ in the right member of inequality (50) and then apply the last inequality, we obtain

$$|G_l(\xi)| \leqslant \min_{k \in n(\xi)} [M'_{k,\varepsilon}(\operatorname{pr}\xi)] \exp\left\{-(a'-\varepsilon)\left[\max_{k \in n(\xi)} \mu_{C_k}(\xi)\right]^{p'}\right\} \leqslant$$
$$\leqslant M'_\varepsilon(\operatorname{pr}\xi) \exp\left\{-(a'-\varepsilon)\left[\frac{\mu_{O(C)}(\xi)}{\rho_C}\right]^{p'}\right\}, \quad \xi \in C_*, \tag{53}$$

where

$$M'_\varepsilon(\operatorname{pr}\xi) = \min_{k \in n(\xi)} [M'_{k,\varepsilon}(\operatorname{pr}\xi)]. \tag{54}$$

For future use, we shall prove the

LEMMA. *Suppose that a cone C'^* is compact in C^*. Then, there exist cones C_{k*} that are compact in C_{k*} such that*

$$C'_* \subset \bigcup_{1 \leqslant k \leqslant t} C'_{k*}. \tag{55}$$

We shall prove this lemma by induction on t. For $t = 1$, it is obvious. Suppose that the lemma is true for $t - 1$ and let us prove it for t.

Since the cone C'^* is compact in C^*, the cone $C'_* \setminus C_{t*}$ is compact in the (open) cone $C_0 = \bigcup_{1 \leqslant k \leqslant t-1} C_{k*}$ because, by virtue of (51), we have the inclusions

$$\operatorname{pr}(\overline{C}'_* \setminus C_{t*}) \subset \operatorname{pr}(C_* \setminus C_{t*}) \subset \operatorname{pr} C_0. \tag{56}$$

(See Fig. 64, for the case of $t = 2$.) Therefore, there exists a cone C'_{t*} that is compact in C_{t*}, and whose complement $C'_* \setminus C'_{t*}$ is also compact in C_0. Now we have assumed that there exist cones C'_{k*} for $k = 1, 2, \ldots t-1$, that are compact in the corresponding C_{k*} such that

$$C'_* \setminus C'_{t*} \subset \bigcup_{1 \leqslant k \leqslant t-1} C'_{k*}. \tag{57}$$

If we add C'_{t*}, to both members of this inclusion, we obtain the desired inclusion (55):

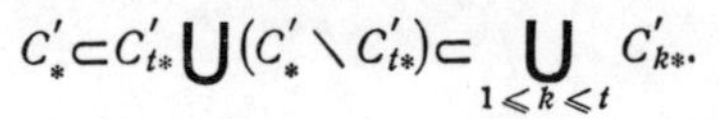

$$C'_* \subset C'_{t*} \bigcup (C'_* \setminus C'_{t*}) \subset \bigcup_{1 \leqslant k \leqslant t} C'_{k*}.$$

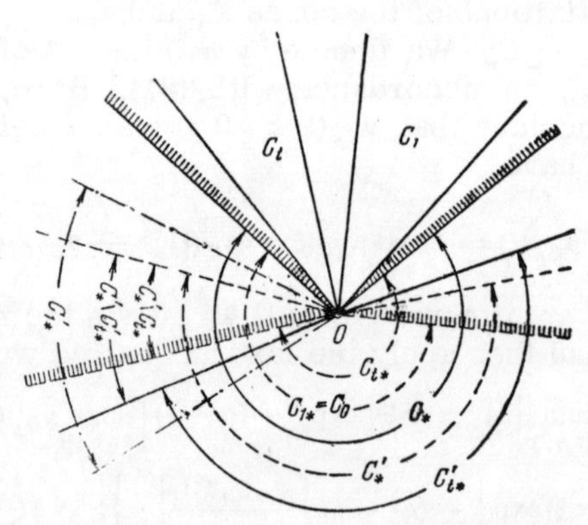

Fig. 64

It follows from this lemma that the function $M'_\varepsilon(\operatorname{pr} \xi)$ is bounded on each cone C'^*, that is compact in C^*. Specifically, from (54) and (55), we have

$$\sup_{\xi \in C'_*} M'_\varepsilon(\mathrm{pr}\,\xi) = \sup_{\xi \in C'_*} \min_{k \in n(\xi)} [M'_{k,\varepsilon}(\mathrm{pr}\,\xi)] \leqslant$$
$$\leqslant \max_{1 \leqslant k \leqslant t} \sup_{\xi \in C'_{k*}} M'_{k,\varepsilon}(\mathrm{pr}\,\xi) = \max_{1 \leqslant k \leqslant t} M'_{k,\varepsilon}(C'_{k*}) = M'_\varepsilon(C'_*). \tag{58}$$

Thus, on every cone C'_* that is compact in $O(C)_* = C_*$, inequalities (53) can be rewritten in the form

$$|G_l(\xi)| \leqslant M'_\varepsilon(C'_*) \exp\left\{-(a'-\varepsilon)\left[\frac{\mu_{O(C)}(\xi)}{\rho_C}\right]^{p'}\right\}, \quad \xi \in C'_*.$$

It follows from this and from Theorem 1 of section 26.4 that

$$f(z) = F[e^{-\xi y} g(\xi)](x) \in H_p(a\rho_C + \varepsilon; O(C)).$$

This completes the proof of the theorem.

27. BOGOLYUBOV'S "EDGE OF THE WEDGE" THEOREM

The "edge of the wedge" theorem, which is a special type of generalization of the principle of holomorphic continuation was discovered and proven by Bogolyubov in connection with the verification of the dispersion relations in quantum field theory (address given at the International Conference in Seattle, Washington, in September, 1956). A complete proof of this theorem was published in the monograph by Bogolyubov, Medvedev, and Polivanov [22] (supplement A, Theorem 1). Other proofs and generalizations of this theorem are contained in the works by Jost and Lehmann [33], Bremermann, Oehme, and Taylor [27, 34], Dyson [35], Epstein [36], and Browder [112] (see also the survey [37]). The proof of the "edge of the wedge" theorem that we shall give is based on ideas of Bogolyubov, Dyson, and Epstein.

We note that the "edge of the wedge" theorem has proven an extremely useful tool not only in quantum field theory but also in other branches of mathematics and mathematical physics.

1. *Statement of the "edge of the wedge" theorem*

*Suppose that a function $f(z)$ is holomorphic in an open set $T^C_R = [z: |z| < R,\ y \in C]$, where C is an open cone such that $C \cap (-C) \neq \emptyset$. Suppose also that an open set $G \subset R^n$ is contained in the sphere $|x| < R$. Let us suppose that the limit**

*Thus defining a generalized function $f \in D^*(G)$ (see section 3.2).

$$\lim_{y \to 0,\, y \in C} \int f(x+iy)\,\varphi(x)\,dx = (f, \varphi), \tag{59}$$

exists for arbitrary $\phi \in D(G)$ *and is independent of the sequence* $y \to 0$, *where* $y \in C$. *Then, the function* $f(z)$ *is holomorphic (and single-valued) in the domain* $T_R^C \cup \widetilde{G}$, *where* $\widetilde{G}$ *is a (complex) neighborhood of the set G of the form*

$$\lim_{y \to 0,\, y \in C} \int f(x+iy)\,\varphi(x)\,dx = (f, \varphi),$$

where the positive number $\theta < 1$ *is independent of* x^0, G, R, *and* f (see Figs. 65 and 66).

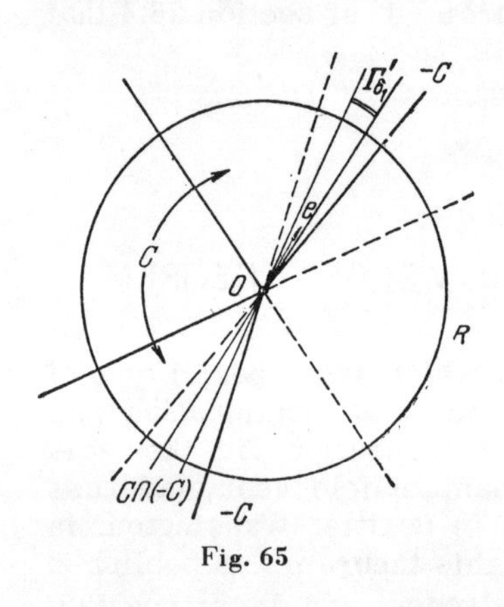

Fig. 65

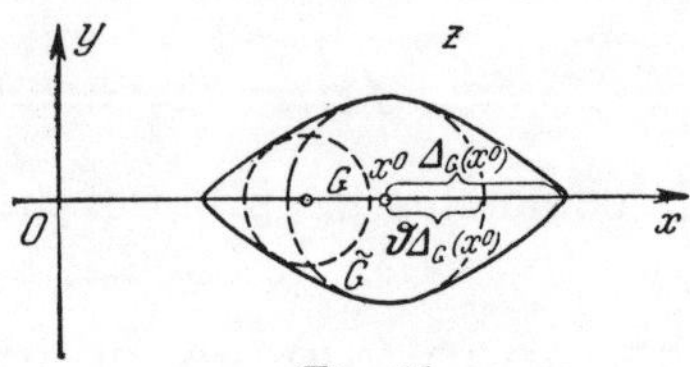

Fig. 66

Remark: If $C \cap (-C) = \varnothing$ in the statement of the theorem, then $f(z)$ is holomorphic only on the portion of the neighborhood $\widetilde{G}$ lying in the tubular domain $T^{O(C)}$ that is, in the domain $T_R^C \cup (\widetilde{G} \cap T^{O(C)})$ (see Epstein [36]).

The "edge of the wedge" theorem is a consequence of the following

LEMMA (special case of the "edge of the wedge" theorem). *Suppose that a function* $f(z)$ *is holomorphic in an open set* $T^{\Gamma_\delta} = [z : |z| < \eta,\ y \in \Gamma_\delta]$, *where* Γ_δ *is a circular cone* $\delta^2 y_1^2 > |y|^2$ *for* $\delta > 1$. *Suppose also that, for an arbitrary function* $\phi \in D[U(0, \eta)]$, *the limit* (59) *exists and is independent of the sequence* $y \to 0$ *for* $y \in \Gamma_\delta$. *Then,* $f(z)$ *is holomorphic in the hypersphere* $|z| < \theta_1 \eta$, *where* $\theta_1 = \theta_1(\delta)$.

Proof: Since $C \cap (-C) \neq \varnothing$, there exists a circular cone (see Fig. 65)

$$\Gamma'_{\delta_1} = [y: \delta_1^2 (ey)^2 > |y|^2 - (ey)^2], \quad |e| = 1, \quad \delta_1 > 0,$$

contained in the cone $C \cap (-C)$. Let us rotate the coordinate axes in R^n in such a way that the vector e will coincide with the vector $(1, \tilde{0})$, $y'' = By$, and $B^* = B^{-1}$. Let us contract the y_1'' axis by taking $\delta = 1 + 1/\delta_1$, $y_1' = 1/\delta\, y_1''$, $\tilde{y}' = \tilde{y}''$. We denote this transformation by C_δ: $y' = C_\delta y''$, $C_\delta^{-1} = C_{1/\delta}$. The product of the last two transformations $y' = C_\delta B y$ maps the cone Γ'_{δ_1} onto the cone $\Gamma_{\delta'}$, where $\delta' = 1 + \delta_1$.

Suppose now that $x^0 \in G$. Let us set $\Delta_G(x^0) = \eta$. Then, $U(x^0, \eta) \subset G \subset U(0, R)$, and, consequently,

$$T^C_\eta(x^0) = [z: |z - x^0| < \eta, \ y \in \Gamma'_{\delta_1}] \subset T^C_R.$$

Thus, the real nonsingular transformation

$$z' = C_\delta B(z - x^0), \quad z = B^* C_{1/\delta} z' + x^0 \tag{60}$$

maps the set $T^C_\eta(x^0)$ onto the set

$$[z': \delta^2 |z'_1|^2 + |\tilde{z}'|^2 < \eta^2, \ y' \in \Gamma_{\delta'}],$$

which, since $\delta > 1$, contains the set $T^{\Gamma_{\delta'}}_{\frac{\eta}{\delta}}$. The transformation (60) maps the function $f(z)$ into the function

$$f_1(z') = f\left(B^* C_{\frac{1}{\delta}} z' + x^0\right),$$

which is holomorphic in $T^{\Gamma_{\delta'}}_{\eta/\delta}$ and satisfies the remaining conditions of this special case of the "edge of the wedge" theorem. Consequently, in accordance with this theorem, $f_1(z')$ is holomorphic in the hypersphere $|z'| < \vartheta_1 \frac{\eta}{\delta} = \vartheta\eta$. Returning to the original variables by using (60), we conclude that the function $f(z) = f_1[C_\delta B(z - x^0)]$ is holomorphic in the domain

$$|C_\delta B(z - x^0)| < \vartheta\eta,$$

which contains the hypersphere $U(x^0, \vartheta\eta)$ since (by virtue of the inequality $\delta > 1$)

$$|C_\delta B(z - x^0)| \leqslant |B(z - x^0)| = |z - x^0|.$$

Continuing in this manner for arbitrary $x^0 \in G$ and remembering that $\eta = \Delta_G(x^0)$, we conclude on the basis of the holomorphic continuation theorem (see section 6.1) that $f(z)$ is holomorphic in the domain $T^C_R \cup \tilde{G}$. Here, we take $\vartheta = \frac{\vartheta_1}{\delta}$. This completes the proof.

Thus, we only need to prove the special case of the "edge of the wedge" theorem. We shall do this in two stages.

2. *The case of continuous boundary values*

Suppose that a function $f(z)$ is holomorphic in T^Γ_η and continuous in $\overline{T}^\Gamma_\eta$, where $\Gamma = \Gamma_1$ is the light cone $y_1^2 > |\tilde{y}|^2$. Then, $f(z)$ is holomorphic in the hypersphere $|z| \leq \frac{\eta}{8}$ and

$$\max_{|z| \leqslant \frac{\eta}{8}} |f(z)| \leqslant \max_{z \in \overline{T}^{\Gamma}_{\eta}} |f(z)|. \tag{61}$$

For $n = 1$, this assertion follows easily from Cauchy's theorem, and it was in fact proven in section 23.2. Therefore, we shall assume that $n \geqslant 2$.

Suppose that $\frac{\eta}{2\sqrt{2}} \leq \eta' < \eta^0 < \frac{\eta}{1+\sqrt{2}}$ and $|x| \leq \frac{\eta'}{2\sqrt{2}}$. Let us show that the function

$$\varphi(x; \zeta) \equiv f[b\zeta + x(1 - \zeta^2)], \quad \zeta = u + iv = \rho e^{i\theta},$$

where $b = (\eta^0, 0, \ldots, 0)$ is holomorphic in the circle $|\zeta| < \frac{\eta^0}{\eta'} = \sigma > 1$.

As a preliminary to this, we note that

$$\begin{aligned} \operatorname{Im}[b\zeta + z(1 - \zeta^2)] &= \\ &= (1 - \rho^2) y + \rho \sin\theta [b - 2\rho(x\cos\theta - y\sin\theta)] \end{aligned} \tag{62}$$

and we write the inclusion

$$D = \left[w\colon\ w = b\zeta + x(1 - \zeta^2),\ |\zeta| < \sigma,\ v \neq 0,\ |x| \leqslant \frac{\eta'}{2\sqrt{2}} \right] \subset T^{\Gamma}_{\eta}. \tag{63}$$

This is true because $v = \rho\sin\theta \neq 0$ and, hence, from (62) we have

$$\begin{aligned} (\operatorname{Im} w_1)^2 - |\operatorname{Im} \tilde{w}|^2 &= \rho^2 \sin^2\theta [(\eta^0 - 2\rho x_1 \cos\theta)^2 - 4\rho^2 |\tilde{x}|^2 \cos^2\theta] = \\ &= \rho^2 \sin^2\theta (\eta^0 - 2\rho x_1 \cos\theta - 2\rho |\tilde{x}| \cos\theta)(\eta^0 - 2\rho x_1 \cos\theta + \\ &+ 2\rho |\tilde{x}| \cos\theta) \geqslant \rho^2 \sin^2\theta (\eta^0 - 2\rho |x_1| - 2\rho |\tilde{x}|)^2 \geqslant \\ &\geqslant \rho^2 \sin^2\theta \left(\eta^0 - 2\sqrt{2}\rho |x|\right)^2 \geqslant \rho^2 \sin^2\theta (\eta^0 - \rho\eta')^2 > 0. \end{aligned} \tag{64}$$

Furthermore, for $|z| \leqslant \eta' / 2\sqrt{2}$ and $|\zeta| < \sigma$,

$$\begin{aligned} |b\zeta + z(1 - \zeta^2)| \leqslant |b||\zeta| + |z||1 - \zeta^2| < \eta^0\sigma + \\ + \frac{\eta'}{2\sqrt{2}}(1 + \sigma^2) \leqslant \eta, \end{aligned} \tag{65}$$

if we take σ sufficiently close to 1. The inclusion (63) follows from inequalities (64) and (65). Furthermore, these inequalities imply that $\overline{D} \subset \overline{T}^{\Gamma}_{\eta}$. Therefore, for every $|x| \leqslant \frac{\eta'}{2\sqrt{2}}$, the holomorphic mapping $w = b\zeta + x(1 - \zeta^2)$ maps the set $|\zeta| < \sigma, v \neq 0$ into T^{Γ}_{η} and it maps the circle $|\zeta| < \sigma$ into $\overline{T}^{\Gamma}_{\eta}$. Consequently, the function $\varphi(x; \zeta)$ is holomorphic for $|\zeta| < \sigma$, $v \neq 0$ and is continuous in the circle $|\zeta| < \sigma$ (see section 4.6). By applying the theorem proven for $n = 1$, we conclude from this that the function $\varphi(x; \zeta)$ is holomorphic in the circle $|\zeta| < \sigma > 1$.

If we apply Cauchy's formula for the circle $|\zeta| \leqslant 1$ to the function $\varphi(x; \zeta)$ and set $\zeta = 0$, we obtain an integral representation for the function $f(x)$ when $|x| \leqslant \frac{\eta'}{2\sqrt{2}}$:

$$f(x)=\varphi(x;\,0)=$$
$$=\frac{1}{2\pi i}\int\limits_{|\zeta|=1}\frac{\varphi(x;\,\zeta)}{\zeta}\,d\zeta=\frac{1}{2\pi}\int\limits_{0}^{2\pi}f\,[be^{i\theta}+x(1-e^{2i\theta})]\,d\theta. \tag{66}$$

Let us now show that the function

$$F(z)=\frac{1}{2\pi}\int\limits_{0}^{2\pi}f\,[be^{i\theta}+z(1-e^{2i\theta})]\,d\theta \tag{67}$$

is holomorphic in the hypersphere $|z|\leqslant\frac{\eta}{2\sqrt{2}}$.

From (62), with $\rho=1$, we have

$$\operatorname{Im}[be^{i\theta}+z(1-e^{2i\theta})]=\sin\theta\,[b-2(x\cos\theta-y\sin\theta)]. \tag{68}$$

Furthermore,

$$\begin{aligned}&[\eta^0-2(x_1\cos\theta-y_1\sin\theta)]^2-4\,|\tilde{x}\cos\theta-\tilde{y}\sin\theta\,|^2\geqslant\\&\quad\geqslant[\eta^0-2\,|\cos\theta\,|\,(|\,x_1|+|\,\tilde{x}\,|)-2\,|\sin\theta\,|\,(|\,y_1|+|\,\tilde{y}\,|)]^2\geqslant\\&\geqslant\left(\eta^0-2\sqrt{2}\,|\cos\theta\,|\,|\,x\,|-2\sqrt{2}\,|\sin\theta\,|\,|\,y\,|\right)^2\geqslant\\&\qquad\qquad\geqslant\left(\eta^0-2\sqrt{2}\,|\,z\,|\right)^2\geqslant(\eta^0-\eta')^2>0.\end{aligned} \tag{69}$$

From inequalities (65) and (69) and from Eq. (68), we have the inclusions

$$D_\varepsilon=\Big[w\colon\ w=be^{i\theta}+z(1-e^{2i\theta}),\ |\,z\,|\leqslant\frac{\eta'}{2\sqrt{2}},\ \varepsilon\leqslant\theta\leqslant\pi-\varepsilon,$$
$$\pi+\varepsilon\leqslant\theta\leqslant 2\pi-\varepsilon\Big]\Subset T^{\Gamma}_{\eta},\quad 0<\varepsilon<\frac{\pi}{2};\quad D_0\subset\overline{T}^{\Gamma}_{\eta}.$$

From this we conclude that the function

$$F_\varepsilon(z)=\frac{1}{2\pi}\left(\int\limits_{\varepsilon}^{\pi-\varepsilon}+\int\limits_{\pi+\varepsilon}^{2\pi-\varepsilon}\right)f\,[be^{i\theta}+z(1-e^{2i\theta})]\,d\theta,\ \ 0<\varepsilon<\frac{\pi}{2}$$

is holomorphic in the hypersphere $|z|\leqslant\frac{\eta'}{2\sqrt{2}}$ (see section 4.6). Since $|f(w)|\leqslant M$ for $w\in D_0\subset\overline{T}^{\Gamma}_{\eta}$, it follows on the basis of (67) that $F_\varepsilon(z)\to F(z)$ uniformly with respect to $|z|\leqslant\frac{\eta'}{2\sqrt{2}}$ as $\varepsilon\to 0$. Consequently, the function $F(z)$ is holomorphic in the hypersphere $|z|\leqslant\eta/8\ \eta'/2\sqrt{2}$ (see section 4.4).

For real values of $z=x$, where x belongs to the hypersphere $|x|\leqslant\eta/8$ this function $F(z)$ that we have constructed coincides with the function $f(x)$, which is the boundary value of $f(x+iy)$ as $y\in\Gamma\to 0$. Consequently, $f(x)$ is holomorphic for $|x|\leqslant\eta/8$. It remains to show that $F(z)$ coincides with $f(z)$ for $z\in T^{\Gamma}_{\frac{\eta}{8}}$, that is, that

$$\psi(z)=F(z)-f(z)=0\quad\text{for}\quad z\in T^{\Gamma}_{\frac{\eta}{8}}.$$

By construction, the function ψ is holomorphic in $T^{\Gamma}_{\frac{\eta}{8}}$ and continuous in $\overline{T}^{\Gamma}_{\frac{\eta}{8}}$, and it vanishes for real $z=x$ in the hypersphere $|x|\leqslant\frac{\eta}{8}$. Let $z=x+iy$ denote an arbitrary point in the set T^{Γ}_{η}. Then, $|z|<\eta/16$ and $y\in\Gamma$. Consider the function $g(\lambda)=\psi(x+\lambda y)$ of a complex variable λ. This function is holomorphic for $|\lambda|<\sqrt{2}$ when $\operatorname{Im}\lambda\neq 0$, it is continuous in the circle $|\lambda|<\sqrt{2}$, and it vanishes for real values of λ in that circle. If we apply the theorem just proven for $n=1$, we conclude that $g(\lambda)\equiv 0$ in the circle $|\lambda|<\sqrt{2}$; that is, $\psi(x+iy)=\psi(z)=0$ at an arbitrary point $z\in T^{\Gamma}_{\frac{\eta}{16}}$. On the basis of the holomorphic continuation theorem, $\psi(z)=0$ for $z\in T^{\Gamma}_{\frac{\eta}{8}}$.

Thus, the function $F(z)$ is indeed a holomorphic continuation of $f(z)$. Inequality (61) follows from the representation (67) and the inclusion $D_0\subset\overline{T}^{\Gamma}_{\eta}$. This completes the proof of the theorem.

3. *Extension to the case of generalized functions*

We shall now prove the particular case of the "edge of the wedge" theorem formulated in section 27.1. For every $y\in\Gamma_\delta\cup\{0\}$ such that $|y|<\sqrt{\frac{3}{8}}\eta$, we define a generalized function $T^y\in D^*\left[U\left(0,\sqrt{\frac{5}{8}}\eta\right)\right]$ by

$$(T^y,\varphi)=\begin{cases}\int f(x'+iy)\varphi(x')\,dx', & y\in\Gamma_\delta,\quad |y|<\sqrt{\frac{3}{8}}\eta;\\ (f,\varphi), & y=0,\end{cases}$$

where

$$(f,\varphi)=\lim_{y\to 0,\, y\in\Gamma_\delta}\int f(x'+iy)\varphi(x')\,dx'.$$

It follows from this definition that, for every $\varphi\in D\left[U\left(0,\sqrt{\frac{5}{8}}\eta\right)\right]$, the function (T^y,φ) is continuous with respect to y and hence bounded on every compact set contained in the set $\left[y\colon y\in\Gamma_\delta\cup\{0\},\ |y|<\sqrt{\frac{3}{8}}\eta\right]$. In particular,

$$|(T^y,\varphi)|\leqslant C_\varphi,\qquad y\in\overline{\Gamma},\qquad |y|\leqslant\frac{\eta}{\sqrt{8}}.$$

This inequality means that the set of generalized functions

$$M^*=\left[T=T^y,\quad y\in\overline{\Gamma},\quad |y|\leqslant\frac{\eta}{\sqrt{8}}\right]$$

is bounded in $D^*\left[U\left(0, \sqrt{\frac{5}{8}}\eta\right)\right]$ (see section 3.2).

Suppose that $\varphi(x) \in D\left[U\left(0, \frac{\eta}{\sqrt{8}}\right)\right]$. Denoting $\dot\varphi(x) = \varphi(-x)$, we introduce the function (see section 3.3)

$$G^{\varphi}(z) = (\dot\varphi * T^y)(x) = (T^y(x'), \varphi(x'-x)),$$

$$|x| \leqslant \frac{\eta}{\sqrt{8}}, \quad |y| \leqslant \frac{\eta}{\sqrt{8}}, \quad y \in \Gamma_\delta \cup \{0\}.$$

Let us show that the function $G^{\varphi}(z)$ is holomorphic in the hypersphere $|z| \leqslant \frac{\eta}{8\sqrt{8}}$ and satisfies the inequality

$$|G^{\varphi}(z)| \leqslant \max_{z \in \overline{T}^{\Gamma}_{\frac{\eta}{\sqrt{8}}}} |G^{\varphi}(z)|, \qquad |z| \leqslant \frac{\eta}{8\sqrt{8}}. \tag{70}$$

Since

$$G^{\varphi}(z) = \int f(z+x') \varphi(x') \, dx', \qquad y \in \Gamma_\delta, \tag{71}$$

$G^{\varphi}(z)$ is holomorphic in $T^{\Gamma_\delta}_{\frac{\eta}{\sqrt{8}}}$. Furthermore, since T^y is a continuous function of y for $|y| \leqslant \eta/\sqrt{8}$ and $y \in \overline{\Gamma}$, in the topology of the space $D^*\left[U\left(0, \sqrt{\frac{5}{8}}\eta\right)\right]$ and since $\varphi(t-x)$ is a continuous function of x for x, $|x| \leqslant \frac{\eta}{\sqrt{8}}$, in the topology of the space $D\left[U\left(0, \sqrt{\frac{5}{8}}\eta\right)\right]$, the function

$$G^{\varphi}(z) = (T^y(x'), \varphi(x'-x))$$

is continuous in $\overline{T}^{\Gamma}_{\frac{\eta}{\sqrt{8}}}$ (see section 3.2). From the theorem in section 27.2, the function $G^{\varphi}(z)$ is holomorphic in the hypersphere $|z| \leqslant \frac{\eta}{8\sqrt{8}}$ and satisfies inequality (70).

Now suppose that $\varphi(x') \to 0$ in $D\left[U\left(0, \frac{\eta}{\sqrt{8}}\right)\right]$. Then, $\varphi(x'-x) \to 0$ in $D\left[U\left(0, \sqrt{\frac{5}{8}}\eta\right)\right]$ uniformly with respect to x for $|x| \leqslant \frac{\eta}{\sqrt{8}}$. Since the set M^* is bounded in $D^*\left[U\left(0, \sqrt{\frac{5}{8}}\eta\right)\right]$, we have

$$G^{\varphi}(z) = (T^y(x'), \varphi(x'-x)) \to 0$$

uniformly with respect to $z \in \overline{T}^{\Gamma}_{\frac{\eta}{\sqrt{8}}}$. But then it follows from inequality (70) that $G^{\varphi}(z) \to 0$ uniformly with respect to z for $|z| \leqslant \frac{\eta}{8\sqrt{8}}$. This means that $G^{\varphi}(z)$ is, for all z such that $|z| \leqslant \frac{\eta}{8\sqrt{8}}$ a generalized function in $D^*\left[U\left(0, \frac{\eta}{\sqrt{8}}\right)\right]$.

We define

$$(T^y, \varphi) = G^{\varphi}(iy), \qquad |y| \leqslant \frac{\eta}{32}.$$

Now let us prove the representation

$$\left.\begin{aligned} G^{\varphi}(z) = (\dot{\varphi} * T^y)(x) = (T^y(x'), \varphi(x' - x)), \\ |x| \leqslant \frac{\eta}{32}, \quad |y| \leqslant \frac{\eta}{32}. \end{aligned}\right\} \tag{72}$$

For $y \in \Gamma_{\delta} \cup \{0\}$, this representation served as a definition of the function $G^{\varphi}(z)$. For $y \in \Gamma_{\delta}$, it coincides with formula (71). From this, we conclude that

$$G^{\varphi}(z) = G^{\psi}(z + x^0), \qquad \psi(x') = \varphi(x' + x^0) \tag{73}$$

for all $|x| \leqslant \frac{\eta}{32}$, $|x^0| \leqslant \frac{\eta}{32}$, $y \in \Gamma, |y| \leqslant \frac{\eta}{32}$. But then, on the basis of the holomorphic continuation theorem (see section 6.1), Eq. (73) is also valid for all $|x| \leqslant \frac{\eta}{32}$, $|x^0| \leqslant \frac{\eta}{32}$ and $|y| \leqslant \frac{\eta}{32}$. In particular, for $x = -x^0$, we have from (73)

$$G^{\varphi}(z) = G^{\psi}(iy) = (T^y, \psi) = (T^y(x'), \varphi(x' - x)),$$

which proves formula (72).

For every $\varphi \in D\left[U\left(0, \frac{\eta}{\sqrt{8}}\right)\right]$, the function $(T^y, \varphi) = G^{\varphi}(iy)$ is continuous with respect to y for $|y| \leqslant \frac{\eta}{32}$. Therefore, on the basis of Schwartz' kernel theorem (see section 3.12), T^y defines a generalized function

$$T(x, y) \in D^*\left[U\left(0, \frac{\eta}{\sqrt{8}}\right) \times U\left(0, \frac{\eta}{32}\right)\right].$$

Thus, the representation (72) can be rewritten in the form

$$G^{\varphi}(z) = (\dot{\varphi}\delta * T)(x, y), \quad \delta = \delta(y). \tag{74}$$

Suppose that a sequence of functions $\varphi_{\alpha}(x)$, for $\alpha = 1, 2, \ldots$, in $D\left[U\left(0, \frac{\eta}{\sqrt{8}}\right)\right]$ approaches $\delta(x)$ as $\alpha \to \infty$. Then,

$$\varphi_{\alpha}(x)\,\delta(y) \to \delta(x)\,\delta(y), \text{ as } \alpha \to \infty,$$

in

$$D^*\left[U\left(0, \frac{\eta}{\sqrt{8}}\right) \times U\left(0, \frac{\eta}{32}\right)\right].$$

In view of the continuity of the convolution operator (see section 3.3), we conclude from this and from (74) that

$$G^{\varphi_\alpha}(z) \to T(x, y) \text{ as } \alpha \to \infty,$$

in

$$D^*\left[U\left(0, \frac{\eta}{\sqrt{8}}\right) \times U\left(0, \frac{\eta}{32}\right)\right].$$

Since the functions $G^{\varphi_\alpha}(z)$ are holomorphic in the hypersphere $|z| < \frac{\eta}{32}$ it follows on the basis of the theorem in section 4.4 that $T(x, y)$ is holomorphic in that hypersphere.

Furthermore, it follows from (71) that

$$G^{\varphi_\alpha}(z) \to f(z) \quad \text{as} \quad \alpha \to \infty \quad \text{in} \quad D^*\left(T^{\Gamma}_{\frac{\eta}{32}}\right).$$

From this, we conclude by virtue of the uniqueness of the limit that the function $T(z) = T(x, y)$ that we have constructed coincides with $f(z)$ in $T^{\Gamma}_{\frac{\eta}{32}}$ and we conclude on the basis of the holomorphic continuation theorem that it coincides with $f(z)$ in $T^{\Gamma_\delta}_{\eta}$. Thus by taking $\vartheta_1 = 1/32$ we complete the proof of the particular case of the "edge of the wedge" theorem.

This in turn completes the proof of the "edge of the wedge" theorem itself.

4. *Strengthening of the "edge of the wedge" theorem*

The "edge of the wedge" theorem that we have just proven is of a local character. Therefore, to determine the dimensions of the neighborhood $\tilde{G}$, we have made no effort to obtain the best constant ϑ. The next step will be to enlarge $\tilde{G}$ globally. For simplicity, we shall take $R = \infty$ in the statement of the theorem and we shall assume that $C = \Gamma = [y: \; y_1^2 > |\tilde{y}|^2]$ is a light cone. (It was shown in section 27.1 that the case of an arbitrary cone C, where $C \cap (-C) \neq \varnothing$ can be reduced to a light cone at the price of poorer approximations.) We recall that a cone Γ consists of two components, namely, a future cone $\Gamma^+ = [y: \; y_1 > |\tilde{y}|]$ and a past cone $\Gamma^- = [y: \; y_1 < -|\tilde{y}|]$.

Thus, the function $f(z)$ is holomorphic in the tubular cone T^{Γ} and the limit (59) exists for arbitrary $\varphi \in D(G)$. According to the "edge of the wedge" theorem (see section 27.1), $f(z)$ is holomorphic in the domain $T^{\Gamma} \cup \tilde{G}$, where

$$\tilde{G} = \bigcup_{x^0 \in G} [z: \; |z - x^0| < \vartheta \Delta_G(x^0)], \quad 0 < \vartheta \leqslant \frac{1}{32}.$$

The question arises of constructing an envelope of holomorphy $H(T^{\Gamma} \cup \tilde{G})$ of the domain $T^{\Gamma} \cup \tilde{G}$. This problem has been solved only for a particular kind of domain G. Examples will be given below.

Let us suppose that $H(T^{\Gamma} \cup \tilde{G})$ is single-sheeted. Then, by using the "disk" theorem (see section 17.2), we can construct the holomorphic extension of the domain $T^{\Gamma} \cup \tilde{G}$. Consider a point $x^0 \in G$. Then, $U(x^0, \eta) \Subset G$ for arbitrary $\eta < \Delta_G(x^0)$. We define $b = (\eta, \tilde{0})$ and $a = (0, \tilde{a})$, where the complex vector $\tilde{a}$ is such that $|\tilde{a}|^2 = \eta^2$. Let us show that the circle $|\zeta| < 1$, which lies on the two-dimensional analytic surface*

$$F_\lambda = [z: \quad z = x^0 + b\zeta + \lambda a (1 - \zeta^2)],$$

is, for sufficiently small λ such that $0 \leqslant \lambda \leqslant \sigma = \sigma(\eta)$, strictly contained in the domain $T^{\Gamma} \cup \tilde{G}$; that is,

$$F_\lambda [|\zeta| \leqslant 1] \Subset T^{\Gamma} \cup \tilde{G}, \quad |\lambda| \leqslant \sigma. \tag{75}$$

It is obviously sufficient to prove the inclusion (75) for $\lambda = 0$. In this case, if $\operatorname{Im} \zeta \neq 0$, we have $\operatorname{Im} z = b \operatorname{Im} \zeta \in \Gamma$. On the other hand, if $\operatorname{Im} \zeta = 0$, then

$$[z: \quad z = x^0 + b\xi, \ -1 \leqslant \xi \leqslant 1] \Subset G.$$

Therefore, the set $F_0[|\zeta| \leqslant 1]$ is contained in $T^{\Gamma} \cup \tilde{G}$. Since this set is closed and bounded, it is strictly contained in $T^{\Gamma} \cup \tilde{G}$.

Let us now show that

$$F_\lambda [|\zeta| = 1] \Subset T^{\Gamma} \cup \tilde{G}, \ 0 \leqslant \lambda \leqslant \frac{1}{2}. \tag{76}$$

If we set $\zeta = e^{i\theta}$, we have, on the basis of (68),

$$\operatorname{Im} z = \operatorname{Im} [x^0 + be^{i\theta} + \lambda a (1 - e^{2i\theta})] =$$
$$= \sin\theta \, [b - 2\lambda (\cos\theta \operatorname{Re} a - \sin\theta \operatorname{Im} a)].$$

From this, we have, for $\sin\theta = 0$ and $0 \leqslant \lambda \leqslant \frac{1}{2}$

$$(\operatorname{Im} z_1)^2 - |\operatorname{Im} \tilde{z}|^2 =$$
$$= \sin^2\theta \, [\eta^2 - 4\lambda^2 |\cos\theta \operatorname{Re} \tilde{a} - \sin\theta \operatorname{Im} \tilde{a}|^2] > 0$$

and, consequently,

$$F_\lambda [|\zeta| = 1, \quad \operatorname{Im} \zeta \neq 0] \subset T^{\Gamma}.$$

---

*By virtue of the choice of vectors a and b, we have on F_λ

$$\frac{\partial z}{\partial \zeta} = b - 2\lambda a \zeta \neq 0.$$

On the other hand, if $\operatorname{Im}\zeta = 0$, then, obviously,

$$F_\lambda[\zeta = \pm 1] \Subset G.$$

Thus, the inclusion (76) is proven.

By virtue of the "disk" theorem (see section 17.2), we conclude from the inclusions (76) and (75) that, for arbitrary $x^0 \in G$ and $\eta < \Delta_G(x^0)$,

$$\left[z:\quad z = x^0 + b\zeta + \lambda a(1 - \zeta^2),\quad |\zeta| \leqslant 1,\ |\lambda| \leqslant \frac{1}{2}\right] \Subset H(T^\Gamma \cup \tilde{G}).$$

If we sum this inclusion over all $x^0 \in G$ and $\eta < \Delta_G(x^0)$, we obtain the inclusion

$$\left[z:\quad z_1 = x_1^0 + \eta\zeta,\ \tilde{z} = \tilde{x}^0 + \tilde{a}(1 - \zeta^2),\ |\zeta| \leqslant 1,\quad 0 \leqslant \eta < \Delta_G(x^0),\ |\tilde{a}| \leqslant \frac{\eta}{2},\ x^0 \in G\right] \subset H(T^\Gamma \cup \tilde{G}).$$

In particular, if we set $\zeta = 0$, we obtain

$$\left[z:\quad z_1 = x_1^0,\ \tilde{z} = \tilde{x}^0 + \tilde{a},\ |\tilde{a}| < \frac{1}{2}\Delta_G(x^0),\quad x^0 \in G\right] \subset H(T^\Gamma \cup \tilde{G}).$$

The left members of these inclusions do not exhaust $H(T^\Gamma \cup \tilde{G})$. However, they are considerably more extensive than the domain $T^\Gamma \cup \tilde{G}$, that is, they are a holomorphic extension.

5. *An example of the construction of an envelope of holomorphy*

Let G be a strip $|x_1| < m$. In this case, Bogolyubov has shown (see [22], p. 167) that every function belonging to the class $H_0(\Gamma)$ (see section 26.4) that is holomorphic in $T^\Gamma \cup \tilde{G}$ is holomorphic in the domain $|\tilde{y}| < \left|\operatorname{Im}\sqrt{z_1^2 - m^2}\right|$. It turns out that this domain coincides with the envelope of holomorphy; that is,

$$H(T^\Gamma \cup \tilde{G}) = \left[z:\quad |\tilde{y}| < \left|\operatorname{Im}\sqrt{z_1^2 - m^2}\right|\right]. \tag{77}$$

Let us prove this formula, following Bremermann, Oehme, and Taylor [27].

Since $\Delta_G(x^0) = m - |x_1^0|$, we have

$$\tilde{G} = \bigcup_{|x_1^0| < m} \left[z:\quad (x_1 - x_1^0)^2 + y_1^2 + |\tilde{y}|^2 < \vartheta^2(m - |x_1^0|)^2,\ \tilde{x}\ \text{ arbitrary}\right].$$

From this it follows that the domain $T^{\Gamma} \cup \tilde{G}$ can be represented in the form

$$T^{\Gamma} \cup \tilde{G} = \\ = [z: z_1 \in B, \ |\tilde{y}| < V(z_1), \\ |\tilde{x}| < +\infty],$$

where B is the complex z_1-plane deleted by the cuts $-\infty < x_1 \leqslant -m$ and $m \leqslant x_1 < \infty$, $y_1 = 0$ (see Fig. 67) and

$$V(z_1) = \max \Big\{ |y_1|, \sup_{|x_1^0| < m} \sqrt{\vartheta^2 (m - |x_1^0|)^2 - (x_1 - x_1^0)^2 - y_1^2} \times \\ \times \theta [\vartheta^2 (m - |x_1^0|)^2 - (x_1 - x_1^0)^2 - y_1^2] \Big\},$$

where $0 < \vartheta \leqslant 1/32$ and $\theta(\xi)$ is a step function whose value is 0 for $\xi < 0$ and 1 for $\xi > 0$.

The domain $T^{\Gamma} \cup \tilde{G}$ is invariant with respect to the group of real rotations and translations of the variables $\tilde{z} = (z_2, z_3, \ldots, z_n)$. According to the Cartan–Thullen theorem (see section 20.5), the envelope of holomorphy $H(T^{\Gamma} \cup \tilde{G})$ possesses the same property. Therefore, to construct $H(T^{\Gamma} \cup \tilde{G})$, it will be sufficient to construct the envelope of holomorphy $H(S)$ of the semitubular domain

B z_1
$-m$ 0 m x_1

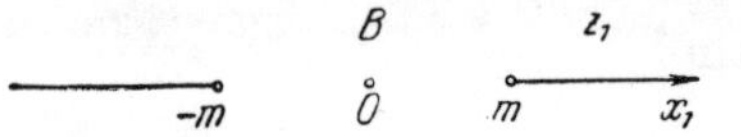

Fig. 67

$$S = [z = (z_1, z_2): \ |y_2| < V(z_1), \ z_1 \in B, \ |x_2| < \infty]$$

in C^2. To obtain the desired domain $H(T^{\Gamma} \cup \tilde{G})$, we now need only construct in $H(S)$ a domain on the base C^n that is invariant with respect to the group of real rotations and translations of the variables $\tilde{z}$.

According to the theorem in section 21.5 the envelope of holomorphy of the semitubular domain S is single-sheeted and is of the form

$$H(S) = [(z_1, z_2): \ |y_2| < V_0(z_1), \ z_1 \in B, \ |x_2| < \infty],$$

where $V_0(z_1)$ is the least superharmonic majorant of the function $V(z)$ in the domain B.

Let us show that

$$V_0(z_1) = \left| \operatorname{Im} \sqrt{z_1^2 - m^2} \right|. \tag{78}$$

The function $\sqrt{z_1^2 - m^2}$ is holomorphic in B and the function $\operatorname{Im} \sqrt{z_1^2 - m^2}$ retains its sign in B. Therefore, the function $\left| \operatorname{Im} \sqrt{z_1^2 - m^2} \right|$ is harmonic and, consequently, superharmonic in B.

Let us show that this function is a superharmonic majorant of the function $V(z_1)$ in the domain B; that is, let us prove the inequality

$$\left|\operatorname{Im}\sqrt{z_1^2-m^2}\right| \geqslant V(z_1), \quad z_1 \in B. \tag{79}$$

The inequality

$$\left|\operatorname{Im}\sqrt{z_1^2-m^2}\right| \geqslant |y_1|, \quad z \in B$$

is verified immediately since

$$\left|\operatorname{Im}\sqrt{z_1^2-m^2}\right|^2 = \\ = \frac{1}{2}\sqrt{(x_1^2+y_1^2-m^2)^2+4m^2y_1^2} - \frac{1}{2}(x_1^2-y_1^2-m^2) \geqslant y_1^2.$$

To conclude the proof of inequality (79), we need only show that

$$\left|\operatorname{Im}\sqrt{z_1^2-m^2}\right| \geqslant \sqrt{\vartheta^2(m-|x_1^0|)^2-(x_1-x_1^0)^2-y_1^2} \tag{80}$$

in the circle

$$(x_1-x_1^0)^2+y_1^2 \leqslant \vartheta^2(m-|x_1^0|)^2 \tag{81}$$

for all $|x_1^0| \leqslant m$. Inequality (80) is equivalent to the inequality

$$\sqrt{(x_1^2+y_1^2-m^2)^2+4m^2y_1^2} \geqslant \\ \geqslant x_1^2-3y_1^2-m^2-2(x_1-x_1^0)^2+2\vartheta^2(m-|x_1^0|)^2 \tag{82}$$

in the circle (81). But this circle lies in the strip $|x_1| \leqslant m$, Therefore, to prove inequality (82), we need only prove that the right member of this inequality is nonpositive for $|x_1^0| \leqslant m$ and $|x_1| \leqslant m$, that is, that

$$x_1^2-3y_1^2-m^2-2(x_1-x_1^0)^2+2\vartheta^2(m-|x_1^0|)^2 \leqslant \\ \leqslant -m^2-x_1^{0^2}+2x_1x_1^0+2\vartheta^2(m-|x_1^0|)^2 \leqslant \\ \leqslant (2\vartheta^2-1)(m-|x_1^0|)^2 < 0.$$

This completes the proof of inequality (79).

Now, let us prove the inequality

$$\left|\operatorname{Im}\sqrt{z_1^2-m^2}\right| - V(z_1) \leqslant \frac{m^2|y_1|}{|z_1|^2}, \quad |z_1| \geqslant \frac{m}{2}(1+\sqrt{5}). \tag{83}$$

First, $V(z_1)=|y_1|$ for $|z_1| \geqslant m$. Therefore, inequality (83) is equivalent to the inequality

$$\sqrt{(|z_1|^2-m^2)^2+4m^2y_1^2} \leqslant x_1^2-y_1^2-m^2+2\left(|y_1|+\frac{m^2|y_1|}{|z_1|^2}\right)^2,$$

Simplifying this inequality, we have

$$|z_1|^4 \leqslant (|z_1|^2 - m^2)(2|z_1|^2 + m^2) + m^2 y_1^2 \left(2 + \frac{m^2}{|z_1|^2}\right)^2.$$

This last inequality is satisfied for $|z_1| \geqslant \frac{m}{2}(1+\sqrt{5})$.

Now, let us prove Eq. (78): that is, let us prove that the function $|\operatorname{Im}\sqrt{z_1^2 - m^2}|$ is the least superharmonic majorant of the function $V(z_1)$ in the domain B (see section 9.7). Here, we assume that a function $V_1(z_1)$, has been found that is superharmonic in B and that satisfies the inequalities

$$V(z_1) \leqslant V_1(z_1) \leqslant |\operatorname{Im}\sqrt{z_1^2 - m^2}|, \qquad z_1 \in B. \tag{84}$$

Let us show that

$$\rho(z_1) = |\operatorname{Im}\sqrt{z_1^2 - m^2}| - V_1(z_1) \equiv 0, \qquad z_1 \in B.$$

The function $\rho(z_1)$ is superharmonic in B and, on the basis of (83) and (84), it satisfies the inequalities

$$0 \leqslant \rho(z_1) \leqslant \frac{m^2|y_1|}{|z_1|^2}, \qquad |z_1| \geqslant 2m, \qquad y_1 \neq 0.$$

Furthermore, it follows from (84) that $\rho(z_1)$ is upper-semicontinuous up to the cuts $|x_1| \geqslant m$, $y_1 = 0$, where it assumes the value 0. Let us take an arbitrary number $R \geqslant 2m$. From what has been said, we conclude that the function $\rho(z_1)$ is subharmonic in $B \cap U(0, R)$ and upper-semicontinuous in $\bar{U}(0, R)$ and that it is bounded above on $\partial[B \cap U(0, R)]$by the number m^2R^{-1} (which $\geqslant m^2|y_1|R^{-2}$). Therefore, on the basis of the maximum principle (see section 9.6), we have

$$0 \leqslant \rho(z_1) \leqslant \frac{m^2}{R}, \qquad z_1 \in B.$$

If we let R approach$+\infty$ in this inequality, we see that $\rho(z_1) \equiv 0$. This completes the proof of Eq. (78).

In view of this equation, we have (cf. the example in section 21.6)

$$H(S) = [(z_1, z_2): |y_2| < |\operatorname{Im}\sqrt{z_1^2 - m^2}|, \ z_1 \in B, \ |x_2| < \infty]$$

from which the desired formula (77) follows.

Remark: (1) It follows from the above considerations that the solution of the Dirichlet problem $\Delta u = 0$ for a domain B with boundary conditions

$$u(x, 0) = 0, \qquad |x| \geqslant m;$$
$$0 \leqslant u(x, y) - |y| = o(1), \qquad |z| \to \infty$$

is unique and is given by the formula

$$u(x, y) = |\operatorname{Im} \sqrt{z^2 - m^2}|, \qquad z = x + iy.$$

(2) For $n = 4$ formula (77) enables us to prove the dispersion relations for a π-meson nucleonic dispersion (see section 33.4) in the interval $0 \leqslant -t < 8\mu^2$ of variation of the square of the impulse transfer-t, where μ is the mass of the π-meson (see Bogolyubov and Vladimirov [39] and Bremermann, Oehme, and Taylor [21].

28. A THEOREM ON C-CONVEX ENVELOPES

1. *Definition of a C-Convex Envelope*

Suppose that an open C consists of connected component cones $C_1, \ldots, C_t$. For the cone C, let us construct new cones C^0 and $\tilde{C}$ as follows: if the cones C_k, for $k = 1, 2, \ldots, t$, are not all convex, let us consider the cone $C^{(1)} = \bigcup_{1 \leqslant k \leqslant t} O(C_k)$ and let us denote its components by $C_1^{(1)}, \ldots, C_{t_1}^{(1)}$, for $t_1 \leqslant t$. If the cones $C_k^{(1)}$ are not all convex, we introduce the components $C_1^{(2)}, \ldots, C_{t_2}^{(2)}$, where $t_2 \leqslant t_1$, of the cone $C^{(2)} = \bigcup_{1 \leqslant k \leqslant t_1} O(C_k^{(1)})$, etc. (see Fig. 68). After a finite number of steps, we obtain an open cone C^0 all the components of which are convex cones. We define $\tilde{C} = C^0 \cap (-C^0)$ (see Fig. 69). Clearly, $\tilde{C} = -\tilde{C}$; If $C \cap (-C) \neq \varnothing$, then $\tilde{C} \neq \varnothing$. The cone $\tilde{C}$ consists of an even number $2l \leqslant t$ of convex components $\tilde{C}_k^{\pm}$, for $k = 1, 2, \ldots, l$, such that $\tilde{C}_k^- = -\tilde{C}_k^+$. We define $\tilde{C}_k = \tilde{C}_k^+ \cup \tilde{C}_k^-$ and $\tilde{C}^{\pm} = \bigcup_{1 \leqslant k \leqslant l} \tilde{C}_k^{\pm}$.

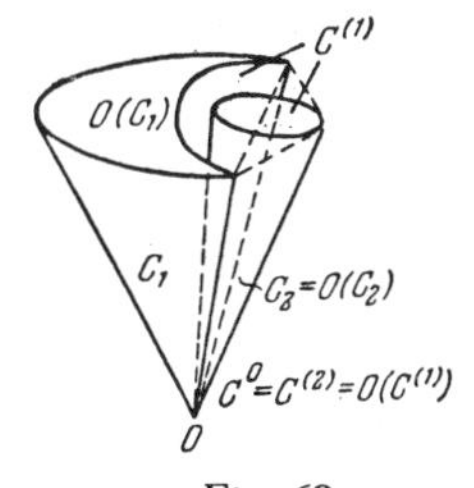

Fig. 68

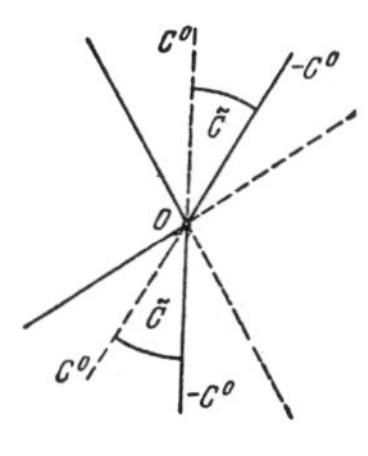

Fig. 69

We shall say that a class $C^{(1)}$ curve is C-similar if, at an arbitrary point x, the tangent vector lies in the cone $\tilde{C} + x$, where the cone $\tilde{C} + x$ is obtained by translation of the cone $\tilde{C}$ by the vector x (see Fig. 70). If $C = \Gamma$ is a light cone, we shall say that the C-similar curve is a time-similar curve.

Let S denote an open set lying in a k-dimensional class $C^{(1)}$ surface (where $1 \leqslant k \leqslant n$). (For $k = n$, this means that $S \subset R^n$.) We shall use the term C-convex envelope of a set S for the smallest open set $B_C(S) \subset R^n$ containing S and possessing the property that if points x' and x'' in $B_C(S)$ can be connected by a C-similar curve lying entirely in $B_C(S)$ then all C-similar curves connecting x' and x'', are contained in $B_C(S)$ (see Fig. 71); that is,

$$B_C(x', x'') = \bigcup_{1 \leqslant k \leqslant l} [\tilde{C}_k^+ + x'] \cap [\tilde{C}_k^- + x''] \subset B_C(S).$$

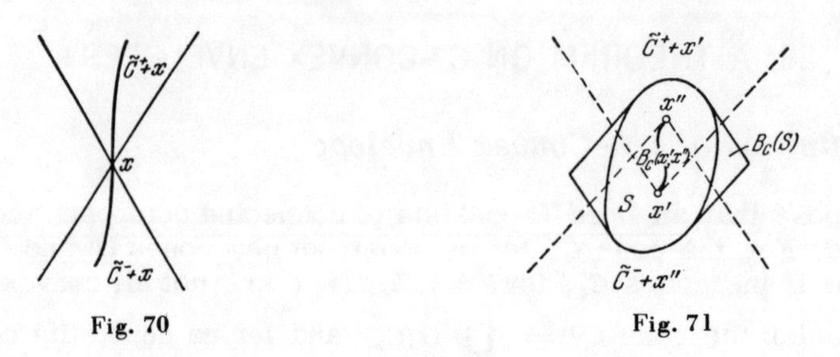

Fig. 70 Fig. 71

It follows from this definition that the envelope $B_C(S)$ can be represented as the sum of an increasing sequence of open sets

$$B_C^{(i)}(S) = \bigcup_{x', x''} B_C(x', x''), \quad B_C^0(S) = S, \quad i = 1, 2, \ldots, \tag{85}$$

where the union is taken over those pairs of points x', x'' in $B_C^{(i-1)}(S)$, which can be connected by a C-similar curve that lies entirely in $C_k = C_k^+$:

$$B_C(S) = \bigcup_{i \geqslant 0} B_C^{(i)}(S). \tag{86}$$

It follows from (85) and (86) that

$$B_C(S) \subset \bigcup_{x' \in S,\, x'' \in S} B_C(x', x'').$$

If

$$B_C(S) = \bigcup_{x' \in S,\, x'' \in S} B_C(x', x''), \tag{87}$$

we shall say that the set S is $\bar{C}$-regular (see Fig. 71). For $n = 2$, every connected set is C-regular. For $n \geqslant 3$, this is not always the case. For example, the spiral shown in Fig. 72 is not a C-regular region.

For $n=1$, there are two and only two open cones without common points, namely, $C^{\pm}=\Gamma^{\pm}=[\pm y>0]$. These constitute the cone $\Gamma=[y\neq 0]$. Therefore $B_{\Gamma}(S)=S$ for $n=1$.

If $C=R^n$, then, obviously, $B_C(S)=R^n$. If $\tilde{C}=\varnothing$, then $B_C(S)=S$. In all cases, either $B_C(S)=S$ or $B_C(S)$ is n-dimensional; $B_C[B_C(G)]=B_C(G)$.

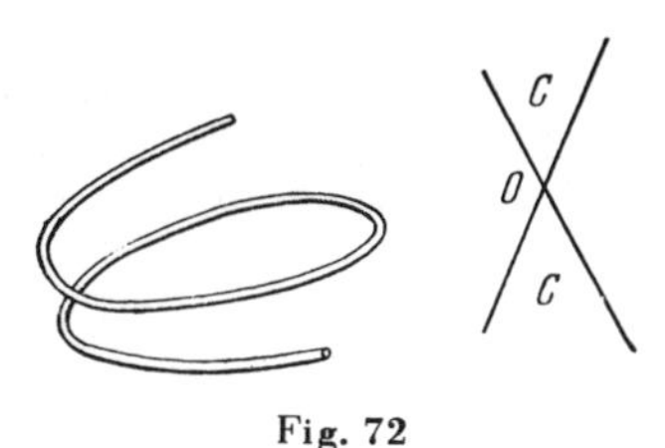

Fig. 72

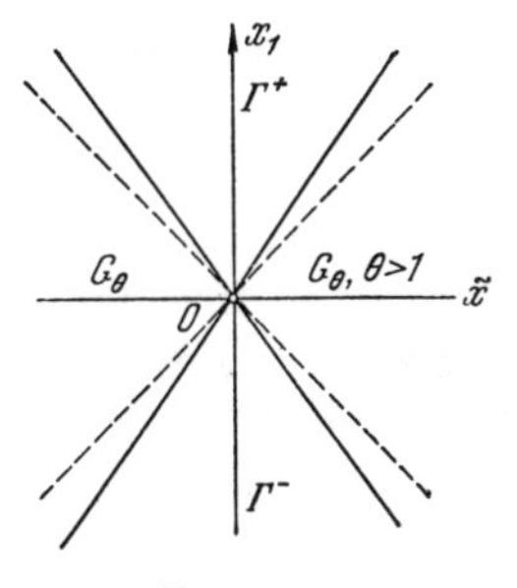

Fig. 73

Suppose that $C=\Gamma$ is a light cone. We shall give three examples of open Γ-regular sets $G\subset R^n$ for which the envelope $B_{\Gamma}(G)$ is essentially greater than G:

$$G_\theta=\left[x:\ x_1^2<\theta|\tilde{x}|^2\right];\quad B_{\Gamma}(G_\theta)=R^n,$$
$$\theta>1,\quad B_{\Gamma}(G_\theta)=G_\theta,\quad 0<\theta\leqslant 1 \text{ (see Fig. 73)};$$

$$G_l=\left[x:\ x_1^2<|\tilde{x}|^2+l\right];\quad B_{\Gamma}(G_l)=G_0,$$
$$l\leqslant 0,\quad B_{\Gamma}(G_l)=G_l,\quad l>0 \text{ (see Fig. 74)};$$

$$G_{a,b}=[x:\ |\tilde{x}|<\varepsilon,\quad a<x_1<b];$$
$$B_{\Gamma}(G_{a,b})=\left[x:\left|x_1-\frac{a+b}{2}\right|+|\tilde{x}|<\frac{b-a}{2}+\varepsilon,\ a<x_1<b\right]$$
$$\text{(see Fig. 75)};$$

in particular,

$$B_{\Gamma}(G_{-\infty,\infty})=R^n,\quad B_{\Gamma}(G_{-\infty,0})=[x:\ x_1+|\tilde{x}|<\varepsilon,\quad x_1<0].$$

LEMMA. *If S_α for $\alpha=1, 2, \ldots,$ is an increasing sequence of open sets lying on a k-dimensional surface of class $C^{(1)}$, the sequence $B_C(S_\alpha)$, for $\alpha=1, 2, \ldots,$ is increasing and*

$$B_C(S)=\bigcup_{\alpha}B_C(S_\alpha),\text{ where } S=\bigcup_{\alpha}S_\alpha.$$

On the basis of Eq. (86), it will be sufficient to prove this lemma for the iterations $B_C^{(i)}(S)$ and $B_C^{(i)}(S_\alpha)$, for $i=0, 1, \ldots,$ defined by formula (85). Since $B_C^{(0)}(S)=S$ and $B_C^{(0)}(S_\alpha)=S_\alpha$, the conclusion for $i=0$ follows immediately from this condition. Let us assume that this lemma is valid for $i-1$, that is, that

$$B_C^{(l-1)}(S_\alpha) \subset B_C^{(l-1)}(S_{\alpha+1}), \qquad B_C^{(l-1)}(S) = \bigcup_\alpha B_C^{(l-1)}(S_\alpha),$$

and let us prove it for l. Obviously,

$$B_C^{(l)}(S_\alpha) \subset B_C^{(l)}(S_{\alpha+1}) \subset B_C^{(l)}(S).$$

Consider any point $x^0 \in B_C^{(l)}(S)$. Then, it follows from (85) that $x^0 \in B_C(x', x'')$, where the points x' and x'' belong to $B_C^{(l-1)}(S)$, and that there exists a C-similar curve l connecting these points that lies

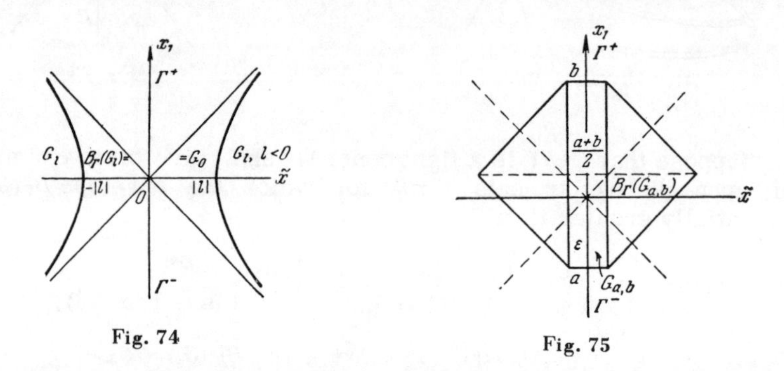

Fig. 74

Fig. 75

in $B_C^{(l-1)}(S)$. Since l is compact, the assumption that the lemma holds for $l - 1$ implies that $l \subset B_C^{(l-1)}(S_N)$ for some N. But then, on the basis of (85), $B_C(x', x'') \subset B_C^{(l)}(S_N)$ and, consequently, $x^0 \in B_C^{(l)}(S_N)$. This means that

$$B_C^{(l)}(S) = \bigcup_\alpha B_C^{(l)}(S_\alpha)$$

so that the lemma is valid for l, which completes the proof.

COROLLARY. *If the S_α are C-regular sets, S is also a C-regular set.*

Proof: By using the lemma just proven, we obtain from (87)

$$B_C(S) = \bigcup_\alpha B_C(S_\alpha) = \bigcup_\alpha \bigcup_{x' \in S_\alpha,\, x'' \in S_\alpha} B_C(x', x'') =$$
$$= \bigcup_{x' \in S,\, x'' \in S} B_C(x', x''), \text{ q.e.d.}$$

We shall say that an open set G is C-convex if $B_C(G) = G$. Clearly, *the interior of the intersection of C-convex sets is itself a C-convex set.*

Finally, we note that $B_C(S_1 \cap S_2) \subset B_C(S_1) \cap B_C(S_2)$.

2. *Simple connectedness of a class of domains*

Suppose that C is an open cone and that G is an open set in R^n. We define $C_R = C \cap U(0, R)$, $T^{C_R} = R^n + iC_R$ and

$$\hat{G} = \bigcup_{x^0 \in G} [z: \quad |z - x^0| < \varepsilon(x^0)],$$

where $\varepsilon(x^0)$ are positive numbers less than $\vartheta\Delta_G(x^0)$.

If a cone C consists of t convex components and G is a domain, then, $T^{C_R} \cup \hat{G}$ is a simply connected domain.

To prove this, we need to show that any two Jordan curves L_1 and L_2 connecting two points z' and z'' in $T^{C_R} \cup \hat{G}$ are homotopic in $T^{C_R} \cup \hat{G}$ (see section 1.5).

Let us suppose first that $t = 1$. Then, T^{C_R} is a convex tubular domain.

As a preliminary, let us prove that if the curves L_1 and L_2 lie in $\hat{G}$, they are homotopic in $T^{C_R} \cup \hat{G}$.

By deforming the curves L_1 and L_2 in the domain $\hat{G}$, we superimpose them on close-lying curves L_1' and L_2' such that terminal sections adjacent to the ends z' and z'', coincide and their middle portions lie in the domain T^{C_R} (see Fig. 76). Since T^{C_R} is convex and hence a simply connected domain, these middle portions are homotopic in T^{C_R}. Therefore, the curves L_1 and L_2 are homotopic in T^{C_R}. Consequently, the curves L_1 and L_2 are homotopic in $T^{C_R} \cup \hat{G}$.

Now, by using this result and the convexity of the domain T^{C_R}, we can easily show that the curves L_1 and L_2 are homotopic wherever they may be located.

Suppose that our assertion holds for $t-1$. Let us prove it for t. Thus, we assume that the domain $T^{C_R} \cup \hat{G}$ is the union of two simply connected domains $T^{C'_R} \cup \hat{G}$ and $T^{C''_R} \cup \hat{G}$, where the cone C' consists of $t-1$ convex components, the cone C'' is convex, and $C' \cap C'' = \varnothing$. We need to show that $T^{C_R} \cup \hat{G}$ is a simply connected domain.

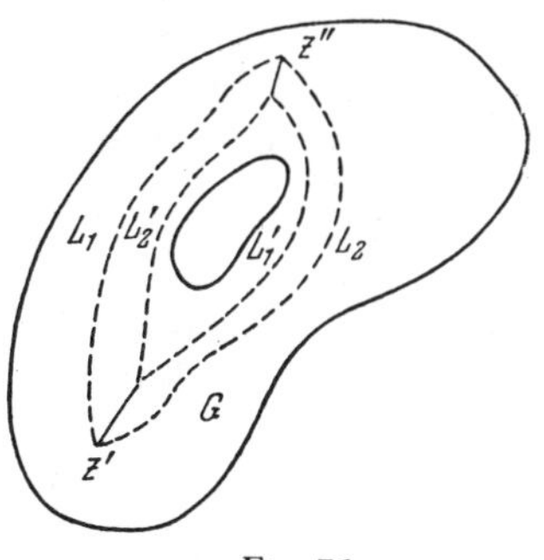

Fig. 76

Let us show first that any two curves L_1 and L_2 connecting two points $z' \in T^{C'_R}$ and $z'' \in T^{C''_R}$ and lying in $T^{C_R} \cup \hat{G}$ are homotopic.

As we move along the curve L_1 (from the point z' to the point z'') we must leave the domain $T^{C'_R} \cup \hat{G}$ and enter the convex domain $T^{C''_R}$. If we again enter the domain $\hat{G}$ as we move further along the curve L_1, then, from what we have proven for $t = 1$, that portion of the curve L_1 lying in $T^{C''_R}$, is homotopic to some curve lying in the domain $\hat{G}$.

Therefore, the curve L_1 can be moved into a curve L_1' that is homotopic to it and that consists of three segments lying respectively in $T^{C'R}$, $\hat{G}$, and $T^{C''R}$ (see Fig. 77). Analogously, for the curve L_2, we can construct a curve L_2' that is homotopic to it. Let z_1 and z_2 be any two curves L_1' and L_2' lying in the domain $\hat{G}$. Let us connect these points with a curve $l \subset \hat{G}$. By construction, that segment of the curve L_1' that lies between the points z' and z_1 is homotopic in $T^{C'R} \cup \hat{G}$ to the curve consisting of l and that segment of the curve L_2', lying between the points z' and z_2. The segment of the curve L_1' lying between z'' and z_1 is homotopic in $T^{C''R} \cup \hat{G}$ to the curve consisting of l and the segment L_2' lying between the points z'' and z_2. This means that the curves L_1' and L_2', and hence the curves L_1 and L_2 are homotopic in $T^{CR} \cup \hat{G}$.

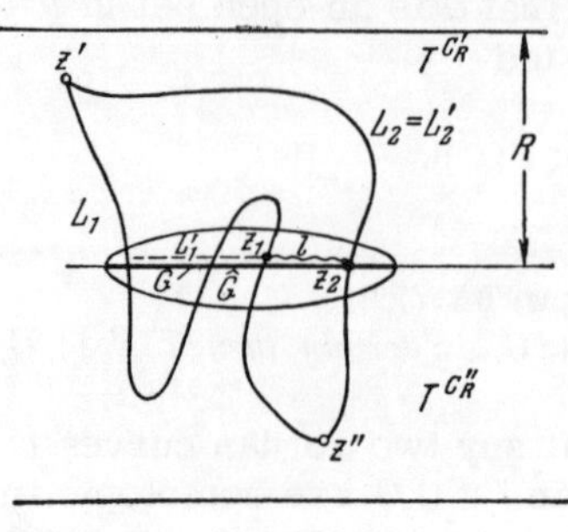

Fig. 77

The cases for other possible locations of the curves L_1 and L_2 in the domain $T^{CR} \cup \hat{G}$ are easily analyzed by the means of the result thus established and the use of induction. On this basis, the proof of our assertion is complete.

3. *The theorem on C-convex envelopes*

Suppose that an open cone C consists of a finite number of components, that $C \cap (-C) \neq \emptyset$ and that the function $f(z)$ is holomorphic in the domain $T^{CR} \cup \hat{G}$. Then, $f(z)$ is holomorphic (and single-valued) in the domain $T^{C^0 R} \cup \tilde{B}_C(G)$. In particular,

$$B_C(G) \subset \operatorname{Re} H\left(T^{CR} \cup \hat{G}\right), \tag{88}$$

where Re H denotes the intersection of H and the plane $y = 0$. The cone C^0 is defined at the end of section 28.1, and the neighborhoods $\hat{G}$ and $\tilde{B}_C$ are defined in section 27.1 and 28.2, respectively.

Remark: The theorem on C-convex envelopes was first proven in our article [92] (see also [30]). Here, it was assumed that the functions belong to the class $H_0(\Gamma)$, though, of course, this was not essential (see [40]). Afterwards, another proof of this theorem for a light cone was given by Borchers [98].

As a preliminary let us prove the following weaker assertion:

LEMMA. *Suppose that a cone C consists of two convex components C^+ and C^- such that $C^- = C^+$. Suppose that a function $f(z)$ is holomorphic in T^{C_R} and in a neighborhood of a C-similar curve connecting two points x' and x''. Then, f is holomorphic (and single-valued)in the domain $T^{C_R} \cup \tilde{B}_C(x', x'')$.*

Proof: On the basis of a theorem of Weierstrass, every C-similar curve $x_j = x_j(\xi)$, for $j = 1, 2, \ldots, n$, where $0 \leqslant \xi \leqslant 1$ (and where the $x_j(\xi)$ are continuously differentiable functions and $[x_1'(\xi), \ldots x_n'(\xi)] \in C$)

can be approximated to an arbitrary degree of accuracy by C-similar polynomial curves $x_j = P_j(\xi)$ for $j = 1, 2, \ldots, n$, where $0 \leqslant \xi \leqslant 1$. (By the degree of the polynomial curve, we mean the highest degree of the polynomials P_j.) The following two facts then follow: (1) there exists a C-similar polynomial curve L_0 (whose degree we note by N) connecting the points x' and x'' in a neighborhood of which $f(z)$ is holomorphic; (2) to prove the lemma, it will be sufficient to show that the function $f(z)$ is holomorphic at points of the set $B_C^{(m)}(x', x'')$, for $m = N, N+1, \ldots$, consisting of all C-similar polynomial curves of degree m that connect the points x' and x''.

Let us consider any polynomial curve of degree m connecting the points x' and x'' as a point in the $(mn-n)$-dimensional space of the coefficients. The set $B_C^{(m)}(x', x'')$ in this space constitutes some domain.

Let us suppose that the function $f(z)$ is not holomorphic at all points of the set $B_C^{(m)}(x', x'')$. Then, there exists a family of curves (see Fig. 78)

$$L_\alpha = [x: \quad x = P_\alpha(\xi), \ 0 \leqslant \xi \leqslant 1], \quad 0 \leqslant \alpha \leqslant 1,$$

in the set $B_C^{(m)}(x', x'')$ that depend continuously on the parameter α and having the property that at points of the curves L_α for $\alpha < 1$ our function $f(z)$ is holomorphic whereas the curve L_1, contains at least one singular point of that function.

Let us extend the vector polynomial $P_\alpha(\xi)$ to complex values $\lambda = \xi + i\eta$. For each $\xi^0 \in [0, 1]$, this polynomial can be represented in the form

$$P_\alpha(\lambda) = d_0(\alpha; \xi^0) + d_1(\alpha; \xi^0)(\lambda - \xi^0) + \ldots \\ \ldots + d_m(\alpha; \xi^0)(\lambda - \xi^0)^m. \tag{89}$$

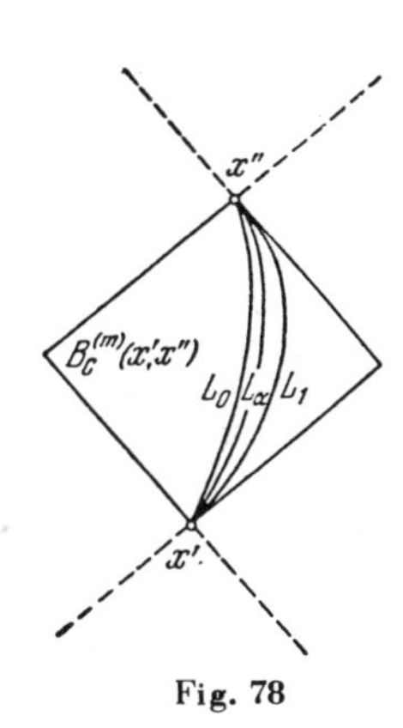

Fig. 78

Here, the coefficients of the vector-valued function $d_k(\alpha; \xi^0)$, for $k = 01, \ldots, m$, are real and continuous with respect to $\alpha \in [0, 1]$ (and are polynomials in ξ^0); also $d_1(\alpha; \xi^0) \in C$ for all $\alpha \in [0, 1]$ and $\xi^0 \in [0, 1]$. Therefore, $P'_\alpha(\lambda) \neq 0$ in a sufficiently small neighborhood of the real interval $0 \leqslant \xi \leqslant 1$. (Obviously, this neighborhood can be chosen independently of $\alpha \in [0, 1]$.)

Consider now the family of "disks":

$$D(\alpha) = [z: \quad z = P_\alpha(\lambda), \ 0 \leqslant \xi \leqslant 1, \ |\eta| \leqslant \sigma], \quad 0 \leqslant \alpha \leqslant 1,$$

where the number σ is yet to be chosen. Let us show that the function $f(z)$ is holomorphic at points of the "disks" $D(\alpha)$, where $0 \leqslant \alpha < 1$ and at points of the boundary $\partial D(1)$ of the limit "disk" $D(1)$.

If $z = P_\alpha(\lambda) \in D(\alpha)$ for $0 \leqslant \alpha \leqslant 1$, from (89) we have for any $\xi^0 \in [0, 1]$

$$\operatorname{Im} z = d_1(\alpha;\ \xi^0)\,\eta + d(\alpha;\ \xi^0,\ \lambda)\operatorname{Im}(\lambda - \xi^0)^2,$$

where, because of the continuity, the vector-valued function $d(\alpha;\ \xi^0,\ \lambda)$ is uniformly bounded for all $\alpha \in [0,\ 1]$, $\xi^0 \in [0,\ 1]$, and $|\lambda - \xi^0| \leqslant \sigma$. Therefore, since the vector-valued function $d_1(\alpha;\ \xi^0)$ is continuous and belongs to the open cone C for all $(\alpha;\ \xi^0) \in [0,\ 1] \times [0,\ 1]$, there exists a sufficiently small number σ, independent of ξ^0 and α, such that $\operatorname{Im} z \in C_R$ for all $z = P_\alpha(\lambda), 0 \leqslant \xi \leqslant 1\, |\eta| \leqslant \sigma, \eta \neq 0,\ 0 \leqslant \alpha \leqslant 1$. This means that $D(\alpha) \subset T^{C_R}$ for $0 \leqslant \alpha \leqslant 1$ if $\eta \neq 0$. If $\eta = 0$, then $D(\alpha) = L_\alpha$. Therefore for $\alpha < 1$, the function $f(z)$ is holomorphic at points of the "disks" $D(\alpha)$. On the other hand, if $\alpha = 1$, the function $f(z)$ is holomorphic at points of $\partial D(1)$ since $\partial D(1)$ reduces to two points x' and x'' when $\eta = 0$.

According to the "disk" theorem (see section 17.2), the function $f(z)$ is holomorphic at all points of the limit "disk" $D(1)$ and, in particular, at points of the curve L_1. But this contradicts our assumption. This contradiction proves that the function $f(z)$ can be holomorphically continued to all points of the set $B_C^{(m)}(x',\ x'')$.

Thus, we have shown that the function $f(z)$ can be holomorphically continued into the domain $T^{C_R} \cup \hat{B}_C(x',\ x'')$, where the neighborhood $\hat{B}_C$ is defined in section 28.2. Since $B_C(x',\ x'')$ is a domain, $T^{C_R} \cup \hat{B}_C(x',\ x'')$ is a simply connected domain (see section 28.2). From the monodromy theorem (see section 6.6), the function $f(z)$ is single-valued in that domain. From the "edge of the wedge" theorem (see section 27.1), the neighborhood $\hat{B}_C(x',\ x'')$ may be replaced by $B_C(x',\ x'')$. This completes the proof of the lemma.

4. *Proof of the theorem on C-convex envelopes*

Let us suppose first that the cone C consists of the convex components C^+ and C^-, where $C^- = -C^+$. Let x' and x'' denote arbitrary points of an open set G and suppose that they can be connected by a C-similar curve lying entirely in G. According to the lemma in section 28.3, the function $f(z)$ is holomorphic in each domain $T^{C_R} \cup \tilde{B}_C(x',\ x'')$. If we take the union of these domains over all admissible pairs x' and x'', we see that $f(z)$ can be holomorphically continued into the domain $T^{C_R} \cup \hat{B}_C^{(1)}(G)$ where $B_C^{(1)}(G)$ is the first approximation to the envelope $B_C(G)$ defined by formula (85) (for $k = 1$) and $\hat{B}_C(G)$ is its complex neighborhood.

Let us show that the function $f(z)$ is single-valued in the domain $T^{C_R} \cup \hat{B}_C^{1)}(G)$. Since $f(z)$ is assumed to be single-valued in $T^{C_R} \cup \tilde{G}$, it will be sufficient for us to show that it is single-valued for real z in the set $B_C^{(1)}(G)$. But $B_C^{(1)}(G)$ is the union of the domains $B_C(x',\ x'')$ over all admissible pairs $x',\ x''$. Therefore, it remains to show that if two domains $B_C(x',\ x'')$ and $B_C(y',\ y'')$ have points in common*

---

*The pairs x', x'' and y', y'' may be different components of the open set G.

(see Fig. 79), the function $f(z)$ is single-valued in their intersection. If this were not the case, the holomorphic function $f(z)$ would be nonsingle-valued in the simply connected domain $T^{C_R} \cup B_C(x', x'') \cup \widetilde{B}_C(y', y'')$ (see section 28.2), and this would contradict the monodromy theorem (see section 6.6). This contradiction proves that the function $f(z)$ is single-valued in the domain $T^{C_R} \cup \widehat{B}_C^{(1)}(G)$.

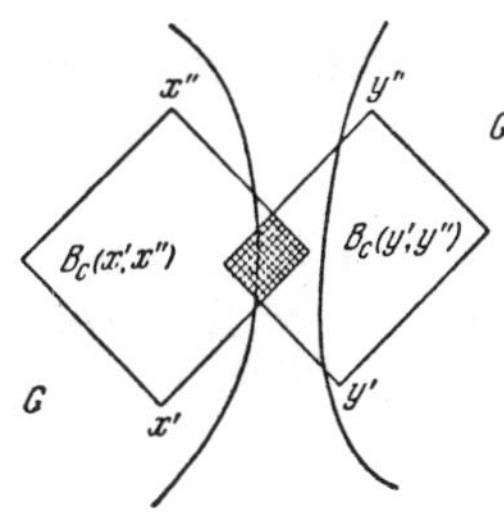

Fig. 79

According to the "edge of the wedge" theorem (see section 27.1), we may take the neighborhood $\widehat{B}_C^{(1)}(G)$ as the neighborhood $\widetilde{B}_C^{(1)}(G)$. Thus, $f(z)$ is holomorphic (and single-valued) in the domain $T^{C_R} \cup \widetilde{B}_C^{(1)}(G)$.

By repeating this process with G replaced by $B_C^{(1)}(G)$, we can show that $f(z)$ is holomorphic in the domain $T^{C_R} \cup \widetilde{B}_C^{(2)}(G)$, etc. Finally, on the basis of (86), we see that $f(z)$ is holomorphic in the domain $T^{C_R} \cup \widetilde{B}_C(G)$.

Suppose now that a cone C consists of connected component cones $C_1, C_2, \ldots, C_t$ and that the function $f(z)$ is holomorphic in the domain $T^{C_R} \cup \widehat{G}$. From Bochner's theorem (see section 17.5), $f(z)$ is holomorphic in the domain $T^{C_R^{(1)}} \cup \widehat{G}$, where $C^{(1)} = \bigcup_{1 \leqslant k \leqslant t} O(C_k)$ and $C_R^{(1)} = C^{(1)} \cap U(0, R)$ (see Fig. 68). It may happen that some of the cones $O(C_k)$ are superimposed one on the other. Nonetheless, the function $f(z)$ remains single-valued in $T^{C_R^{(1)}} \cup \widehat{G}$ (see Remark in section 21.2). (Let us denote the components of the cone $C^{(1)}$ by $C_1^{(1)}, \ldots, C_{t_1}^{(1)}$, where $t_1 \leqslant t$). Again applying Bochner's theorem, we see that $f(z)$ is holomorphic in the domain $T^{C_R^{(2)}} \cup \widehat{G}$, where $C^{(2)} = \bigcup_{1 \leqslant k \leqslant t_1} O(C_k^{(1)})$, etc. Finally, we conclude that $f(z)$ is holomorphic (and single-valued) in $T^{C_R^0} \cup \widehat{G}$ and that all the components of C^0 are convex cones. The cone $\widetilde{C} = C^0 \cap (-C^0)$ consists of an even number $2l \leqslant t$ of convex components $\widetilde{C}_k^{\pm}$, for $k = 1, 2, \ldots, l$, where $\widetilde{C}_k^- = -\widetilde{C}_k^+$ (see section 28.1).

Let x' and x'' be arbitrary points in G that can be connected by a C-similar curve lying entirely in G. If we apply the theorem that we have just proven to each of the cones $\widetilde{C}_k = \widetilde{C}_k^+ \cup \widetilde{C}_k^-$, where $k = 1$ $2, \ldots l$, we conclude that $f(z)$ is holomorphic (and single-valued) in the domain

$$\bigcup_{1 \leqslant k \leqslant l} \left[T^{C_R^0} \cup \widetilde{B}_{\widetilde{C}_k}(x', x'')\right] = T^{C_R^0} \cup \widehat{B}_C(x', x'')$$

and, consequently, in the domain $T^{C_R^0} \cup \widehat{B}_C^{(1)}(G)$. Using the recursion formulas (86), we conclude, as above, that $f(z)$ is holomorphic (and single-valued) in the domain $T^{C_R^0} \cup \widehat{B}_C(G)$. According to the "edge of the wedge" theorem, we may take the neighborhood $\hat{B}_C$ as the neighborhood $\tilde{B}_C$. This completes the proof of the theorem.

29. SOME APPLICATIONS OF THE PRECEDING RESULTS

1. *Entire functions*

A function is said to be *entire* if it is holomorphic in C^n. The results of this section are contained in our articles [42] and [40].

THEOREM 1. *Suppose that a function $f(z)$ is holomorphic in a tubular cone T^C, where C consists of a finite number of components and $C \cap (-C) \neq \emptyset$. Suppose, that, for arbitrary $\phi \in D(G)$, the limit*

$$\lim_{y \to 0,\, y \to C} \int f(x+iy)\,\varphi(x)\,dx = (f, \varphi),$$

exists and is independent of the sequence $y \to 0$, where $y \in C$. Then, $f(z)$ is holomorphic in $T^{C^0} \cup \tilde{B}_C(G)$. If, in addition, the open set G is such that $B_C(G) = R^n$, then $f(z)$ is an entire function.

Proof: From the "edge of the wedge" theorem (see section 27.1), $f(z)$ is holomorphic in the domain $T^C \cup \tilde{G}$. From the theorem on C-convex envelopes (see section 28.3), $f(z)$ is holomorphic in the domain $T^{C^0} \cup \tilde{B}_C(G)$. If $B_C(G) = R^n$, then

$$\tilde{B}_C(G) = \tilde{R}^n = \bigcup_{x^0 \in R^n} [z\colon\ |z - x^0| < \vartheta \cdot \infty] = C^n$$

which completes the proof.

THEOREM 2. *Suppose that $f(z) \in H_p(a + \epsilon; C)$, where the cone C consists of the components $C_1, C_2, \ldots, C_t$ and $C \cap (-C) \neq \emptyset$. Suppose also that the limiting values $f_k(x) \in S^*$, for $k = 1, 2, \ldots, t$, of f coincide in an open set G and that $\tilde{B}_C(G) = R^n$. Then, $f(z)$ is an entire function and*

$$|f(z)| \leqslant \begin{cases} M_\epsilon (1 + |z|)^\beta e^{(a+\epsilon)\rho_C^p |y|^p}, & p > 1,\ a > 0; \\ M (1 + |z|)^\beta e^{a\rho_C |y|}, & p = 1,\ a \geqslant 0. \end{cases} \tag{90}$$

Proof: From Theorem 1, the function $f(z)$ is an entire function. Therefore, the generalized functions $f_k(x)$ coincide in R^n. From the theorem in section 26.7, $f(z)$ satisfies inequalities (90), which completes the proof.

COROLLARY (a generalization of Liouville's theorem). *If $p = 1$ and $a = 0$ in the conditions of Theorem 2, the function $f(z)$ is a polynomial.*

Proof: It follows from Theorem 2 that $f(z)$ is an entire polynomially bounded function. From Liouville's classical theorem (see section 6.3), $f(z)$ is a polynomial of degree not exceeding β.

Remarks: (1) On the basis of the remark to the "edge of the wedge" theorem made in section 27.1, the condition that $C \cap (-C) \neq \emptyset$ in Theorems 1 and 2 can be weakened and replaced with the condition that $O(C) = R^n$.

(2) The condition that $B_C(G) = R^n$ is sufficient but not necessary for Theorems 1 and 2. The problem arises as to how we may characterize all open sets G that verify these theorems. In section 32.4, this will be done for a light cone Γ, for functions belonging to the class $H_0(\Gamma)$, and for Γ-regular open sets C (see section 28.1).

2. *The class $H_1(a; C)$*

For the definition of functions in the class $H_1(a; C)$, see section 26.4. Suppose that a cone G is connected. According to Theorem 2 (section 26.4), for a function $f(z)$ to belong to $H_1(a; C)$, it is necessary and sufficient that its spectral function $g(\xi)$ vanish for $\mu_C(\xi) > a$. From Corollary 1 to this theorem, the class $H_1(a; C)$ coincides with the class $H_1(a\rho_C; O(C))$.

From this and from Theorem 1 in section 29.1, we have the following

THEOREM.* *For a function $f(z)$ to belong to $H_1(a; C)$, where the cone C consists of the components $C_1, C_2, \ldots, C_t$ and $C \cap (-C) \neq \emptyset$, to be holomorphic in the domain $T^C \cup \tilde{G}$, it is necessary and sufficient that its boundary values $f_k(x)$, for $k = 1, 2, \ldots, t$, coincide in G and that its spectral functions $g_k(\xi) = F^{-1}[f_k]$ vanish when $\mu_{C_k}(\xi) > a$. Here, it is necessary that $f(z)$ be holomorphic in $T^{C^0} \cup \tilde{B}_C(G)$ and hence that the $f_k(x)$ coincide in $B_C(G)$.*

In what follows, we shall assume for simplicity that the cone C consists of two convex components in $C^\pm$ such that $C^- = -C^+$. We shall denote the corresponding boundary values and spectral functions by

$$f^\pm(x \pm i0) = \lim_{y \to 0,\, y \in C^\pm} f^\pm(x + iy), \quad g^\pm(\xi) = F^{-1}[f^\pm(x \pm i0)]. \tag{91}$$

Consider the case of $a = 0$. In this case, the class $H_1(0; C) = H_0(C)$ constitutes a class in the sense of section 16.1 (see remark in section 26.4). Therefore, the set of class $H_0(C)$ functions that are holomorphic in the domain $T^C \cup \tilde{G}$ also constitutes a class $K = K_{T^C \cup \tilde{G}}$ (in the sense of section 16.1).

We shall refer to the largest domain in which every class K function is holomorphic (and single-valued) as the envelope of $T^C \cup \tilde{G}$ and shall denote it by $K(T^C \cup \tilde{G})$. The following questions arise in connection with this definition:

---

*See Vladimirov [30].

(1) When do the envelopes $H(T^C \cup \tilde{G})$ and $K(T^C \cup \tilde{G})$ exist, that is, when are they single-sheeted?
We note that these envelopes always exist in the class of nonsingle-sheeted domains (see section 20.1) and that

$$H(T^C \cup \tilde{G}) \subset K(T^C \cup \tilde{G}). \tag{92}$$

(2) When does equality hold in the inclusion (92)?
(3) How can we construct the envelopes $H(T^C \cup \tilde{G})$ and $K(T^C \cup \tilde{G})$?

In the example that we considered in section 27.5 (where $C=\Gamma$ and $G=[|x_1|<m]$), the envelopes H and K were single-sheeted and they coincided. A second example of the construction of the envelopes H and K is given by Theorem 2 of section 29.1: if $B_C(G)=R^n$, then $H=K=C^n$. In section 33.2, the envelope $K(T^\Gamma \cup \tilde{G})$ has been constructed under certain hypotheses regarding G.

Let us suppose that the envelope $K(T^C \cup \tilde{G})$ exists. Then, it follows from the lemma in section 16.3 and from the Cartan-Thullen theorem (see section 16.1) that it is a K-convex domain (cf. section 20.3) and hence is a domain of holomorphy (see section 16.7).

Even so, we still do not know the answer to the following question:

(4) Is $K(T^C \cup \tilde{G})$ a K-convex envelope of the domain $T^C \cup \tilde{G}$, and, if so, is it the smallest one?

The answer to such questions is not always affirmative since not every K-convex domain is a domain of holomorphy of some class K function. For example, if B is a complete circular domain and K is the P-class of polynomials, then $K(B)=C^n$ although, according to Weil's theorem (see section 24.9), a P-convex envelope of a domain B coincides with its envelope of holomorphy [which is equal to the logarithmically convex envelope (see section 21.2)].

From our constructions and from inclusions (88) and (92), we derive the inclusions

$$G \subset B_C(G) \subset \operatorname{Re} H(T^C \cup \tilde{G}) \subset \operatorname{Re} K(T^C \cup \tilde{G}). \tag{93}$$

Below, we shall see that the envelope $B_C(G)$ does not always exhaust $\operatorname{Re} K(T^C \cup \tilde{G})$.

3. *Quasi-analyticity of a certain class of generalized functions*

According to the Paley-Wiener-Schwartz theorem (see section 26.4), generalized functions with a compact spectrum are holomorphic and polynomially bounded. The question arises as to what the situation is if the spectrum is not compact.

Suppose that a cone C consists of two open convex components C^+ and C^- such that $C^- = -C^+$. Let us denote by $L_a(C)$, where $a \geqslant 0$, the class of generalized functions $f=F[g]$ in S^* whose spectra S_g are contained in the closed set (see Fig. 80)

$$F = F^+ \cup F^-, \qquad F^{\pm} = [\xi : \mu_{C^{\pm}}(\xi) \leqslant a]. \tag{94}$$

The set F consists of convex closed sets $F^{\pm}$ (see section 26.6). It is a regular set (see section 3.5). In addition

$$\mu_{C^-}(\xi) = \mu_{C^+}(-\xi), \quad F^- = -F^+. \tag{95}$$

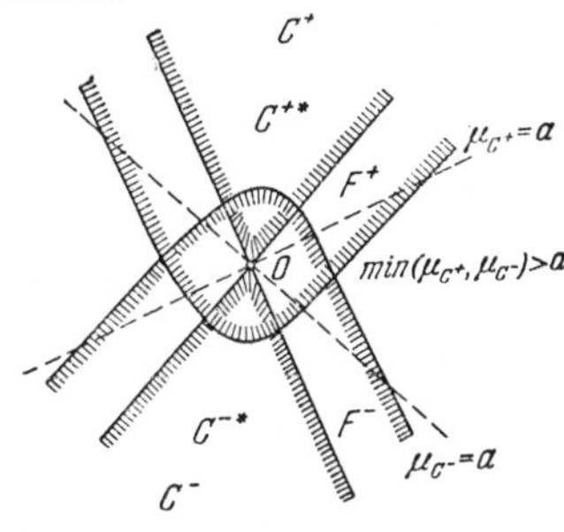

Fig. 80

Remark 1: Suppose that the spectrum $f \in S^*$ is contained in the set $(L^+ + \xi^+) \cup \cup (L^- + \xi^-)$, where L^+ and L^- are closed convex cones not containing any entire straight line such that $L^+ = -L^-$ and where ξ^+ and ξ^- are fixed points (see Fig. 81). Then, $f \in L_a(C)$, where $C = C^+ \cup C^-$, $C^{\pm} =$ int $L^{\pm *}$, and $a = \max[0, \mu_{C^+}(\xi^+), \mu_{C^-}(\xi^-)]$.

To see this, note that, from Lemma 1 of section 25.2, int $L^{\pm *} \neq \varnothing$. Furthermore,

$$L^{\pm} = L^{\pm **} = \bar{C}^{\pm *} = C^{\pm *} = [\xi : \mu_{C^{\pm}}(\xi) \leqslant 0]$$

(see section 25.1). Since the functions $\mu_{C^{\pm}}(\xi)$ are convex and homogeneous of degree one, we have

$$\mu_{C^{\pm}}(\xi) \leqslant \sup_{\xi \in L^{\pm} + \xi^{\pm}} \mu_{C^{\pm}}(\xi) = \sup_{\xi \in L^{\pm}} \mu_{C^{\pm}}(\xi + \xi^{\pm}) \leqslant$$

$$\leqslant \sup_{\xi \in L^{\pm}} \mu_{C^{\pm}}(\xi) + \mu_{C^{\pm}}(\xi^{\pm}) \leqslant \mu_{C^{\pm}}(\xi^{\pm}),$$

for all $\xi \in L^{\pm} + \xi^{\pm}$. The assertion made then follows.

Remark 2: For $n = 1$, every generalized function $f \in S^*$ belongs to $L_0(\Gamma)$, where $\Gamma = [\xi \neq 0] = \Gamma^+ \cup \Gamma^-$, where in turn $\Gamma^{\pm} = [\pm \xi > 0]$.

LEMMA. Every generalized function $g \in S^*$ with carrier $S_g \subset F$ can be represented in the form of the difference $g = g^+ - g^-$ of the generalized function $g^{\pm} \in S^*$ with carrier $S_{g^{\pm}} \subset F^{\pm}$. This representation is not unique: the $g^{\pm}$ are defined up to an arbitrary generalized function η (the same for g^+ and g^-) with carrier contained in the compact set $F^+ \cap F^-$. In particular, for $a = 0$, this η is an arbitrary finite combination of the δ-function and its derivatives at zero.

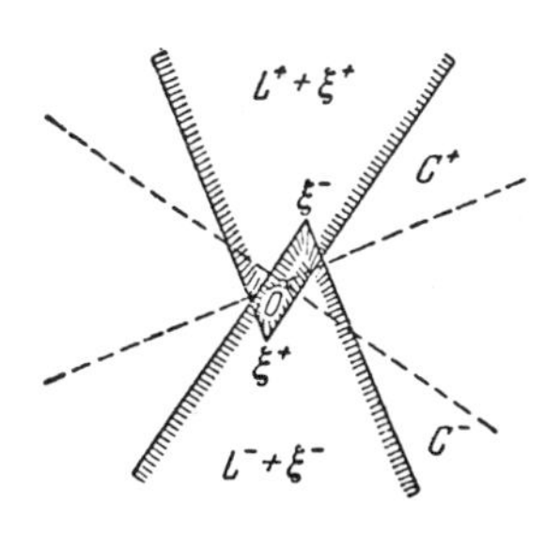

Fig. 81

Since F is a regular set, the generalized function g can, by the theorem in section 3.8, be represented in the form

$$g = \sum_{|\alpha| \leqslant M} D^{\alpha}_{\mu_{\alpha}}(\xi),$$

where the $\mu_{\alpha}(\xi)$ are measures of power increase with carrier in F. If we set

$$\mu_\alpha^+(\varepsilon) = \mu_\alpha(\varepsilon \cap F^+),$$
$$\mu_\alpha^-(\varepsilon) = \mu_\alpha^+(\varepsilon) - \mu_\alpha(\varepsilon),$$

for an arbitrary bounded Borel set ε, we obtain the desired generalized functions in the form

$$g^\pm = \sum_{|\alpha| \leqslant M} D^\alpha \mu_\alpha^\pm.$$

Let us suppose that there are two representations:

$$g = g_1^+ - g_1^- = g_2^+ - g_2^-,$$
$$S_{g_k^+} \subset F^+, \qquad S_{g_k^-} \subset F^-, \qquad k = 1, 2.$$

Then, the carrier of the generalized function

$$\eta = g_1^+ - g_2^+ = g_1^- - g_2^-$$

is contained in

$$F^+ \cap F^- = [\xi : \mu_{C^+}(\xi) \leqslant a \text{ and } \mu_{C^-}(\xi) \leqslant a] = [\xi : \mu_C(\xi) \leqslant a].$$

But the closed set $F^+ \cap F^-$ is compact because it is contained in the sphere $|\xi| \leqslant \rho_C a$ since $\mu_C(\xi) \geqslant \frac{1}{\rho_C} \mu_{O(C)}(\xi) = \frac{1}{\rho_C}|\xi|$ and, on the basis of Lemma 3 of section 25.1, $1 \leqslant \rho_C < +\infty$. For $a = 0$, $F^+ \cap F^- = \{0\}$ and, on the basis of section 3.5,

$$\eta = \sum_{|\alpha| \leqslant m} c_\alpha D^\alpha \delta.$$

THEOREM. *For a function f to belong to $L_a(C)$, it is necessary and sufficient that it have a representation in the form of the difference*

$$f(x) = f^+(x + i0) - f^-(x - i0) \tag{96}$$

of the boundary values of the functions $f^\pm(z) \in H_1(a;\ C^\pm)$. Here, the functions $f_\pm(z)$ are defined up to an arbitrary entire function with compact spectrum in $F^+ \cap F^-$ the same entire function for both functions $f^\pm$. In particular, for $a = 0$, the functions $f^\pm(z)$ are determined up to an arbitrary polynomial.

If in addition $f(x) = 0$ in an open set G, it can be represented in the form of the difference

$$f(x) = f(x + i0) - f(x - i0) \tag{97}$$

of the boundary values of some function $f(z) \in H_1(a;\ C)$ that is holomorphic in the domain $T^C \cup \widetilde{B}_C(G)$. In particular,

$$f(x) = 0, \quad x \in B_C(G). \tag{98}$$

Proof: Suppose that $f(x)$ can be represented in the form (96). Then, according to the theorem in section 29.2 the spectral functions

$$g^{\pm}(\xi)=F^{-1}\left[f^{\pm}(x\pm i0)\right](\xi)=0 \text{ for } \mu_{C^{\pm}}(\xi)>a.$$

Therefore,

$$F^{-1}[f](\xi)=g^{+}(\xi)-g^{-}(\xi)=0 \text{ for } \min\left[\mu_{C^{+}}(\xi), \mu_{C^{-}}(\xi)\right]>a.$$

This means that $f\in L_a(C)$.

Conversely, suppose that $f\in L_a(C)$. From the preceding lemma, we can represent the generalized function $F^{-1}[f]\in S^*$ in the form of the difference

$$F^{-1}[f]=g^{+}-g^{-} \tag{99}$$

of generalized functions $g^{\pm}\in S^*$ such that $S_{g^{\pm}}\subset F^{\pm}$. According to the theorem in section 29.2,

$$f^{\pm}(z)=F\left[g^{\pm}(\xi)e^{-\xi y}\right](x)\in H_1\left(a;\ C^{\pm}\right).$$

By considering formulas (91), we obtain from this equation the desired representation (96). Since the functions $g^{\pm}$ are defined up to an arbitrary generalized function with compact carrier in $F^{+}\cap F^{-}$, the functions $f^{\pm}(z)$ are defined up to the corresponding arbitrary entire function. In particular, for $f^{\pm}(z)$, the functions $a=0$ are defined up to an arbitrary polynomial.

Let us now prove the second part of the theorem. Suppose that $f(x)=0$. Then, on the basis of (96), we have

$$f^{+}(x+i0)=f^{-}(x-i0), \quad x\in G.$$

By applying the theorem in section 29.2, we then obtain the remaining assertions of the theorem.

It follows from this theorem that generalized functions belonging to the class $L_a(C)$ possess the property of *quasi-analyticity*: vanishing of a function $f(x)$ in this class in a domain G implies that it vanishes in the C-convex envelopes $B_C(G)$ of the domain G.

This kind of quasi-analyticity for a causal commutator in quantum field theory was predicted by Jost and proven by Dyson in [41] (see also Bogolyubov and Vladimirov [97], Vladimirov [92, 30, 40, 42], Borchers [98], and Araki [99]).

The reason for this phenomenon of quasi-analyticity consists in the fact that although $f(x)\in L_a(C)$ is a generalized function, it can nonetheless be represented in the form of the difference (97) of the values $f(x\pm i0)$ of some function $f(z)$ belonging to the class $\bigcup_{a'\geq a} H_1(a';\ C)$ (see section 26.4) that is holomorphic in the domain $T^C\cup\tilde{G}$ and hence holomorphic in the corresponding K-convex envelope, which cannot be arbitrary (it must in any case be pseudoconvex [see section 17.1]. Therefore, at its real points (in which, by virtue of (97), $f(x)$ must vanish), it cannot be arbitrary.

4. *Quasi-analyticity of the solutions of a certain class of linear equations*

Let us use the results of the preceding section to establish the quasi-analytic nature of solutions $u(x)$ in the space S^* of the convolution equation

$$f_0 * u = f \tag{100}$$

with the hypotheses

$$f \in L_a(C), \quad f_0 \in \overset{*}{\theta}_C, \quad F[f_0](\xi) \neq 0$$
$$\text{for } \min\left[\mu_{C^+}(\xi), \mu_{C^-}(\xi)\right] > a. \tag{101}$$

We note that, since f_0 is a convolutor in S^*, the function $F[f_0]$, is a multiplier in S^*, and all its derivatives are infinitely differentiable functions of power increase (see section 3.9).

Convolution equations include differential equations with constant coefficients

$$P(iD)\, u = f(x), \qquad f_0 = P(iD)\, \delta(x), \tag{102}$$

difference equations

$$\sum_{1 \leqslant k \leqslant N} c_k u(x - x_k) = f(x), \quad f_0 = \sum_{1 \leqslant k \leqslant N} c_k \delta(x - x_k), \tag{103}$$

integral equations

$$\int K(x - x')\, u(x')\, dx' = f(x), \quad f_0 = K(x) \tag{104}$$

and various combinations of these: differential equations with displaced arguments, integro-differential equations, and others. The function $F[f_0](\xi)$ takes the forms

$$P(\xi), \quad \sum_{1 \leqslant k \leqslant N} c_k e^{ix_k\xi}, \quad \int K(x)\, e^{ix\xi}\, dx.$$

for Eqs. (102)-(104) respectively.

By applying formula (24) of Chapter 1 to Eq. (100), we get the Fourier transformations

$$F[f_0]\, F[u] = F[f].$$

It follows from this, from (95), and from conditions (101) that $F^{-1}[u](\xi) = 0$ for $\min\left[\mu_{C^+}(\xi), \mu_{C^-}(\xi)\right] > a$,; that is, $u \in L_a(C)$. If we now apply the theorem in section 29.3, we get the following result (see [30, 93]).

THEOREM 1. *Every solution $u \in S^*$ of the equation $f_0 * u = f$ that satisfies conditions* (101) *and vanishes in an open set G can be represented in the form of the difference* (97) *of the boundary values $f(z \pm i0)$ of a function $f(z) \in H_1(a; C)$ that is holomorphic in the domain $T^C \cup \widetilde{B}_C(G)$. This solution therefore vanishes in the larger domain constituting the envelope $B_C(G)$.*

COROLLARY 1. *If two solutions of Eq.* (100) *that belong to S^* coincide in the domain G, they also coincide in $B_C(G)$.*

From Remark 1 in section 29.3, we obtain the

COROLLARY 2. *Suppose that $F[f_0](\xi) = 0$ outside two closed convex cones L^+ and L^- (where $L^- = -L^+$) that do not contain any entire straight line. Then, in the representation* (97) *of each solution $u \in S^*$ of the equation $f_0 * u = 0$, the function $f(z) \in H_0(C)$, where $C = C^+ \cup C^-$ and $C^\pm = \operatorname{int} L^{\pm *}$*

For example, every solution $u \in S^*$ of the iterated wave equation $\square^m u = 0$ (for $m \geqslant 1$ that vanishes in an open set G also vanishes in $B_\Gamma(G)$, that is, in the convex envelope of G with respect to timelike curves.

Proof: In this case, $f_0 = \square^m \delta(x)$,

$$F[f_0](\xi) = (-\xi_1^2 + \xi_2^2 + \dots + \xi_n^2)^m$$

and $F^\pm = \overline{\Gamma}^\pm$, that is, $u \in L_0(\Gamma)$. Noting that $\operatorname{int} \overline{\Gamma}^{\pm *} = \Gamma^\pm$ and $\Gamma = \Gamma^+ \cup \Gamma^-$ and using Corollary 2, we obtain the assertion made.

THEOREM 2. *Suppose that the equation*

$$i^k \frac{\partial^k u}{\partial x_1^k} = P(iD)\, u, \tag{105}$$

can be solved for the highest derivative with respect to x_1 and that this equation satisfies condition (101):

$$\xi_1^k \neq P(\xi), \qquad \min\left[\mu_{C^+}(\xi),\ \mu_{C^-}(\xi)\right] > a. \tag{106}$$

Let $u \in S^*$ denote a solution of Eq. (105). *Suppose that this solution satisfies the conditions*

$$u(0, \tilde{x}) = \frac{\partial u(0, \tilde{x})}{\partial x_1} = \dots = \frac{\partial^{k-1} u(0, \tilde{x})}{\partial x_1^{k-1}} = 0, \quad \tilde{x} \in g, \tag{107}$$

where g is an open set lying on the hyperplane $x_1 = 0$. Then, $u(x)$ vanishes in the envelope $B_C(g)$ and, in particular, Eqs. (107) *are valid in a larger set than g, namely, in the intersection of $B_C(g)$ and the plane $x_1 = 0$.*

Proof: Since an arbitrary derivative $D^\alpha u(x)$ belongs to S^* and satisfies Eq. (105) and since the differential operator $i^k D_1^k - P(iD)$ is hypoelliptic with respect to x_1, we conclude that $D^\alpha u(a, \tilde{x}) \in S^*$ for all a (see section 3.11). Therefore, conditions (107) are meaningful.

Suppose that the function $\varphi_\varepsilon(x) \in D(U(0, \varepsilon))$ approaches $\delta(x)$ as $\varepsilon \to +0$. Since $u(x) \in S^*$ satisfies Eq. (105) and conditions (107), it follows that

$$u_{\varphi_\varepsilon}(x) = (u(x_1, \tilde{x}'), \varphi_\varepsilon(\tilde{x} - \tilde{x}'))$$

satisfies this equation in the ordinary sense and satisfies the conditions

$$\frac{\partial^\alpha}{\partial x_1^\alpha} u_{\varphi_\varepsilon}(0, \tilde{x}) = 0, \ \alpha = 0, 1, \ldots, k-1, \qquad \tilde{x} \in g(\varepsilon), \tag{108}$$

where $g(\varepsilon)$ is the set of points in the domain g that lie at a distance from ∂g greater than ε (see Fig. 82). By using Holmgren's theorem (see Petrovskiy [105], p. 49), we conclude from (108) that $u_{\varphi_\varepsilon}(x) = 0$ in some n-dimensional neighborhood $g'(\varepsilon)$ of the $(n-1)$-dimensional set $g(\varepsilon)$. From this, we conclude on the basis of Theorem 1 that $u_{\varphi_\varepsilon}(x) = 0$ in the envelope $B_C[g'(\varepsilon)]$ and *a fortiori* in the envelope $B_C[g(\varepsilon)] \subset B_C[g'(\varepsilon)]$.

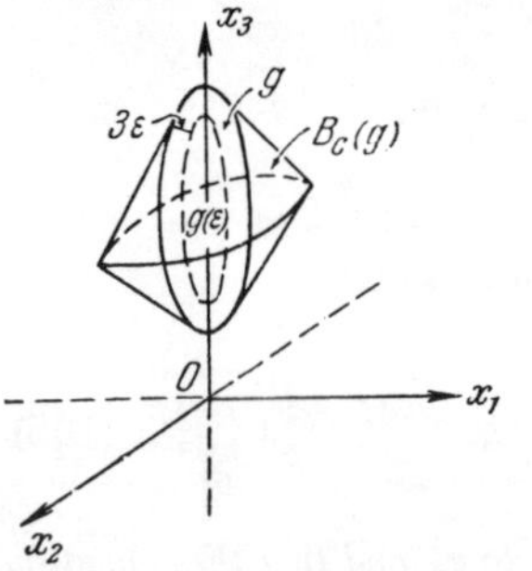

Now let $\varepsilon \to +0$. Then, in S^*,

$$u_{\varphi_\varepsilon}(x) \to u(x), \quad \frac{\partial^\alpha u_{\varphi_\varepsilon}(0, \tilde{x})}{\partial x_1^\alpha} \to \\ \to \frac{\partial^\alpha u(0, \tilde{x})}{\partial x_1^\alpha}, \alpha = 0, 1, \ldots. \tag{109}$$

Here, the sequence of open sets $g(\varepsilon)$ increases monotonically and $g = \bigcup_{\varepsilon > 0} g(\varepsilon)$.

According to the lemma in section 28.1, the sequence $B_C[g(\varepsilon)]$ also increases monotonically and

$$B_C(g) = \bigcup_{\varepsilon > 0} B_C[g(\varepsilon)]. \tag{110}$$

Fig. 82

Since $u_{\varphi_\varepsilon}(x) = 0$ in $B_C[g(\varepsilon)]$, it follows from (109) and (110) that $u_{\varphi_\varepsilon}(x) = 0$ for $B_C[g(\varepsilon)]$. Analogously, it follows from (109) and (110) that Eqs. (107) are valid in the intersection of $B_C(g)$ and the plane $x_1 = 0$, which completes the proof.

The quasi-analytic nature of the solutions of the wave equation was pointed out in the book by Courant and Hilbert [88] (Chapter VI, § 8). For example, if a solution vanishes identically in a neighborhood of the interval $|x_1| \leqslant 1$, $\tilde{x} = 0$, then it must also vanish in the double cone $|x_1| + |\tilde{x}| < 1$ (see Fig. 75). In connection with this, it is stated in that book that we encounter the remarkable phenomenon of the existence of nonanalytic functions whose values in some arbitrarily thin region uniquely determine the behavior of the function in a much larger region.

Similar phenomena of quasi-analyticity have been studied by John [89] Nirenberg [90] and Broda [91] in connection with questions on the uniqueness of solutions of partial differential equations (in particular, those with constant coefficients). Nirenberg [90] offered the hypothesis that if a sufficiently smooth solution $u(x)$ of a differential equation $P(iD)u=0$ vanishes in a domain G, it also vanishes in the domain $B^P(G)$, that is, the convex envelope with respect to timelike curves. A curve of class $C^{(1)}$ is said to be timelike with respect to the operator $P(iD)$ if, at an arbitrary point ξ^0 of the curve, the equations $\xi t=0$, $P_0(\xi)=0$ are simultaneously satisfied only by $\xi=0$, where t is the tangent vector to that curve at ξ^0 (see Fig. 83) and $P_0(\xi)$ is the principal part of the polynomial $P(\xi)$. We note that, for the wave operator $\Box$, timelike and Γ -like curves coincide, so that $B^{\Box}(G)=B_\Gamma(G)$.

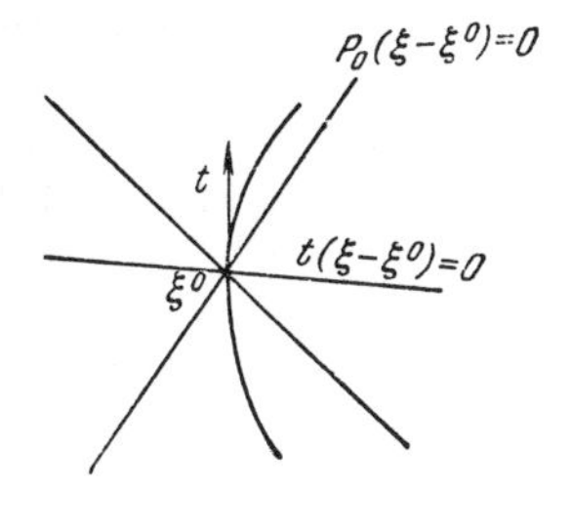

Fig. 83

These results prove that Nirenberg's hypothesis remains true for solutions in the space S^* when the envelope $B^P(G)$ is replaced by the envelope $B_C(G)$. It would be interesting to extend these results to the space $D^*(O)$.

The method of studying regions of uniqueness of solutions of differential equations with constant coefficients that we have expounded here is based on the property of pseudoconvexity of domains of holomorphy. Therefore, it is quite different from previously used methods of studying such questions (see [88-91]).

5. *A boundary-value problem for class $H_0(C)$ holomorphic functions*

The results obtained in section 29.3 enable us to solve the following problem concerning the conjugacy of two holomorphic functions.

Suppose that an open cone $C\subset R^n$ consists of two convex components C^+ and C^-, where $C^-=-C^+$. Let $f(x)$ denote a function belonging to $S^*_\cdot$. Find two functions $f^\pm(z)$ belonging respectively to the classes $H_0(C^\pm)$ such that

$$f^+(x+i0)=f^-(x-i0)+f(x). \tag{111}$$

This problem is a generalization of the well-known classical problem of determining a piecewise-holomorphic function from a given jump on the real line (see, for example, Muskhelishvili [107] to the case of many complex variables and to a broader class of admissible jumps in f. For $n=1$, problem (111) and its analogs have been studied in various spaces of generalized functions by Köthe [130], Tillman [131-133], Parasyuk [108], Sato [134, 135], Bremermann and Durand [106], Luszczki and Zielezny (136), and Rogozhin [137]. The similar problem (for $n\geqslant 1$) on the representation of an arbitrary generalized function in the form of the sum of 2^n boundary values of a function that is holomorphic in the open set

$$(C^1 \setminus R^1) \times (C^1 \setminus R^1) \times \ldots \times (C^1 \setminus R^1)$$

has been studied by Tillman (131-133).

The theorem in section 29.3 provides the following condition for solvability of problem (111). For problem (111) to have a solution, it is necessary and sufficient that $f(x) \in L_0(C)$; this solution is not unique but is determined up to an arbitrary polynomial.

Thus, the conditions under which the problem has a solution are quite different in the spaces C^1 and C^n, where $n \geqslant 2$; in the space C^1, we know from Remark 2 of section 29.3 that the problem has a solution for all $f \in S^*$ whereas this is not the case in C^n for $n \geqslant 2$.

We shall now give an algorithm for solving this conjugacy problem. Suppose that $f \in L_0(C)$ and $g = F^{-1}[f]$. By hypothesis, $S_g \subset C^{+*} \cup C^{-*}$, where $C^{-*} = -C^{+*}$ and $C^* = \{0\} = C^{+*} \cap C^{-*}$. Therefore, there exists a hyperplane $\xi e = 0$ (where $e \in C^+$) such that $\pm \xi e > 0$ for $\xi \in C^{\pm *}$ and $\xi \neq 0$. Let N denote the order of $g \in S^*$. Then,

$$G^{\pm}(\xi) = \pm (\xi e)^{N+1} \theta(\pm \xi e) g(\xi) \in S^*, \quad S_{G^{\pm}} \subset C^{\pm *}$$

and

$$G^+(\xi) - G^-(\xi) = (\xi e)^{N+1} g(\xi). \tag{112}$$

Since division by a polynomial is always possible in the space S^* (see section 3.10), let us divide Eq. (112) by $(\xi e)^{N+1}$. This gives us a representation of g in the form

$$g(\xi) = g^+(\xi) - g^-(\xi), \tag{113}$$

where the generalized functions $g^{\pm} \in S^*$ are chosen in such a way that

$$S_{g^{\pm}} \subset C^{\pm *} \tag{114}$$

Setting

$$f^{\pm}(z) = F\left[g^{\pm}(\xi) e^{-\xi y}\right](x),$$

we obtain, by virtue of (113), the solution to our problem.

The spectral functions $g^{\pm}$ in the representation (113) subject to conditions (114) are not unique. Let us find their general form. Let $g = g_1^+ - g_1^-$ be another such representation. Then, $\eta = g^+ - g_1^+ = g^- - g_1^-$. Since the carrier $g^{\pm} - g_1^{\pm}$ is contained in $C^{\pm *}$, the carrier η is

$$C^{+*} \cap C^{-*} = \{0\} \text{ or } \varnothing.$$

From this, we conclude on the basis of formula (16) of Chapter 1 that the generalized function η can be represented in the form

$$\eta = \sum_{|\alpha| \leqslant M} c_\alpha D^\alpha \delta(\xi),$$

where the c_α are arbitrary numbers (see section 3.5). Therefore, the functions $f^{\pm}(z)$ are determined up to an arbitrary polynomial.

It would be interesting to generalize this method for solving the more general problem in which condition (111) is replaced with

$$f^+(x + i0) = g(x) f^-(x - i0) + f(x), \quad g \in \theta_M, \quad f \in S^*. \tag{115}$$

(For $n = 1$, see Parasyuk [108].)

In the case in which $g(x) = P(x)$ is a polynomial such that $P(z) \neq 0$ in T^{C^-} the problem (115) can also be solved under the condition that $f \in L_0(C)$. Here, $f^+(z)$ and $f^-(z)$ are respectively determined up to expressions of the form $Q(z)$ and $\frac{Q(z)}{P(z)}$, where Q is an arbitrary polynomial.

This is true because if $\Phi^{\pm}(z)$ is a solution of the problem (111), then

$$f^+(z) = \Phi^+(z), \qquad f^-(z) = \frac{\Phi^-(z)}{P(z)}$$

is a solution of the problem (115).

To generalize what has been said, it will be sufficient to show that it is always possible to divide by a polynomial $P(z)$ such that $P(z) \neq 0$ (for $z \in T^{C^-}$ in the ring $H_0(C^-)$, that is, that the equation $P(z)\chi(z) = \Phi(z)$ has a (unique) solution $\chi(z)$ in the ring $H_0(C^-)$ for any right member $\Phi(z)$ in $H_0(C^-)$. This assertion follows from the

LEMMA. *If the polynomial $P(z) \neq 0$ in the tubular radial domain $T^C = R^n + iC$, then $1 / P(z) \in H_0(C)$.*

Proof: If $P(z) \neq 0$ in T^C then $P(z) \neq 0$ in its envelope of holomorphy $T^{O(C)}$ (see section 20.3). Therefore we may assume C to be a convex cone. Since $P(z) \neq 0$ in T^C, the function $\frac{1}{P(z)}$ is holomorphic in T^C. Thus, it remains only to show that this function satisfies inequality (18) (for $a = 0$). For this, we use the second algebraic lemma in the article by Hörmander [81]: Let $Q(x)$ be a polynomial with real coefficients, let N be the set of its real zeros, and denote the comment of N by $CN = R^n \setminus N$. Then, for certain values of a, b, and c (where $c > 0$),

$$|Q(x)| \geqslant c(1 + |x|)^a \Delta^b_{CN}(x).$$

(For $N = \varnothing$, instead of $\Delta^b_{CN}(x)$, we may take 1.)

Now let $P(z)$ denote an arbitrary polynomial and let M denote the set of its zeros. Again, $CM = C^n \setminus M$. If we apply Hörmander's lemma to the polynomial $Q(x, y) = |P(z)|^2$ (with real coefficients) the real zeros of which coincide with M, we see that this lemma remains valid in this case also:

$$|P(z)| \geqslant d(1 + |z|)^\alpha \Delta^\beta_{CM}(z), \quad (\Delta^\beta_{C\varnothing}(z) = 1). \tag{116}$$

Suppose that the polynomial $P(z) \neq 0$ in T^C. If $M = \varnothing$, then, on the basis of inequality (116), we have $\frac{1}{P(z)} \in H_0(C)$. Suppose that $M \neq \varnothing$. If $\beta \leqslant 0$, in (116), we conclude, on the basis of the inequality

$$\Delta_{CM}(z) \leqslant |z| + \Delta_{CM}(0) = A + |z|, \quad A \geqslant 0$$

that $\frac{1}{P(z)} \in H_0(C)$.

We still need to consider the case in which $M \neq \varnothing$ and $\beta > 0$. Since the cone C is convex, for all $z \in T^C$ we have (see section 25.1) $C^{**} = \bar{C}$ $\operatorname{int} \bar{C} = C$, and

$$\Delta_{CM}(z) \geqslant \Delta_{TC}(z) = \Delta_C(y) = \inf_{\xi \in C^*, |\xi| = 1} \xi y = -\mu_{C^*}(y). \tag{117}$$

Let C' be a cone that is compact in C. It follows from Lemma 2 of section 25.2 that

$$-\mu_{C^*}(y) \geqslant \sigma |y|, \qquad y \in C', \qquad \sigma = \sigma(C') > 0.$$

From this inequality and inequalities (117) and (116), we conclude that

$$|P(z)| \geqslant d\sigma^\beta (1 + |z|)^\alpha |y|^\beta, \qquad z \in T^{C'}$$

that is, $\frac{1}{P(z)} \in H_0(C)$. This completes the proof of the lemma.

6. *The impossibility of nonlocal theories of a certain type*

Wightman posed the question [24]: Does vanishing of the commutator $[A(x), B(y)]$ for $(x - y)^2 < -l^2$ imply* that it vanishes for $(x - y)^2 < 0$. This question was answered in the affirmative by Petrina [43] by means of the theorem of a Γ-convex envelope [92]. Another proof of this result appears in Wightman's article [32]. We shall present the solution of this question, following Petrina.

---

*Here, we use the following notations, which are employed in quantum mechanics: (337m) (see section 30).

We are dealing with quantum field theory, in which the following axioms are assumed satisfied:

(a) In the Hilbert space $\mathfrak{H}$ of amplitudes of state $|\rangle$, the linear operators of the field $A(x)$, $B(y)$, ..., are given. These are operator-valued generalized functions in S^* with a common domain of definition D that is dense in $\mathfrak{H}$. Here, $AD \subset D$.

(b) Invariance under the group of translations. This means that in the space $\mathfrak{H}$ there is a unitary representation

$$U(a) = e^{-iPa} = \int e^{-ipa} \, d\varepsilon(p), \qquad U(a) D \subset D$$

of the group of translations $\{a\}$, where $P = (P_0, P_1, P_2, P_3)$ is the energy-momentum operator and $\varepsilon(p)$ is the corresponding spectral measure. (The operators P_j are self-adjoint in $\mathfrak{H}$ and they commute with each other.) This representation transforms the operators of the field $A(x)$ according to the rule

$$U(a) A(x) U(-a) = A(x-a).$$

(c) Completeness of the system of (generalized)* eigenamplitudes of the state $|p\rangle$ of the energy-momentum operator P, $P|p\rangle = p|p\rangle$, $p = (p_0, p_1, p_2, p_3) = (p_0, \mathbf{p})$ and nonnegativeness of the spectrum of the energy operator P_0. Thus, the carrier of the spectral measure $\varepsilon(p)$ is contained in the closure of the future light cone $\bar{\Gamma}^+ = [p: p_0 \geqslant 0, \; p^2 \geqslant 0]$; that is, $\varepsilon(p)$ is a retarded function.

(d) Nonlocal commutativity: For any two field operators $A(x)$ and $B(y)$, the commutator

$$[A(x), B(y)] = 0 \qquad \text{if} \qquad (x-y)^2 < -l^2.$$

Then, *conditions* (a)-(d) *imply local commutativity*:

$$[A(x), B(y)] = 0 \text{ for } (x-y)^2 < 0.$$

Proof: Since the system of eigenamplitudes of the state $|p\rangle$ is complete in $\mathfrak{H}$, it will be sufficient to show that the generalized function in S^*

$$f(x-y) = \langle p' | [A(x), B(y)] | p \rangle \, e^{-i \frac{p'-p}{2}(x+y)}$$

*This completeness will exist if $\mathfrak{H}$ is a rigged Hilbert space, that is, if there exists a countably Hilbert kernel space Φ with nondegenerate scalar product such that $\mathfrak{H}$ is the completion of Φ with respect to this scalar product (see Gel'fand and Vilenkin [44], Chapter 1, section 4.

vanishes for $(x-y)^2<0$ for arbitrary eigenamplitudes of the state $|p'\rangle$ and $|p\rangle$. By using (b), we obtain the chain of equations

$$\begin{aligned}
f(x-y) &= \\
&= e^{-i\frac{p'-p}{2}(x+y)}[\langle p'|A(x)B(y)|p\rangle-\langle p'|B(y)A(x)|p\rangle]= \\
&= e^{-i\frac{p'-p}{2}(x+y)}[e^{ip'x-ipy}\langle p'|U(x)A(x)U(-x)U(x-y)\times \\
&\times U(y)B(y)U(-y)|p\rangle-e^{ip'y-ipx}\langle p'|U(y)B(y)U(-y)\times \\
&\times U(-x+y)U(x)A(x)U(-x)|p\rangle]= \\
&= e^{i\frac{p'+p}{2}(x-y)}\langle p'|A(0)\int e^{-i(x-y)q}\,d\varepsilon(q)B(0)|p\rangle- \\
&- e^{-i\frac{p'+p}{2}(x-y)}\langle p'|B(0)\int e^{i(x-y)q}\,d\varepsilon(q)A(0)|p\rangle= \\
&= \int e^{i(x-y)t}\Big[\Big\langle p'\Big|A(0)\,d\varepsilon\Big(-t+\frac{p'+p}{2}\Big)B(0)\Big|p\Big\rangle- \\
&- \Big\langle p'\Big|B(0)\,d\varepsilon\Big(t+\frac{p'+p}{2}\Big)A(0)\Big|p\Big\rangle\Big].
\end{aligned}$$

It follows from this and from (c) that the spectrum of $f(x-y)$ is contained in

$$\left(\bar{\Gamma}^+-\frac{p'+p}{2}\right)\cup\left(\bar{\Gamma}^-+\frac{p'+p}{2}\right).$$

But the light cones Γ^+ and $\Gamma^-=-\Gamma^+$ are self-adjoint (see section 25.1) and $\frac{1}{2}(p'+p)\in\bar{\Gamma}^+$. Therefore, by using Remark 1 of section 29.3, we conclude that $f(x-y)$ belongs to the class $L_a(\Gamma)$ where $\Gamma=\Gamma^+\cup\Gamma^-$ is a light cone and

$$a=\max\left[0,\ \mu_{\Gamma^+}\left(\frac{p'+p}{2}\right),\ \ \mu_{\Gamma^-}\left(\frac{p'+p}{2}\right)\right]=\frac{1}{2}|p'+p|.$$

Furthermore, we conclude from (d) that

$$f(x-y)=0 \text{ for } x-y\in G_l=[t:\ t^2<-l^2].$$

Keeping the theorem of section 29.3 in mind, we conclude from this that $f(x-y)=0$ if $x-y\in B_\Gamma(G_l)$. Noting now that $B_\Gamma(G_l)=G_0=[t:\ t^2<0]$ (see section 28.1), we conclude that $f(x-y)=0$ for $(x-y)^2<0$, which completes the proof.

7. *Products of generalized functions*

It is known (see Schwartz [138]), that there does not exist a multiplication operation (commutative or not) of all generalized functions that is associative and bilinear, that coincides with ordinary multiplication for continuous functions, and that satisfies the conditions $x\delta(x)=0$ and $P\,1/x\,x=1$, where

$$\left(P\frac{1}{x},\ \varphi\right)=\mathrm{Vp}\int\frac{\varphi(x)}{x}\,dx \qquad (n=1).$$

(If there were, we would have

$$0=P\frac{1}{x}\,[x\delta(x)]=\left[P\frac{1}{x}\,x\right]\delta(x)=1\cdot\delta(x)=\delta(x),$$

which is impossible.)

However, if we confine ourselves to particular classes of generalized functions or if we relax some of the conditions enumerated, such a multiplication operation can exist. Along these lines, see Bogolyubov and Parasyuk [139] and König [157, 158].

Here, we shall mention one of the possible ways of defining a multiplication of certain generalized functions which are the differences between boundary values of functions that are holomorphic in the tubular domains $T^{C_R^+}$ and $T^{C_R^-}$ and that satisfy inequality (13). Here, $C_R^{\pm}=C^{\pm}\cap U(0,R)$. We note in passing that, by virtue of the theorem in section 29.3, every class $L_a(C)$ generalized function satisfies the conditions listed. (For $n=1$, every generalized function in S^* belongs to $L_0(\Gamma)$ where $\Gamma=[y\neq 0]$ [see Remark 2 in section 29.3].)

Thus, suppose* that, in the sense of convergence in the space S^*,

$$f_k(x)=\lim_{y\to 0,\ y\in C^+} F_a^y(x), \qquad k=1,\ 2,\ \dots,\ m\ ^*)$$

where

$$F_k^y(x)=f_k^+(x+iy)-f_k^-(x-iy).$$

Here the functions $f_k^{\pm}(z)$ are holomorphic in $T^{C_R^{\pm}}$ respectively and they satisfy inequality (13).

Let us suppose now that, for arbitrary φ in S there exists a finite limit

$$\lim_{y\to 0,\ y\in C^+}\int F_1^y(x)\,F_2^y(x)\,\dots\,F_m^y(x)\,\varphi(x)\,dx, \tag{118}$$

that is independent of the sequence $y\to 0$, $y\in C^+$. Then, since the space S^* is dense (see section 3.7), this limit defines a generalized function in S^*, which we shall call the *product* $f_1f_2\dots f_m$ of the generalized functions $f_1,\ f_2,\ \dots,\ f_m$. Thus,

---

*According to our agreement (see section 26.3) the notation $y\to 0,\ y\in C^+$ means that the sequence $y\to 0$ is contained in some cone that is compact in C^+.

$$f_1 f_2 \ldots f_m = \lim_{y \to 0,\ y \in C^+} F_1^y F_2^y \ldots F_m^y \quad (\text{in } S^*), \tag{119}$$

if the limit on the right exists and is independent of the sequence $y \to 0$, $y \in C^+$. Obviously, this product is commutative and associative.

In connection with this product, the very natural question arises: For what generalized functions belonging to the class in question does this product exist? First of all, we should note two extremely important subclasses of this class in which such a product is known to exist for an arbitrary number of factors and to belong to the same subclass, namely, the set of boundary values of functions that are holomorphic in $T^{C_R^+}$ (resp. in $T^{C_R^-}$) and that satisfy inequality (13).

Suppose, for example, that certain functions $f_k(x)$ $k = 1, 2, \ldots, m$, are limiting values of functions $f_k^+(z)$ that are holomorphic in $T^{C_R^+}$ and that satisfy inequality (13). Then,

$$F_k^y(x) = f_k^+(x + iy), \quad f_k^-(x + iy) = 0, \quad k = 1, 2, \ldots, m$$

and, on the basis of the theorem in section 26.3, the product $f_1 f_2 \ldots f_m$ exists, belongs to S^*, and is a boundary value of the function $f_1(z) f_2(z) \ldots f_m(z)$, which is holomorphic in $T^{C_R^+}$ and satisfies inequality (13).

Thus, *the set of boundary values of functions that are holomorphic in* $T^{C_R^+}$ *and that satisfy inequality* (13) *constitute a commutative ring with unity without zero divisors with respect to the multiplication defined above.*

We note that, for suitable spectral functions $g = F^{-1}[f]$ this multiplication becomes convolution: $g_1 * g_2$. This set of spectral functions constitutes a subring of the ring considered in section 26.2.

Since the sets of holomorphic functions

$$\bigcup_{a' > a} H_p(a'; C) \text{ and } H_0(C)$$

constitute rings with respect to ordinary multiplication, the sets of their boundary values also constitute rings with respect to the multiplication just defined. By using Theorems 1 and 2 of section 26.4, we can describe these rings in terms of spectral functions.

As an example, we shall do this for the important case of light cones.

Suppose that Γ^+ and Γ^- are future and past light cones respectively (see section 25.1). We shall call a generalized function $g \in S^*$ an advanced or a retarded function according as $S_g \subset \bar{\Gamma}^-$ or $S_g \subset \bar{\Gamma}^+$ respectively. On the basis of Theorem 2 of section 26.4, we conclude that, for a generalized function g in S^* to be an advanced (resp. retarded) function, it is necessary and sufficient that its Fourier-Laplace transform belong to the class $H_0(\Gamma^-)$ [resp. $H_0(\Gamma^+)$]. From this and from the preceding construction, it follows that *it is possible to define a multiplication on the set of Fourier transforms of advanced (resp. retarded) functions such that the product of these functions is the Fourier transform of some advanced (retarded) function.*

As an application of this result, we note that multiplication is always possible for generalized functions of the type of Pauli-Iordan transition functions of equal frequency:

$$P\left(\frac{\partial}{\partial x}\right)\int e^{ikx}\theta(\pm k_0)\,\delta(k^2-m^2)\,dk. \tag{120}$$

Products of such functions result when the products of linear operators in quantum field theory are reduced to normal form. We note that the product of two, let us say, advanced functions of the type (120) will not in general be a function of the same type. Nonetheless, we may now assert that this product will be the Fourier transform of some advanced function.

Another construction of the operation of multiplication of identical-frequency transposition functions (120) is given in the book by Bogolyubov and Shirkov [87] (section 16) on the basis of the Pauli-Villars regularization.

Turning to the general case, we note that the existence of the limit (118) for arbitrary $\varphi \in S$ implies existence of the limit in (119) with respect to the norm of functionals in the space $S^{(N)*}$ for some N, which depends on $f_1, \ldots, f_m$ (since weak convergence in the space S^* implies strong convergence [see section 3.7]). Let us now suppose that the limit (118) does not exist for all $\varphi \in S$ but that it exists for all φ in a closed subspace M of the space $S^{(N)}$ for some N. (Since M is closed in $S^{(N)}$, it is a Banach space with norm $\| \;\|_N$.) From the Banach-Steinhaus theorem (see section 3.7), the limit (118) defines a continuous linear functional χ on M. We shall use the term product $f_1 f_2 \ldots f_m$ of the generalized functions $f_1, f_2, \ldots, f_m$ for any continuous linear functional in the space $S^{(N)*} \subset S^*$ that is a continuation of $\bar{\chi}$ from M onto $S^{(N)}$. (According to the Hahn-Banach theorem [see section 3.4], such an extension always exists but it is not in general unique.)

The case in which M consists of those functions φ in $S^{(N)}$ that vanish together with all derivatives of order $p \leqslant N$ inclusively at $x=0$ is of interest. In this case, all continuations $f_1 f_2 \ldots f_m$ of the functional χ from M onto $S^{(N)}$ are given by the formula

$$(f_1 f_2 \ldots f_m, \varphi) = \\ = \left(\chi, \varphi - \sum_{|\alpha| \leqslant p} \frac{D^\alpha \varphi(0)}{\alpha!}\varphi_0(x)\,x^\alpha\right) + \sum_{|\alpha| \leqslant p} c_\alpha (\delta^{(\alpha)}, \varphi), \tag{121}$$

where φ_0 is an arbitrary function in S that is identically equal to unity in a neighborhood of the point 0 and the c_α are arbitrary constants. *Here, the extensions defined by formula (121) are actually independent of the function* φ_0.

Proof: Let us represent an arbitrary function φ in $S^{(N)}$ in the form $\varphi = \varphi_1 + \varphi_2$, where

$$\varphi_2 = \sum_{|\alpha| \leqslant p} \frac{D^\alpha \varphi(0)}{\alpha!} \varphi_0(x)\, x^\alpha \in S, \qquad \varphi_1 = \varphi - \varphi_2 \in M,$$

and let us apply the functional $f_1 f_2 \ldots f_m$ to it. Since this functional coincides with χ on M, we obtain formula (121) with

$$c_\alpha = \frac{(-1)^{|\alpha|}}{\alpha!} (f_1 f_2 \ldots f_m, \varphi_0(x)\, x^\alpha). \tag{122}$$

Let us show that formula (121) is independent of φ_0. Suppose that we have two representations of the form (121) corresponding to functions φ_0 and φ_0^*. Then, on the basis of (122), their difference is equal to

$$\left(\chi, \sum_{|\alpha| \leqslant p} \frac{D^\alpha \varphi(0)}{\alpha!} (\varphi_0^* - \varphi_0)\, x^\alpha \right) + \sum_{|\alpha| \leqslant p} (c_\alpha^* - c_\alpha)(\delta^{(\alpha)}, \varphi) =$$

$$= \left(f_1 f_2 \ldots f_m, \sum_{|\alpha| \leqslant p} \frac{D^\alpha \varphi(0)}{\alpha!} (\varphi_0^* - \varphi_0)\, x^\alpha \right) +$$

$$+ \sum_{|\alpha| \leqslant p} c_\alpha' (\delta^{(\alpha)}, \varphi) = \sum_{|\alpha| \leqslant p} d_\alpha (\delta^{(\alpha)}, \varphi),$$

which proves our assertion.

It should be noted that holomorphic functions (of a single variable) were also used by Bremermann and Durand [106] for defining the analogous product of generalized functions of certain classes. The product of causal functions of quantum field theory was defined by Bogolyubov and Parasyuk [139,140] (see also Stepanov [141-143], Shcherbina [144,145], Taylor [146], Bremermann [147], and Steinmann [148]).

In conclusion, let us look at two examples.

(1) Find $(n = 1)$

$$\left(\frac{1}{x \pm i0}\right)^k = \frac{1}{x \pm i0} \frac{1}{x \pm i0} \cdots \frac{1}{x \pm i0}.$$

The inequality $\left|\frac{1}{z^k}\right| \leqslant |y|^{-k}$ implies that $1/z^k \in H_0(\Gamma^\pm)$ for $\pm y > 0$ respectively. Therefore, the boundary values of the function $1/z^k$ as $y \to \pm 0$ exists in S^* (see section 26.3). We denote them respectively by $1/(x \pm i0)^k$. From the definition (119) of the product of generalized functions, it follows that

$$\left(\frac{1}{x \pm i0}\right)^k = \frac{1}{(x \pm i0)^k}.$$

Suppose that $k=1$. Using the definition of the Cauchy principal value (at the points $x=0$ and $x=\infty$), we obtain for all $\varphi\in S$,

$$\lim_{\varepsilon\to+0}\int_{-\infty}^{\infty}\frac{\varphi(x)}{x\pm i\varepsilon}\,dx=\lim_{\varepsilon\to+0}\mathrm{Vp}\int_{-\infty}^{\infty}\frac{\varphi(x)-\varphi(0)}{x\pm i\varepsilon}+\varphi(0)\lim_{\varepsilon\to+0}\mathrm{Vp}\int_{-\infty}^{\infty}\frac{dx}{x+i\varepsilon}=$$

$$=\lim_{N\to+\infty}\int_{-N}^{N}\frac{\varphi(x)-\varphi(0)}{x}\,dx+\varphi(0)\lim_{\varepsilon\to+0}\int_{-\infty}^{\infty}\frac{-i\varepsilon\,dx}{x^2+\varepsilon^2}=$$

$$=\mathrm{Vp}\int_{-\infty}^{\infty}\frac{\varphi(x)}{x}\,dx\mp i\pi\varphi(0).$$

These are the famous formulas of Sokhotskiy. Let us rewrite them in terms of generalized functions

$$\frac{1}{x\pm i0}=\mp i\pi\delta(x)+P\frac{1}{x}.$$

Then, by making use of the continuity of the operation of differentiating generalized functions (see section 3.3), we conclude for ($k=1$ 2, ...) that

$$\left(\frac{1}{x\pm i0}\right)^k=\frac{1}{(x\pm i0)^k}=$$
$$=\frac{(-1)^{k-1}}{(k-1)!}\frac{d^{k-1}}{dx^{k-1}}\frac{1}{x\pm i0}=\frac{\pm(-1)^k}{(k-1)!}i\pi\delta^{(k-1)}(x)+P\frac{1}{x^k},\tag{123}$$

where

$$P\frac{1}{x^k}=\frac{(-1)^{k-1}}{(k-1)!}\frac{d^{k-1}}{dx^{k-1}}P\frac{1}{x}.$$

(2) Find $\delta^2(x)=\delta(x)\delta(x)$ for $n=1$. It follows from (123) that

$$\delta(x)=\frac{1}{2\pi i}\left(\frac{1}{x-i0}-\frac{1}{x+i0}\right).$$

Therefore,

$$F^y(x)=\frac{1}{2\pi i}\left(\frac{1}{x-iy}-\frac{1}{x+iy}\right)=\frac{y}{\pi(x^2+y^2)}.$$

Suppose that $\varphi\in S^{(1)}$ and that $\varphi(0)=0$. Then, calculating the corresponding limit (118), we obtain

$$\lim_{y\to+0}\int_{-\infty}^{\infty}[F^y(x)]^2\varphi(x)\,dx=$$
$$=\frac{1}{\pi^2}\lim_{y\to+0}\int_{-\infty}^{\infty}\frac{y^2}{(x^2+y^2)^2}\varphi(x)\,dx=0.$$

Thus, on the closed subspace $M \subset S^{(1)}$ of functions that vanish at the point 0, the function χ is equal to zero. But then, on the basis of (121), we have

$$\delta^2(x) = c\delta(x),$$

where c is an arbitrary constant.

30. GENERALIZED FUNCTIONS ASSOCIATED WITH LIGHT CONES

In this and the following section, we shall use different notations, which are especially suited for light cones. We denote points in the spaces R^{n+1} and C^{n+1} respectively by

$$x = (x_0, x_1, \ldots, x_n) = (x_0, \boldsymbol{x}),$$

$$z = x + iy = (z_0, \boldsymbol{z}); \quad dx = dx_0\, d\boldsymbol{x}.$$

We denote the scalar product of points ξ and z by

$$\xi z = \xi_0 z_0 - \boldsymbol{\xi z}, \qquad \boldsymbol{\xi z} = \xi_1 z_1 + \ldots + \xi_n z_n,$$

so that

$$x^2 = x_0^2 - |\boldsymbol{x}|^2, \qquad |\boldsymbol{x}|^2 = x_1^2 + \ldots + x_n^2.$$

We recall that a light cone $\Gamma = [x\colon x^2 > 0]$ consists of two convex components, namely, a future light cone $\Gamma^+ = [x : x_0 > |\boldsymbol{x}|]$ and a past light cone $\Gamma^- = [x\colon x_1 < -|\boldsymbol{x}|]$. We denote $T^{\pm} = T^{\Gamma^{\pm}} = R + i\Gamma^{\pm}$; $T = T^+ \cup T^-$.

We define the generalized functions $\theta(\pm x_0)\delta^{(k)}(x^2)$, for $0 \leqslant k <$, $(< n-1)/2$, (where $n \geqslant 2$), by the formula

$$\int \theta(\pm x_0)\delta^{(k)}(x^2)\varphi(x)\,dx = \\ = (-1)^k \int \left(\frac{\partial}{2x_0\,\partial x_0}\right)^k \left(\frac{\varphi(x_0, \boldsymbol{x})}{2x_0}\right)\Bigg|_{x_0 = \pm|\boldsymbol{x}|} d\boldsymbol{x}, \quad \varphi \in S. \tag{124}$$

For $k < n - 1/2$ this integral always converges, We set

$$\left.\begin{aligned} \delta^{(k)}(x^2) &= \theta(x_0)\delta^{(k)}(x^2) + \theta(-x_0)\delta^{(k)}(x^2), \\ \varepsilon(x_0)\delta^{(k)}(x^2) &= \theta(x_0)\delta^{(k)}(x^2) - \theta(-x_0)\delta^{(k)}(x^2). \end{aligned}\right\} \tag{125}$$

Let us prove the following formula for $n \geqslant 2$:

$$\Box_n[\theta(\pm x_0)\theta(x^2)] = 2(n-1)\theta(\pm x_0)\delta(x^2). \tag{126}$$

where $\Box_n$ is the wave operator

$$\Box_n = \frac{\partial^2}{\partial x_0^2} - \frac{\partial^2}{\partial x_1^2} - \ldots - \frac{\partial^2}{\partial x_n^2}.$$

For $\varphi \in S$, we have

$$\int \Box_n [\theta(x_0)\theta(x^2)]\varphi(x)\,dx = \int \theta(x_0)\theta(x^2)\Box_n\varphi(x)\,dx =$$
$$= \int\limits_{x_0 > |x|} \Box_n \varphi(x)\,dx_0\,dx_1 \ldots dx_n =$$
$$= \int\limits_{\Gamma^+} d\left(\frac{\partial\varphi}{\partial x_0} dx_1 \ldots dx_n + \frac{\partial\varphi}{\partial x_1} dx_0\,dx_2 \ldots dx_n - \ldots \right.$$
$$\left. \ldots + (-1)^n \frac{\partial\varphi}{\partial x_n} dx_0\,dx_1 \ldots dx_{n-1}\right).$$

We now use Stokes' formula (see section 22.4). Since the point 0, the vertex of the cone Γ^+, is singular, we need to draw a sphere of radius ε around it and integrate over the region $\Gamma_\varepsilon^+ = \Gamma^+ \setminus U(0, \varepsilon)$ (see Fig. 84). We obtain

$$\int \Box_n [\theta(x_0)\theta(x^2)]\varphi(x)\,dx = \lim_{\varepsilon \to +0} \int\limits_{\partial\Gamma_\varepsilon^+} \frac{\partial\varphi}{\partial x_0} dx_1 \ldots dx_n +$$
$$+ \frac{\partial\varphi}{\partial x_1} dx_0\,dx_2 \ldots dx_n - \ldots + (-1^n)\frac{\partial\varphi}{\partial x_n} dx_0\,dx_1 \ldots dx_{n-1}.$$

The boundary $\partial\Gamma_\varepsilon^+$ consists of two portions, namely, $\partial\Gamma^+ \setminus U(0, \varepsilon)$ and $\Gamma^+ \cap \partial U(0, \varepsilon)$. On the first of these, since $x_0^2 = x^2$, we have

$$x_0\,dx_0 = x_1\,dx_1 + \ldots + x_n\,dx_n, \quad x_0 \geqslant \varepsilon. \tag{127}$$

From this relation, we have the identity

$$\frac{\partial\varphi}{\partial x_0} dx_1 \ldots dx_n + \frac{\partial\varphi}{\partial x_1} dx_0\,dx_2 \ldots dx_n + \ldots$$
$$\ldots + (-1)^n \frac{\partial\varphi}{\partial x_n} dx_0\,dx_1 \ldots dx_{n-1} =$$
$$= d\left[\frac{\varphi}{x_0}((-1)^{n-1} x_1\,dx_2 \ldots dx_n + \ldots + x_n\,dx_1 \ldots dx_{n-1})\right] -$$
$$- 2(n-1)\varphi \frac{dx_1 \ldots dx_n}{2x_0}.$$

By using Stokes' formula and remembering that the integral over the second portion approaches 0 as $\varepsilon \to 0$, we obtain

$$\int \Box_n [\theta(x_0)\theta(x^2)]\varphi(x)\,dx =$$
$$= \lim_{\varepsilon \to +0} \left[-2(n-1) \int\limits_{\partial\Gamma_\varepsilon^+} \varphi \frac{dx_1 \ldots dx_n}{2x_0} + \right.$$
$$+ \int\limits_{\partial\Gamma^+ \cap \partial U(0,\varepsilon)} \frac{\varphi}{x_0}((-1)^{n-1} x_1\,dx_2 \ldots dx_n + \ldots$$
$$\left. \ldots + x_n\,dx_1 \ldots dx_{n-1})\right].$$

As $\varepsilon \to +0$, the second term in the square brackets in the right member approaches zero. In the first term, the integration is over the outer side of the surface $\partial\Gamma_\varepsilon^+$. Therefore, on the basis of (124), we conclude that it approaches

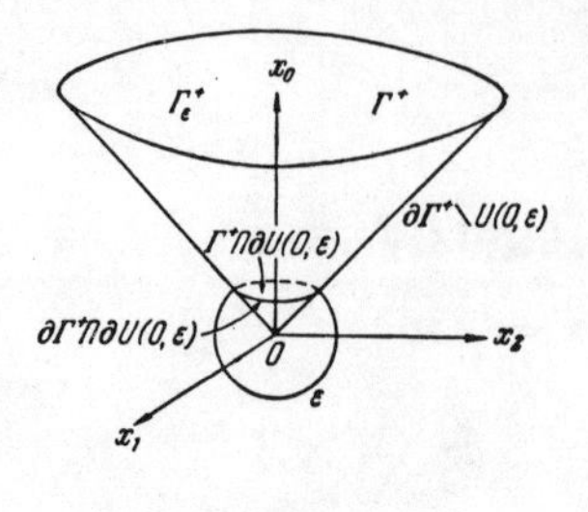

Fig. 84

$$2(n-1)\int \theta(x_0)\delta(x^2)\varphi(x)\,dx$$

as $\varepsilon \to +0$. This proves formula (126).

We could obtain formula (126) more simply by formally differentiating the left member of Eq. (126) and using the properties of the one-dimensional delta-function. Proceeding in this way, we obtain the formula

$$\square_n^k[\theta(\pm x_0)\theta(x^2)] = \frac{4^k\Gamma\left(\frac{n+1}{2}\right)}{\Gamma\left(\frac{n+1}{2}-k\right)}\theta(\pm x_0)\delta^{(k-1)}(x^2), \tag{128}$$

for $k < n+1/2$, where n is odd. For $k \geqslant n+1/2$, where n is even, this formula can be used to define the generalized functions

$$\theta(\pm x_0)\delta^{(k)}(x^2).$$

We now introduce the following generalized functions:

$$\square_n^k \ln|x^2| = (-1)^{k-1}\frac{4^k\Gamma(k)\Gamma\left(\frac{n+1}{2}\right)}{\Gamma\left(\frac{n+1}{2}-k\right)}P\frac{1}{(x^2)^k} \tag{129}$$

($1 \leqslant k < n+1/2$ if n is odd; $k \geqslant 1$ is arbitrary if n is even);

$$\left.\begin{aligned}
&\square_n^k\left\{\theta(\pm x_0)[\theta(x^2)x^2]^{-\frac{1}{2}}\right\} = \\
&= (-1)^k\frac{4^k}{\sqrt{\pi}}\Gamma\left(k+\frac{1}{2}\right)\frac{\Gamma\left(\frac{n}{2}\right)}{\Gamma\left(\frac{n}{2}-k\right)}\theta(\pm x_0)[\theta(x^2)x^2]^{-k-\frac{1}{2}}, \\
&\square_n^k[-\theta(-x^2)x^2]^{-\frac{1}{2}} = \\
&= \frac{4^k}{\sqrt{\pi}}\Gamma\left(k+\frac{1}{2}\right)\frac{\Gamma\left(\frac{n}{2}\right)}{\Gamma\left(\frac{n}{2}-k\right)}[-\theta(-x^2)x^2]^{-k-\frac{1}{2}}
\end{aligned}\right\} \tag{130}$$

($0 \leqslant k < n/2$ if n is ever; $k \geqslant 0$ is arbitrary if n is odd);

$$\left.\begin{aligned}&[\theta(x^2)\,x^2]^{-k-\frac{1}{2}}=\\&\quad=\theta(x_0)\,[\theta(x^2)\,x^2]^{-k-\frac{1}{2}}+\theta(-x_0)\,[\theta(x^2)\,x^2]^{-k-\frac{1}{2}},\\&\varepsilon(x_0)\,[\theta(x^2)\,x^2]^{-k-\frac{1}{2}}=\\&\quad=\theta(x_0)\,[\theta(x^2)\,x^2]^{-k-\frac{1}{2}}-\theta(-x_0)\,[\theta(x^2)\,x^2]^{-k-\frac{1}{2}}.\end{aligned}\right\} \tag{131}$$

Let us now prove the following auxiliary

LEMMA. *For a point z to belong to $T = T^+ \cup T^-$, it is necessary and sufficient that $(z - x') \neq \rho$, where $\rho \geq 0$, for all $x' \in R^{n+1}$*

Proof: From the condition that

$$(x-x'+iy)^2=(x-x')^2-y^2+2iy\,(x-x')\neq\rho\geqslant 0 \qquad x'\in R^n \tag{132}$$

it follows that $y^2 \neq 0$. If y^2 were negative, we could choose a vector x' such that

$$x_0'=x_0,\quad y\,(x-x')=0,\quad |\mathbf{x}-\mathbf{x}'|^2=-y^2,$$

and then, from (132) we would have $(z-x')^2=0$, which is impossible. Therefore, $y^2>0$, which means that $z\in T$.

Conversely, if $y^2>0$, then either $y\,(x-x')\neq 0$, or, for $y\,(x-x')=0$,

$$(x-x')^2-y^2=\left[\frac{\mathbf{y}\,(\mathbf{x}-\mathbf{x}')}{y_0}\right]^2-|\mathbf{x}-\mathbf{x}'|^2-y^2\leqslant$$
$$\leqslant -y^2\left[1+\frac{|\mathbf{x}-\mathbf{x}'|^2}{y_0^2}\right]<0,$$

which means that condition (132) is satisfied. This completes the proof.

It follows from this lemma that $z^2\neq\rho\geqslant 0$ for all $z\in T$. Therefore, the functions $\ln(-z^2)$, and $(-z^2)^p$ (where p is an arbitrary real number) are holomorphic (and single-valued*) in T, and, on the basis of the lemma in section 29.5, belong to the class $H_0(\Gamma)$. This means that their boundary values as $y\to 0$, where $y\in\Gamma^\pm$, exist and belong to S^* (see section 26.3). Since these boundary values are independent of the sequence $y\to 0$, where $y\in\Gamma^\pm$, we may take sequences of the particular form $(\pm\varepsilon,\,\mathbf{0})$, where $\varepsilon\to+0$, for such sequences.

---

*The single-valuedness of these functions in T likewise follows from the monodromy theorem (see section 6.6).

Thus, we may introduce generalized functions of the form

$$\lim_{\varepsilon\to+0} [|\boldsymbol{x}|^2-(x_0\pm i\varepsilon)^2]^{-\frac{s}{2}} = [|\boldsymbol{x}|^2-(x_0\pm i0)^2]^{-\frac{s}{2}}, \quad s=0,\ \pm 1,\ \ldots,$$

$$\lim_{\varepsilon\to+0} \ln[|\boldsymbol{x}|^2-(x_0\pm i\varepsilon)^2] = \ln[|\boldsymbol{x}|^2-(x_0\pm i0)^2] = \pm i\pi\varepsilon(x_0)\,\theta(x^2)+\ln|x^2|. \tag{133}$$

For $s\leqslant 1$, these generalized functions are locally integrable functions of power increase at infinity and they may be calculated without difficulty. For example,

$$[|\boldsymbol{x}|^2-(x_0\pm i0)^2]^{-\frac{1}{2}} = [-\theta(-x^2)\,x^2]^{-\frac{1}{2}} \pm i\varepsilon(x_0)[\theta(x^2)\,x^2]^{-\frac{1}{2}}. \tag{134}$$

Therefore, it will be more interesting to consider the case in which $s>1$. Suppose that $s=2k$ — is an even number. Then,

$$[|\boldsymbol{x}|^2-(x_0\pm i0)^2]^{-k} = \frac{-\Gamma\left(\frac{n+1}{2}-k\right)\Box_n^k \ln[|\boldsymbol{x}|^2-(x_0\pm i0)^2]}{4^k\Gamma(k)\,\Gamma\left(\frac{n+1}{2}\right)} \tag{135}$$

($1\leqslant k<n+1/2$ if n is odd; $k\geqslant 1$ is arbitrary if n is even.) By using formulas (125), (128), (129), and (135), we obtain in this case

$$[|\boldsymbol{x}|^2-(x_0\pm i0)^2]^{-k} = \frac{\pm i\pi}{(k-1)!}\,\varepsilon(x_0)\,\delta^{(k-1)}(x^2)+(-1)^k P\frac{1}{(x^2)^k}. \tag{136}$$

We note that formulas (136) may be regarded as a generalization of the well-known formulas (123) for a single variable:

$$\frac{1}{(x\pm i0)^k} = \frac{(-1)^{k-1}}{(k-1)!}\frac{d^k}{dx^k}\ln(x\pm i0) = \frac{(-1)^{k-1}}{(k-1)!}\frac{d^k}{dx^k}(\pm i\pi\theta(-x)+\ln|x|) = \pm i\pi\frac{(-1)^k}{(k-1)!}\delta^{(k-1)}(x)+P\frac{1}{x^k}.$$

Suppose now that $s=2k+1$ is an odd number. Then,

$$[|\boldsymbol{x}|^2-(x_0\pm i0)^2]^{-k-\frac{1}{2}} = \frac{\sqrt{\pi}\,\Gamma\left(\frac{n}{2}-k\right)\Box_n^k[|\boldsymbol{x}|^2-(x_0\pm i0)^2]^{-\frac{1}{2}}}{4^k\Gamma\left(k+\frac{1}{2}\right)\Gamma\left(\frac{n}{2}\right)} \tag{137}$$

($0 \leqslant k < n/2$ if n is even; $k \geqslant 0$ is arbitrary if n is odd). In this case, by using formulas (130), (131), (134), and (137), we obtain

$$[|x|^2-(x_0\pm i0)^2]^{-k-\frac{1}{2}}= \\ =[-\theta(-x^2)x^2]^{-k-\frac{1}{2}}\pm i(-1)^k\varepsilon(x_0)[\theta(x^2)x^2]^{-k-\frac{1}{2}}. \tag{138}$$

Let us now evaluate the Fourier-Laplace transformation of the function $\theta(\xi_0)\theta(\xi^2)$. Since this function vanishes outside the cone $\bar{\Gamma}^+$, it follows from Theorem 2 of section 26.4 that its Fourier-Laplace transformation is a holomorphic function belonging to the class $H_0(\Gamma^+)$. Therefore, on the basis of the holomorphic continuation theorem (see section 6.1), it will be sufficient to evaluate this transformation at points of the form $(x_0+i\varepsilon, x)$, where $\varepsilon>0$, and then to continue it holomorphically to all values of z in the domain T^+. We have

$$\int\theta(\xi_0)\theta(\xi^2)e^{i(x_0+i\varepsilon)\xi_0-ix\xi}d\xi= \\ =\sigma_{n-1}\int_0^\infty\int_0^\pi\rho^{n-1}\sin^{n-2}\theta e^{-i|x|\rho\cos\theta}d\theta\times \\ \times\int_\rho^\infty e^{i(x_0+i\varepsilon)\xi_0}d\xi_0\,d\rho=\frac{i\sigma_{n-1}}{x_0+i\varepsilon}\int_0^\infty\rho^{n-1}e^{i(x_0+i\varepsilon)\rho}\times \\ \times\int_0^\pi\sin^{n-2}\theta e^{-i|x|\rho\cos\theta}d\theta= \\ =\frac{i\sigma_{n-1}}{x_0+i\varepsilon}\sqrt{\pi}\,\Gamma\left(\frac{n-1}{2}\right)\left(\frac{2}{|x|}\right)^{\frac{n-2}{2}}\times \\ \times\int_0^\infty\rho^{\frac{n}{2}}J_{\frac{n-2}{2}}(|x|\rho)e^{i(x_0+i\varepsilon)\rho}d\rho= \\ =2^n\pi^{\frac{n-1}{2}}\Gamma\left(\frac{n+1}{2}\right)[|x|^2-(x_0+i\varepsilon)^2]^{-\frac{n+1}{2}}.$$

In carrying out these calculations, we used formula *6.413.1* and *4.432.2* of the book of tables [109]. Here, σ_n is the area of the surface of a unit sphere in R^n.

Thus,

$$\int\theta(\xi_0)\theta(\xi^2)e^{iz\xi}d\xi=2^n\pi^{\frac{n-1}{2}}\Gamma\left(\frac{n+1}{2}\right)(-z^2)^{-\frac{n+1}{2}},\quad z\in T^+. \tag{139}$$

Therefore, by using formula (126), we obtain

$$\int\theta(\pm\xi_0)\delta(\xi^2)e^{iz\xi}d\xi=\frac{1}{2(n-1)}\int\Box_n[\theta(\pm\xi_0)\theta(\xi^2)]e^{iz\xi}d\xi= \\ =-\frac{z^2}{2(n-1)}\int\theta(\pm\xi_0)\theta(\xi^2)e^{iz\xi}d\xi= \\ =\frac{2^{n-1}}{n-1}\pi^{\frac{n-1}{2}}\Gamma\left(\frac{n+1}{2}\right)[-z^2]^{-\frac{n-1}{2}} \tag{140}$$

if $z \in T^{\pm}$

We now introduce the generalized functions

$$\left.\begin{aligned} D_n^{(\pm)}(x) &= \frac{1}{(2\pi)^n i} \int \theta(\pm \xi_0)\, \delta(\xi^2)\, e^{ix\xi}\, d\xi, \\ D_n(x) &= D_n^{(+)}(x) - D_n^{(-)}(x) = \frac{1}{(2\pi)^n i} \int \varepsilon(\xi_0)\, \delta(\xi^2)\, e^{ix\xi}\, d\xi \end{aligned}\right\} \tag{141}$$

which will be important for us later on.

Explicit expressions for $D_n^{(\pm)}(x)$ follow directly from (140)

$$D_n^{(\pm)}(x) = \frac{\pi^{-\frac{n+1}{2}}}{2(n-1)i} \Gamma\left(\frac{n+1}{2}\right) [\,|\boldsymbol{x}|^2 - (x_0 \pm i0)^2]^{-\frac{n-1}{2}}.$$

Finally, by using formulas (136) and (138), we obtain from the above expression

$$D_n^{(\pm)}(x) =$$

$$= \begin{cases} \frac{1}{4}\pi^{-m-1}\Big[\pm \pi \varepsilon(x_0)\, \delta^{(m-1)}(x^2) + \\ \qquad + i(-1)^{m-1}(m-1)!\, P\frac{1}{(x^2)^m}\Big], \quad n = 2m+1; \\ \frac{1}{4}\pi^{-m-\frac{1}{2}}\Gamma\left(m-\frac{1}{2}\right)\Big\{\mp(-1)^m \varepsilon(x_0)\, [\theta(x^2)\, x^2]^{-m+\frac{1}{2}} - \\ \qquad - i[-\theta(-x^2)\, x^2]^{-m+\frac{1}{2}}\Big\}, \quad n = 2m. \end{cases} \tag{142}$$

Thus, on the basis of (140)-(142), the function $D_n(x)$ can be represented in the form of the difference

$$\begin{aligned} D_n(x) &= \frac{1}{2i(n-1)} \pi^{-\frac{n+1}{2}} \Gamma\left(\frac{n+1}{2}\right) \times \\ &\times \left\{ [\,|\boldsymbol{x}|^2 - (x_0 + i0)^2]^{-\frac{n-1}{2}} - [\,|\boldsymbol{x}|^2 - (x_0 - i0)^2]^{-\frac{n-1}{2}} \right\} \end{aligned} \tag{143}$$

between boundary values of the function $b_n(-z^2)\ n{-}1/2$, which is holomorphic in the domain T^+ and T^-, and it can be expressed in explicit form:

$$D_n(x) =$$

$$= \begin{cases} \frac{1}{2}\pi^{-m}\varepsilon(x_0)\, \delta^{(m-1)}(x^2), \quad \text{if} \quad n = 2m+1,\ m \geqslant 1, \\ \frac{1}{2}(-1)^{m-1}\pi^{-m-\frac{1}{2}}\, \Gamma\left(m-\frac{1}{2}\right) \varepsilon(x_0)\, [\theta(x^2)\, x^2]^{-m+\frac{1}{2}}, \\ \qquad\qquad \text{if} \quad n = 2m. \end{cases} \tag{144}$$

Let us show that the generalized function* $D_n(x)$ satisfies the wave equation $\square_n D_n = 0$. It follows from Eqs. (124) and (125) that $\xi^2 \varepsilon(\xi_0)\delta(\xi^2) = 0$ and, hence, on the basis of (141), $D_n(x)$ satisfies the wave equation.

Let us show that the function $D_n(x)$ satisfies the conditions

$$\left.\begin{aligned} &\lim_{x_0 \to \pm 0} D_n(x_0, \boldsymbol{x}) = 0, \\ &\lim_{x_0 \to \pm 0} \frac{\partial}{\partial x_0} D_n(x_0, \boldsymbol{x}) = \delta(\boldsymbol{x}), \\ &\lim_{x_0 \to \pm 0} \frac{\partial^2}{\partial x_0^2} D_n(x_0, \boldsymbol{x}) = 0. \end{aligned}\right\} \tag{145}$$

It follows from formula (143) that, for every x_0, the generalized functions

$$\frac{\partial^k D_n(x_0, \boldsymbol{x})}{\partial x_0^k} \in S^*, \quad k = 0, 1, \ldots$$

are meaningful (see also section 3.11). Consider a function $\varphi(\boldsymbol{x}) \in S$. We define $\tilde{\varphi}(\boldsymbol{\xi}) = F^{-1}[\varphi]$. By using the definition (141) of the function $D_n(x)$, we obtain

for $x_0 \to \pm 0$

$$(D_n(x_0, \boldsymbol{x}), \varphi(\boldsymbol{x})) = \frac{1}{i} \int \varepsilon(\xi_0)\delta(\xi_0^2 - |\boldsymbol{\xi}|^2) e^{i\xi_0 x_0} \tilde{\varphi}(\boldsymbol{\xi})\, d\xi_0\, d\boldsymbol{\xi} =$$

$$= 2 \int \theta(\xi_0)\delta(\xi_0^2 - |\boldsymbol{\xi}|^2) \sin(\xi_0 x_0)\tilde{\varphi}(\boldsymbol{\xi})\, d\xi_0\, d\boldsymbol{\xi} =$$

$$= \int \frac{\sin(|\boldsymbol{\xi}| x_0)}{|\boldsymbol{\xi}|} \tilde{\varphi}(\boldsymbol{\xi})\, d\boldsymbol{\xi} \to 0,$$

$$\frac{\partial}{\partial x_0}(D_n(x_0, \boldsymbol{x}), \varphi(\boldsymbol{x})) =$$

$$= \int \cos(|\boldsymbol{\xi}| x_0)\tilde{\varphi}(\boldsymbol{\xi})\, d\boldsymbol{\xi} \to \int \tilde{\varphi}(\boldsymbol{\xi})\, d\boldsymbol{\xi} = \varphi(0),$$

$$\frac{\partial^2}{\partial x_0^2}(D_n(x_0, \boldsymbol{x}), \varphi(\boldsymbol{x})) = -\int \sin(|\boldsymbol{\xi}| x_0)|\boldsymbol{\xi}|\tilde{\varphi}(\boldsymbol{\xi})\, d\boldsymbol{\xi} \to 0,$$

as $x_0 \to \pm 0$, which proves relations (145).

Finally, let us show that the generalized functions

$$E_n^{(\pm)}(x) = \begin{cases} \frac{1}{2}\pi^{-\frac{n-1}{2}}\theta(\pm x_0)\delta^{\left(\frac{n-3}{2}\right)}(x^2), & n \text{ odd} \\ \frac{1}{2}(-1)^{\frac{n}{2}-1}\Gamma\left(\frac{n-1}{2}\right)\pi^{-\frac{n+1}{2}}\theta(\pm x_0)[\theta(x^2)x^2]^{-\frac{n-1}{2}}, & n \text{ even} \end{cases} \tag{146}$$

*For $n = 1$ we set $D_1 = \frac{1}{2}\varepsilon(x_0)\theta(x_0^2 - x_1^2)$.

are elementary solutions of the wave equation $\Box_n E_n^{(\pm)} = \delta$ (see section 3.10).

To prove this, it will be sufficient to show that

$$-z^2 \tilde{E}_n^{(\pm)}(z) = 1, \quad z \in T^{\pm}$$

where $\tilde{E}_n^{(\pm)}$ is the Fourier-Laplace transform of $E_n^{(\pm)}$. For odd n, this equation follows from formulas (128) and (139):

$$-z^2 \int E_n^{(\pm)}(x) e^{izx} dx =$$

$$= \frac{-z^2}{2} \pi^{-\frac{n-1}{2}} \int \theta(\pm x_0) \delta^{\left(\frac{n-3}{2}\right)}(x^2) e^{izx} dx =$$

$$= -\frac{z^2}{\Gamma\left(\frac{n+1}{2}\right)} 2^{-n} \pi^{-\frac{n-1}{2}} \int \Box_n^{\frac{n-1}{2}} [\theta(\pm x_0) \theta(x^2)] e^{izx} dx =$$

$$= \frac{(-z^2)^{\frac{n+1}{2}}}{\Gamma\left(\frac{n+1}{2}\right)} 2^{-n} \pi^{-\frac{n-1}{2}} \int \theta(\pm x_0) \theta(x^2) e^{izx} dx = 1.$$

Suppose now that n is an even number $n \geqslant 2$. By using formula (130), we obtain

$$-z^2 \tilde{E}_n^{(\pm)}(z) =$$

$$= \frac{-z^2 (-1)^{\frac{n}{2}-1}}{2} \Gamma\left(\frac{n-1}{2}\right) \pi^{-\frac{n+1}{2}} \int \theta(\pm x_0) [\theta(x^2) x^2]^{-\frac{n-1}{2}} \times$$

$$\times e^{izx} dx =$$

$$= -z^2 \frac{\pi^{-\frac{n}{2}} 2^{-n+1}}{\Gamma\left(\frac{n}{2}\right)} \int \Box_n^{\frac{n}{2}-1} \theta(\pm x_0) [\theta(x^2) x^2]^{-\frac{1}{2}} e^{izx} dx =$$

$$= (-z^2)^{\frac{n}{2}} \frac{\pi^{-\frac{n}{2}} 2^{-n+1}}{\Gamma\left(\frac{n}{2}\right)} \int \theta(\pm x_0) [\theta(x^2) x^2]^{-\frac{1}{2}} e^{izx} dx = 1,$$

since direct evaluation of the last integral yields (for even n)

$$\int \theta(\pm x_0) [\theta(x^2) x^2]^{-\frac{1}{2}} e^{izx} dx = 2^{n-1} \pi^{\frac{n}{2}} \Gamma\left(\frac{n}{2}\right) (-z^2)^{-\frac{n}{2}}.$$

This integral can be evaluated in a manner analogous to the integral (139) by using the same tables of integrals. It follows from (146) that the generalized function

$$E_n(x) = \frac{1}{2} [E_n^{(+)}(x) + E_n^{(-)}(x)] =$$

$$= \begin{cases} \frac{1}{4} \pi^{-\frac{n-1}{2}} \delta^{\left(\frac{n-3}{2}\right)}(x^2), & n \text{ odd} \\ \frac{(-1)^{\frac{n}{2}-1}}{4} \Gamma\left(\frac{n-1}{2}\right) \pi^{-\frac{n+1}{2}} [\theta(x^2) x^2]^{-\frac{n-1}{2}}, & n \text{ even} \end{cases} \tag{147}$$

is also an elementary solution of the wave equation. Finally, we note the following relationship between $E_n^{(\pm)}$ and D_n:

$$D_n(x) = E_n^{(+)}(x) - E_n^{(-)}(x) \tag{148}$$

which follows from (146) and (144).

With regard to generalized functions associated with light cones, see also Bogolyubov and Shirkov [87], Gel'fand and Shilov [38] and Steinmann [23].

31. REPRESENTATIONS OF THE SOLUTIONS OF THE WAVE EQUATION

1. *Representation I*

In the preceding section, it was shown that the generalized function $D_n(x)$ satisfies the wave equation and conditions (145). This means that it is a fundamental solution of the Cauchy problem for the semispaces $x_0 > 0$ and $x_0 < 0$ simultaneously. Furthermore, it is clear from formulas (144) that the carrier of the function $D_n(x)$ is contained in the closed light cone $\bar{\Gamma}$. (For odd n, this carrier coincides with $\partial\Gamma$ and, for even n, it coincides with $\bar{\Gamma}$.) Therefore, for every $x_0 = a$, the generalized functions

$$D_n(a, \boldsymbol{x}) \text{ and } \frac{\partial}{\partial x_0} D_n(a, \boldsymbol{x})$$

are finite.

Let $u \in S^*$ be an arbitrary solution of the wave equation. Then, by virtue of the hypoellipticity of the wave operator with respect to x_0 (see section 3.11), we conclude that, for every a,

$$u(a, \boldsymbol{x}) \in S^*, \quad \frac{\partial}{\partial x_0} u(a, \boldsymbol{x}) \in S^*.$$

In view of the properties of the function $D_n(x)$ and the uniqueness of the solution to the Cauchy problem for the wave equation, we conclude from this that our solution $u(x)$ can be represented in the form

$$\begin{aligned} u(x) = D_n(x_0 - a, \boldsymbol{x}') * \frac{\partial u(a, \boldsymbol{x}')}{\partial x_0} + \\ + \frac{\partial D_n(x_0 - a, \boldsymbol{x}')}{\partial x_0} * u(a, \boldsymbol{x}'). \end{aligned} \tag{149}$$

This formula expresses the solution $u(x)$ in terms of its value and the value of its conormal derivative on an arbitrary spacelike hyperplane $x_0 = a$.

By introducing the generalized functions in S^*

$$(\delta(x_0 - a)\, u(x_0, x), \varphi) = \int u(a, x)\, \varphi(a, x)\, dx,$$

$$\left(\delta(x_0 - a) \frac{\partial}{\partial x_0} u(x_0, x), \varphi\right) = \int \frac{\partial}{\partial x_0} u(a, x)\, \varphi(a, x)\, dx$$

we rewrite formula (149) as the sum of two $(n+1)$-dimensional convolutions

$$u(x) = D_n(x) * \delta(x_0 - a) \frac{\partial u(x)}{\partial x_0} + \frac{\partial D_n(x)}{\partial x_0} * \delta(x_0 - a)\, u(x).$$

Shifting the differentiation with respect to x_0 in the second term on the right to the second member of the convolution in accordance with formula (14) of section 3.3, we obtain

$$u(x) = D_n(x) * \left[\delta(x_0 - a) \frac{\partial u(x)}{\partial x_0} + \frac{\partial}{\partial x_0} (\delta(x_0 - a)\, u(x))\right].$$

By denoting

$$\psi_a(x) = \delta(x_0 - a) \frac{\partial u(x)}{\partial x_0} + \frac{\partial}{\partial x_0} (\delta(x_0 - a)\, u(x)),$$

we conclude from this that, *for any hyperplane $x_0 = a$, an arbitrary solution $u \in S^*$ of the wave equation can be represented in the form*

$$u = D_n * \psi_a, \quad \text{and} \quad \psi_a \in S^*, \quad S_{\psi_a} \subset [x_0 = a]. \tag{150}$$

2. *Spacelike hypersurfaces*

A closed class $C^{(1)}$ hypersurface Σ (see section 1.4) is said to be spacelike if, at an arbitrary point x^0 of this hypersurface, the conormal $\gamma = \gamma(x^0)$ lies in the cone $\Gamma^+ + x^0$, that is, if,

$$\gamma_0\, dx_0 = \gamma_1\, dx_1 + \ldots + \gamma_n\, dx_n, \quad \gamma_0 > |\gamma|, \quad |\gamma| = 1, \tag{151}$$

where $(dx_0, dx_1, \ldots, dx_n)$ is the displacement vector on Σ at the point x^0. If the hypersurface Σ is defined by the equation $x_0 = \omega(x)$, then, on the basis of (151), $|\text{grad}\ \omega| < 1$, $\gamma_0 = (1 + |\text{grad}\ \omega|^2)^{-\frac{1}{2}}$ and $= \gamma_0\, \text{grad}\ \omega$.

By fixing the direction of the conormal on Σ (on the side on which $x_0 \to +\infty$), we define an orientation of the hypersurface Σ (see section 22.2). We shall keep this orientation in mind in what follows. Unless the contrary is stated, we shall assume that the hypersurface Σ is sufficiently smooth.

For a hypersurface Σ of class $C^{(1)}$ to be spacelike, it is necessary and sufficient that

$$\Sigma \cap \{\overline{\Gamma} + x^0\} = \{x^0\}$$

for an arbitrary point $x^0 \in \Sigma$. If the point x^0 does not lie on Σ, the intersection of the cone $\Gamma + x^0$ with the hypersurface Σ is along a bounded simply connected domain lying on Σ. The spacelike hypersurface Σ partitions the space R^{n+1} into two domains $x > \Sigma$ and $x < \Sigma$ lying on the sides $x_0 \to +\infty$ and $x_0 \to -\infty$ respectively (see Fig. 85).

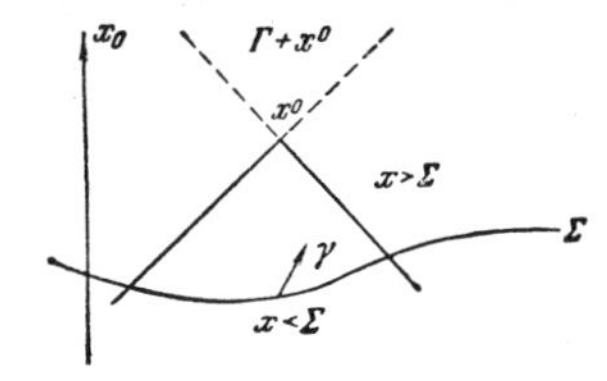

Fig. 85

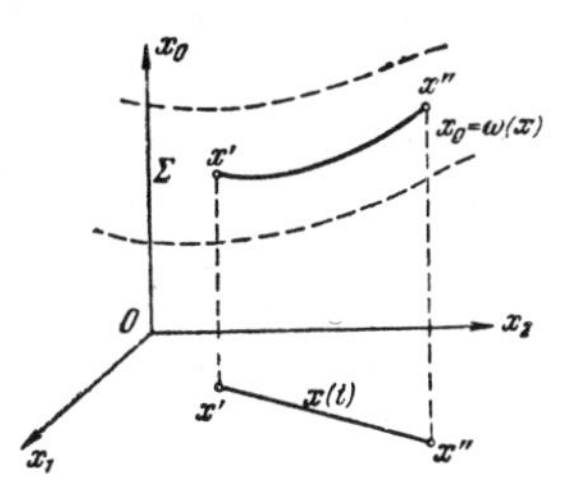

Fig. 86

We note that *a spacelike hypersurface is a regular set* (in the sense of section 3.5).

Proof: If we connect two points $x' \in \Sigma$ and $x'' \in \Sigma$ with a curve

$$x_0(t) = \omega[x(t)], \quad x(t) = tx' + (1-t)x'', \quad 0 \leqslant t \leqslant 1$$

lying $\Sigma = [x:\ x_0 = \omega(x)]$ (see Fig. 86), we see that the length of this curve does not exceed $\sqrt{2}\,|x' - x''|$ since

$$l = \int_0^1 \sqrt{|x_0'(t)|^2 + |x'(t)|^2}\, dt =$$

$$= \int_0^1 \sqrt{[\operatorname{grad} \omega,\ (x' - x'')]^2 + |x' - x''|^2}\, dt \leqslant$$

$$\leqslant \int_0^1 \sqrt{(|\operatorname{grad} \omega|^2 + 1)|x' - x''|^2}\, dt \leqslant$$

$$\leqslant \sqrt{2}\,|x' - x''| \leqslant \sqrt{2}\,|x' - x''|, \quad \text{q.e.d.}$$

We shall say that a spacelike hypersurface Σ is *strictly spacelike* if, for all $x \in \Sigma$, the inequality $|x_0 - a| \leqslant \rho |x|$ holds for some $\rho < 1$ and a (see Fig. 87).

Suppose that f is some (generalized) function. We denote by

$$\delta(\Sigma) f, \quad \delta(\Sigma) \frac{\partial f}{\partial \gamma}$$

and

$$\frac{\partial}{\partial \gamma} (\delta(\Sigma) f)$$

the generalized functions (layers) connected with a spacelike hypersurface Σ that act according to the rules

$$\left.\begin{aligned}(\delta(\Sigma) f, \varphi) &= \int\limits_{\Sigma} f(x)\varphi(x)\,d\Sigma, \\ \left(\delta(\Sigma)\frac{\partial f}{\partial\gamma}, \varphi\right) &= \int\limits_{\Sigma} \frac{\partial f}{\partial\gamma}\varphi(x)\,d\Sigma, \quad \left(\frac{\partial}{\partial\gamma}(\delta(\Sigma) f), \varphi\right) = \\ &= -\int\limits_{\Sigma} f(x)\frac{\partial\varphi(x)}{\partial\gamma}\,d\Sigma,\end{aligned}\right\} \qquad (152)$$

where $d\Sigma = \gamma_0^{-1}\,dx$ is the element of area on the surface Σ and the operator

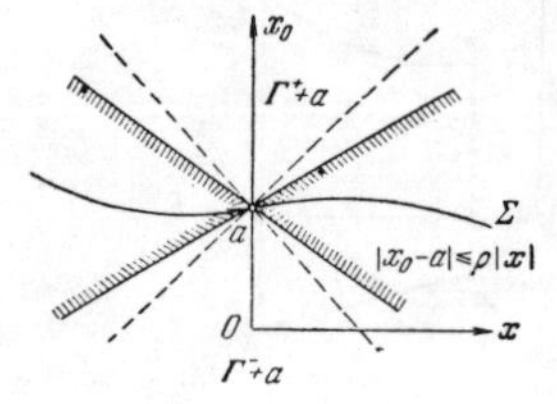

Fig. 87

$$\frac{\partial}{\partial\gamma} = \gamma_0\frac{\partial}{\partial x_0} + \gamma_1\frac{\partial}{\partial x_1} + \ldots + \gamma_n\frac{\partial}{\partial x_n}$$

is the conormal derivative on Σ, so that, by virtue of (151),

$$\frac{\partial}{\partial\gamma}\,d\Sigma = \frac{\partial}{\partial x_0}\,dx_1\ldots dx_n + \frac{\partial}{\partial x_1}\,dx_0\,dx_2\ldots$$
$$\ldots dx_n + \frac{\partial}{\partial x_n}\,dx_2\ldots dx_{n-1}\,dx_0.$$

Obviously, the generalized functions (152) do not always exist: for example, for $f \in C^{(1)}$, they exist in D^*. If $f \in \theta_M$, then

$$\delta(\Sigma)D_n, \quad \delta(\Sigma)\frac{\partial D_n}{\partial\gamma}, \quad \frac{\partial}{\partial\gamma}(\delta(\Sigma)D_n) \qquad (153)$$

also exist in S^*.

Let us show now that if $f = u$ is a solution in S^* of the wave equation, the generalized functions (152) exist in S^* for an arbitrary strictly spacelike hypersurface Σ. This assertion follows from the fact that the generalized functions (153) exist in S^* and from the representation (150). Specifically,

$$\left.\begin{aligned}\delta(\Sigma)u &= \delta(\Sigma)(D_n * \psi_0) = \delta(\Sigma)D_n * \psi_0, \\ \delta(\Sigma)\frac{\partial u}{\partial\gamma} &= \delta(\Sigma)\frac{\partial}{\partial\gamma}(D_n * \psi_0) = \delta(\Sigma)\frac{\partial D_n}{\partial\gamma} * \psi_0, \\ \frac{\partial}{\partial\gamma}(\delta(\Sigma)u) &= \frac{\partial}{\partial\gamma}[\delta(\Sigma)(D_n * \psi_0)] = \frac{\partial}{\partial\gamma}(\delta(\Sigma)D_n) * \psi_0,\end{aligned}\right\} \qquad (154)$$

and the right members of these equations exist in S^*.

3. *Representation II*

In this section, we shall show that the hyperplane $x_0 = a$ in formula (150) can be replaced with an arbitrary strictly spacelike hypersurface Σ.

Let us assume first that $u(x)$ is a sufficiently smooth solution of the wave equation. We introduce the functions $\bar{u}(x)$ and $\underline{u}(x)$ defined by

$$\bar{u}(x)=\begin{cases} u(x), & x>\Sigma, \\ 0, & x<\Sigma, \end{cases} \quad \underline{u}(x)=u(x)-\bar{u}(x). \tag{155}$$

Let us show that the functions $\bar{u}(x)$ and $\underline{u}(x)$ satisfy the equations

$$\square_n \bar{u}(x)=\psi_\Sigma(u;\ x), \quad \square_n \underline{u}(x)=-\psi_\Sigma(u;\ x), \tag{156}$$

where the generalized function $\psi_\Sigma \in D^*$ is defined by the formula (cf. section 31.1)

$$\psi_\Sigma(u;\ x)=\delta(\Sigma)\frac{\partial u}{\partial \gamma}+\frac{\partial}{\partial \gamma}(\delta(\Sigma)\, u). \tag{157}$$

Remembering that

$$d\left[\left(u\frac{\partial\varphi}{\partial\gamma}-\varphi\frac{\partial u}{\partial\gamma}\right)d\Sigma\right]=d\left[\left(u\frac{\partial\varphi}{\partial x_0}-\varphi\frac{\partial u}{\partial x_0}\right)dx_1\ldots dx_n+\right.$$
$$+\left(u\frac{\partial\varphi}{\partial x_1}-\varphi\frac{\partial u}{\partial x_1}\right)dx_0\,dx_2\ldots dx_n+\ldots$$
$$\left.\ldots+\left(u\frac{\partial\varphi}{\partial x_n}-\varphi\frac{\partial u}{\partial x_n}\right)dx_1\ldots dx_{n-1}\,dx_0\right]=$$
$$=(u\square_n\varphi-\varphi\square_n u)\,dx_0\,dx_1\ldots dx_n,$$

and using Stokes' formula (see section 22.4), we obtain Green's formula

$$(\square_n\bar{u},\ \varphi)=(\bar{u},\ \square_n\varphi)=\int_{x>\Sigma} u\square_n\varphi\, dx_0\,dx_1\ldots dx_n=$$
$$=\int_{x>\Sigma}\varphi\square_n u\, dx_0\,dx_1\ldots dx_n-$$
$$-\int_\Sigma\left(u\frac{\partial\varphi}{\partial\gamma}-\varphi\frac{\partial u}{\partial\gamma}\right)d\Sigma, \quad \varphi\in D.$$

Here, we took the orientation on Σ as opposite to the orientation defined by the outer normal to the domain $x>\Sigma$ (see section 31.2). Since $\square_n u=0$, the first of formulas (156) follows from the formula just obtained and from (154). The second of formulas (156) follows from the first and from the equation

$$u(x)=\bar{u}(x)+\underline{u}(x). \tag{158}$$

Since the carriers of the elementary solutions $E_n^{(\pm)}$ of the wave equation [see section 30, formulas (146)] are contained in $\bar{\Gamma}$ and since the carrier of ψ_Σ is contained in Σ, their convolutions $E_n^{(\pm)} * \psi_\Sigma$ exist in D^*. This is a consequence of the following property of a strictly spacelike hypersurface Σ: For any compact set K, the set of points

$$[(x, y): x \in \overline{\Gamma}, \quad y \in \Sigma, \quad x + y \in K]$$

is bounded (see Schwartz [11], Vol. II, p. 26). Therefore, the representations

$$\overline{u} = E_n^{(+)} * \psi_\Sigma, \quad \underline{u} = -E_n^{(-)} * \psi_\Sigma \tag{159}$$

follow from Eqs. (156).

If we add Eqs. (159) and apply relations (148) and (158), we obtain a representation for a sufficiently smooth solution $u(x)$ of the wave equation

$$u = D_n * \psi_\Sigma. \tag{160}$$

We note that formulas (159) and (160) are simply Kirchhoff's classical formulas written in terms of generalized functions (see for example, Riesz [110]). Formula (150) follows from formula (160) for $\Sigma = (x_0 = a)$.

If our solution $u(x)$ belongs to θ_M (and consequently $\psi_\Sigma \in S^*$) and if Σ is a strictly spacelike hypersurface, the convolutions (159) and (160) exist in S^*. This follows from the following

LEMMA. *Suppose that* $f \in S^*$, $G \in S^*$, $S_f \subset \overline{\Gamma}$, *where*

$$F = [x : |x_0 - a| \leqslant \rho\,|\boldsymbol{x}|] \text{ and } 0 \leqslant \rho < 1.$$

Then, the convolution $f * g$ *exists in* S^* *and is continuous with respect to* g (*in in the sense that* $f * g$ *in* S^* *as* $g \to 0$, *where* $S_g \subset F$).

Proof: Since F is a regular set, g can be extended in accordance with the theorem in section 3.8 to be a continuous linear functional on a Banach space with norm $\| \ \|_{m,F}$ for some $m = m(g) \geqslant 0$. Therefore, to prove the asserted properties of the convolution

$$(f * g, \varphi) = (g, f * \varphi), \quad \varphi \in S$$

it will be sufficient to show the existence of numbers $N \geqslant 0$ and $K > 0$ such that, for all $\varphi \in S$,

$$\| f * \varphi \|_{m,F} \leqslant K \, \| \varphi \|_N. \tag{161}$$

Remembering that $\overline{\Gamma}$ is regular, we use the theorem of section 3.8 to represent f in the form

$$f = \sum_{|\alpha| \leqslant M} D^\alpha \mu_\alpha,$$

where the measures μ_α possess the properties

$$\int (1+|x|)^{-p} |d\mu_\alpha(x)| < C, \quad S_{\mu_\alpha} \subset \bar{\Gamma}.$$

By setting $N = 2p + m$, we then obtain the following chain of inequalities:

$$\begin{aligned}
&\|(f * \varphi)(x)\|_{m, F} = \\
&= \sup_{|\beta| \leqslant m,\ x \in F} (1+|x|)^m |D^\beta (f(x'), \varphi(x+x'))| \leqslant \\
&\leqslant \sum_{|\alpha| \leqslant M} \sup_{|\beta| \leqslant m,\ x \in F} (1+|x|)^m \left| \int_{\bar{\Gamma}} D^{\alpha+\beta} \varphi(x+x')\, d\mu_\alpha(x') \right| \leqslant \\
&\leqslant \sum_{|\alpha| \leqslant M} \sup_{|\beta| \leqslant m,\ x \in F} (1+|x|)^{m+p} \times \\
&\times \int_{\bar{\Gamma}} \frac{|D^{\alpha+\beta} \varphi(x+x')(1+|x+x'|)^{2p+m}}{(1+|x+x'|)^{p+m}(1+|x'|)^p} |d\mu_\alpha(x')| \leqslant \\
&\leqslant K_1 \|\varphi\|_N \sup_{x \in F,\ x' \in \bar{\Gamma}} \left(\frac{1+|x|}{1+|x+x'|} \right)^{p+m} \leqslant \\
&\leqslant K_2 \|\varphi\|_N \sup_{\substack{|x_0| \leqslant \rho |x| \\ |x_0'| \geqslant |x'| \geqslant 1}} \left(\frac{1+|x|}{1+|x+x'|} \right)^{p+m}.
\end{aligned} \tag{162}$$

Let us now show that there exists a number $\sigma = \sigma(\rho) < 1$ such that

$$xx' \geqslant -\sigma |x| |x'| \quad \text{for} \quad |x_0| \leqslant \rho |x|, \quad |x_0'| \geqslant |x'|. \tag{163}$$

If such a σ did not exist, there would be points x' and x'' satisfying the conditions

$$x'x'' = -|x'| |x''|, \quad |x_0'| \leqslant \rho |x'|, \quad |x_0''| \geqslant |x''|,$$

which is impossible for $\rho < 1$.

Keeping inequality (163) in mind, we continue our chain of inequalities (162):

$$\begin{aligned}
&\|f * \varphi\|_{m, F} \leqslant \\
&\leqslant K_2 \|\varphi\|_N \sup_{\substack{|x_0| \leqslant \rho |x| \\ |x_0'| \geqslant |x'| \geqslant 1}} \left(\frac{1+|x|}{1+\sqrt{|x|^2+|x'|^2-2\sigma |x| |x'|}} \right)^{p+m} \leqslant \\
&\leqslant K_2 \|\varphi\|_N \sup_{x,\ |x'| \geqslant 1} \left(\frac{1+|x|}{1+\sqrt{2(1-\sigma)|x| |x'|}} \right)^{p+m} \leqslant K \|\varphi\|_N, \quad \text{q.e.d.}
\end{aligned}$$

Let us suppose now that the solution $u(x)$ of the wave equation belongs to S^*. Then, its regularization

$$u_\varepsilon(x) = (u(x_0, \boldsymbol{x}), \omega_\varepsilon(\boldsymbol{x}-\boldsymbol{x}')) \quad (D \ni \omega_\varepsilon(\boldsymbol{x}) \to \delta(\boldsymbol{x}), \ \varepsilon \to +0)$$

belongs to θ_M and also satisfies the wave equation. Here, the following limit relationships are valid in the sense of convergence in S^* as $\varepsilon \to +0$:

$$\left. \begin{aligned} u_\varepsilon(x) \to u(x), \quad u_\varepsilon(0, \boldsymbol{x}) = u(0, \boldsymbol{x}), \\ \frac{\partial}{\partial x_0} u_\varepsilon(0, \boldsymbol{x}) \to \frac{\partial}{\partial x_0} u(0, \boldsymbol{x}). \end{aligned} \right\} \tag{164}$$

Let us show that in the space S^*,

$$\psi_\Sigma(u_\varepsilon; x) \to \psi_\Sigma(u; x), \qquad \varepsilon \to +0. \tag{165}$$

For $\Sigma = (x_0 = 0)$ relation (165) follows from (164):

$$\begin{aligned} \psi_0(u_\varepsilon; x) = \delta(x_0)\frac{\partial u_\varepsilon}{\partial x_0} + \frac{\partial}{\partial x_0}(\delta(x_0)u_\varepsilon) \to \\ \to \delta(x_0)\frac{\partial u}{\partial x_0} + \frac{\partial}{\partial x_0}(\delta(x_0)u) = \psi_0(u; x), \quad \varepsilon \to +0. \end{aligned} \tag{166}$$

Furthermore, from formulas (154) and (157), we obtain

$$\begin{aligned} \psi_\Sigma(u_\varepsilon; x) = \delta(\Sigma)\frac{\partial u_\varepsilon}{\partial \gamma} + \frac{\partial}{\partial \gamma}(\delta(\Sigma)u_\varepsilon) = \\ = \left[\delta(\Sigma)\frac{\partial D_n}{\partial \gamma} + \frac{\partial}{\partial \gamma}(\delta(\Sigma)D_n)\right] * \psi_0(u_\varepsilon; x). \end{aligned} \tag{167}$$

Since the carrier of the generalized function in the square brackets (in 167) is contained in $\overline{\Gamma}$ and since the carrier $\psi_0(u_\varepsilon; x)$ is contained in the hyperplane $x_0 = 0$ we use the lemma just proven and the limit relation (166) to derive the limit relation (165) from (167).

Suppose that Σ is a strictly spacelike hypersurface. If we apply formula (160) to the function u_ε,

$$u_\varepsilon = D_n * \psi_\Sigma(u_\varepsilon; x),$$

and use the lemma and the limit relations (164) and (165), we conclude that formula (160) remains valid for an arbitrary solution $u(x)$ in S^*. Thus, we have proven the

THEOREM. *For any strictly spacelike hypersurface Σ, an arbitrary solution $u(x)$ in S^* of the wave equation can be represented in the form*

$$u = D_n * \psi_\Sigma, \tag{168}$$

where

$$\psi_\Sigma(u; x) = \delta(\Sigma)\frac{\partial u}{\partial \gamma} + \frac{\partial}{\partial \gamma}(\delta(\Sigma)u) \in S^*, \qquad S_{\psi_\Sigma} \subset \Sigma. \tag{169}$$

Conversely, if a generalized function $u(x)$ has the representation (168), where $\psi_\Sigma \in S^*$ and $S_{\psi_\Sigma} \subset \Sigma$, then u belongs to S^* and satisfies the wave equation.

4. *Representation III*

Every solution $u(x) \in S^*$ of the wave equation belongs to the class $L_0(\Gamma)$. Therefore, from the theorem in section 29.3, it can be represented in the form of the difference (96)

$$u(x) = f^+(x_0 + i0, \, \boldsymbol{x}) - f^-(x_0 - i0, \, \boldsymbol{x}) \tag{170}$$

of the boundary values of the functions $f^\pm(z)$ belonging to the classes $H_0(\Gamma^\pm)$ respectively. We recall that if $f^\pm(z)$ belongs to the class $H_0(\Gamma^\pm)$ (see section 26.4) this means that these functions are holomorphic in the corresponding tubular radial domain $T^\pm = R^{n+1} + i\Gamma^\pm$ and that, in an arbitrary subdomain $R^{n+1} + i[\pm y_0 > (1+\delta)|\boldsymbol{y}|]$, of it, where $\delta > 0$, they satisfy the inequality

$$|f^\pm(z)| \leqslant C_\delta (1 + |z|)^\beta (1 + |y_0|^{-\alpha}) \tag{171}$$

for certain nonnegative α and β (which do not depend on δ).

Our problem is to find an integral representation for the functions $f^\pm(z)$. We assume $n > 1$. (The case in which $n = 1$ is easily examined, but we do not need it.) As a preliminary, let us find an integral representation for all derivatives $D^l f^\pm(z)$ of sufficiently high order l such that $|l| \geqslant s$, where s is an integer to be chosen later. If we differentiate the representation (168) and use formula (143), we obtain

$$\begin{aligned} D^l u(x) = D^l D_n * \psi_\Sigma = \\ = c_n D^l \Big\{ [|\boldsymbol{x}|^2 - (x_0 + i0)^2]^{-\frac{n-1}{2}} - \\ - [|\boldsymbol{x}|^2 - (x_0 - i0)^2]^{-\frac{n-1}{2}} \Big\} * \psi_\Sigma, \end{aligned} \tag{172}$$

where

$$c_n = \frac{1}{2i(n-1)} \pi^{-\frac{n+1}{2}} \Gamma\left(\frac{n+1}{2}\right).$$

In what follows, we shall need the following

LEMMA. 1. *Suppose that a spacelike hypersurface Σ is such that $-\infty < a \leq x_0 \leq b < +\infty$ for all $x \in \Sigma$. Then, for arbitrary $m \geq 0$, there exists a nonnegative number s such that*

$$\left\| D^l [-(z-x')^2]^{-\frac{n-1}{2}} \right\|_{m, \Sigma} \leqslant K_{\delta, l} (1+|z|)^{\beta} (1+|y_0|^{-\alpha}) \tag{173}$$

for all y in an arbitrary subcone $|y_0| \geq (1+\delta)|y|$, where $\delta > 0$ and all l such that $|l| \geq s$. Here, $\alpha \geq 0$ and $\beta \geq 0$ do not depend on δ. Also, for all $\phi \in S$,

$$\left\| D^l (-z^2)^{-\frac{n-1}{2}} * \varphi - D^l [|x|^2 - (x_0 \pm i0)^2]^{-\frac{n-1}{2}} * \varphi \right\|_{m, \Sigma} \to 0, \tag{174}$$

as $y \to 0$, where $y \in \Sigma$.

Proof: According to the lemma in section 30, the function $(z-x')^2 \neq 0$ for all $z \in T$ and $x' \in R^{n+1}$. Therefore, the function $[-(z-x')^2]^{-1}$ is infinitely differentiable with respect to x'. Suppose that $y_0 > 0$. To prove inequality (173), we use (141). Assuming $s = s(m)$ sufficiently large and remembering that

$$-y_0|\xi| + y\xi \leqslant -\eta|\xi|y_0, \qquad \eta = \frac{\delta}{1+\delta},$$

for $y_0 \geqslant (1+\delta)|y|$, we obtain a chain of inequalities for $|l| \geqslant s$:

$$\left\| D^l [-(z-x')^2]^{-\frac{n-1}{2}} \right\|_{m, \Sigma} =$$

$$= K_1 \sup_{|\gamma| \leqslant m,\ x' \in \Sigma} (1+|x'|)^m \left| \int \theta(\xi_0) \delta(\xi^2) \xi^{\gamma+l} e^{i(z-x')\xi} d\xi \right| \leqslant$$

$$\leqslant K_2 \sup_{\substack{|\gamma| \leqslant m,\ x' \\ a \leqslant x_0' \leqslant b}} \left| (1+|x'|^2)^{\left[\frac{m}{2}\right]+1} \times \right.$$

$$\left. \times \int |\xi|^{\gamma_0+l_0-1} (\xi)^{\gamma+l} e^{i(z_0-x_0')|\xi| - i(z-x')\xi} d\xi \right| =$$

$$= K_2 \sup_{\substack{|\gamma| \leqslant m,\ x' \\ a \leqslant x_0' \leqslant b}} \left| \int e^{ix'\xi} (1-\Delta)^{\left[\frac{m}{2}\right]+1} \times \right.$$

$$\left. \times \left[|\xi|^{\gamma_0+l_0-1} \xi^{\gamma+l} e^{i(z_0-x_0')|\xi| - iz\xi} \right] d\xi \right| \leqslant$$

$$\leqslant K_3 (1+|z|)^{\beta} \int |\xi|^p e^{-y_0|\xi| + y\xi} d\xi =$$

$$= K_4 (1+|z|)^{\beta} \int_0^{\infty} \rho^p e^{-\eta y_0 \rho} d\rho \leqslant K_{\delta, l} (1+|z|)^{\beta} (1+y_0^{-\alpha}).$$

This completes the proof of inequality (173). Let us prove relation (174). For $\varphi \in S$,

$$\left(D^l (-z^2)^{-\frac{n-1}{2}} * \varphi \right)(x') = \int D^l [-(x+iy)^2]^{-\frac{n-1}{2}} \varphi(x+x') dx =$$

$$= c_n' \left(F[\theta(\xi_0) \delta(\xi^2) \xi^l e^{-y\xi}](x), \varphi(x+x') \right) =$$

$$= c_n' \int \theta(\xi_0) \delta(\xi^2) \xi^l e^{-y\xi - i\xi x'} F[\varphi](\xi) d\xi =$$

$$= \frac{c_n'}{2} \int |\xi|^{l_0-1} \xi^l e^{-y_0|\xi| + y\xi - ix_0|\xi| + ix\xi} F[\varphi](|\xi|, \xi) d\xi.$$

Therefore,

$$\left\| \left\{ D^l [-(x+iy)^2]^{-\frac{n-1}{2}} - D^l [|\mathbf{x}|^2 - (x_0+i0)^2]^{-\frac{n-1}{2}} \right\} * \varphi \right\|_{m,\Sigma} \leqslant$$

$$\leqslant \frac{c_n'}{2} \sup_{|\gamma| \leqslant m,\, x' \in \Sigma} (1+|x'|)^m \left| \int |\xi|^{l_0+\gamma_0-1} \xi^{l+\gamma} e^{-ix_0'|\xi|+ix'\xi} \times \right.$$

$$\left. \times (e^{-y_0|\xi|+y\xi} - 1) F[\varphi](|\xi|, \xi)\, d\xi \right| \leqslant$$

$$\leqslant K_5 \sup_{\substack{|\gamma| \leqslant m,\, x' \\ a \leqslant x_0' \leqslant b}} \left| \int e^{ix'\xi} (1-\Delta)^{\left[\frac{m}{2}\right]+1} \times \right.$$

$$\times |\xi|^{l_0+\gamma_0-1} \xi^{l+\gamma} e^{-ix_0'|\xi|} (e^{-y_0|\xi|+y\xi} - 1) \times$$

$$\left. \times F[\varphi](|\xi|, \xi)\, d\xi \right| \to 0 \quad \text{for } y \to 0,\ y \in \Gamma^+ \text{ q.e.d.}$$

Let us show now that if a spacelike hypersurface Σ satisfies the condition of Lemma 1, the right member of (172) can, for sufficiently large s (depending only on the order of ψ_Σ), be broken into two terms, so that each derivative $D^l u(x)$, for $|l| \geqslant s$, can be represented in the form of the difference

$$D^l u(x) = f_l^+(x_0 + i0, \mathbf{x}) - f_l^-(x_0 - i0, \mathbf{x}) \tag{175}$$

between the boundary values of the functions

$$f_l^\pm(z) = c_n \int \psi_\Sigma(u; x') D^l [-(z-x')^2]^{-\frac{n-1}{2}} dx' \tag{176}$$

belonging to the classes $H_0(\Gamma^\pm)$ respectively. The integral in (176) should be understood symbolically as the value of the functional ψ_Σ applied to the corresponding function.

To prove the preceding assertion, note that, since the carrier ψ_Σ is contained in the regular set Σ (see section 31.2), it follows from the theorem in section 3.8 that ψ_Σ can be extended to a continuous linear functional on a Banach space with norm $\| \;\|_{m,\Sigma}$ for some $m \geqslant 0$. According to the lemma in section 30, the function $[-(z-x')^2]^{-1}$ is holomorphic in $T^\pm$ for all x'. Suppose that $s = s(m) \geqslant 0$ is the integer defined in Lemma 1. Then, it follows from inequalities (173) that, for all l and j (where $|l| \geqslant s$, and $0 \leqslant j \leqslant n$,

$$\left\| \frac{\partial}{\partial z_j} D^l [-(z-x')^2]^{-\frac{n-1}{2}} - \frac{1}{\Delta z_j} \left\{ D^l [-(z+\Delta z_j - x')^2]^{-\frac{n-1}{2}} - \right.\right.$$

$$\left.\left. - D^l [-(z-x')^2]^{-\frac{n-1}{2}} \right\} \right\|_{m,\Sigma} \to 0$$

uniformly with respect to z, which belongs to an arbitrary set that is compact in T. Therefore, the right members in (176) have all first-order derivatives in $T^{\pm}$, and these derivatives can be obtained by differentiating under the "integral" sign. Therefore, on the basis of Hartogs' fundamental theorem (see section 4.2), we conclude that the functions $f_l^{\pm}(z)$, where $|l|\geqslant s$ are holomorphic in T^+ and T^- respectively. Furthermore, on the basis of (173), they satisfy inequality (171):

$$|f_l^{\pm}(z)| \leqslant |c_n| \|\psi_\Sigma\|_{-m,\Sigma} \left\| D^l [-(z-x')^2]^{-\frac{n-1}{2}} \right\|_{m,\Sigma} \leqslant$$
$$\leqslant c'_{\delta,l}(1+|z|)^{\beta_l}(1+|y_0|^{-\alpha_l}), \quad |y_0| \geqslant (1+\delta)|y|$$

that is, they belong to the classes $H_0(\Gamma^{\pm})$. Finally, it follows from (174) that

$$f_l^{\pm}(x_0 \pm i0, x) = c_n D^l [|x|^2 - (x_0 \pm i0)^2]^{-\frac{n-1}{2}} * \psi_\Sigma \tag{177}$$

since, for all $\varphi \in S$,

$$\int f_l^{\pm}(x+iy)\varphi(x)\,dx =$$
$$= c_n \left(\psi_\Sigma(x'), \int D^l [-(x+iy)^2]^{-\frac{n-1}{2}} \varphi(x+x')\,dx \right) \to$$
$$\to c_n \left(\psi_\Sigma, D^l [|x|^2 - (x_0 \pm i0)^2]^{-\frac{n-1}{2}} * \varphi \right) =$$
$$= c_n \left(D^l [|x|^2 - (x_0 \pm i0)^2]^{-\frac{n-1}{2}} * \psi_\Sigma, \varphi \right)$$

as $y \in \Gamma^{\pm}$, where $y \to 0$. If we note now that the representation (175) follows from formulas (177) and (172), all the assertions made above are proven.

On the other hand, if we differentiate Eq. (170), we obtain

$$D^l u(x) = D^l f^+(x_0 + i0, x) - D^l f^-(x_0 - i0, x).$$

Since the derivative of the boundary value coincides with the boundary value of the derivative, the last equation can be rewritten in the form

$$D^l u(x) = (D^l f^+)(x_0 + i0, x) - (D^l f^-)(x_0 - i0, x).$$

If we compare this equation with Eq. (175), we conclude on the basis of the theorem in section 29.3 that the $f_l^{\pm}(z)$ coincides with the $D^l f^{\pm}(z)$ up to the polynomials $P_l^{\pm}$, that is, on the basis of (176)

$$D^l f^{\pm}(z) = c_n \int \psi_\Sigma(u; x') D^l [-(z-x')^2]^{-\frac{n-1}{2}} dx' + P_l^{\pm}(z), \quad |l| \geqslant s. \tag{178}$$

To determine the functions $f^{\pm}(z)$ from their derivatives of order s, we need the following

LEMMA 2. *Suppose that a function $f(z)$ is holomorphic in a domain D_1 and that all its first-order derivatives $\partial f/\partial z_j$ have holomorphic (and single-valued) continuations f_j to a domain $D \supset D_1$, where D is a simply connected or a Runge domain. Then, $f(z)$ can be holomorphically (and single-valuedly) continued into the domain D, and its continuation $F(z)$ is given by the formula*

$$F(z) = f(a) + \int_a^z \sum_j f_j(z')\,dz'_j \equiv L_0\{f_j\}, \tag{179}$$

where the integration is over an arbitrary piecewise-smooth path in the domain D that connects an arbitrary point $a \in D_1$ with a variable point $z \in D$.

Proof: For $f_j = \partial f/\partial z_j$ and $z \in D_1$, we have, in D_1,

$$\frac{\partial f_k}{\partial z_j} = \frac{\partial f_j}{\partial z_k}. \tag{180}$$

Since the functions f_j are holomorphic in D, it follows on the basis of the holomorphic continuation theorem (see section 6.1) that Eqs. (180) remain valid in the domain D. These equations are equivalent to the relation

$$d\Big(\sum_j f_j\,dz_j\Big) = 0, \qquad z \in D.$$

But then, it follows from Stokes' formula (see section 22.4) that the line integral in (179) is independent of the path of integration for all homotopic paths in D that connect the points a and z. Since $\partial F/\partial z_j = f_j$, it follows from the fundamental theorem of Hartogs (see section 4.2) that the function $F(z)$ is holomorphic in D. If the domain D is simply connected, then, from the monodromy theorem (see section 6.6), the function $F(z)$ is single-valued in D. If D is a Runge domain, then $F(z)$ is also single-valued in D (see section 24.8). Since $f_j = \partial f/\partial z_j$ in the domain D_1, it follows on the basis of section 6.7 that the functions $F(z)$ and $f(z)$ coincide in the domain D_1, which completes the proof.

Let us apply Lemma 2 to the functions (178) for $|l| = s-1$ and $D = D_1 = T^{\pm}$. We can, on the basis of formula (179) determine all the derivatives $D^l f^{\pm}(z)$ of order $s-1 = |l|$ from the right members of Eqs. (178) for $|l| = s$. From these derivatives, we can determine all derivatives of order $s-2 = |l|$, etc. Finally, after s steps, we obtain the functions $f^{\pm}(z)$ themselves. Let us denote this operation by L_0^s. Thus,

$$f^{\pm}(z) = L_0^s\Big\{c_n \int \psi_\Sigma(u, x') D^l [-(z-x')^2]^{-\frac{n-1}{2}}\,dx' + P_l^{\pm}(z)\Big\}, \quad z \in T^{\pm}, \tag{181}$$

where the $P_l^{\pm}$ are polynomials, $|s|=l$, and ψ_Σ is given by formula (169).

Remark: For $s=2$ the operator L_0^s takes the form

$$L_0^2\{f_{ij}\}=f(a)+\int_a^z \sum_k \left[\frac{\partial f(a)}{\partial z_k}+\int_a^{z'} \sum_j f_{kj}(z'')\,dz_j''\right] dz_k'.$$

Let us summarize these results in the form of the

Theorem: Suppose that a spacelike hypersurface Σ is such that $-\infty<a\leqslant x_0\leqslant b<+\infty$ for all $x\in\Sigma$. Then, every solution $u(x)\in S^*$ of the wave equation for $n>1$ can be represented in the form of the difference (170) of the boundary values of functions $f^{\pm}(z)$ of the class $H_0(\Gamma^{\pm})$ which are represented by formula (181) for certain $s\geqslant 0$, $\psi_\Sigma\in S^*$, $S_{\psi_\Sigma}\subset\Sigma$, and $P_l^{\pm}$, where $|l|=s$. Here, the functions $f^{\pm}(z)$ are determined up to an arbitrary polynomial.

32. THE JOST–LEHMANN–DYSON INTEGRAL REPRESENTATION

In connection with the proof of the dispersion relations in quantum field theory, Jost and Lehmann [33] and Dyson [41] obtained a general representation for functions of the class $L_0(\Gamma)$ that vanish in a domain G that is bounded, crudely speaking, by two spacelike hypersurfaces. This representation leads to an integral representation for functions of the class $H_0(\Gamma)$ that are holomorphic in the domain $T\cup\tilde{G}$. This last representation makes it possible to construct the envelope $K(T\cup\tilde{G})$ (see section 29.2).

1. *The envelope $B_\Gamma(G)$*

Suppose that $G\subset R^{n+1}$ is a Γ-regular open set (see section 28.1). This means that the envelope $B_\Gamma(G)$ can be represented in the form

$$\begin{aligned} B_\Gamma(G)&=\bigcup_{x'\in G,\,x''\in G} B_\Gamma(x',x'')=\\ &=\bigcup_{x'\in G,\,x''\in G}[x:\ x_0'+|\mathbf{x}-\mathbf{x}'|<x_0<x_0''-|\mathbf{x}-\mathbf{x}''|]=\\ &=[x:\ A^-(\mathbf{x})<x_0<A^+(\mathbf{x})], \end{aligned} \tag{182}$$

where (see Fig. 88)

$$\left.\begin{aligned} A^-(\mathbf{x})&=\inf_{x'\in G}[x_0'+|\mathbf{x}-\mathbf{x}'|],\\ A^+(\mathbf{x})&=\sup_{x'\in G}[x_0'-|\mathbf{x}-\mathbf{x}'|]. \end{aligned}\right\} \tag{183}$$

Let us note certain properties of the functions $A^{\pm}(x)$. It follows from the definition of $B_{\Gamma}(G)$ that these functions are unchanged if we replace G in (183) with $B_{\Gamma}(G)$.

Let x_1 and x_2 be points such that $|x_1 - x_2| < \varepsilon$. Then, it follows from the inequalities

$$x_0' - |x_2 - x'| - \varepsilon < x_0' - |x_1 - x'| < x_0' - |x_2 - x'| + \varepsilon$$

that

$$|A^+(x_1) - A^+(x_2)| \leqslant \varepsilon.$$

Therefore, either the function $A^+(x)$ is identically equal to $+\infty$ or it is continuous. Analogously, either $A^-(x) \equiv -\infty$ or $A^-(x)$ is continuous. Here, we always have $A^+(x) \geqslant c - |x|$ and $A^-(x) \leqslant d + |x|$.

Thus, all Γ-regular open sets can be divided into three classes, which we classify as rank zero if $A^- \equiv -\infty$ and $A^+ \equiv +\infty$, as rank 1 if $A^- \equiv -\infty$ and $A^+ < +\infty$ or $A^- > -\infty$ and $A^+ \equiv +\infty$, and as rank 2 if $A^- > -\infty$ and $A^+ < +\infty$ (see Fig. 88).

We shall say that a hyperboloid $(x - x')^2 = \lambda^2$, where $x' = (x_0', \ldots, x_n')$ and λ are real parameters defining this hyperboloid, is admissible for an open set G if the set $(x - x')^2 \geqslant \lambda^2$ has no points in common with G (and, consequently, with the envelope $B_{\Gamma}(G)$). We denote by $N(G)$ (see Fig. 89) the set of parameters $(x', \lambda) \subset R^{n+2}$ corresponding to admissible hyperboloids for G. Thus,

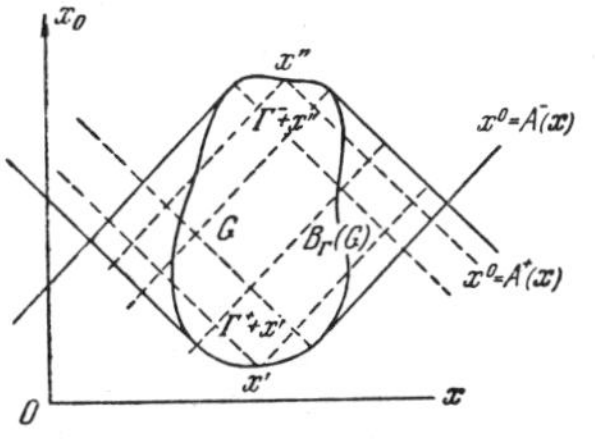

Fig. 88

$$N(G) = \bigcap_{x \in G} [(x', \lambda) : (x - x')^2 \leqslant \lambda^2] = [(x', \lambda) : \varphi(x') \leqslant \lambda^2], \tag{184}$$

where

$$\varphi(x') = \sup_{x \in G} (x - x')^2 = \sup_{x \in \partial G} (x - x')^2$$

is a lower-semicontinuous function (see section 2.3). It follows from (184) that

$$N(G) = [(x', \lambda): \quad B^+(x', \lambda) \leqslant x_0' \leqslant B^-(x', \lambda)], \tag{185}$$

$$\left.\begin{aligned} B^-(x', \lambda) &= \inf_{x \in G} \left[x_0 + \sqrt{|x - x'|^2 + \lambda^2}\right], \\ B^+(x', \lambda) &= \sup_{x \in G} \left[x_0 - \sqrt{|x - x'|^2 + \lambda^2}\right] \end{aligned}\right\} \tag{186}$$

are continuous functions. We shall say that $N(G)$ is *mutual* to the set G. It is closed and $N(G) = N[B_\Gamma(G)]$.

Suppose that $N(G) \neq \varnothing$ where G is a Γ-regular open set. We shall call the open set obtained by enveloping G by all admissible hyperboloids (see Fig. 89) the *envelope* of G and shall denote it by $B(G)$. Thus, on the basis of (184),

$$B(G) = \operatorname{int} \bigcap_{(x', \lambda) \in N(G)} [x: \ (x - x')^2 < \lambda^2] = R^{n+1} \setminus \overline{\bigcup_{(x', \lambda) \in N(G)} [x: \ (x - x')^2 = \lambda^2]}. \tag{187}$$

If a Γ-regular open set G is such that $N(G) = \varnothing$ (for example, if G is of rank 0 or 1), let us consider the increasing sequence of Γ-regular open sets (see Fig. 90)

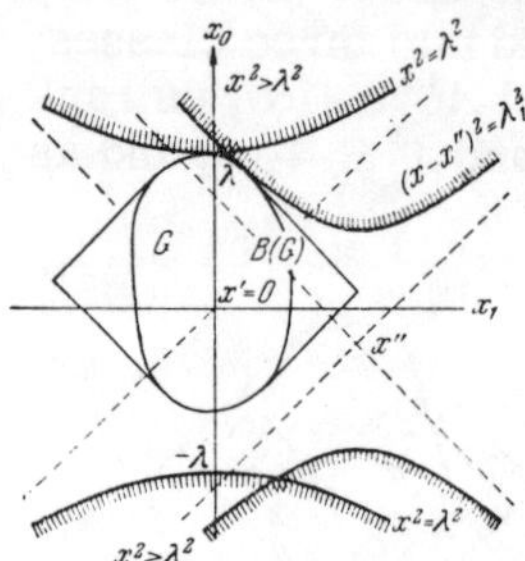

Fig. 89

$$G_\alpha^\Gamma = B_\Gamma(G) \cap (|x_0| < \alpha),$$

$$\alpha = 1, 2, \ldots, \quad B_\Gamma(G) = \bigcup G_\alpha^\Gamma$$

and let us construct their envelopes $B(G_\alpha^\Gamma)$. The sequence $B(G_\alpha)$, for $\alpha = 1, 2, \ldots$, is also an increasing sequence, and we define

$$\hat{B}(G) = \bigcup_\alpha B(G_\alpha^\Gamma). \tag{188}$$

We note that, on the basis of (184), the sequence of closed sets $N(G_\alpha^\Gamma)$, for $\alpha = 1, 2, \ldots$, is a decreasing sequence and $\lim N(G_\alpha^\Gamma) = \varnothing$ since

$$\lim_{\alpha \to \infty} N(G_\alpha^\Gamma) = \bigcap_\alpha N(G_\alpha^\Gamma) = \bigcap_\alpha \bigcap_{x \in G_\alpha^\Gamma} [(x', \lambda): \ (x - x')^2 \leqslant \lambda^2] = \bigcap_{x \in \bigcup_\alpha G_\alpha^\Gamma} [(x', \lambda): \ (x - x')^2 \leqslant \lambda^2] = N[B_\Gamma(G)] = N(G). \tag{189}$$

On the basis of (187), formula (188) may be rewritten in the form

$$B(G) = R^{n+1} \setminus \left\{ \lim_{\alpha \to +\infty} \overline{\bigcup_{(x', \lambda) \in N(G_\alpha^\Gamma)} [x: (x - x')^2 = \lambda^2]} \right\}. \tag{190}$$

If it is possible to take the limit in formula (190) as $\alpha \to \infty$ in the set $N(G_\alpha^\Gamma)$ (for example, if G is of rank 0), then, since $\lim N(G_\alpha^\Gamma) = \varnothing$, we have $B(G) = R^{n+1}$. However, taking the limit is not always possible since, when we envelop the set $B_\Gamma(G) \cap (|x_0| < \alpha)$ with all admissible hyperboloids $(x - x')^2 = \lambda^2$ as $\alpha \to +\infty$, one or both branches

$$x_0 = x_0' \pm \sqrt{|\mathbf{x} - \mathbf{x}'|^2 + \lambda^2}$$

of these hyperboloids become infinite, so that they reduce to hyperplanes of the form $\lambda(x-x')=0$, where $\lambda\in\bar{\Gamma}$ (that is, they become infinitely distant admissible hyperboloids [see Fig. 90]).

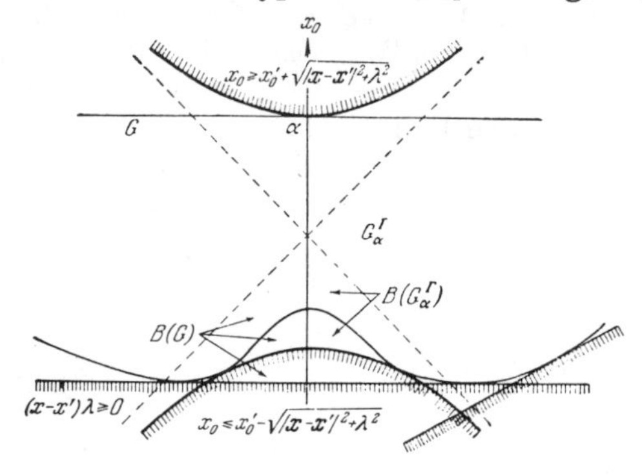

Fig. 90

We shall say that a hyperplane $\lambda(x-x')=0$ is admissible for an open set G if $\lambda\in\bar{\Gamma}$ and if the semispace $\lambda(x-x')\geqslant 0$ has no points in common with G (and hence with $B_\Gamma(G)$)(see Fig. 90). The set of parameters (x', λ) of the corresponding admissible hyperplanes for G will be denoted by $N_\infty(G)$, so that

$$N_\infty(G)=\bigcap_{x\in G}[(x', \lambda): \lambda(x-x')\leqslant 0, \lambda\in\bar{\Gamma}]. \tag{191}$$

Thus, if a Γ -regular open set G is such that $N(G)=\varnothing$, then, by virtue of (190),

$$B(G)=R^{n+1}\setminus\overline{\bigcup_{(x', \lambda)\in N_\infty(G)}[x: \lambda(x-x')=0]}=O(G). \tag{192}$$

In other words, $B(G)$ is obtained by enveloping G by the admissible hyperplanes (see Fig. 90); that is, $B(G)$ coincides with the convex envelope of G.

It follows from the construction that

$$B_\Gamma(G)\subset B(G). \tag{193}$$

In the examples considered in section 28.1, $B_\Gamma(G)=B(G)$. There exist cases, however, in which $B_\Gamma(G)\neq B(G)$. For example:

(1) $G=[x: |x_0|<a+|\boldsymbol{x}|]$, $a>0$; $B_\Gamma(G)=G$, $B(G)=R^{n+1}$, $N(G)=N_\infty(G)=\varnothing$ (Fig. 91).

(2) $G=[x: x_0+\theta|\boldsymbol{x}|>0]$, $\theta<1$; $B_\Gamma(G)=G$, $B(G)=R^{n+1}$, $N(G)=N_\infty(G)=\varnothing$ (Fig. 92).

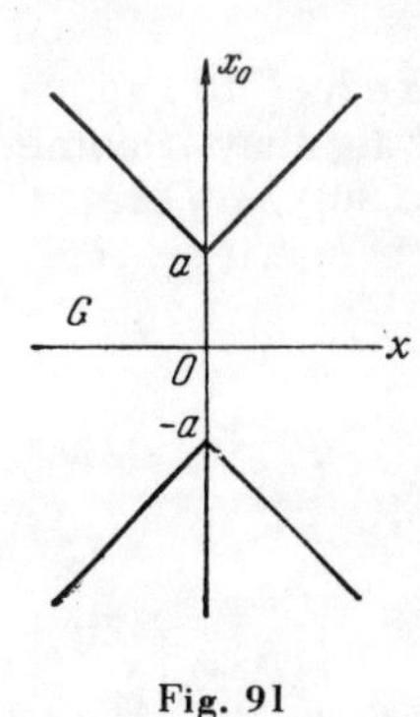

Fig. 91

(3) $G = [x: a-\sqrt{|x|^2+b^2} < x_0 < -a+\sqrt{|x|^2+c^2}]$, $a \geqslant 0$, $0 \leqslant b \leqslant c$; $B_r(G) = B(G) = G$, if $b \geqslant c-2a$; $B_r(G) = G$ and $B(G) = [x: a-\sqrt{|x|^2+(c-2a)^2} < x_0 < -a+\sqrt{|x|^2+c^2}]$, if $b < c-2a$; $N(G) = [(x', \lambda): |x'_0| + |\mathbf{x}'| \leqslant a, |\lambda| \geqslant \max\{b - \sqrt{(a-x'_0)^2-|\mathbf{x}'|^2}, c - \sqrt{(a+x'_0)^2-|\mathbf{x}'|^2}\}]$ (see Fig. 93; $b=a$, $c=4a$) (see Dyson [41]).

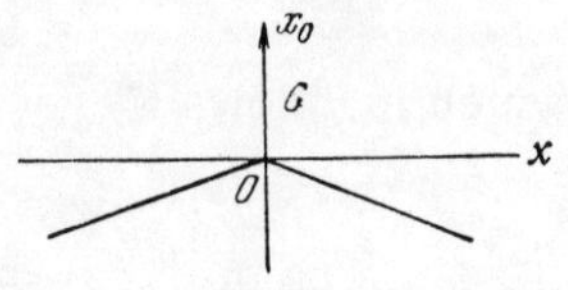

Fig. 92

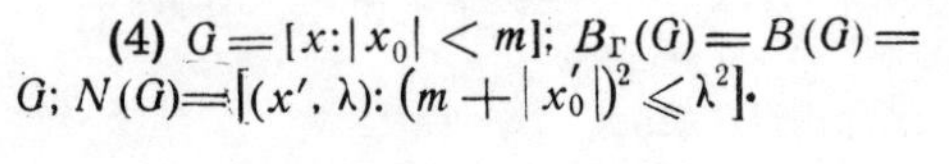

(4) $G=[x:|x_0|<m]$; $B_\Gamma(G)=B(G)=G$; $N(G)=[(x',\lambda): (m+|x'_0|)^2 \leqslant \lambda^2]$.

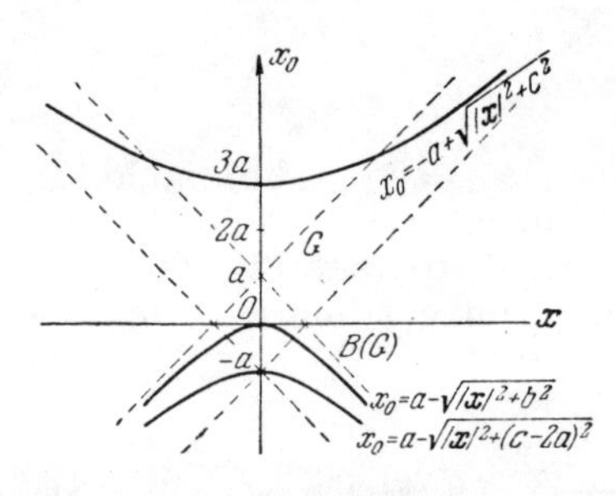

Fig. 93

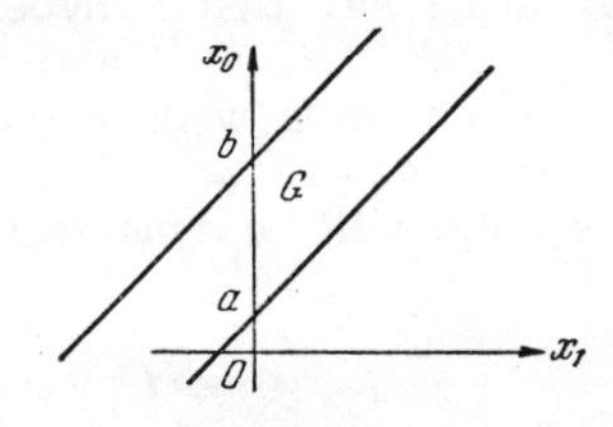

Fig. 94

(5) $G=[x: |x_0|+|\mathbf{x}|<a]$; $B_\Gamma(G)=B(G)=G$; $N(G)=[(x',\lambda): (a+|x'_0|)^2 \leqslant \lambda^2+|\mathbf{x}'|^2]$.

(6) $G=[x: a<x_0-x_1<b]$; $B_\Gamma(G)=B(G)=G$; $N_\infty(G) \neq \varnothing$ (see Fig. 94).

2. *Integral representations of functions belonging to the class* $(L_0\Gamma)$

THEOREM 1.* *For a generalized function $f(z)$ in S^* to belong to the class $L_0(\Gamma)$, it is necessary and sufficient that it have a representation of the form*

$$f(x)=u(x, 0), \tag{194}$$

where $u(x,\lambda) \in S^*$ satisfies the wave equation $\square_{n+1}u = 0$ and the condition

$$\frac{\partial}{\partial\lambda} u(x, 0)=0. \tag{195}$$

*See Dyson [41] and Wightman and Gårding [61].

The representation (194) *under conditions* (195) *is unique.*

Proof of the sufficiency: Suppose that $u(x, \lambda)$ belongs to $S^*_{(x,\lambda)}$ and satisfies the wave equation $\square_{n+1}u = 0$. Since the wave operator is hypoelliptic with respect to λ, $u(x, 0) \in S^*$ (see section 3.11). By performing the Fourier transformation on the wave equation $\square_{n+1}u = 0$, we obtain

$$(\xi^2 - \sigma^2) F^{-1}[u](\xi, \sigma) = 0, \tag{196}$$

from which it follows that

$$F^{-1}[u](\xi, \sigma) = 0 \quad \text{for all } \sigma \text{ and } \xi^2 < 0. \tag{197}$$

If we perform the inverse Fourier transformation with respect to the variable σ in Eq. (197), we obtain

$$F^{-1}[u(x, \lambda)](\xi) = 0 \quad \text{and} \quad \xi^2 < 0. \tag{198}$$

In particular, by setting $\lambda = 0$ in (198), we see that $u(x, 0) \in L_0(\Gamma)$.

Proof of the necessity: Suppose that $f(x) \in L_0(\Gamma)$. We define the generalized function $u_1(\xi, \lambda) \in S^*_{(\xi,\lambda)}$ of the variables (ξ_0, ξ, λ) by

$$u_1(\xi, \lambda) = F^{-1}[f](\xi) \cos(\lambda \sqrt{\xi^2}),$$

by which we mean the functional

$$\begin{aligned} u_1(\xi, \lambda) = {} & F^{-1}[f](\xi) \sum_{k=0}^{N} (-1)^k \frac{(\xi^2)^k \lambda^{2k}}{(2k)!} + \\ & + F^{-1}[f](\xi)\, \theta(\xi^2) \left[\cos(\lambda \sqrt{\xi^2}) - \sum_{k=0}^{N} (-1)^k \frac{(\xi^2)^k \lambda^{2k}}{(2k)!} \right], \end{aligned} \tag{199}$$

where N is the order of $F^{-1}[f]$. The function

$$\psi(\xi, \lambda) = \theta(\xi^2) \left[\cos(\lambda \sqrt{\xi^2}) - \sum_{k=0}^{N} \frac{(-1)^k (\xi^2)^k \lambda^{2k}}{(2k)!} \right]$$

is N times continuously differentiable and is of polynomial increase. Therefore, every term in (199) is meaningful and defines a generalized function in S^*. Furthermore, for every fixed λ, we have $u_1(\xi, \lambda) \in S^*$. Also, for arbitrary $\varphi(\xi) \in S$, the following limit relations hold (in the sense of convergence in the space $S^{(N)}$):

$$\psi(\xi, \lambda)\varphi(\xi) \to 0 \text{ and } \frac{\partial}{\partial \lambda} \psi(\xi, \lambda)\varphi(\xi) \to 0 \quad \text{and} \quad \lambda \to 0.$$

Therefore, $u_1(\xi, \lambda)$ possesses the following limit properties (in the sense of convergence in the space S^*):

$$u_1(\xi, \lambda) \to F^{-1}[f](\xi) \text{ and } \frac{\partial}{\partial \lambda} u_1(\xi, \lambda) \to 0 \quad \text{and} \quad \lambda \to 0.$$

From this, it follows that the generalized function in $S^{*}_{(x,\lambda)}$

$$u(x, \lambda) = \int u_1(\xi, \lambda)\, e^{i\xi x}\, d\xi \equiv F[u_1(\xi, \lambda)](x)$$

satisfies conditions (194) and (195). Let us show that it satisfies the wave equation. In view of (199), we obtain

$$\begin{aligned}\square_{n+1} u &= -\int \left(\xi^2 + \frac{\partial^2}{\partial \lambda^2}\right) u_1(\xi, \lambda)\, e^{i\xi x}\, d\xi = \\ &= \frac{(-1)^{N+1}}{(2N)!} \lambda^{2N} \int F^{-1}[f](\xi)\,[1 - \theta(\xi^2)]\,(\xi^2)^{N+1} e^{i\xi x}\, d\xi.\end{aligned} \tag{200}$$

Since the carrier $F^{-1}[f]$ is contained in the cone $\xi^2 \geqslant 0$, since its order is N, and since the function

$$[1 - \theta(\xi^2)]\,(\xi^2)^{N+1}$$

and all its first N derivatives vanish in the cone $\xi^2 \geqslant 0$, it follows (see section 3.5) that

$$F^{-1}[f](\xi)\,[1 - \theta(\xi^2)]\,(\xi^2)^{N+1} = 0.$$

It follows from this and from (200) that $u(x, \lambda)$ satisfies the wave equation $\square_{n+1} u = 0$.

Let us prove that this function $u(x, \lambda)$ that we have constructed is unique. Suppose that a second such function existed. Then, on the basis of (194) and (195), their difference $v(x, \lambda)$ would satisfy the wave equation and the conditions

$$v(x, 0) = \frac{\partial v(x, 0)}{\partial \lambda} = 0.$$

By using Theorem 2 of section 29.4, we conclude from this that $v(x, \lambda) = 0$, which completes the proof.

Remark: The generalized function $u(x, \lambda)$ of Theorem 1 can be expressed in terms of f and represented symbolically in the form

$$u(x, \lambda) = \int F^{-1}[f](\xi) \cos\left(\lambda \sqrt{\xi^2}\right) e^{i\xi x}\, d\xi, \tag{201}$$

where $F^{-1}[f](\xi) \cos\left(\lambda \sqrt{\xi^2}\right)$ is understood to mean the functional defined by the right member of Eq. (199).

Let us suppose now that a function $f(x)$ of class $L_0(\Gamma)$ vanishes in a Γ-regular open set G. According to Theorem 1, $f(x)$ can be represented in the form (194), where the function $u(x, \lambda)$ satisfies the wave equation $\square_{n+1} u = 0$ and condition (195). From this, we conclude on the basis of Theorem 2 of section 29.4 that $u(x, \lambda)$ vanishes in the Γ-convex envelope $B_\Gamma(G)$ of the set G, which lies in the plane $\lambda = 0$. (The envelope $B_\Gamma(G)$ is $(n+2)$-dimensional.) But the open set G is assumed to be Γ-regular in R^{n+1}. Then, it will also be Γ -regular in the space R^{n+2} of the variables (x, λ) (see Fig. 95). Therefore, when we apply formulas (182) and

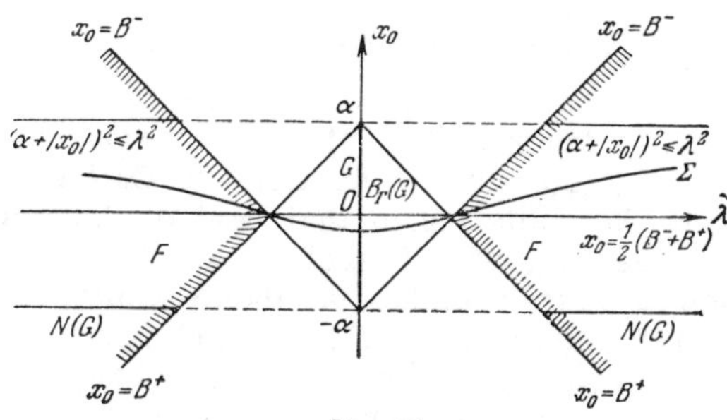

Fig. 95

(183) to this case, we conclude that

$$u(x, \lambda)=0, \quad (x, \lambda)\in B_\Gamma(G)= \\ =[(x, \lambda):\quad B^-(x, \lambda)<x_0<B^+(x, \lambda)], \tag{202}$$

where the functions $B^\pm(x, \lambda)$ are defined by formulas (186) and are continuous.

We shall say that a Γ -regular open set G satisfies condition (A) if there exists a strictly spacelike hypersurface Σ contained in the set (see Fig. 95)

$$B_\Gamma(G)\cup N(G)=[(x, \lambda):\ B^-<x_0<B^+, \text{ or } B^+\leqslant x_0\leqslant B^-].$$

For example, condition (A) is satisfied if the hypersurface $x_0=\frac{1}{2}(B^++B^-)$ is spacelike and satisfies the condition

$$|B^+(x, \lambda)+B^-(x, \lambda)-a|<2\rho\sqrt{|\boldsymbol{x}|^2+\lambda^2} \tag{203}$$

for some $\rho<1$ and a. Then, we may take this hypersurface as the Σ of the definition. For example, if a set G is symmetric with respect to the hyperplane $x_0=0$, then $B^+=-B^-$, so that $\Sigma=(x_0=0)$.

Condition (203) will be satisfied, for example, if G lies in the strip $|x_0|<\alpha$ (see Fig. 95) because then, on the basis of (186), we have

$$\left.\begin{aligned} &-\alpha\leqslant B^-, \quad B^+\leqslant\alpha, \\ &\frac{1}{2}|B^++B^-|\leqslant\frac{1}{2}\Big|\sup_{x'\in G}[x_0'-\sqrt{|\boldsymbol{x}-\boldsymbol{x}'|^2+\lambda^2}]+ \\ &+\inf_{x'\in G}[x_0'+\sqrt{|\boldsymbol{x}-\boldsymbol{x}'|^2+\lambda^2}]\Big|\leqslant\sup_{x'\in G}|x_0'|\leqslant\alpha. \end{aligned}\right\} \tag{204}$$

In this case, the set $N(G)$ contains the set $(\alpha+|x_0|)^2\leqslant\lambda^2$ (see Fig. 95 and cf. Example 4 of section 32.1).

Therefore, for large $|\lambda|$ we may take the hyperplane $x_0=0$ as Σ.

On the basis of the results of section 31.3, the function $u(x, \lambda)$ can be represented by formula (168):

$$u=D_{n+1}*\psi, \quad \psi\in S^*, \quad S_\psi\subset\Sigma,$$

where $D_{n+1}(x, \lambda)$ is a fundamental solution of the Cauchy problem for the wave equation $\Box_{n+1}u=0$ (see section 30). On the basis of (202), the function $u(x, \lambda)$ vanishes in that part of the hyperplane ψ in which $B^-<B^+$ (see Fig. 95). Therefore, on the basis of (169), the weight function Σ also vansihes at these points. The remaining points of Σ satisfy the inequality $B^+\leqslant B^-$ and, hence, on the basis of (185), belong to $N(G)$ (see Fig. 95). Therefore, $S_\psi\subset\Sigma\cap N(G)$ and the preceding representation may be rewritten in the form

$$u=D_{n+1}*\psi, \quad \psi\in S^*, \quad S_\psi\subset\Sigma\cap N(G). \tag{205}$$

If we set $\lambda=0$ in (205) and use Theorem 1, we obtain the following result:

THEOREM 2*. *For a generalized function $f(x)$ of class $L_0(\Gamma)$ to vanish in an open set G that satisfies condition (A), it is necessary and sufficient that it have a representation of the form*

$$f=D_{n+1}*\psi|_{\lambda=0}, \quad \psi\in S^*, \quad S_\psi\subset N(G)\cap\Sigma. \tag{206}$$

3. *Remarks*

(1) Let us rewrite the representation (206) symbolically in integral form:

$$f(x)=\int\limits_{N(G)\cap\Sigma} D_{n+1}(x-x', \lambda')\psi(x', \lambda')\,dx'\,d\lambda'. \tag{207}$$

Let us transform formula (207) for $n=3$. By setting $m=2$ in (144), we obtain

$$D_4(x, \lambda)=-\frac{1}{4\pi^2}\varepsilon(x_0)[\theta(x^2-\lambda^2)(x^2-\lambda^2)]^{-\frac{3}{2}}=$$

$$=-\frac{1}{4\pi^2}\varepsilon(x_0)\int\limits_{\lambda^2}^{\infty}(k^2-\lambda^2)^{-\frac{3}{2}}\delta(x^2-k^2)\,dk^2.$$

By substituting this expression into (207), formally reversing the order of "integration," and using formula (184) for $N(G)$, we obtain the well-known Jost-Lehmann-Dyson integral representation

$$f(x) = \int \varepsilon(x_0 - x'_0)\, dx' \int\limits_{\max[0,\, \varphi(x')]}^{\infty} \delta[(x-x')^2 - k^2]\, dk^2 \times$$

$$\times \int\limits_{\sqrt{\max[0,\, \varphi(x')]}}^{\sqrt{k^2}} \frac{(k^2-\lambda'^2)^{-3/2}}{-4\pi^2} [\psi(x', \lambda') + \psi(x', -\lambda')]\, d\lambda',$$

that is,

$$f(x) = \int\limits_{N(G)} \varepsilon(x_0 - x'_0)\, \delta[(x-x')^2 - k^2]\, \chi(x', k^2)\, dx'\, dk^2. \tag{208}$$

We note that formula (208) was obtained from the representation (207) by formal manipulations. We shall not have occasion to use this formula in what follows. A formula analogous to (208) holds in the space of an arbitrary number of dimensions.

(2) The representation (208) was obtained under the assumption that G satisfies condition (A). Dyson [41] conjectured that this representation was valid for an arbitrary open set G. The following examples show that in general this is not true even for Γ -regular domains (see Vladimirov and Nikitin [111]). Suppose that G is either the semispace $x_0 < 0$ or the strip $0 < x_0 - x_1 < a$ (see Fig. 94). (These domains are of rank 1 and 2, respectively.) In either case, $N(G) = \varnothing$. If the representation (208) were valid for these domains, the generalized functions $\delta(x_0)$ and $\delta(x_0 - x_1)$ of class $L_0(\Gamma)$ corresponding to them would vanish throughout all space, which is impossible.

(3) For Γ -regular open sets G that do not satisfy condition (A), we can obtain the corresponding integral representation by applying the representation (208) to a sequence of Γ-regular sets $G_\alpha^\Gamma = B_\Gamma(G) \cap [\,|x_0| < \alpha]$ (assuming that these sets satisfy condition (A)) and then taking the limit as $\alpha \to +\infty$. For example, if we proceed in this manner for a set G of rank 1, we obtain

$$f(x) = \int\limits_{N_\infty(G)} \varepsilon(x_0 - x'_0)\, \delta[\lambda(x - x')]\, \chi(x', \lambda)\, dx'\, d\lambda, \tag{209}$$

where the set $N_\infty(G)$ is defined by formula (191). But this reasoning is purely formal. It does show, however, that the representation (208) evidently remains valid for an arbitrary Γ -regular open set G if, to the set $N(G)$ of admissible hyperboloids, we add the set $N_\infty(G)$ of infinitely distant admissible hyperboloids, that is, the admissible hyperplanes $\lambda(x - x') = 0$.

4. *The quasi-analytic nature of functions of class $L_0(\Gamma)$*

For functions of class $L_0(\Gamma)$ the results of section 29.3 can be strengthened as follows:

THEOREM. *Suppose that a generalized function $f(x)$ of class $L_0(\Gamma)$ vanishes in a Γ-regular open set G. Then, $f(x) = 0$ in the envelope $B(G)$.*

Proof: Let us suppose first that $N(G) \neq \varnothing$. According to the theorem in section 32.2, $f(x)$ can be represented in the form (194): $f(x) = u(x, 0)$. Here, on the basis of (202), the function $u(x, \lambda)$ vanishes in the envelope

$$B_\Gamma(G) = [(x, \lambda): B^-(x, \lambda) < x_0 < B^+(x, \lambda)],$$

where $B^\pm(x, \lambda)$ are determined by formulas (186). But then $u(x, \lambda)$ vanishes outside the sphere of influence of that portion of the hypersurface $\Sigma = \left[x: x_0 = \frac{1}{2}(B^- + B^+)\right]$ where $B^+ \leqslant B^-$ (see Fig. 95); that is,

$$u(x, \lambda) = 0 \text{ for } (x, \lambda) \overline{\in} \overline{\bigcup_{\substack{(x', \lambda') \in \Sigma \\ B^+ \leqslant x_0' \leqslant B^-}} [\overline{\Gamma} + (x', \lambda')]}.$$

Therefore, *a fortiori*

$$u(x, \lambda) = 0 \text{ for } (x, \lambda) \overline{\in} \overline{\bigcup_{B^+ \leqslant x_0' \leqslant B^-} [(x - x')^2 - (\lambda - \lambda')^2 \geqslant 0]}.$$

If we set $\lambda = 0$ and apply formulas (194), (185), and (187), we then obtain

$$\text{for } x \in R^{n+1} \setminus \overline{\bigcup_{(x', \lambda') \in N(G)} [(x - x')^2 - \lambda'^2 \geqslant 0]} = B(G).$$

Suppose now that $N(G) = \varnothing$. According to the theorem in section 29.3, $f(x) = 0$ in $B_\Gamma(G)$. By applying the theorem that we have proven to the increasing sequence of Γ-regular open sets

$$G_\alpha^\Gamma = B_\Gamma(G) \cap (|x_0| < \alpha), \quad \alpha = 1, 2, \ldots,$$

$$\bigcup_\alpha G_\alpha^\Gamma = B_\Gamma(G), \quad N(G_\alpha^\Gamma) \neq \varnothing,$$

we see that

$$f(x) = 0 \text{ for } x \in B(G_\alpha^\Gamma), \quad \alpha = 1, 2, \ldots .$$

The assertion of the theorem follows from this on the basis of (188).

COROLLARY. *If G is a Γ-regular open set, then the carrier of an arbitrary function $f(x)$ of class $L_0(\Gamma)$ that vanishes in G is connected in the union of the (admissible) hyperboloids for G including the infinitely distant hyperboloids, that is, the hyperplanes, $\lambda(x - x') = 0$* (see Dyson [41]).

This follows from the preceding theorem and formulas (187) and (192).

In particular, if $N(G) \neq \varnothing$, the carrier f is contained in the union of admissible hyperboloids. If G is of rank 1, the carrier of f is contained in the union of admissible hyperboloids.

For functions of class $H_0(\Gamma)$ and $L_0(\Gamma)$ and for Γ -regular open sets G, Theorems 1 and 2 of sections 29.1, the theorem of section 29.3, and Theorems 1 and 2 of section 29.4 can be strengthened in the sense that the envelope $B_\Gamma(G)$ in these theorems can be replaced with the larger envelope $B(G)$. For example, the corollary to Theorem 2 of section 29.1 in the strengthened forms reads (see Vladimirov [40]):

Suppose that $f(z) \in H_0(\Gamma)$ and $f(x_0+i0, x) = f(x_0-i0, x)$ for $x \in G$, where G is a Γ -regular open set. Then, for an arbitrary such function $f(z)$ to be a polynomial, it is necessary and sufficient that $B(G) = R^{n+1}$.

Proof of the sufficiency: Suppose that $B(G) = R^{n+1}$. According to the theorem in section 29.3, the function

$$f(x_0+i0, x) - f(x_0-i0, x) \in L_0(\Gamma)$$

belongs to $L_0(\Gamma)$. Then, by applying the theorem proven above, we conclude that

$$f(x_0+i0, x) = f(x_0-i0, x), \qquad x \in B(G) = R^{n+1}.$$

According to the corollary to Theorem 2 of section 29.1 (the generalization of Liouville's theorem), $f(z)$ is a polynomial.

Proof of the necessity: Suppose that every function satisfying the conditions of the theorem is a polynomial. Let us show that $B(G) = R^{n+1}$. Let us suppose that $B(G) \neq R^{n+1}$. Then, on the basis of (187) and (192), there is either an admissible hyperboloid $(x-x')^2 - \lambda^2 = 0$ or an admissible hyperplane $\lambda(x-x') = 0$ lying outside $B(G)$. But then, the functions

$$f_1(z) = [(z-x')^2 - \lambda^2]^{-1}, \qquad f_2(z) = [\lambda(z-x')]^{-1}$$

satisfy the conditions of the theorem but they are not polynomials, which is impossible. (The holomorphy of the function f_1 in T follows from the lemma in section 30; the holomorphy of f_2 in T is obvious since $\operatorname{Im}[\lambda(z-x')] = \lambda y \neq 0$ for $y \in \Gamma$ and $\lambda \in \bar{\Gamma}$, where $\lambda \neq 0$; that these are functions of class $H_0(\Gamma)$ follows from the lemma of section 29.5.)

33. CONSTRUCTION OF AN ENVELOPE OF HOLOMORPHY $K(T \cup \tilde{G})$

1. *The envelope* $K'(T \cup \tilde{G})$

We shall say that a complex hyperboloid $(z-x')^2 = \lambda^2$ or a complex plane $\lambda(z-x') = 0$ is admissible for the domain $T \cup \tilde{G}$, is the corresponding real hyperboloid $(x-x')^2=\lambda^2$ or the real hyperplane $\lambda(x-x')=0$ is admissible for G (see section 32.1).

Suppose that G is a Γ-regular open set. We denote by $K'(T\cup\tilde{G})$ the envelope obtained by enveloping the domain $T\cup\tilde{G}$ with admissible complex hyperboloids for G (if $N(G)\neq\varnothing$) or admissible complex planes (if $N(G)=\varnothing$), so that

$$K'(T\cup\tilde{G})=C^{n+1}\setminus\overline{\bigcup_{(x',\lambda)\in N(G)}[z:\ (z-x')^2=\lambda^2]},\qquad N(G)\neq\varnothing; \tag{210}$$

or

$$K'(T\cup\tilde{G})=C^{n+1}\setminus\overline{\bigcup_{(x',\lambda)\in N_\infty(G)}[z:\ \lambda(z-x')=0]},\qquad N(G)=\varnothing, \tag{211}$$

respectively.

Reasoning as in section 32.1, we represent formula (211) in the form of the union of the increasing sequence of envelopes $K'(T\cup\tilde{G}_\alpha^\Gamma)$, for $\alpha=1, 2, \ldots$, of the form (210), where $G_\alpha^\Gamma=B_\Gamma(G)\cap(|x_0|<\alpha)$:

$$K'(T\cup\tilde{G})=\bigcup_\alpha K'(T\cup\tilde{G}_\alpha^\Gamma),\qquad N(G)=\varnothing. \tag{212}$$

We conclude from (187), (192), (210), and (211) that

$$\operatorname{Re} K'(T\cup\tilde{G})=B(G). \tag{213}$$

For definiteness, let us study the envelope $K'(T\cup\tilde{G})$, which corresponds to the most important case, that in which $N(G)\neq\varnothing$. The case in which $N(G)=\varnothing$ can be examined in an analogous fashion.

Keeping in mind the definition of the set $N(G)$ (see section 32.1), we conclude from (210) that $K'(T\cup\tilde{G})$ consists entirely of those interior points z^0 for which every complex hyperboloid $(z-x')^2=\lambda^2$ passing though z^0 possesses the property that the set $(x-x')^2\geqslant\lambda^2$ for real $z=x$ has at least one point in common with the open set G. Furthermore, for every point z^0 lying outside $K'(T\cup\tilde{G})$, there exists a sequence of points $z^{(k)}$, for $k=1, 2, \ldots$, such that $z^{(k)}\to z^0$, where $z^{(k)}\overline{\in} K'(T\cup\tilde{G})$, and the analytic surface $(z-x')^2=\lambda^2$ lying entirely outside $K'(T\cup\tilde{G})$ passes through each point $z^{(k)}$. But, for every $(x',\lambda)\in N(G)$, the function

$$[(z-x')^2-\lambda^2]^{-1}$$

is holomorphic in $T\cup\tilde{G}$ (see lemma of section 30), it belongs to the class $H_0(\Gamma)$ (see lemma of section 29.5), and, consequently, it belongs to the class of functions $K=K_{T\cup\tilde{G}}$ (see sections 29.2 and 16.1).

Therefore, on the basis of the first criterion for K-convexity (see section 16.3), we conclude that the envelope $K'(T \cup \tilde{G})$ is P-convex. Therefore,

$$K(T \cup \tilde{G}) \subset K'(T \cup \tilde{G}), \tag{214}$$

where the envelope $K(T \cup \tilde{G})$ is defined in section 29.2.

The envelope $K(T \cup \tilde{G})$ is a Runge domain.

To prove this assertion, it will be sufficient to show, on the basis of Weil's theorem (see section 24.9) that the envelope $K'(T \cup \tilde{G})$ is polynomially convex. On the basis of (210) and (184), it can be represented in the form

$$K'(T \cup \tilde{G}) = \text{int} \bigcap_{(x', \lambda) \in N(G)} \{C^{n+1} \setminus [z: (z - x')^2 = \lambda^2]\} =$$

$$= \text{int} \bigcap_{x'} \{C^{n+1} \setminus [z: (z - x')^2 = \rho, \; \rho \geqslant \max[0, \varphi(x')]]\},$$

that is, it is in the interior of the intersection of the exteriors of the analytic hypersurfaces (see section 18.7)

$$(z - x')^2 = \rho, \quad \rho \geqslant \max[0, \varphi(x')].$$

Therefore, we need only show that a domain of the form

$$\sigma = [z: \quad z^2 \neq \mu, \; \mu \geqslant \mu_0 \geqslant 0]$$

is polynomially convex (see section 16.2). To do this, we denote by $K = K_\sigma$ the smallest class (see section 16.1) of functions that are holomorphic in σ containing the functions*

$$\varphi_\mu(z^2) = (z^2 - \mu)^{-1}, \quad \mu_0 \leqslant \mu.$$

The class K obviously consists of functions of the form

$$P(z, \mu)[\varphi_\mu(z^2)]^m, \quad m = 1, 2, \ldots, \; \mu_0 \leqslant \mu,$$

where the P are polynomials. On the basis of the first criterion for K-convexity (see section 16.3), the domain σ is K-convex.

The function $\varphi_\mu(\zeta)$ is holomorphic in the ζ-plane cut along the line $\text{Re}\,\zeta \geqslant \mu_0$, $\text{Im}\,\zeta = 0$, that is, in a simply connected domain. Therefore, it can be approximated to any desired degree of accuracy by means of polynomials in this cut plane (see section 24.8). But then, the function $\varphi_\mu(z^2)$ can also be approximated by polynomials to

---

*The idea of introducing such a class is due to Shirinbekov.

any desired degree of accuracy in the domain σ. It follows from this that the class of polynomials is dense in the class K (in the sense of section 16.2). According to the lemma in section 16.2, the domain σ is polynomially convex, which completes the proof.

2. Integral representation of functions of class $H_0(\Gamma)$

We shall say that a Γ-regular open set G satisfies condition (A′) if $N(G) \neq \varnothing$ and there exists a spacelike hypersurface Σ contained in the set $B_\Gamma(G) \cup N(G)$ and in some strip $a \leqslant x_0 \leqslant b$ (where $(a > -\infty$ and $b < +\infty$ [see Fig. 95]). Here, the set $F = N(G) \cap (a \leqslant x_0 \leqslant b)$ is regular (in the sense of section 3.5). Obviously, if G satisfies condition (A′), it satisfies condition (A) (see section 32.2).

THEOREM.* *Every function $f(z)$ of class $H_0(\Gamma)$ that is holomorphic in a domain $T \cup \widetilde{G}$, where the open set G satisfies condition (A′), is holomorphic in the envelope $K'(T \cup \widetilde{G})$ and, for all $z \in K'(T \cup \widetilde{G})$, it can be represented in the form*

$$f(z) = L_0^s \left\{ c_{n+1} \int \psi(x', \lambda) D_z^l [-(z - x')^2 + \lambda^2]^{-\frac{n}{2}} dx' \, d\lambda + \right.$$
$$\left. + P_l(z) \right\} + P_0(z), \tag{215}$$

where $\psi \in S^*$ with carrier in $N(G) \cap \Sigma$, where P_0 and the P_l, for $|l| = s$, are polynomials, where s is an integer (here (ψ, P_0, P_l) and s depend on f), where Σ is a spacelike hypersurface defined by condition (A′), and where L_0^s is an integration operator in $K'(T \cup \widetilde{G})$ (see section 31.4).

Proof: Suppose that $f(z)$ belongs to $H_0(\Gamma)$ and is holomorphic in the domain $T \cup \widetilde{G}$. According to the theorem of section 29.3, the generalized function

$$f_0(x) = f(x_0 + i0, x) - f(x_0 - i0, x) \tag{216}$$

belongs to the class $L_0(\Gamma)$ and vanishes in the Γ-regular open set G. According to Theorem 1 of section 32.2, f_0 can be represented in the form (194): $f_0(x) = u(x, 0)$, where $u(x, \lambda)$ belongs to S^*, satisfies the wave equation $\Box_{n+1} u = 0$ and, by virtue of (202), vanishes in the envelope $B_\Gamma(G)$, where Γ is an $(n+2)$-dimensional light cone. According to the theorem of section 29.3, $u(x, \lambda)$ can be represented in the form of the difference

$$u(x, \lambda) = F(x_0 + i0, x, \lambda) - F(x_0 - i0, x, \lambda) \tag{217}$$

---

*See Jost and Lehmann [33], Vladimirov and Logunov [29], Omnes [31] and Vladimirov [30].

of the boundary values of the function $F(z, \zeta)$ of class $H_0(\Gamma)$. It follows from the results of section 31.4 that all derivatives $D_z^l F(z, \zeta)$ of sufficiently high order $|l| \geqslant s$ can be represented for $z \in R^{n+2} + i\Gamma$ in the form

$$D_z^l F(z, \zeta) = \\ = c_{n+1} \int \psi(x', \lambda) D_z^l [-(z-x')^2 + (\zeta-\lambda)^2]^{-\frac{n}{2}} dx' d\lambda + P_l(z, \zeta) \tag{218}$$

where the P_l are polynomials, $\psi \in S^*$, $S_\psi \subset \Sigma \cap N(G)$, and Σ is a spacelike hypersurface lying in the strip $a \leqslant x_0 \leqslant b$ and in the set $B_\Gamma(G) \cup N(G)$. (Since G possesses the property (A′), such a hypersurface exists; here, $S_\psi \subset F$, where

$$F = N(G) \cap (a \leqslant x_0 \leqslant b) \tag{219}$$

is a regular set.)

Let us show now that the functions

$$F_l(z) = c_{n+1} \int \psi(x', \lambda) D_z^l [-(z-x')^2 + \lambda^2]^{-\frac{n}{2}} dx' d\lambda + \\ + P_l(z, 0), \ |l| \geqslant s \tag{220}$$

are holomorphic in $K'(T \cup \tilde{G})$ if s is sufficiently great.

It follows from the definition of the envelope $K'(T \cup \tilde{G})$ (see section 33.1) that

$$(z - x')^2 - \lambda^2 \neq 0, \quad z \in K'(T \cup \tilde{G}), \quad (x', \lambda) \in N(G).$$

Therefore, the functions

$$\varphi_l(z; x', \lambda) \equiv D_z^l [-(z - x')^2 + \lambda^2]^{-\frac{n}{2}}$$

are holomorphic with respect to z in $b < +\infty$) and are infinitely differentiable with respect to $(x', \lambda) \in N(G)$.

We now need the

LEMMA. *For an arbitrary nonnegative integer m and subdomain* $G' \Subset K'(T \cup \tilde{G})$,

$$\|\varphi_l(z; x', \lambda)\|_{m, F} \leqslant c_{m, l}(G'), \ z \in G' \tag{221}$$

for all l such that $|l| \geq s$, *where* $s = \max[0, m - n]$ *and the set F is defined in* (219).

Proof: Let z range over $G' \Subset K'(T \cup \tilde{G})$. Then, for (x', λ) in the strip $a \leqslant x_0' \leqslant b$ (where $(a > -\infty$ and $K'(T \cup \tilde{G})$, we have the inequality

$$|-(z - x')^2 + \lambda^2| \geqslant |-(x - x')^2 + y^2 + \lambda^2| \geqslant |x'|^2 + \lambda^2 - C,$$

where C is a nonnegative number depending on G', a, and b. Using this inequality, we obtain

$$\|\varphi_l(z;\ x',\ \lambda)\|_{m,\ F} =$$

$$= \sup_{\substack{|\alpha| \leqslant m \\ (x',\ \lambda) \in F}} \left(1+\sqrt{|x'|^2+\lambda^2}\right)^m \left| D^\alpha_{(x',\ \lambda)} D^l_z [-(z-x')^2+\lambda^2]^{-\frac{n}{2}} \right| \leqslant$$

$$\leqslant c_1 \sup_{\substack{|\alpha| \leqslant m,\ x',\ \lambda \\ a \leqslant x'_0 \leqslant b}} \left(1+\sqrt{|x'|^2+\lambda^2}\right)^{m+|\alpha|+|l|} \times$$

$$\times \left| -(z-x')^2+\lambda^2 \right|^{-\frac{n}{2}-|\alpha|-|l|} \leqslant$$

$$\leqslant c_2 + c_3 \sup_{\substack{|\alpha| \leqslant m \\ |x'|^2+\lambda^2 \geqslant 2c}} \frac{\left(1+\sqrt{|x'|^2+\lambda^2}\right)^{m+|\alpha|+|l|}}{\left(|x'|^2+\lambda^2-C\right)^{\frac{n}{2}+|\alpha|+|l|}} =$$

$$= c_2 + c_3 \sup_{r \geqslant \sqrt{2C}} \frac{(1+r)^{m+|l|}}{(r^2-C)^{\frac{n}{2}+|l|}} = c_{m,\ l}(G'),\ \text{q.e.d.}$$

The carrier of the generalized function $\psi(x',\ \lambda)$ in (220) is contained in the regular set F. According to the theorem of section 3.8, ψ can be continued so as to be a continuous linear functional on a Banach space with norm $\|\ \|_{m,\ F}$ for some nonnegative m. Let us set $s = \max[0,\ m-n]$. It follows from inequality (221) and the structure of the functions $\varphi_l(z;\ x',\ \lambda)$ that, for all l and j (where $(|l| \geqslant s$ and $0 \leqslant j \leqslant n)$,

$$\left\| \frac{\partial \varphi_l(z;\ x',\ \lambda)}{\partial z_j} - \frac{\varphi_l(z+\Delta z_j;\ x',\ \lambda) - \varphi_l(z;\ x',\ \lambda)}{\Delta z_j} \right\|_{m,\ F} \underset{|\Delta z_j| \to 0}{\longrightarrow} 0,$$

uniformly as $z \in G' \Subset K'(T \cup \tilde{G})$. Therefore, the right members in (220) have all first derivatives in $K'(T \cup \tilde{G})$, and these derivatives can be obtained by differentiating under the "integral" sign. On the basis of Hartogs' fundamental theorem (see section 4.2), the functions $F_l(z)$, for $|l| \geqslant s$, are holomorphic in $K'(T \cup \tilde{G})$.

We note now that, on the basis of (218) and (220),

$$D^l_z F(z,\ 0) = F_l(z),\quad z \in T = R^{n+1} + i\Gamma.$$

Therefore, by using Lemma 2 of section 31.4 (since $D_1 = T^+$ and $D = K'(T \cup \tilde{G})$ is a Runge domain [see section 33.1]), we conclude that the function $F(z,\ 0)$ is holomorphic in $K'(T \cup \tilde{G})$ and can be represented in the form

$$F(z,\ 0) = L^s_0 \Big\{ c_{n+1} \int \psi(x',\ \lambda) \times \\ \times D^l_z [-(z-x')^2+\lambda^2]^{-\frac{n}{2}}\, dx'\, d\lambda + P_l(z,\ 0) \Big\}. \tag{222}$$

Furthermore, since $F(z,\ \zeta) \in H_0(\Gamma)$, it follows that $F(z,\ 0) \in H_0(\Gamma)$. Finally, it follows from (217) that

$$f_0(x) = u(x,\ 0) = F(x_0 + i0,\ \mathbf{x},\ 0) - F(x_0 - i0,\ \mathbf{x},\ 0).$$

By comparing this result with Eq. (216) and using the theorem in section 29.3, we conclude that the functions $f(z)$ and $F(z, 0)$ can differ only by a polynomial, which we denote by $P_0(z)$. From this and from (222), we obtain the representation (215). This completes the proof of the theorem.

3. *Remarks and examples*

(1) It follows from the preceding theorem and the inclusion (214) that

$$K(T \cup \tilde{G}) = K'(T \cup \tilde{G}). \tag{223}$$

From this and from (213), we conclude that

$$B(G) = \operatorname{Re} K(T \cup \tilde{G}).$$

It is shown in the article by Bros, Messiah, and Stora [26] that if a domain G is bounded by two spacelike hypersurfaces,* the envelopes $H(T \cup \tilde{G})$ and $K'(T \cup \tilde{G})$ must coincide. Thus, if, in addition the domain G satisfies condition (A′), then, on the basis of (223),

$$H(T \cup \tilde{G}) = K(T \cup \tilde{G}). \tag{224}$$

Furthermore, since $K'(T \cup \tilde{G})$ is a polynomially convex domain (see section 33.1), the envelope $K(T \cup \tilde{G})$ is a (in fact, the smallest) polynomially convex (and *a fortiori* K-convex) envelope of the domain $T \cup \tilde{G}$. Consequently, in this case, we have obtained answers to all the questions posed in section 29.2.

(2) In the representation (215), the weight function $\psi(x', \lambda)$ can, by means of formulas (201) and (169), be expressed linearly in terms of the difference (216) of the boundary values $f(x_0 \pm i0, x)$ of the function $f(z)$. Therefore, function $f(z)$ of every class $H_0(\Gamma)$ that is holomorphic in the domain $T \cup \tilde{G}$, where G satisfies condition (A′), is completely determined by its boundary values $f(x_0 \pm i0, x)$ for $x \in R^{n+1} \setminus B(G)$.

(3) On the one hand, the representation (215) is a generalization of the well-known Källén-Lehmann representation (see [22]) to the case of noninvariant functions. On the other hand, for $G = \varnothing$, it can be regarded as a modification of Bochner's representation (see section 25.4). Specifically, as is easy to see from the proof of the preceding theorem, in this case it makes the form

---

*According to the theorem on a Γ-convex envelope (see section 28.3), this condition can be weakened by assuming that the envelope $B_\Gamma(G)$ is bounded by two spacelike hypersurfaces.

$$f^{\pm}(z) = L_0^s\Big\{c_{n+1}\int \rho(x', \lambda)\, D_z^l[-z_0^2 + (z_1 - x_1')^2 + \ldots$$
$$\ldots + (z_n - x_n')^2 + \lambda^2]^{-\frac{n}{2}}\, dx'\, d\lambda + P_l^{+}(z)\Big\} + P_0^{\pm}(z), \qquad (225)$$

where the hyperplane $x_0' = 0$ is used for Σ. In particular, for $f^- \equiv 0$ the right member of (225) yields a representation of the function f^+ of $H_0(\Gamma^+)$ and vanishes in T^-.

(4) All the results obtained for a light (circular) cone $x_0^2 > |x|^2$ can be carried over in an obvious manner to the case of an elliptic cone

$$x_0^2 > \sum_{1 \leqslant j,\, k \leqslant n} a_{jk} x_j x_k, \quad \sum_{j,\, k} a_{jk} x_j x_k \geqslant \sigma |x|^2, \; \sigma > 0.$$

Here, the wave operator becomes the corresponding hyperbolic operator with constant coefficients.

(5) Let G be the strip $a < x_0 < b$. Then, (see example 4 of section 32.1 and Fig. 95),

$$N(G) = \left[(x', \lambda)\colon \left(\frac{b-a}{2} + \left|x_0' - \frac{a+b}{2}\right|\right)^2 \leqslant \lambda^2\right],$$

so that for Σ we may take the hyperplane

$$x_0' = \frac{a+b}{2}$$

and

$$F = \left[(x', \lambda)\colon \quad x_0' = \frac{a+b}{2},\right.$$
$$\left.\lambda^2 \geqslant \left(\frac{b-a}{2}\right)^2, \quad |x| < \infty\right]$$

is a regular set. Thus,

$$K(T \cup \tilde{G}) = [z\colon \quad (z - x')^2 \neq \lambda^2, \; (x', \lambda) \in F],$$

or

$$K(T \cup \tilde{G}) = \left[z\colon \, |\operatorname{Im} z| < \left|\operatorname{Im} \sqrt{\left(z_0 - \frac{a+b}{2}\right)^2 - \left(\frac{b-a}{2}\right)^2}\right|\right].$$

This result is due to Bogolyubov (see section 27.5; see also Sorokina [96]).

(6) Let us suppose that an open set G is such that $N(G) = \varnothing$ and that a sequence G_α^Γ, for $\alpha = 1, 2, \ldots$, possesses property (A′) beginning with sufficiently large α. Then, by using the theorem proven in the preceding section, we conclude that function $f(z)$ of

function $f(z)$ *of every class* $H_0(\Gamma)$ *that is holomorphic in* $T \cup \tilde{G}$, *where* $N(G) = \varnothing$, *is holomorphic in the union of the increasing sequence of envelopes* $K'(T \cup \tilde{G}_\alpha^\Gamma)$, *as* $\alpha \to \infty$, *that is (on the basis of* (212), *in the envelope* $K'(T \cup \tilde{G})$ *defined by formula* (211). Consequently, Eq. (223) also holds in this case.

Suppose, for example, that G is the halfspace $x_0 < 0$ and that $N(G) = \varnothing$. By using Example (5), we obtain (for $b = 0$ and $a = -\alpha$)

$$K(T \cup \tilde{G}) = \bigcup_\alpha K(T \cup \tilde{G}_\alpha^\Gamma) = \bigcup_\alpha [z: \ |\mathrm{Im}\, z| < \\ < |\mathrm{Im}\sqrt{(z_0 + \alpha) z_0}|] = [z: \ z_0 \neq \rho, \ \rho \geqslant 0].$$

In this case, since $K(T \cup \tilde{G}_\alpha^\Gamma) = H(T \cup \tilde{G}_\alpha^\Gamma)$, Eq. (224) remains valid.

4. *Applications to the justification of the dispersion relations*

As we know (see, for example, Bogolyubov, Medvedev and Polivanov [22]), the amplitude of the dispersion T corresponding to the process

$$p + k = p' + k',$$

where

$$p^2 = m^2, \ p'^2 = m'^2, \ k^2 = \mu^2, \ k'^2 = \mu'^2$$

are the squares of the masses, where $s = (p + k)^2$ is the square of the total energy, and where $-t = -(p - p')^2$ is the square of the transfer of momentum, is of the form

$$T_{p,\,p'}(q) = -i \int e^{iqx}\theta(x_0) \langle p' \left| \left[A\left(\frac{x}{2}\right),\ B\left(-\frac{x}{2}\right) \right] \right| p \rangle \, dx, \quad q = \frac{k + k'}{2}, \tag{226}$$

where the commutator (see section 29.6)

$$g_{p,\,p'}(x) = \langle p' \left| \left[A\left(\frac{x}{2}\right),\ B\left(-\frac{x}{2}\right) \right] \right| p \rangle$$

belongs to S^* and vanishes for $x^2 < 0$ (local commutativity) and its Fourier transform $\tilde{g}_{p,\,p'}(q)$ vanishes in the domain

$$G = \left[q: \ \left(q + \frac{p + p'}{2}\right)^2 < c^2 \text{ or } q_0 + \frac{p_0 + p_0'}{2} < 0 \right] \cap \\ \cap \left[q: \ \left(q - \frac{p + p'}{2}\right)^2 < b^2 \text{ or } q_0 - \frac{p_0 + p_0'}{2} < 0 \right]$$

(the spectral condition with masses considered). Taking $\frac{1}{2}(p + p') = (a, 0)$, let us rewrite the expression for the domain G in the form (see Fig. 93)

$$G=\left[q:\quad a-\sqrt{|q|^2+b^2}<q_0<-a+\sqrt{|q|^2+c^2}\right],$$
$$a\geqslant 0,\quad 0\leqslant b\leqslant c.$$

It follows from these properties and the results of section 29.3 that the amplitude $T_{p,\,p'}(q)$ is the boundary value of the function $f_{p,\,p'}(\zeta)$, which is the Fourier-Laplace transform of the function $-i\theta(x_0)\,g_{p,\,p'}(x)$, and which is holomorphic in the domain $T\cup\tilde{G}$:

$$T_{p,\,p'}(q)=f_{p,\,p'}(q_0+i0,\ q).$$

According to the theorem in section 33.2, the function $f_{p,\,p'}(\zeta)$ is holomorphic in the envelope

$$K(T\cup\tilde{G})=\text{int}\ [\zeta:\ (\zeta-x')^2\neq\lambda^2,\quad (x',\ \lambda)\in N(G)],$$

where (see Example 3 of section 32.1)

$$N(G)=\left[(x',\ \lambda):\quad |x_0'|+|\boldsymbol{x}'|\leqslant a,\right.$$
$$\left.|\lambda|\geqslant\max\left\{b-\sqrt{(a-x_0')^2-|\boldsymbol{x}'|^2},\ c-\sqrt{(a+x_0')^2-|\boldsymbol{x}'|^2}\right\}\right].$$

This result proves the dispersion relations.

Without going into details, let us point out certain steps in the proof. For simplicity, let us assume that we are dealing with a scalar theory, so that the amplitude T depends on the six scalar variables $p^2,\ p'^2,\ k^2,\ k'^2,\ s$, and t, two of which, namely, $p^2=m^2$ and $p'^2=m'^2$ we assume fixed. Therefore, $T_{p,\,p'}(q)=T(s,\ k^2,\ k'^2,\ t)$.

For $t<0$, the integral in (226) becomes meaningless in the so-called nonphysical region of energies. This difficulty was first overcome by Bogolyubov in 1956 in his proof of the dispersion relations. His idea consists in the fact that, for negative values of $k^2=\tau$ and $k'^2=\tau'$ that satisfy the inequality

$$\lambda(\tau,\ \tau',\ t)\equiv\tau^2+\tau'^2+t^2-2\tau\tau'-2\tau t-2\tau' t<0,$$

the nonphysical region disappears and the dispersion relation for $T(s,\ \tau,\ \tau',\ t)$ is easily derived. With accuracy up to insignificant simplifications, it takes the form*

---

*Equation (227) is simply Cauchy's formula for the s-plane with the cuts

$$\operatorname{Im}s=0,\quad \operatorname{Re}s\leqslant l_1,\quad \operatorname{Re}s\geqslant l_2.$$

Here, in the "slits" $\operatorname{Im}s=0$ and $l_1<\operatorname{Re}s<l_2$ the function T is real, so that, by virtue of Schwartz' symmetry principle the following equation is valid on the cuts:

$$\operatorname{Im}T(s,\ \tau,\ \tau',\ t)=\frac{1}{2i}\left[T(s+i0,\ \tau,\ \tau',\ t)-T(s-i0,\ \tau,\ \tau',\ t)\right].$$

The integral in (227) should be understood in the sense of generalized functions. For more information, see Taylor [34].

$$T(s, \tau, \tau', t) = \frac{(s+s_0)^n}{\pi}\left(\int_{-\infty}^{l_1} + \int_{l_2}^{\infty}\right)\frac{\operatorname{Im} T(s', \tau, \tau', t)}{(s'-s)(s'+s_0)^n}\,ds' + \sum_{0 \leqslant k \leqslant n-1} c_k(\tau, \tau', t)\, s^k. \tag{227}$$

To obtain the dispersion relation for physical values of $\tau=\mu^2$ and $\tau'=\mu'^2$, the right member of (227) must be holomorphically continued for every $t \leqslant 0$ as a function of the complex variables (s, τ, τ') up to the values $\tau=\mu^2$ and $\tau'=\mu'^2$. (Here, s ranges over the plane with the cuts.) To achieve such a continuation, we need only continue the function $\operatorname{Im} T(s', \tau, \tau', t)$ holomorphically into a complex neighborhood of the (real) intervals of variation of the variables (τ, τ', t) up to the values $\tau=\mu^2$ and $\tau'=\mu'^2$.

This continuation can be done by means of the established analyticity properties of the function $f_{p,p'}(\zeta)$. It turns out that it is not always possible to achieve such a continuation and, if it is possible, it is only for values of the impulse transfer $-t$ lying in a bounded interval. For example, for a π-meson-nucleonic scattering,

$$\left(m=m',\ \mu=\mu',\ a=\frac{m-2\mu}{2},\quad b=3\mu,\quad c=m+\mu\right)$$

this interval is

$$0 \leqslant -t \leqslant \frac{32}{3}\,\frac{2m+\mu}{2m-\mu}\,\mu^2$$

(cf. section 27.5; see Lehmann [28] and Vladimirov and Logunov [29]).

In conclusion, we make two remarks:

(1) Suppose that $a=0$ in the spectral conditions. In this case,

$$N(G) = [(x', \lambda):\quad x'=0,\quad \lambda^2 \geqslant \max(b^2, c^2)].$$

Let us suppose also that the function $f_{p,p'}(\zeta)$ decreases sufficiently fast at infinity that we may set $s=0$ and $P_0=0$ in the representation (215). This representation then takes the form

$$f_{p,p'}(\zeta) = \int\limits_{\max(b^2, c^2)} \frac{\psi(\lambda)\,d\lambda}{(\zeta^2-\lambda^2)^{n/2}}. \tag{228}$$

From this it follows that the envelope $K(T \cup \tilde{G})$ is bounded by the analytic hypersurface $z^2=\rho$, where $\max(b^2, c^2) \leqslant \rho < +\infty$, and that the function $f(\zeta)$ does in fact depend on ζ^2. However, if $f_{p,p'}(\zeta)$ does not decrease sufficiently rapidly at infinity, the representation (228) is valid up to

$$f_{p,\,p'}(\zeta) = \sum_{1 \leqslant k \leqslant N} P_k(\zeta)\,\Phi_k(\zeta^2),$$

where the P_k are polynomials and the functions $\Phi(w)$ are holomorphic and polynomially bounded in the w-plane with cut along the line $\operatorname{Im} w = 0$, $\max(b^2, c^2) \leqslant \operatorname{Re} w$. By virtue of the theorem in section 32.4, the case in which $G = [q: \; q^2 < -b^2]$ reduces to the case that we have considered (where $b = c = 0$). Since $B(G) = B_\Gamma(G) = [q^2 < 0]$. These results were established by a different method by Bogolyubov and Vladimirov [97].

The three conditions

(α) $g_{p,\,p'}(x) = 0$ for $x^2 < 0$ (local commutativity),

(β) $\tilde{g}_{p,\,p'}(q) = 0$ for $q^2 < m^2$ (spectralness),

(γ) Sufficiently rapid decrease of the function $f_{p,\,p'}(\zeta)$ at infinity imply Lorentz invariance and the Källén-Lehmann representation [228]. This result shows obviously that the axioms of quantum field theory are not independent (see also Streater [114]).

(2) The spectral conditions cannot be given arbitrarily: the domain G in which $\tilde{g}_{p,\,p'}(q)$ vanishes must be such that $B(G) \neq R^4$. To see this, note that in the opposite case, we would, by virtue of the quasi-analytic nature of class $L_0(\Gamma)$ functions (see section 32.4), have $\tilde{g}_{p,\,p'}(q) \equiv 0$, so that $f_{p,\,p'}(\zeta)$ would be a polynomial and the theory would become trivial.

References

1. B. A. Fuks, Introduction to the Theory of Analytic Functions of Several Complex Variables, Providence, American Mathematical Society, 1963.
2. B. A. Fuks, Spetsial'nyye glavy teorii analiticheskikh funktsiy mnogikh kompleksnykh peremennykh. (Special chapters in the theory of analytic functions of several complex variables), Fizmatgiz, 1963.
3. S. Bochner and W. T. Martin, Several Complex Variables, Princeton University Press, 1948.
4. H. Behnke and P. Thullen, Theorie der Funktionen mehrerer komplexer Veränderlichen, Ergebnisse der Mathem., Berlin, 1934.
5. G. M. Goluzin, Geometricheskaya teoriya funktsiy kompleksnogo peremennogo (Geometric theory of functions of a complex variable), Gostekhizdat, 1952.
6. F. Hartogs, "Zur Theorie der analytischen Funktionen mehrerer unabhängiger Veränderlichen, insbesondere über die Darstellung derselben durch Reihen, welche nach Potenzen einer Veränderlichen fortschreiten," Math. Ann., 62 (1906), 1-88.
7. H. Cartan and P. Thullen, "Zur Theorie der Singularitäten der Funktionen mehrerer komplexen Veränderlichen," Math. Ann., 106 (1932), 617-647.
8. I. P. Natanson, Teoriya funktsiy veshchestvennoy peremennoy (Theory of functions of a real variable), Gostekhizdat, 1950.
9. I. I. Privalov, Subgarmonicheskiye funktsii (Subharmonic functions), ONTI-NKTP, 1937.
10. G. H. Hardy, J. E. Littlewood, and G. Pólya, Inequalities, Cambridge University Press, 1934.
11. L. Schwartz, Théorie des distributions, t. I-11, Paris, 1950-1951.
12. P. Lelong, Fonctions plurisousharmoniques; mesures de Radon associées. Applications aux fonctions analytiques, Colloque sur les fonctions de plusieurs variables, Brussels, 1953, 21.40.
13. H. J. Bremermann, "Complex convexity," Trans. Amer. Math. Soc. 82 (1956), 17-51.
14. P. S. Alexsandrov, Kombinatornaya topologiya (Combinatorial topology), Gostekhizdat, 1947.

15. H. J. Bremermann, "Über die Äquivalenz der pseudokonvexen Gebiete under der Holomorphiegebiete im Raum von n Komplexen Veränderlichen", Math. Ann., 128 (1954), 63-91.
16. H. J. Bremermann, "On the conjecture of the equivalence of the plurisubharmonic functions and the Hartogs functions," Math. Ann., 131 (1956), 76-86.
17. H. J. Bremermann, "Note on plurisubharmonic and Hartogs functions," Proc. Amer. Math. Soc., 7 (1956), 771-775.
18. H. J. Bremermann, "Die Holomorphiehüllen der Tuben- und Halbtubengebiete," Math. Ann., 127 (1954), 406-423.
19. K. Oka, "Sur les fonctions analytiques de plusieurs variables complexes. VI. Domaines pseudoconvexes," Tokohu Math. J., 49 (1942), 15-52.
20. K. Oka, "Sur les fonctions analytiques de plusieurs variables complexes. IX. Domaines finis sans point critique intérieur," Ja. J. Math., *23* (1953), 97-155.
21. F. Norquet, "Sur les domaines d'holomorphie des fonctions uniformes de plusieurs variables complexes (Passage du local au global)," Bull. Soc. Mathém. de France, 82 (1954), 137-159.
22. N. N. Bogolyubov, B. V. Medvedev, and M. K. Polivanov, Voprosy teorii dispersionnykh sootnosheniy (Questions in the theory of dispersion relations), Fizmatgiz, 1958.
23. O. Steinmann, "Structure of the two-point function," J. Math. Phys., 4 (1963), 583-588.
24. A. S. Wightman, "Nekotoryye matematicheskiye problemy relativistskoy kvantovoy teorii" (Certain mathematical problems of relativistic quantum theory), Matematika, 6:4 (1962), 96-133.
25. G. Källén and A. Wightman, "The analytic properties of the vacuum expectation value of a product of three scalar local fields," Math Fys. Scr. Dan. Vid. Selsk., 1, No. 6 (1958).
26. J. Bros, A. Messiah, and R. Stora, "A problem of analytic completion related to the Jost-Lehmann-Dyson formula," J. Math. Phys., 2 (1961), 639-651.
27. H. J. Bremermann, R. Oehme, and J. G. Taylor, "A proof of dispersion relations in quantized field theories," Phys. Rev., 109 (1958), 2178-2190.
28. H. Lehmann, "Scattering matrix and field operators," Nuovo Cimento, 14 (1959), Suppl. No. 1, 153-176.
29. V. S. Vladimirov and A. A. Logunov, "Ob analiticheskikh svoystvakh obobshchennykh funktsiy kvantovoy teorii polya" (Analytic properties of the generalized functions of quantum field theory), Izv. Akad. Nauk SSSR, seriya matem., 23 (1959), 661-676.
30. V. S. Vladimirov, "O postroyenii obolochek golomorfnosti dlya oblastey spetsial'nogo vida i ikh primeneniya" (Construction of envelopes of holomorphy for domains of a special

type and their applications). Trudy Matem. in-ta im. V. A. Steklova, 60 (1961), 101-144.

31. R. Omnes, "Démonstration des relations de dispersion", in the book: Dispersion Relations and Elementary Particles, ed. C. de Witt and R. Omnes, New York, Wiley, 1960, pp. 317-384.
32. A. S. Wightman, "Quantum field theory and analytic functions of several complex variables," J. Indian Math. Soc., 24 (1960-61), 625-677.
33. R. Jost and H. Lehmann, "Integral-Darstellung kausaler Kommutatoren," Nuovo Cimento, 5 (1957), 1598-1610.
34. J. G. Taylor, "Dispersion relations and Schwartz's distributions," Ann. of Phys., 5 (1958), 391-398.
35. F. J. Dyson, "Connection between local commutativity and regularity of Wightman functions," Phys. Rev., 110 (1958), 579-581.
36. H. Epstein, "Generalization of the edge-of-the-wedge theorem," J. Math. Phys., 1, (1960), 524-531.
37. V. S. Vladimirov, "O teoreme ostriya klina Bogolyubova" (On Bogolyubov's "edge of the wedge" theorem), Izv. Akad. Nauk SSSR, ser. matem. 26 (1962), 825-838.
38. I. M. Gel'fand and G. Ye. Shilov, Obobshchennyye funktsii (Generalized functions), Vols. 1, 2, Fizmatgiz, 1958.
39. N. N. Bogolyubov and V. S. Vladimirov, "Ob analiticheskom prodolzhenii obobshchennykh funktsiy" (Analytic continuation of generalized functions), Izv. Akad. Nauk SSSR, ser. matem., 22 (1958), 15-48.
40. V. S. Vladimirov, "Ob odnom obobshchenii teoremy Liuvillya" (A generalization of Liouville's theorem), Trudy Matem. in-ta im. V. A. Steklova, 64 (1961), 9-27.
41. F. J. Dyson, "Integral representation of causal commutators," Phys. Rev., 110 (1958), 1460-1464.
42. V. S. Vladimorov, "O funktsiyakh, golomorfnykh v trubchatykh konusakh" (Functions that are holomorphic in tubular cones), Izv. Akad. Nauk SSSR, ser. matem., 27 (1963), 75-100, 1186.
43. D. Ya. Petrina, "O nevozmozhnosti postroyeniya nelokal'noy teorii polya s polozhitel'nym spektrom operatura energii-impul'sa" (On the impossibility of constructing a nonlocal field theory with positive spectrum of the energy-impulse operator). Ukr. mat. Zhur., 13 (1961), No. 4, 109-111.
44. I. M. Gel'fand and N. Ya. Vilenkin, Obobshchennyye funktsii (Generalized functions), Vol. 4, Fizmatgiz, 1961.
45. L. Schwartz, "Transformation de Laplace des distributions," Medd. Lunds. Univ. mat. Semin. (Supplementband), 1952, 196-206.
46. J. L. Lions, "Supports dans la transformation de Laplace," J. Analyse Math., 2 (1952-1953), 369-380.

47. J. G. Taylor, "A theorem of continuation for functions of several complex variables," Proc. Cambr. Phil. Soc., 54 (1958), 377-382.
48. H. Behnke and K. Stein, "Konvergente Folgen von Regularitätsbereichen und die Meromorphie-Konvexität, Math. Ann., 116 (1938), 204-216.
49. P. Lelong, "La convexité et les fonctions analytiques de plusieurs variables complexes," J. Math. pures et appl., 31 (1952), 191-219.
50 D. Ruelle. "Domain of holomorphy of the three-point function," Helv. Phys. Acta, 34 (1961), 587-592.
51. H. Behnke and F. Sommer, "Analytische Funktionen mehrerer Veränderlichen. Über die Voraussetzungen des Kontinuitätssatzes", Math. Ann., 121 (1950), 356-378.
52. S. Hitotsumatsu, "On some conjectures concerning pseudo-convex domains," J. Math. Soc. Jap., 6 (1954), 177-195.
53. V. V. Stepanov, Kurs differentsial'nykh uravneniy (Course of differential equations), Fizmatgiz, 1958.
54. H. Kneser, "Die singulären Kanten bei analytischen Funktionen mehrerer Veränderlichen," Math. Ann., 106 (1932), 656-660.
55. S. Bochner, "Analytic and meromorphic continuation by means of Green's formula," Annals of Math., 44 (1943), 652-673.
56. F. Sommer, "Uber die Integralformeln in der Funktionentheorie mehrerer komplexer Veränderlichen," Math. Ann., 125 (1952), 172-182.
57. S. Bochner, "Group invariance of Cauchy's formula in several variables," Annals of Math., 45 (1944), 686-707.
58. G. de Rham, Variétés différentiables, Actualités no. 1222, Paris, Hermann, 1935.
59. H. Hefer, "Zur Funktionentheorie mehrerer Veränderlichen. Über eine Zerlegung analytischer Funktionen und die Weilsche Integraldarstellung," Math. Ann., 122 (1950), 276-278.
60. H. Grauert and K. Remmert, "Komplexe Räume," Math. Ann., 136 (1958), 245-318.
61. A. S. Wightman, "Analytic functions of several complex variables." In the book: Dispersion Relations and Elementary Particles, ed. C. de Witt and R. Omnes, New York, Wiley, 1960, 227-315.
62. L. A. Ayzenberg and B. S. Mityagin, "Prostranstva funktsiy, analitcheskikh v kratno-krugovykh oblastyakh" (Spaces of functions that are analytic in multiple-circular domains), SMZh, 1 (1960), 153-170.
63. J. Bros, "Les problèmes de construction d'enveloppes d'holomorphie en théorie quantique des champs," Séminaire d'Analyse de P. Lelong, 13 March 1962.
64. W. S. Brown, "Analyticity properties of the momentum-space vertex function," J. Math. Phys., 3 (1962), 221-235.

65. P. Thullen, "Zur Theorie der Singularitäten der Funktionen zweier komplexer Veränderlichen," Math. Ann., 106 (1932), 64-76.
66. W. Rothstein, "Ein neuer Beweis des Hartogsschen Hauptsatzes und seine Ausdehnung auf meromorphe Funktionen," Math. Zeit., 53 (1950), 84-95.
67. V. S. Vladimirov and M. Shirinbekov, "O postroyenii obolochek golomorfnosti dlya oblastey Gartogsa" (Construction of envelopes of holomorphy for Hartogs' domains), Ukr. Mat. Zhur., 15 (1963), No. 2, 189-192.
68. S. Bergmann, "Über eine Integraldarstellung von Funktionen von zwei komplexen Veränderlichen," Matem. sb., 1, (1936), 242-257.
69. A. Weil, "L'intégral de Cauchy et les fonctions de plusieurs variables," Math. Ann., 111 (1935, 178-182.
70. A. I. Markushevich, Teoriya analiticheskikh funktsiy (Theory of analytic functions), Gostekhizdat, 1950.
71. A. Weil, "Sur les séries des polynomes de deux variables complexes," C. R. Acad. Sc., 194 (1932), 1304-1307.
72. K. Oka, "Sur les fonctions analytiques des plusieurs variables complexes," I. J. Sci. Hiroshima Univ., ser. A, 6 (1936), 245-255.
73. H. J. Bremermann, "Die Charakterisierung Rungescher Gebiete durch plurisubharmonische Funktionen," Math. Ann., 136 (1958), 173-186.
74. M. Shirinbekov, "Ob oblastyakh Runge v prostranstve mnogikh kompleksnykh peremennykh" (Runge domains in the space of several complex variables), Doklady Akad. Nauk SSSR, 145 (1962), 45-47.
75. K. Oka, "Sur les fonctions analytiques des plusieurs variables complexes, II, Domaines d'holomorphie," J. Sci. Hiroshima Univ., ser. A, 7 (1937), 115-130.
76. S. G. Gindikin, "Analiticheskiye funktsii v trubchatykh oblastyakh " (Analytic functions in tubular domains), Doklady Akad. Nauk SSSR, 145 (1962), 1205-1208.
77. S. L. Sobolev, "Méthode nouvelle à résoudre le problème de Cauchy pour les équations linéaires hyperboliques normales," Matem. sb., 1 (43) (1936), 39-72.
78. N. N. Bogolyubov and O. S. Parasyuk, "Ob analiticheskom prodolzhenii obobshchennykh funktsiy" (Analytic continuation of generalized functions), Doklady Akad. Nauk SSSR, 109 (1956), 717-719.
79. L. Gårding and J. L. Lions, "Functional analysis," Nuovo Cimento, 14 (1959), Suppl. 1, 9-66.
80. H. Whitney, Analytic extensions of differentiable functions defined in closed sets," Trans. Amer. Math. Soc., 36 (1934), 63-89.

81. L. Hörmander, "On the division of distributions by polynomials," Arkiv mat., 3 (1958), 555-568.
82. S. Lojasiewicz, "Division d'une distribution par une fonction analytique de variables réelles," C. R. Acad. Sci., 246 (1958), 683-686.
83. L. Gårding and B. Malgrange, "Opérateurs différentiels partiellement hypoelliptiques," C. R. Acad. Sci., 247 (1958), 2083-2085.
84. L. Gårding and B. Malgrange, "Opérateurs différentiels partiellement hypoelliptiques et partiellement elliptiques," Math. Scand., 9 (1961), 5-21.
85. L. Hörmander, "On the theory of general partial differential operators," Acta Math., 94 (1955), 161-248.
86. G. I. Eskin, "Obobshcheniye teoremy Paleya-Vinera-Shvartsa" (Generalization of the Paley-Weiner-Schwartz theorem), Usp. mat. nauk, XVI, no. 1 (97) (1961), 185-188.
87. N. N. Bogolyubov and D. V. Shirkov, Vvedeniye v teoriyu kvantovannykh poley (Introduction to quantum field theory), II, GTTI, 1957.
88. R. Courant and D. Hilbert, Methods of Mathematical Physics, Vol. II, New York and London, Interscience, 1962.
89. F. John, "On linear partial differential equations with analytic coefficients. Unique continuation of data," Communs. pure and appl. mathem., 2 (1949), 209-253.
90. L. Nirenberg, "Uniqueness in Cauchy problems for differential equations with constant leading coefficients," Communs pure and appl. mathem., 10 (1957), 89-105.
91. B. Broda, "On uniqueness theorems for differential equations with constant coefficients," Math. Scand., 9 (1961), 55-68.
92. V. S. Vladimirov, "O postroyenii obolochek golomorfnosti dlya oblastey spetsial'nogo vida" (The construction of envelopes of holomorphy for domains of a special type), Doklady Akad. Nauk SSSR, 134 (1960), 251-254.
93. V. S. Vladimirov, "O primenenii svoystv oblastey golomorfnosti k izucheniyu resheniy differentsial'nykh uravneniy" (Application of the properties of domains of holomorphy to the study of the solutions of differential equations), Doklady Akad. Nauk SSSR, 134 (1960), 511-513.
94. M. S. Lavrent'yev and B. V. Shabat, Metody teorii funktsiy kompleksnogo peremennogo (Methods in the theory of functions of a complex variable), Fizmatgiz, 1958.
95. L. V. Kantorovich and G. P. Akilov, Functional Analysis in Normed Spaces, Oxford and New York, Pergamon, 1964.
96. N. G. Sorokina, "Ob odnoy teoreme N. N. Bogolyubova" (On a theorem of N. N. Bogolyubov), Ukr. mat. Zhur., 11, No. 2 (1959), 220-222.

97. N. N. Bogolyubov and V. S. Vladimirov, "Odna teorema ob analiticheskom prodolzhenii obobshchennykh funktsiy (A theorem on the analytic continuation of generalized functions), NDVSh, fiz.-matem. nauki, No. 3 (1958), 26-35; No. 2 (1959), 179.

98. H. J. Borchers, "Über die Vollständigkeit lorentzinvarianter Felder in einer zeitartigen Röhre," Nuovo Cimento, 19 (1961), 781-793.

99. H. Araki, "A generalization of Borchers' theorem," Helv. Phys. Acta, 36 (1963), 132-139.

100. A. A. Logunov and B. M. Stepanov, "Dispersionnyye sootnosheniya dlya reaktsiy fotorozhdeniya π-mezonov" (Dispersion relations for reactions involving the photogeneration of π-mesons), Doklady Akad. Nauk SSSR, 110 (1956), 368-370.

101. I. T. Todorov, "Analytical properties of the scattering amplitude for inelastic processes involving strange particles," Nuclear Phys., 18 (1960), 521-528.

102. R. F. Streater, "Some integral representations in field theory," Nuovo Cimento, 15 (1960), 936-948.

103. R. F. Streater, "Special methods of analytic completion in field theory," Proc. Roy. Soc., Ser. A., 256 (1960), 39-52.

104. R. F. Streater, "The double commutator in quantum field theory," J. Math. Phys., 1 (1960), 231-233.

105. I. G. Petrovskiy, Lectures on Partial Differential Equations (in press).

106. H. J. Bremermann and L. Durand, "On analytic continuation, multiplication and Fourier transformations of Schwartz distributions," J. Math. Phys., 2 (1961), 240-258.

107. N. I. Muskhelishvili, Singulyarnyye integral'nyye uravneniya (Singular integral equations), Fizmatgiz, 1962.

108. O. S. Parasyuk, "O 'parnykh' integral'nykh uravneniyakh v klasse obobshchennykh funktsiy" ("Pairs" of integral equations in the class of generalized functions), Doklady Akad. Nauk SSSR, 110 (1956), 957-958.

109. I. M. Ryzhik and I. S. Gradshteyn, Tables of Integrals, Sums, Series, and Products (in press).

110. M. Riesz, "L'intégral de Riemann-Liouville et le problème de Cauchy," Acta. Math., 81 (1949), 1-222.

111. V. S. Vladimirov and V. F. Nikitin, "Ob integral'nom predstavlenii Iosta-Lemana-Daysona" (On the Jost-Lehmann-Dyson integral representation), Doklady Akad. Nauk SSSR, 138 (1961), 809-812.

112. F. E. Browder, "On the edge-of-the-wedge theorem", Canad. J. of Math., 15 (1963), 125-131.

113. R. Jost, "Eine Bemerkung zum CTP-theorem," Helv. Phys. Acta, 30 (1957), 409-416.

114. R. F. Streater, "Analytic properties of products of field operators," J. Math. Phys., 3 (1962), 256-261.
115. E. E. Levi, "Studii sui punti singolari essenziali delle funzioni analitiche di due o più variabili complesse," Ann. di Mat., ser. III, 17 (1910), 61-87.
116. E. E. Levi, "Sulle ipersuperfici dello spazio a 4 dimensioni che possono essere frontiera del campo di esistenza di una funzione analitica di due variabili complesse," Ann. di Mat., ser. III, 18 (1911), 69-79.
117. S. Stoilow, Teoria funcţiilor de o variabilă complexă, Bucharest, 1958-1962, Vols. I and II. (This has been translated into Russian as Teoriya funktsiy kompleksnogo premennogo.)
118. H. Cartan, "Les fonctions de deux variables complexes et le problème de la représentation analytique," J. Math. pures et appl., 10 (1931), 1-114.
119. F. Hartogs, "Über die aus den singulären Stellen einer analytischen Funktion mehrerer Veränderlichen bestehenden Gebilde," Acta Math., 32 (1909), 59-79.
120. H. Behnke and K. Stein, "Konvergente Folgen nichtschlichter Regularitätsbereiche," Ann. di Mat. ser. IV, 28, (1949), 317-326.
121. T. Bonnesen and W. Fenchel, "Theorie der konvexen Körper," Ergebnisse der Mathem., Berlin, 1934.
122. L. A. Ayzenberg, "Integral'noye predstavleniye funktsiy, golomorfnykh v vypuklykh oblastyakh prostranstva C^n (Integral representation of functions that are holomorphic in convex domaines of the space C^n), Doklady Akad. Nauk, 151 (1963), 1247-1249.
123. E. Cartan. "Sur les domaines bornés hologènes de l'espace de n variables complexes," Abhandl. Math. Semin. Hamburg. Univ., 11 (1935), 116-162.
124. J. Bros, H. Epstein, and V. Glaser, "Some rigorous analyticity properties of the four-point function in momentum space," CERN-Geneva, 1963, preprint.
125. G. Källén, "The analyticity domain of the four point function," Nuclear Phys., 25 (1961), 568-603.
126. H. Araki, "Generalized retarded functions and analytic functions in momentum space in quantum field theory," J. Math. Phys., 2 (1961), 163-177.
127. D. Ruelle, "Connection between Wightman functions and Green functions," Nuovo Cimento, 19 (1961), 356-376.
128. O. Steinmann, "Über den Zusammenhang zwischen den Wightman-Funktionen und den retardierten Kommutatoren," Helv. Phys. Acta, 33 (1960), 257-298.
129. O. Steinmann, "Wightman-Funktionen und retardierte Kommutatoren, II," Helv. Phys. Acta, 33 (1960), 347-362.
130. G. Köthe, "Die Randverteilungen analytischer Funktionen," Math. Zeit., 57 (1952), 13-33.

131. H. G. Tillman, "Randverteilungen analytischer Funktionen und Distributionen," Math. Zeit.,59 (1953), 61.83.
132. H. G. Tillman, "Distributionen als Randverteilungen analytischer Funktionen, II," Math. Zeit., 76 (1961), 5-21.
133. H. G. Tillman, "Darstellung der Schwartzschen Distributionen durch analytische Funktionen," Math. Zeit., 77 (1961), 106-124.
134. M. Sato, "On a generalization of the concept of functions," Proc. Japan. Acad., 34 (1958), 126-135.
135. M. Sato, "Theory of hyperfunctions," I. J. Fac. Sc. Tokyo, 8 (1959), 139-193.
136. Z. Luszczki and Z. Zielezny, "Distributionen der Räume D'_{Lp} als Randverteilungen analytischer Funktionen," Colloq. Math., 8 (1961), 125-131.
137 V. S. Rogozhin, "Krayevaya zadacha Rimana v prostranstve obobshchennykh i mnogochleny Fabera" (Riemann's boundary-value problem in the space of generalized functions and Faber's polynomials), Doklady Akad. Nauk SSSR, 152 (1963), 1308-1311.
138. L. Schwartz, "Sur l'impossibilité de la multiplication des distributions," C. R. Acad. Sci., 239 (1954), 847-848.
139. N. N. Bogolyubov and O. S. Parasyuk, "Über die Multiplikation der Kausalfunktionen in der Quantentheorie der Felder," Acta Math., 97 (1957), 227-266.
140. O. S. Parasyuk, "K teorii R-operatsii Bogolyubova" (On the theory of Bogolyubov's R-operation), Ukr. mat. zhurn., XII, No. 3 (1960), 287-307.
141. B. M. Stepanov, "Abstraktnaya teoriya R-operatsii" (The abstract theory of the R-operation), Izv. Akad. Nauk SSSR, ser. matem., 27 (1963), 819-830.
142. B. M. Stepanov, "K voprosy o postroyenii S-matritsy" (On the construction of an S-matrix), Doklady Akad. Nauk SSSR, 151 (1963), 84-86.
143. B. M. Stepanov, "Struktura nerelyativistskikh kontrchlenov" (The structure of nonrelativistic counterterms), Doklady Akad. Nauk SSSR, 133 (1960), 547-549.
144. V. A. Shcherbina, "Regulyarizatsiya proizvedeniy obobshchennykh funktsiy tipa prichinnykh" (Regularization of the products of generalized functions of the causal type), Doklady Akad. Nauk SSSR, 143 (1962), 815-817.
145. V. A. Shcherbina, "O lokal'nykh regulyarizatsiyakh koeffitsiyentnykh funktsiy matritsy rasseyaniya" (On local regularizations of the coefficients of functions of the dispersion matrix), Doklady Akad. Nauk SSSR, 143 (1962), 1075-1077.
146. J. G. Taylor, "The renormalization constants in perturbation theory," Nuovo Cimento, 17 (1960), 695-702.

147. H. J. Bremermann, On finite renormalization constants and the multiplication of causal functions in perturbation theory, Technical report, No. 8, Berkeley, 1959.
148. O. Steinmann, "Zur Definition der retardierten und zeitgeordneten Produkte," Helv. Phys. Acta, 36 (1963), 90-112.
149. S. Bochner, "Bounded analytic functions in several variables and multiple Laplace integrals," Amer. J. Math., 59 (1937), 732-738.
150. H. J. Bremermann, "Construction of the envelopes of holomorphy of arbitrary domains," Revista Math. Hisp.-Amer. (4), 17 (1957), 175-200.
151. E. Martinelli, "Sulle estensioni della formula integrale di Cauchy alle funzioni analitiche di piú variabili complesse," Ann. di Mat., ser. IV, 34 (1953), 277-347.
152. A. A. Temlyakov, "Integral'nyye predstavleniya" (Integral representations), "Uchenyye zapiski," Mosk. obl. ped. in-ta, 96 (1960), 3-14.
153. L. A. Ayzenberg, "Integral'nyye predstavleniya funkysiy, golomorfnykh v kratno-krugovykh oblastyakh" (Integral representations of functions that are holomorphic in multiple-circular domains), Doklady Akad. Nauk SSSR, 138 (1961), 9-12.
154. S. Bochner, Lectures on Fourier Integrals, Princeton University Press, 1959.
155. R. E. A. C. Paley and N. Wiener, Fourier transform in the complex domain, Colloq. Amer. Math. Soc., XIX, 1934.
156. A. Plancherel and G. Pólya, "Fonctions entières et intégrales de Fourier multiples," Comm. Math. Helv., 9 (1937), 224-248; 10 (1938), 110-163.
157. H. König, "Multiplikation und Variablentransformation in der Theorie der Distributionen," Arch. Math., 6 (1955), 391-396.
158. H. König, "Multiplikation von Distributionen, I," Math. Ann., 128 (1955), 420-452.
159. P. Lelong, "Domaines convexes par rapport aux fonctions plurisousharmoniques," J. d'Analyse Mathématique, II (1952-53), 178-208.
160. L. Hörmander, Linear partial differential operators, Springer-Verlag, 1963.

Index